Energy Applications of 2D Nanomaterials

Energy Applications of 2D Nanomaterials

Edited by
Ram K. Gupta

CRC Press
Taylor & Francis Group
Boca Raton London New York

CRC Press is an imprint of the
Taylor & Francis Group, an **informa** business

First edition published 2022
by CRC Press
6000 Broken Sound Parkway NW, Suite 300, Boca Raton, FL 33487-2742

and by CRC Press
2 Park Square, Milton Park, Abingdon, Oxon, OX14 4RN

© 2022 Taylor & Francis Group, LLC

CRC Press is an imprint of Taylor & Francis Group, LLC

Library of Congress Cataloging-in-Publication Data
Names: Gupta, Ram K., editor.
Title: Energy applications of 2D nanomaterials / edited by Ram K. Gupta.
Description: First edition. | Boca Raton : CRC Press, 2022. |
Includes bibliographical references and index. |
Identifiers: LCCN 2021058600 (print) | LCCN 2021058601 (ebook) |
ISBN 9781032013879 (hardback) | ISBN 9781032013909 (paperback) | ISBN 9781003178422 (ebook)
Subjects: LCSH: Two-dimensional materials. | Electric batteries—Materials. |
Supercapacitors—Materials.
Classification: LCC TA418.9.T96 E54 2022 (print) | LCC TA418.9.T96 (ebook) |
DDC 620.1/12—dc23/eng/20220204
LC record available at https://lccn.loc.gov/2021058600
LC ebook record available at https://lccn.loc.gov/2021058601

ISBN: 978-1-032-01387-9 (hbk)
ISBN: 978-1-032-01390-9 (pbk)
ISBN: 978-1-003-17842-2 (ebk)

DOI: 10.1201/9781003178422

Typeset in Times
by codeMantra

I would like to dedicate this book to my parents who taught me the importance of electricity (energy).

Contents

Preface

Energy serves as a focal point for every activity that is carried out in our daily lives. The human brain, for instance, is considered the most efficient processing system, faster than that of any supercomputer ever developed to date and, consumes energy as low as 20 W. "Aurora" the supercomputer being developed by Intel is claimed to hit a computing speed of 1 exaflop – equivalent to one quintillion floating-point computations per second. The United States Department of Energy estimates electricity target to power "Aurora" will be 40 MW. Although we are trying to hit the extravagant efficiencies closer to the brain, the devices that are as efficient to the low energy consumption of the brain remain a challenge. The answer to all these problems is hidden in the material science of electronic devices that are being fabricated. The development of materials has allowed us to fit billions of transistors in small-sized chips and allow progress per Moore's law. The major contribution to these materials includes two-dimensional nanomaterials, which are widely used materials in energy applications due to their unique structure at low cost.

2D nanomaterials have emerged as promising candidates for energy devices owing to their superior electrochemical properties, surface area, nano-device integration, multifunctionality, printability, and mechanical flexibility. In this book, we summarize research aspects for different materials in the 2D category such as graphene and its derivatives, transition metal chalcogenides, and transition metal oxides/hydroxides for energy applications such as in batteries, supercapacitors, solar cells, and fuel cells. Apart from conventional techniques, this book explores new aspects of synthesizing 2D nanomaterials beyond traditionally layered structures such as 2D metal oxides and polymers, broadening the vision for readers to explore novel material systems for enhanced energy applications.

The book is divided into two volumes to cover synthesis, properties, and applications of 2D nanomaterials for energy. The first volume covers basic concepts, chemistries, the importance of 2D nanomaterials for energy along theoretical consideration in designing new 2D nanomaterials. The effect of doping, structural variation, phase, and exfoliation on structural and electrochemical properties of 2D nanomaterials are discussed for their applications in the energy sector. The second volume (this volume) covers a wide range of applications of 2D nanomaterials for energy. Many advanced 2D materials such as graphene, chalcogenides, phosphides, MXene, and their nanocomposites are covered for their applications in fuel cells, photovoltaics, batteries, and supercapacitors. The future applications and challenges in fabricating flexible energy generation and storage devices are also covered.

Ram K. Gupta, Associate Professor
Department of Chemistry
Kansas Polymer Research Center
Pittsburg State University
Pittsburg, Kansas, United States

Editor

Dr. Ram K. Gupta is an Associate Professor at Pittsburg State University. Dr. Gupta's research focuses on conducting polymers and composites, green energy production and storage using biowastes and nanomaterials, optoelectronics and photovoltaics devices, organic-inorganic hetero-junctions for sensors, bio-based polymers, flame-retardant polymers, biocompatible nanofibers for tissue regeneration, scaffold and antibacterial applications, corrosion inhibiting coatings, and bio-degradable metallic implants. Dr. Gupta has published over 235 peer-reviewed articles, made over 300 national, international, and regional presentations, chaired many sessions at national/international meetings, edited many books, and written several book chapters. He has received over two and a half million dollars for research and educational activities from many funding agencies. He is serving as Editor-in-Chief, Associate Editor, and editorial board member for numerous journals.

Contributors

Bincy George Abraham
Department of Chemical Engineering
Indian Institute of Technology-Madras
Chennai, Tamil Nadu, India

Arpana Agrawal
Department of Physics
Shri Neelkantheshwar Government
    Post-Graduate College
Khandwa, India

Junaid Ahmad
Department of Physics
Division of Science and Technology
University of Education Lahore
Lahore, Pakistan

Khuram Shahzad Ahmad
Department of Environmental Sciences
Fatima Jinnah Women University
Rawalpindi, Pakistan

Farman Ali
Department of Chemistry
Hazara University
Mansehra, Pakistan

Jazib Ali
Electronic Engineering Department
University of Rome Tor Vergata
Rome, Italy

Nisar Ali
Key Laboratory for Palygorskite
    Science and Applied Technology of
    Jiangsu Province
National & Local Joint Engineering
    Research Center for Deep Utilization
    Technology of Rock-salt Resource
Faculty of Chemical Engineering
Huaiyin Institute of Technology
Huaian, China

Ghulam Abbas Ashraf
Department of Physics
Zhejiang Normal University
Jinhua, Zhejiang

Mohd Asyadi Azam
Fakulti Kejuruteraan Pembuatan
Universiti Teknikal Malaysia Melaka
Hang Tuah Jaya, Malaysia

M. F. Aziz
Centre for Research and Innovation
    Management (CRIM)
Universiti Teknikal Malaysia Melaka
Hang Tuah Jaya, Malaysia
and
Fakulti Kejuruteraan Pembuatan
Universiti Teknikal Malaysia Melaka
Hang Tuah Jaya, Malaysia

Manahil Bakhtiar
Department of Chemistry
Hazara University
Mansehra, Pakistan

Kaushalya Bhakar
School of Chemical Sciences
Central University of Gujarat
Gandhinagar, India

Emre Bicer
Battery Research Laboratory
Faculty of Engineering and
    Natural Sciences
Sivas University of Science and
    Technology
Sivas, Turkey

Muhammad Bilal
School of Physics
College of Physical Science and
    Technology
Yangzhou University
Yangzhou, P. R. China
and
School of Environmental Science and
    Engineering
Yangzhou University
Yangzhou, P. R. China

Faheem K. Butt
Department of Physics
Division of Science and Technology
University of Education Lahore
Lahore, Pakistan

Raghuram Chetty
Department of Chemical Engineering
Indian Institute of Technology-Madras
Chennai, Tamil Nadu, India

Felipe M. de Souza
Department of Chemistry
Kansas Polymer Research Center
Pittsburg State University
Pittsburg, Kansas

P. Dhanasekaran
CSIR-Scientist Pool Scheme,
    Fuel Cells Division
CSIR-Central Electrochemical Research
    Institute-Madras
Chennai, India

Zeliha Ertekin
Department of Chemistry
Hacettepe University
Beytepe, Turkey

Sehrish Gull
School of Physics and Optoelectronics
Shenzhen University
Shenzhen, China

Ram K. Gupta
Department of Chemistry
Kansas Polymer Research Center
Pittsburg State University
Pittsburg, Kansas

Jianhua Hou
School of Environmental Science and
    Engineering
Yangzhou University
Yangzhou, P. R. China

Linrui Hou
School of Materials Science &
    Engineering
University of Jinan
Jinan, China

Asif Hussain
School of Physics
College of Physical Science and
    Technology
Yangzhou University
Yangzhou, P. R. China
and
School of Environmental Science and
    Engineering
Yangzhou University
Yangzhou, P. R. China

Misbah Jabeen
Department of Chemistry
COMSATS University Islamabad
Islamabad, Pakistan

Shaan Bibi Jaffri
Department of Environmental Sciences
Fatima Jinnah Women University
Rawalpindi, Pakistan

Solen Kinayyigit
Nanocatalysis and Clean Energy
    Technologies Laboratory
Institute of Nanotechnology
Gebze Technical University
Gebze, Turkey
and
NANOTerial Technology Corporation
Istanbul, Turkey

Anuj Kumar
Department of Chemistry
GLA University
Mathura, India

Dinesh Kumar
School of Chemical Sciences
Central University of Gujarat
Gandhinagar, India

Chuanbo Li
School of Science
Minzu University of China
Beijing, P.R. China
and
Optoelectronics Research Center
Minzu University of China
Beijing, P.R. China

Yang Liu
School of Materials Science &
    Engineering
University of Jinan
Jinan, China

Mohammad Mazloum-Ardakani
Department of Chemistry
Faculty of Science
Yazd University
Yazd, Iran

Neeraj Mehta
Physics Department
Institute of Science
Banaras Hindu University
Varanasi, India

**Hamideh
Mohammadian-Sarcheshmeh**
Department of Chemistry
Faculty of Science
Yazd University
Yazd, Iran

Syed Ali Raza Naqvi
Department of Chemistry
Government College University
Faisalabad, Pakistan

Meena Nemiwal
Department of Chemistry
Malaviya National Institute of
    Technology
Jaipur, India

Demet Ozer
Department of Chemistry
Hacettepe University
Beytepe, Turkey

Shiv Kumar Pal
Physics Department
Institute of Science
Banaras Hindu University
Varanasi, India

Alok Kumar Rai
Department of Chemistry
University of Delhi (North Campus)
Delhi, India

Naresh A. Rajpurohit
School of Chemical Sciences
Central University of Gujarat
Gandhinagar, India

Tahir Rasheed
Interdisciplinary Research Center for
    Advanced Materials
King Fahd University of Petroleum and
    Minerals (KFUPM)
Dhahran, Saudi Arabia

Sajid Ur Rehman
School of Science
Minzu University of China
Beijing, P. R. China
and
Optoelectronics Research Center
Minzu University of China
Beijing, P. R. China

Zia Ur Rehman
School of Physics
College of Physical Science and
  Technology
Yangzhou University
Yangzhou, P. R. China
and
School of Environmental Science and
  Engineering
Yangzhou University
Yangzhou, P. R. China

Nur Ezyanie Safie
Fakulti Kejuruteraan Pembuatan
Universiti Teknikal Malaysia Melaka
Hang Tuah Jaya, Malaysia

Syed Sakhwat Shah
Department of Chemistry
Quaid-i-Azam University Islamabad
Islamabad, Pakistan

Tauqir A. Sherazi
Department of Chemistry
COMSATS University Islamabad
Abbottabad Campus, Pakistan

Jay Singh
Department of Chemistry
University of Delhi (North Campus)
Delhi, India

Rajesh Kumar Singh
Department of Physics &
  Astronomical Sciences
Central University of Himachal Pradesh
Dharamshala, India

Zhaolin Tan
School of Materials Science &
  Engineering
University of Jinan
Jinan, China

Zeeshan Tariq
School of Science
Minzu University of China
Beijing, P.R. China
and
State Key Laboratory on Integrated
  Optoelectronics
Institute of Semiconductors
Chinese Academy of Sciences
Beijing, P.R. China
and
University of Chinese Academy of
  Sciences
Beijing, P.R. China

Sami Ullah
Department of Physics
Division of Science and Technology
University of Education Lahore
Lahore, Pakistan

Abdullah Uysal
Nanocatalysis and Clean Energy
  Technologies Laboratory
Institute of Nanotechnology
Gebze Technical University
Gebze, Turkey

S. Vinod Selvaganesh
Department of Chemical Engineering
Indian Institute of Technology-Madras
Chennai, India

Jingxuan Wei
School of Materials Science &
  Engineering
University of Jinan
Jinan, China

Sachin Kumar Yadav
Physics Department
Institute of Science
Banaras Hindu University
Varanasi, India

Changzhou Yuan
School of Materials Science &
    Engineering
University of Jinan
Jinan, China

Maryam Zaqa
School of Physics
College of Physical Science and
    Technology
Yangzhou University
Yangzhou, P. R. China

Jinyang Zhang
School of Materials Science &
    Engineering
University of Jinan
Jinan, China

Yamin Zhang
School of Materials Science &
    Engineering
University of Jinan
Jinan, China

1 Importance of 2D Nanomaterials for Energy

Sami Ullah, Faheem K. Butt, and Junaid Ahmad
University of Education Lahore

CONTENTS

DOI: 10.1201/9781003178422-1

1.1   INTRODUCTION

Energy scavenging is growing in popularity across the board as a source of electricity for everything from the house to industry. In order to assure a consistent energy supply in the future, scientists are striving to develop green, affordable, and sustainable energy systems at a commercial scale. The discovery of graphene by exfoliating graphite has opened an exclusive research gate for scientists, and the study of two-dimensional nanomaterials (2D NMs) has become a growing area [1]. Researchers all around the globe including material scientists, chemists, engineers, and physicists have been working on economical and environmentally benign techniques to meet future energy demand. Materials having Van der Waals bonding and covalent bonding in the inter-layer and intra-layer, respectively, in atomically thick layered crystal structure are known as 2D NMs. Owing to their atomically thick nature and elevated surface area-to-volume ratio, 2D NMs display substantially different behaviors than their bulk counterparts, with unique properties related to quantum confinement [2]. Researchers are now working to develop practical applications for these fascinating nanomaterials. 2D NMs are being applied in (i) chemical energy scavenging, i.e., osmotic power generation; (ii) thermal energy scavenging, i.e., pyroelectric systems and thermoelectric; (iii) mechanical energy scavenging, i.e., piezoelectric and triboelectric devices; and (iv) solar energy scavenging, i.e., photocatalysis, photovoltaic cells, and perovskites [3]. Nanogenerators based on 2D NMs could be a viable solution for large-scale power generation from renewable sources including rolling wheels, ocean waves, and wind [4]. Furthermore, the energy provided by these nanogenerators can be used to power portable electronic gadgets, allowing for multifunctional applications such as code transmission and body motion sensing using a single generator that scavenges energy from the human body [5].

In this chapter, we devote our attention to fundamentals, rational design, synthesis routes, versatile properties, and potential of 2D NMs such as graphene, MXenes, graphitic carbon nitride (g-C_3N_4), transition metal dichalcogenides (TMDCs), and layered transition metal oxides (LTMOs) in energy-related applications. We hope that this chapter will not only encourage the developments of new 2D NMs with boosted utilization in energy applications, but will also help to address the challenges that these materials face in the production and storage of sustainable and renewable energy.

1.2   2D NANOMATERIALS: THE CENTER OF ATTENTION

What is the significance of 2D NMs? Do they have any unique characteristics that distinguish them from their bulk counterparts, or other nanomaterials, i.e., three-dimensional (3D) networks, one-dimensional (1D) nanostructures, and zero-dimensional (0D) nanoparticles? In this section, we are going to throw light on the unique properties of 2D NMs by giving a few examples. The first thing to notice in these materials is the electron confinement in two dimensions without interlayer interactions, particularly single-layer nanosheets. This allows highly fascinating electronic properties than other nanomaterials and makes them attractive nominees for electronic devices and condensed matter research field. The second important property that we are taking into account is their atomic thickness. This property makes 2D NMs ideal for transparent and ultra-flexible electrical and optoelectronic devices due to their outstanding optical transparency and mechanical flexibility. Last, lateral sizes with ultrathin thickness of these materials provide ultra-elevated specific surface areas that allow applying them in surface-active applications [6].

The exclusive electrical characteristics of graphene were the focus of early research. The so-called massless Dirac Fermions are one of the most notable characteristics that benefit from the electrical band structure and morphology. So, graphene is great for electrical devices because of its high conductivity and large charge carrier mobility.

Another well-researched 2D TMD nanosheet is MoS_2. When the thickness of this material is reduced to monolayer from bulk, then its indirect bandgap (1.3 eV) may be transformed to a direct bandgap (1.8 eV) [7].

The sensitivity of electronic structure to external stimuli (like mechanical deformation, chemical modification, doping, adsorption of other materials, and external electric fields) in 2D NMs is another appealing feature, which allows precise tuning of their electronic properties at a low cost [6]. Because of their in-plane covalent bonding and ultrathin thickness, 2D NMs have outstanding mechanical characteristics. Graphene has a breaking strength of 42 N/m and a Young's modulus of 1.0 TPa, according to experimental findings [8]. This strength indicates that graphene is the strongest as well as the thinnest material ever measured. More notably, graphene can withstand elastomeric deformations of up to 20%. Single-layer MoS_2 nanosheets have been shown to have a higher value of Young's modulus (270 GPa) compared to steel (205 GPa) and bulk MoS_2 (240 GPa) [9]. According to the calculations, covalently bonded 2D NMs exhibit outstanding mechanical properties [10]. Additionally, the 2D NMs' atomic thickness ensures their great optical transparency. 2D NMs are particularly appealing in the development of highly flexible, next-generation wearable, and transparent electrical/optoelectronic devices because they combine exceptional light transmittance, mechanical, and electronic properties. It's worth noting that graphene has a white light absorbance of about 2.3% and a negligible reflectance of 0.1%, representing outstanding light transmission.

In addition, 2D NMs have extraordinary ratios of exposed surface atoms and ultrahigh specific surface areas due to their huge lateral size and atomic thickness. This property is particularly advantageous in numerous surface-active applications including supercapacitors, photocatalysis, organic catalysis, and electrocatalysis where a large specific surface area is a vital need. For instance, graphene shows a theoretical specific surface area of 2,630 m²/g. Because all other 2D NMs have a 2D nanosheet structure, it is fair to expect that they will have high specific surface areas as well. 2D NMs are also promising building blocks for functional composites because of their elevated specific surface area. They can be utilized as reinforced fillers to strengthen composites as well as templates for directly preparing various forms of nanostructures on their surfaces [6].

1.3 GRAPHENE: AN OVERVIEW

1.3.1 BRIEF INTRODUCTION

Among the various 2D NMs, graphene is the most researched material because of its unique structure, zero bandgap, high thermal conductivity, great charge carrier mobility at room temperature, low-cost fabrication, impermeability to gases, large surface area, and tunable surface chemistry [11]. Owing to these incredible properties, graphene has become a spotlight for scientists in many applications including energy harvesting and conversion devices, optoelectronics, chemical sensors, coatings and nanocomposites, membranes, and various biomedical technologies [6]. Graphene is an atomically thick single-layered allotropic form of carbon having lattice-like honeycomb. Amazingly, in 1 mm of graphite, there are ~3 million graphene layers. Graphene, the strongest material ever discovered having hardness greater than steel, elasticity more than rubber, toughness higher than steel, and weight lighter than aluminum [12].

1.3.2 CRYSTAL STRUCTURE

Graphene takes the shape of a two-dimensional, atomically thick hexagonal lattice with one atom forming each vertex via sp^2 hybridization. The C–C bond has a length of 0.142 nm as depicted in Figure 1.1a [13]. Each lattice has three σ bonds with strong connections, producing a sturdy hexagonal structure. The vertical position of π-bond to the lattice plane is primarily responsible for graphene's electrical conductivity [14]. Because of its sp^2 orbital hybridization (s, p_x, and p_y orbitals combination which forms the σ-bond) and closely packed carbon atoms—graphene is extremely stable. The π-bond is formed by the last p_z electron. The π-band and π^*-bands are formed through the π-bonds hybridization. The partially filled band, which allows electrons free movement, is responsible for graphene's remarkable electrical characteristics. We can say that graphene is a unit structure that includes graphite, fullerene, and carbon nanotubes, as well as aromatic molecules of unlimited size, such as very planar polycyclic aromatic hydrocarbons. To put it another way, graphene is a honeycomb lattice plane made up of a tightly packed single layer of carbon atoms. In its single layer, carbon atoms form a benzene ring in which each atom gives an unpaired electron through sp^2 hybridization with neighboring carbon atoms.

Furthermore, comparable to CNTs, graphene edges may be categorized as zigzag or armchair based on various carbon chains, as illustrated in Figure 1.1b. The distinct types of edges result in a wide range of conducting behaviors. A graphene nanoribbon with a zigzag edge generally conducts electricity same as metal, but a nanoribbon with an armchair edge might conduct energy like a metal or a semiconductor [15].

1.3.3 FABRICATION

Graphene's quality plays an important role in determining its optical and electrical characteristics as impurities, defects, grain boundaries, structural disorders, numerous domains, and wrinkles all affect these properties. The demand for large-size samples in electronic applications is only achievable with the chemical vapor deposition (CVD) method; however, it is challenging to create good-quality single crystalline

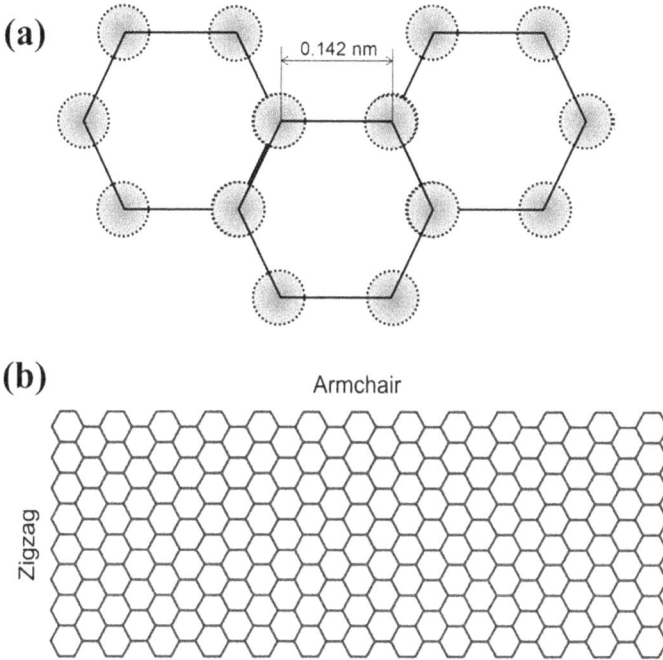

FIGURE 1.1 Hexagonal structure of graphene and graphene nanoribbon (a, b, respectively). (Adapted with permission from Ref. [14]. Copyright (2018) Elsevier.)

graphene thin films with extremely boosted thermal as well as electrical conductivities and outstanding optical transparency. Another bottleneck in the traditional synthesis of graphene is the employment of harmful chemicals, which generally result in the creation of hazardous waste and dangerous fumes. As a result, ecologically acceptable low-cost techniques for producing graphene must be developed.

The most common graphene synthesis techniques currently accessible are chemical vapor deposition, micromechanical cleavage, chemical reduction of exfoliated graphene oxide, unzipping of carbon nanotubes, epitaxial growth on SiC substrates, and liquid-phase exfoliation of graphite [16]. Nevertheless, depending on the application, each of these approaches might have its own set of benefits and drawbacks. To overcome these barriers to commercializing graphene, researchers from various companies, universities, and R&D institutes from all over the orb are working together to introduce novel routes for its large-scale production with high quality via facile and environmentally friendly methods.

1.3.4 PROPERTIES

One of the contributing factors why graphene is so exciting to the scientists working on molecular electronics is because of its electronic characteristics as it is the finest electrical conductor on the planet. The exclusive arrangement of carbon atoms in graphene permits electrons to move at extremely fast speeds without scattering, conserving energy that would otherwise be wasted in conventional conductors. Researchers discovered that graphene can transmit electricity even at supposedly zero carrier concentrations because electrons do not slow down or localize. The periodic potential of graphene's honeycomb lattice interacts with the electrons traveling around carbon atoms, resulting in new quasi-particles that have lost their mass (named as massless Dirac fermions). As a result, graphene is always conducting. They also move far quicker than electrons in other semiconductors [17].

The most fascinating features that make graphene an amazing reinforcing agent in nanocomposites are its intrinsic mechanical characteristics including toughness, stiffness, and strength. sp^2 bonds stability creates the hexagonal lattice and resists a range of in-plane deformations cause them [18]. Fracture toughness is considered as the significant mechanical characteristics of graphene, which was measured as a critical stress intensity factor of 4 ± 0.6 MPa. Fracture toughness is a feature that is extremely relevant to engineering applications [18]. The experimental and simulated breaking forces were nearly comparable, with an experimental value of 340 ± 50 N/m for second-order elastic stiffness. Supposing 0.335 nm effective thickness, this value translates to Young's modulus of 1 ± 0.1 TPa [18]. Monolayer graphene, which is defect-free, is thought to be the strongest material ever discovered, with a strength of 42 N/m and equivalent to the intrinsic strength of 130 GPa [18].

1.3.5 IMPORTANCE

Graphene is a game-changing material that has the potential to open up advanced industries and perhaps replace existing materials and technologies. The whole potential of graphene may be understood when it is employed both to improve existing materials and to transform them. When all of graphene's incredible qualities are combined, it might have an impact comparable to that of the Industrial Revolution. The incredible qualities of graphene allow it to have a substantial impact on a wide range of processes, products, and industries. No other substance possesses as many superlatives as graphene, making it perfect for wide-ranging applications. Transportation, electronics, medical, defence, energy, sensors, composites, coatings, membranes, and desalination are just some of the fields where graphene research is having an extensive impact.

Graphene-based nanomaterials have several intriguing uses in energy-related fields. For instance, they improve the charge rate and energy capacity of rechargeable batteries; used to develop highly efficient supercapacitors; graphene-based electrodes can make possible an encouraging route for developing lightweight, flexible, and inexpensive solar cells; and multifunctional-graphene mats are favorable substrates for catalytic mechanisms [19]. We can summarize importance of graphene in four primary energy-related fields such as supercapacitors, solar cells, graphene batteries, and catalysis.

1.4 MXENES: AN OVERVIEW

1.4.1 BRIEF INTRODUCTION

Since the discovery of Ti_3C_2 in 2011, the class of 2D transition-metal nitrides, carbonitrides, and carbides (MXenes) has grown fast. Surface terminations, such as fluorine (F_2), hydroxyl ($-OH$), or oxygen (O_2), are always present in the materials studied to date, and these terminations confer hydrophilicity to their surfaces. A total of 30 distinct MXenes have been prepared in laboratory and dozen more have been anticipated theoretically. The solid solutions availability, controlled surface terminations, and the novel finding of multi-transition-metal layered MXenes all open the door to the creation of a broad range of novel structures. Thermoelectric, optical, electrical, and plasmonic characteristics have also been demonstrated by these materials which make them promising for a wide spectrum of applications [20].

1.4.2 Crystal Structure

MXenes, a special type of 2D layered transition metal nitride or/and carbide formed by selective etching of raw MAX phases. Their general representation is $M_{n+1}AX_n$ ($n = 1, 2,$ or 3). Here M stands for the transition metal (i.e., V, Cr, Ti, Nb, and so on), A refers to element from group IVA or IIIA (e.g., Si, Sn, Al, In, so on), and X is nitrogen or/and carbon (Figure 1.2a). Crystal structures for three different MAX phases of MXenes are shown in Figure 1.2b. MAX phases show P63/mmc symmetry with a layered hexagonal structure, where X atoms fill the octahedral positions and M layers are almost hexagonally close-packed together. In the $M_{n+1}X_n$ layers, the element A is metallically bonded to the M element and interleaved. Using powerful etching solutions, such as HF, the A layer can be selectively etched from the MAX phases, resulting in MXenes with three distinct structures (M_4X_3, M_2X, or M_3X_2).

1.4.3 Fabrication

Several recent investigations have examined the fabrication of MXenes. In a nutshell, researchers at Drexel University initially disclosed multilayered MXenes ($Ti_3C_2T_x$) fabricated by etching an A layer from a MAX phase of Ti_3AlC_2 in 2011 [21]. The word "MXene" was introduced to differentiate this novel family of 2D NMs from graphene, and it refers to both the original MAX phases and MXenes made from them. Wet etching with HCl-NaF, hydrofluoric acid (HF), or HCl-LiF produced multilayered MXene flakes. As a result various MXenes such as $Mo_2Ti_2C_3$, Mo_2ScC_2, Cr_2TiC_2, Mo_2TiC_2, $(Nb_{0.8}Zr_{0.2})_4C_3$, Mo_2C, $(Nb_{0.8}Ti_{0.2})_4C_3$, Nb_4C_3, Nb_2C, TiNbC, Ti_3CN, Ta_4C_3, Ti_2C, Ti_4N_3, $(Ti_{0.5}Nb_{0.5})_2C$, Ti_3C_2, V_2C, Zr_3C_2, and $(V_{0.5}Cr_{0.5})_3C_2$ were yielded [24].

A historical timeframe for the fabrication of MXene is shown in Figure 1.2c. MXene ($Ti_3C_2T_x$) was the first multilayered MXene to be fabricated in 2011. Ti_2CT_x, $Ta_4C_3T_x$, Ti_3CNT_x, (V, Cr)$_3C_2T_x$, and $(TiNb)_2CT_x$ were among the MXenes discovered in 2012. Intercalation and delamination using organic molecules were used to isolate single-layer MXenes in 2013. MXenes were first synthesized in situ using HF etchants such as NH_4HF_2 or HCl-LiF in 2014. Large-scale delamination of different MXenes was shown in 2015 using an amine-assisted approach or Bu_4NOH. In 2017, $(Mo_{2/3}Sc_{1/3})_2Al$ was used to synthesize MXene with ordered divacancies ($Mo_{1.3}CT_x$) [24].

1.4.4 Properties

MXenes are exciting materials because of their promising features such as elevated surface areas, plenty of surface functionalities, exceptional hydrophilicity, easily

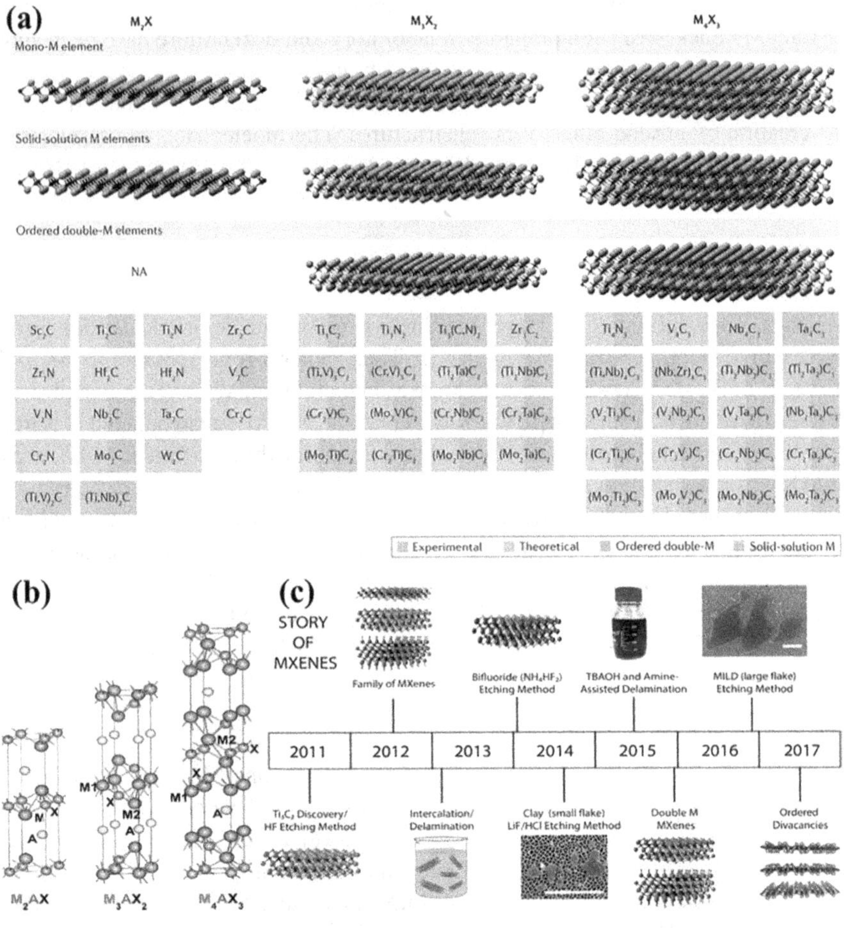

(a)

M$_2$X
Mono-M element

M$_3$X$_2$

M$_4$X$_3$

Solid-solution M elements

Ordered double-M elements

NA

Sc$_2$C	Ti$_2$C	Ti$_2$N	Zr$_2$C
Zr$_2$N	Hf$_2$C	Hf$_2$N	V$_2$C
V$_2$N	Nb$_2$C	Ta$_2$C	Cr$_2$C
Cr$_2$N	Mo$_2$C	W$_2$C	
(Ti,V)$_2$C	(Ti,Nb)$_2$C		

Ti$_3$C$_2$	Ti$_3$N$_2$	Ti$_3$(C,N)$_2$	Zr$_3$C$_2$
(Ti,V)$_3$C$_2$	(Cr,V)$_3$C$_2$	(Ti,Ta)C$_2$	(Ti,Nb)C$_2$
(Cr,V)C$_2$	(Mo,V)C$_2$	(Cr,Nb)C	(Cr,Ta)C$_2$
(Mo,Ti)$_2$C$_2$	(Cr,Ti)C$_2$	(Mo,Nb)C	(Mo,Ta)C$_2$

Ti$_4$N$_3$	V$_4$C$_3$	Nb$_4$C$_3$	Ta$_4$C$_3$
(Ti,Nb)$_4$C$_3$	(Nb,Zr)$_4$C$_3$	(Ti,Nb)$_4$C$_3$	(Ti,Ta)$_4$C$_3$
(V,Ti)$_4$C$_3$	(V,Nb)$_4$C$_3$	(V,Ta)$_4$C$_3$	(Nb,Ta)$_4$C$_3$
(Cr,Ti)$_4$C$_3$	(Cr,V)$_4$C$_3$	(Cr,Nb)$_4$C$_3$	(Cr,Ta)$_4$C$_3$
(Mo,Ti)$_4$C$_3$	(Mo,V)$_4$C$_3$	(Mo,Nb)$_4$C$_3$	(Mo,Ta)$_4$C$_3$

Experimental   Theoretical   Ordered double-M   Solid-solution M

(b)

M$_2$AX   M$_3$AX$_2$ M$_4$AX$_3$

(c) STORY OF MXENES

Family of MXenes

Bifluoride (NH$_4$HF$_2$) Etching Method

TBAOH and Amine-Assisted Delamination

MILD (large flake) Etching Method

| 2011 | 2012 | 2013 | 2014 | 2015 | 2016 | 2017 |

Ti$_3$C$_2$ Discovery/ HF Etching Method

Intercalation/ Delamination

Clay (small flake) LiF/HCl Etching Method

Double M MXenes

Ordered Divacancies

FIGURE 1.2 (a) MXenes family with three different MAX phases. (Adapted with permission from Ref. [21]. Copyright (2017) Springer Nature.) (b) Sketch of crystal structures of MXenes (MAX phases). (Adapted with permission from Ref. [22]. Copyright (2015) Elsevier.) (c) Historical timeframe for the fabrication of MXenes. (Adapted with permission from Ref. [23]. Copyright (2017) American Chemical Society.)

producing large quantities in water, outstanding chemical stability, high electrical conductivity, desired tunable electrical and optical properties, activated metallic hydroxide sites, synergistic effects of intercalants, efficient absorption of electromagnetic waves, high negative zeta-potential, ready to bond to various species, nontoxicity, and environmental benign characteristics [25].

1.4.5 Importance

MXenes are interesting materials for a wide range of applications due to their diverse chemistry and structure. MXenes were first used for energy storage, and this application has received the greatest attention. MXenes, on the other hand, may be able to surpass other materials in a variety of applications such as water purification, electromagnetic interference shielding, composites reinforcement, photothermal therapy, lubricants, electrocatalysts, gas- and bio-sensors, photocatalysts, and many more [20].

1.5 G-C$_3$N$_4$: AN OVERVIEW

1.5.1 BRIEF INTRODUCTION

In recent years, the development of the visible light responsive (VLR) photocatalysts has gained much attention of scientists for the remarkable usage of the visible light that is the major part of the solar spectrum. In this regard, a polymeric semiconductor known as g-C$_3$N$_4$ urged the researchers for the new class of VLR photocatalysts owing to its simple fabrication, attractive electronic band formation, and easy availability in nature. This material belongs to the 18th century when Liebig and Berzelius developed the embryonic type of melon [26].

1.5.2 CRYSTAL STRUCTURE

g-C$_3$N$_4$ has various phases of C$_3$N$_4$ including α-C$_3$N$_4$, g-h-triazine, β-C$_3$N$_4$, g-o-triazine, cubic C$_3$N$_4$, pseudocubic C$_3$N$_4$, and g-h-heptazine, and g-C$_3$N$_4$ is regarded as the most stable form. It was observed that the triazine and tri-s-triazine/heptazine rings are the fundamental tectonic units to develop the allotropes of g-C$_3$N$_4$ as shown in Figure 1.3a. This was according to the first principle density functional theory studies done by Kroke et al. [27]. Furthermore, the development of melon polymer from the melem by the polycondensation of urea, melamine or dicyanamide showed that the tecton was the most stable. Therefore, tri-s-triazine is considered as the basic building unit for the development of the g-C$_3$N$_4$ as shown in Figure 1.3a [28,29].

1.5.3 FABRICATION

g-C$_3$N$_4$ is a well-structured polymer, which comprises nitrogen and carbon elements. It can be synthesized by using the various nitrogen-containing precursors, which are easily available and economically suitable such as melamine, cyanamide, urea, thiourea, and dicyandiamide. Various approaches have been employed for the production of g-C$_3$N$_4$ including solvothermal, plasma sputtering deposition, chemical vapor deposition, and thermal polycondensation. Among all these methods, thermal polycondensation has gained much attention owing to its economic and facile approach. The nitrogen containing little molecules can polymerize into g-C$_3$N$_4$ while in the process of simple calcination method at 450°C–650°C. Electronic band framework and structural properties of developed g-C$_3$N$_4$ mainly depend upon the type of precursor. For example, g-C$_3$N$_4$ developed from thiourea shows shorter bandgap as compared to developed from urea. Similarly, g-C$_3$N$_4$ obtained from urea exhibits larger specific surface area as compared to that developed from melamine [31].

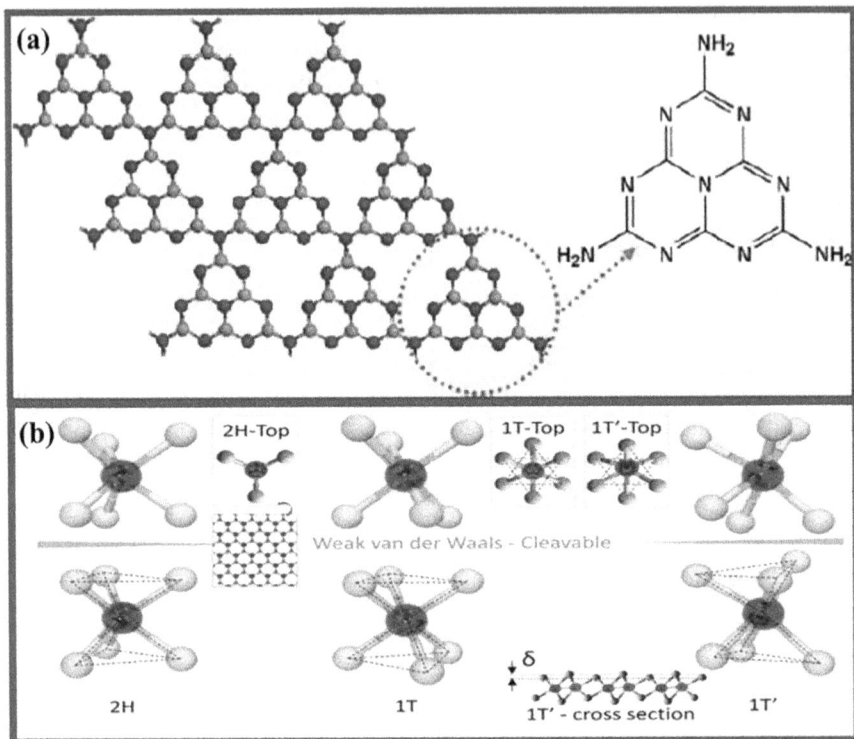

FIGURE 1.3 (a) Structure of g-C$_3$N$_4$. (Adapted with permission [28]. Copyright (2014) American Chemical Society.) (b) Framework of layered TMDCs. (Adapted with permission from Ref. [30]. Copyright (2017) Elsevier.)

1.5.4 PROPERTIES

g-C$_3$N$_4$ has grasped worldwide attraction as VLR photocatalyst owing to its enthralling properties such as adequate and tunable bandgap (~2.7 eV), suitable electronic band framework, safe, excellent stability, cheap, excellent thermal and chemical stability, and facile preparation. Furthermore, it consists of two non-metal and earth-abundant elements like carbon and nitrogen. Moreover, it can be easily prepared by utilizing the nitrogen-rich materials, which are low cost and easily available [29].

1.5.5 IMPORTANCE

Owing to the above-mentioned features, g-C$_3$N$_4$ gained much attention and considered as an effective VLR photocatalyst. It is appeared as a novel class of safe, next-generation, metal-free, and VLR polymer semiconductor material, which is easily available on earth. It is playing a pivotal role in various applications in different fields, e.g., energy storage and conversion and environmental remediation. Some of the important applications of this fascinating material are organic pollutant decontamination from water, hydrogen production, imaging, sensing, and energy storage as well as energy conversion, and many more [29].

1.6 TRANSITION METAL DICHALCOGENIDES (TMDCs): AN OVERVIEW

1.6.1 BRIEF INTRODUCTION

2D TMDCs are appeared as a novel family of materials with exceptional properties due to which these materials are suitable for various applications ranging from nano-electronics and nano-photonics to sensing applications. In current years, 2D TMDCs have grasped much consideration of researchers owing to the exposition of the first transistor and the detection of powerful photoluminescence (PL) in monolayers of MoS_2. TMDCs have very old and useful history. The structure of first TMDCs was observed in 1923 by Linus Pauling. At the end of the 1960s, almost 60 TMDCs were discovered which of them 40 have a layered framework. In 2004 rapid development in the research of graphene urged the development of technologies that are suitable for layered materials opened the new pathways for the study of TMDCs [32].

1.6.2 CRYSTAL STRUCTURE

TDMCs are well-defined layered materials and its every unit (MX_2) consists of a layer of transition metal (M), which is sandwiched between two layers of chalcogenides (X). The framework of TDMCs can be classified on the basis of the order of atoms into trigonal prismatic (hexagonal, H), octahedral (tetragonal, T), and their disordered phase (T′) as illustrated in Figure 1.3b. Generally, one transition metal and two chalcogen atoms are utilized to design MX_2, but in some cases, 2:3 is utilized to design (M_2X_3) [33] and 1:1 to design metal chalcogenides (MX) [34]. In H-phase material, every metal atom has six terminals out to two tetrahedrons in the direction of –z and +z and the hexagonal arrangement can easily be analyzed in Figure 1.3b. T-phase material has a trigonal chalcogen surface on the upper side and structure is rotated at 180° at the lower side in a single surface. If the atoms of metal are further mangled, then, the T′ phase is formed due to which atomic displacements occur in the atoms of chalcogen in z-direction [30].

1.6.3 FABRICATION

2D TMDCs exhibit excellent characteristics, but the development of these materials at large scale and defect-free atomic coatings with controlled thickness on required substrates is a difficult task. Mechanical exfoliation method can develop TMDCs of good quality monolayers, but this method is unable to use at a large scale. CVD is considered as a good approach to develop the TDMCs because this method has ability to produce TDMCs at a large scale and can control the morphology of the material. Latest modifications in the CVD process improved the standard of TMDCs layers. Moreover, atomic layer deposition and metal-organic CVD are also gaining attention owing to the development of wafer-scale and high-standard TMDCs films. TMDCs developed by these approaches have unique features such as uniformity on the whole substrate and enhanced performance as compared to those developed by the exfoliation technique [35].

1.6.4 PROPERTIES

TMDCs have a structure like graphene such as thin, flexible, and transparent. In contrast with graphene, 2D TMDCs have semiconductor behavior and are potentially used for the design of ultra-small and low-power transistors that are more proficient as compared to traditional silicon-based transistors that are fighting with the decreasing size of devices. Moreover, TMDCs have high carrier movement, on/off fraction with universal silicon and can be grown on flexible substrates and can endure stress as well as strain [36].

2D TMDCs show distinctive physiochemical, electrical, and optical features, which are due to the surface and quantum confinement effects that arise during the conversion of indirect bandgap into a direct bandgap when the pristine material is converted into monolayers. Furthermore, these materials exhibit higher surface/volume ratio, Van der Waals gap between every adjacent layer, high specific surface area, mechanically strong, and flexibility [30].

1.6.5 IMPORTANCE

As a result of tunable bandgap, these materials exhibit a powerful PL and higher exciton binding energy and considered as a potential element for different kinds of optoelectronic gadgets such as light-emitting diodes, solar cells, photo-detectors, and phototransistors [37]. For example, MoS_2 exhibits distinctive characteristics such as direct bandgap (~1.8 eV), better mobility, excellent current on/off fraction, high optical absorption, and huge PL emission due to direct bandgap in monolayer makes it suitable for electronics and optoelectronic devices [37]. Furthermore, due to sheet-like morphology and excellent unique properties, 2D TMDCs have grasped much attention toward energy storage applications including batteries, sensors, and supercapacitors [30]. Various new electronic and optoelectronic applications like tunneling transistors, photodetectors, flexible electronics, barristers, and LEDs can be designed by exploring promising features of 2D TMDCs.

1.7 LAYERED TRANSITION METAL OXIDES (LTMOs): AN OVERVIEW

1.7.1 BRIEF INTRODUCTION

The history of LTMOs is longer than that of other atomically thin materials. LTMOs are made up of a variety of naturally occurring minerals and have been used in lubricants, building materials, heat management, paints, and a variety of more purposes. Because O_2 strongly pulls the transition metal s electrons in LTMOs, the strongly correlated d electrons are accountable for the majority of the physical, chemical, and structural features. Because of the plenty of crystal structures, chemical composition, and relative simplicity of creating O_2 defects, LTMOs are also extremely tunable. When compared to their bulk counterparts, 2D LTMOs frequently have differing chemical and physical properties. These discrepancies result in a broad spectrum of exceptional electrical properties, including multiferroicity and high-temperature superconductivity, as well as unique optical, thermal, and mechanical phenomena [38]. LTMOs' catalytic and chemical properties can also be altered by lowering their thickness. Compared to TMDCs, LTMOs have gotten comparatively less research despite their unique features.

1.7.2 CRYSTAL STRUCTURE

Intercalation of a wide range of guest species, including anions, cations, and polymers, is possible due to the weak interlayer bonding and wide interlayer spacing of layered oxide materials. The transition metal–oxygen clusters that make up these multilayer structures are typically octahedral, or six-fold, coordinated by O_2 ligands. The strong interaction between the electrons of the O_2 ligand and the transition metal cation in layered oxides causes O_2 in adjacent layers to connect poorly to transition metals [39]. Through sharing edges, rarely, faces, and corners the octahedra are constructed into extended structures. The intralayer bonding is substantially stronger as compared to the interlayer bonding in LTMOs, which is a key property. Transition metals in high oxidation states, i.e., +4, +5, and +6—typically produce layered oxides.

The layered crystal structures of many redox-active layered oxides are presented in Figure 1.4a–f [40].

1.7.3 FABRICATION

Many LTMOs in nature are made up of negatively charged slices with alkaline-cations (such as Cs^+, K^+, and Rb^+) occupying the interlayer gaps. These slices are generally formed by edge- or corner-shared octahedral units of MO_6 (M = Mn, Ti,

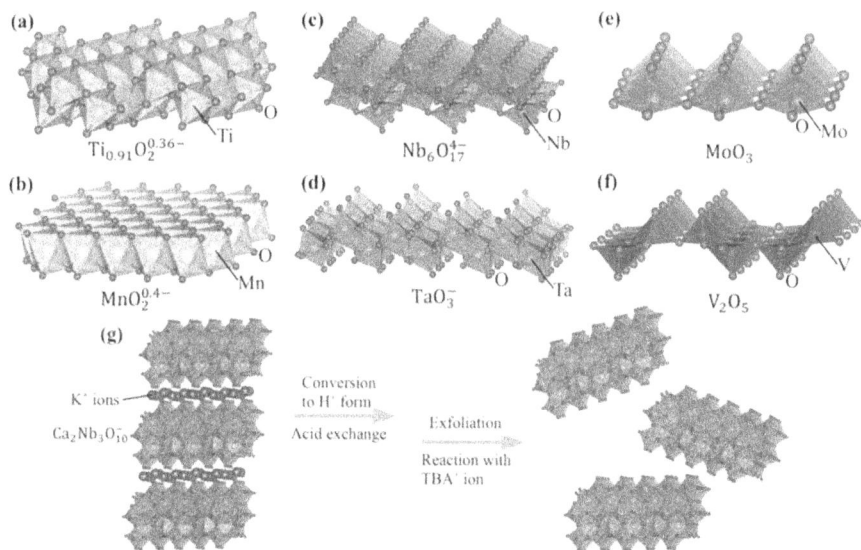

FIGURE 1.4 Crystal structures of $Ti_{0.91}O_2^{0.36-}$, $MnO_2^{0.4-}$, $Nb_6O_{17}^{4-}$, TaO_3^-, MoO_3, and V_2O_5 (a, b, c, d, e, and f, respectively). (g) The cation exchange–assisted liquid exfoliation method was used to exfoliate layered $KCa_2Nb_3O_{10}$ into $Ca_2Nb_3O_{10}^-$ nanosheets. (Adapted with permission from Ref. [40]. Copyright (2019) Wiley Online Library.)

Nb, Ta, W, Mo, Ru, and so on) that form ionic bondings with the surrounding alkaline cations, as seen in Figure 1.4. To convert these materials into 2D nanosheets, cation exchange–assisted liquid exfoliation was devised (Figure 1.4g). When treated with an acid solution, alkaline cations in the interlayer can be swapped for H^+ cations, resulting in hydrated protonic compounds. Because of their Bronsted solid acidity, the interlayer protons can be further substituted with organo-ammonium ions in an aqueous base solution, $(C_4H_9)_4NOH$. Such a process frequently introduces a huge water volume, resulting in a substantial reduction in interlayer electrostatic interaction and interlayer expansion. Exfoliation of the expanded compounds into metal oxide nanosheets using mechanical shaking or sonication is simple, i.e., $Ti_{0.91}O_2^{0.36-}$, $MnO_2^{0.4-}$, $Nb_6O_{17}^{4-}$, TaO_3^-, or $Ca_2Nb_3O_{10}^-$. Some LTMOs, such as α-MoO_3 and V_2O_5 are bounded by the weak Van der Waals forces also. Each layer of MoO_3 is composed of edge-shared distorted MoO_6 octahedra as shown in Figure 1.4e; whereas V_2O_5's layered anisotropic structure is made up of distorted trigonal-bipyramidal-polyhedral O atoms surrounding V atoms as depicted in Figure 1.4f. The 2D monolayers in these oxides can be created via liquid exfoliation or micro-mechanical cleavage methods [41].

1.7.4 PROPERTIES

LTMOs can be made out of a wide range of transition metals. The cationic species and their flexibility in changing oxidation state dictate many of the essential features of LTMOs. Because the cations attain multiple binding configurations and charge states, a vast range of structures can be stabilized, especially in 2D LTMOs. For the same LTMO layer at various stoichiometries, distinct oxidation states caused various electrical properties ranging from metallic to insulating behavior. Because of the confined nature of the d-electronic states, various metal–insulator transitions (i.e., Verwey and Mott transitions) occur as a function of temperature and pressure. LTMOs with different bandgaps are found in 2D, allowing for the engineering of electrical and optical characteristics nearly at any wavelength [42]. The redox characteristics of nanostructured LTMOs are different, and several of them exhibit reversible trends. Many 2D LTMOs also have outstanding chemical and thermal stability. A variety of well-known methods can be used to manipulate O_2 vacancies in nanostructured LTMOs, which allows for a great deal of customization. In addition, the formation energy of O_2 vacancies in LTMOs may decrease dramatically as the reducibility of metal cations increases, causing the surface chemical characteristics that differ significantly from the bulk structure [43]. In short, LTMOs are the intriguing class of materials with unique properties such as increased optical transparency, mechanical stress endurance, excellent carrier mobility, and so on [44].

1.7.5 IMPORTANCE

The features of LTMOs have aided in the creation of high energy density electrochemical storage. High ionic and electronic conductivity, the ability to undergo redox processes, and the availability of interlayer sites for cations intercalation are among these properties. We can say that LTMOs have an exclusive combination of chemical and physical features, as well as structural diversity, resulting their use in optics, catalysis, electronics, energy units (supercapacitors, fuel cells, solar cells, and batteries), biosystems, thermal units, sensors, superconductors, and piezoelectronics [44]. So far, a wide spectrum of 2D LTMOs has been synthesized, as well as nanocomposites, and superlattices of these planar oxides. Owing to extraordinary features, LTMOs have been used in plenty of important applications and they have capability to apply in many other novel applications in near future.

1.8 CONCLUDING REMARKS AND FUTURE OUTLOOK

From foundational research to the development of next-generation technology, 2D NMs research has progressed at a breakneck pace during the last decade. Undoubtedly, this significant advancement has transformed the distinctive function of dimensionality in influencing the intrinsic features of nanomaterials as well as their broad variety of possible applications. In this chapter, we categorized current advancements based on a variety of factors, including crystal structures and composition, fabrication routes, importance properties, and possible applications. After graphene, 2D NMs have developed as a novel class of nanomaterials that allows scientists to select materials with appropriate features for specific purposes. These 2D NMs can be made using a variety of well-established synthetic processes, each with its own set of benefits and drawbacks. Various fabrication routes may be utilized to generate 2D NMs with diverse morphological and structural properties like crystallinity, size, defect, thickness, crystal phase, strain, surface property, and doping, which are advantageous for a variety of applications. Atomically thick 2D NMs demonstrate high performance and the potential to replace current commercial technologies. Their application performance has been shown to be a function of their meticulously constructed structural properties.

Extensive research into atomically thick 2D NMs also presents some obstacles. First, in terms of material synthesis, quantity, production rate, quality, and yield of 2D NMs are still far from the industry or commercialization standards. To address industry demands, one of the primary difficulties is to develop improved yield and mass production of these 2D NMs. The structural features of 2D NMs determine their chemical, physical, and electrical properties, as well as their performance in applications. Most existing well-developed technologies, however, make it challenging to achieve a controlled production of atomically thick 2D NMs with desired morphologies. As a result, one of the problems in this sector is to prepare atomically thick 2D NMs with desirable structural features in a highly controllable manner. Second, knowing the growth mechanism of atomically thick 2D NMs is vital from a characterization standpoint, but it is not straightforward. As a result, finding or developing appropriate characterization tools to investigate the growth mechanism of atomically thick 2D NMs is another issue in this sector. Some promising in situ characterization techniques, including in situ TEM, in situ XPS, and in situ Raman spectroscopy, have been developed in recent years, having potential importance for researching the growth mechanism of atomically thick 2D NMs. Third, because of their limited chemical or/and physical stability, most contemporary atomically thick 2D NMs lack longstanding durability and stability, limiting their potential uses.

Their lack of stability is primarily due to the following circumstances: (i) Long-term storage of 2D NMs distributed in liquid solution is impossible. Due to their high proclivity for irreversible aggregation, they lose a large amount of progress resulting from their 2D structural traits. (ii) In ambient settings, most 2D NMs are easily oxidized, resulting in structural decomposition/degradation. (iii) In applications like LIBs and electrocatalysis, structural decomposition, collapse, or change may occur during the chemical process. As a result, the 2D NM's stability is obviously crucial, and it must be sustained not only throughout processing and storage but also during use. As a result, one of the most pressing concerns in this sector is to find simple but effective ways to stabilize these 2D NMs in order to greatly extend their stability.

Despite significant advances, research on 2D NMs, particularly those beyond graphene, is still in its early stages. First, because atomically thick 2D NMs are defined by their dimensionality rather than their compositions, we consider that any type of atomically thick 2D NMs could be made if their growth could be restricted into 2-dimensions and down to single- or few-atomic layers under the precise conditions. Consequently, the key purpose is to apply those well-established technologies and discover innovative approaches to fabricate novel 2D NMs that should exhibit novel functionalities and properties. Ion intercalation-assisted liquid exfoliation micromechanical cleavage, and sonication-assisted liquid exfoliation, are some examples of well-developed exfoliation technologies. Second, it's important to note that each material has its own set of drawbacks. Hybridizing a material with other materials to make heteronanostructures or composites is the simplest technique to overcome a material's disadvantage. Graphene has been frequently employed as a extremely conductive matrix to incorporate with other low conductive materials, for instance metal oxides and TMDs, to improve their electrical conductivity and hence increase their performance in desirable applications such as LIBs, supercapacitors, electrocatalysis, and so on.

More fascinatingly, the synergistic action of many components may result in some new desirable functionalities or features. Both cobalt and cobalt oxide nanosheets have been shown to have moderate electrocatalytic activities, however, their composite produced outstanding activity for CO_2 reduction. Considering this, one of the most interesting future research approaches is to produce nanocomposites or distinct heterostructures employing atomically thick 2D NMs as building blocks, thereby improving their characteristics as well as capabilities. Third, material preparation research is always conducted before the prospective uses of the materials are investigated. Even though a variety of atomically thick 2D NMs have been developed in recent years, the majority of their potential uses have yet to be investigated. Even though MoS_2 research dates back decades, its excellent Hydrogen evolution reaction (HER) activity was only discovered around 10 years ago. Given that these freshly produced atomically thick 2D NMs may have exceptional performance in some unknown applications, identifying the best appropriate use for each atomically thick 2D NMs is another intriguing future direction.

ACKNOWLEDGEMENT

The authors acknowledge HEC for grant 7435/Punjab/NRPU/R&D/HEC/2017.

REFERENCES

1. Novoselov, K.S., A.K. Geim, S.V. Morozov, D.-E. Jiang, Y. Zhang, S.V. Dubonos, I.V. Grigorieva, and A.A. Firsov, Electric field effect in atomically thin carbon films. *Science*, 2004. **306**(5696): pp. 666–669.
2. Hu, J., Z. Guo, P.E. Mcwilliams, J.E. Darges, D.L. Druffel, A.M. Moran, and S.C. Warren, Band gap engineering in a\ 2D material for solar-to-chemical energy conversion. *Nano letters*, 2016. **16**(1): pp. 74–79.
3. Zhang, H., Ultrathin two-dimensional nanomaterials. *ACS Nano*, 2015. **9**(10): pp. 9451–9469.
4. Han, S.A., J. Lee, J. Lin, S.-W. Kim, and J.H. Kim, Piezo/triboelectric nanogenerators based on 2-dimensional layered structure materials. *Nano Energy*, 2019. **57**: pp. 680–691.
5. Dong, Y., S.S.K. Mallineni, K. Maleski, H. Behlow, V.N. Mochalin, A.M. Rao, Y. Gogotsi, and R. Podila, Metallic MXenes: A new family of materials for flexible triboelectric nanogenerators. *Nano Energy*, 2018. **44**: pp. 103–110.
6. Geim, A.K. and K.S. Novoselov, The rise of graphene. In: P. Rodgers (Ed.) *Nanoscience and Technology: A Collection of Reviews from Nature Journals*. 2010, Singapore: World Scientific, pp. 11–19.
7. Meyer, J.C., A.K. Geim, M.I. Katsnelson, K.S. Novoselov, T.J. Booth, and S. Roth, The structure of suspended graphene sheets. *Nature*, 2007. **446**(7131): pp. 60–63.
8. Lee, C., X. Wei, J.W. Kysar, and J. Hone, Measurement of the elastic properties and intrinsic strength of monolayer graphene. *Science*, 2008. **321**(5887): pp. 385–388.
9. Ganatra, R. and Q. Zhang, Few-layer MoS_2: A promising layered semiconductor. *ACS Nano*, 2014. **8**(5): pp. 4074–4099.
10. Ataca, C., H. Sahin, and S. Ciraci, Stable, single-layer MX_2 transition-metal oxides and dichalcogenides in a honeycomb-like structure. *The Journal of Physical Chemistry C*, 2012. **116**(16): pp. 8983–8999.
11. Geim, A.K., Graphene: Status and prospects. *Science*, 2009. **324**(5934): pp. 1530–1534.
12. Allen, M.J., V.C. Tung, and R.B. Kaner, Honeycomb carbon: A review of graphene. *Chemical Reviews*, 2010. **110**(1): pp. 132–145.
13. Yan, Y., W.I. Shin, H. Chen, S.-M. Lee, S. Manickam, S. Hanson, H. Zhao, E. Lester, T. Wu, and C.H. Pang, A recent trend: Application of graphene in catalysis. *Carbon Letters*, 2020. **31**: pp. 1–23.
14. Zhen, Z. and H. Zhu, Structure and properties of graphene. In: H. Zhu (Ed.) *Graphene*. 2018, New York: Elsevier. pp. 1–12.
15. Son, Y.-W., M.L. Cohen, and S.G. Louie, Energy gaps in graphene nanoribbons. *Physical Review Letters*, 2006. **97**(21): p. 216803.
16. Choi, W., I. Lahiri, R. Seelaboyina, and Y.S. Kang, Synthesis of graphene and its applications: A review. *Critical Reviews in Solid State and Materials Sciences*, 2010. **35**(1): pp. 52–71.
17. Bøggild, P., D.M. Mackenzie, P.R. Whelan, D.H. Petersen, J.D. Buron, A. Zurutuza, J. Gallop, L. Hao, and P.U. Jepsen, Mapping the electrical properties of large-area graphene. *2D Materials*, 2017. **4**(4): p. 042003.
18. Papageorgiou, D.G., I.A. Kinloch, and R.J. Young, Mechanical properties of graphene and graphene-based nanocomposites. *Progress in Materials Science*, 2017. **90**: pp. 75–127.
19. Choi, H.-J., S.-M. Jung, J.-M. Seo, D.W. Chang, L. Dai, and J.-B. Baek, Graphene for energy conversion and storage in fuel cells and supercapacitors. *Nano Energy*, 2012. **1**(4): pp. 534–551.
20. Anasori, B., M.R. Lukatskaya, and Y. Gogotsi, 2D metal carbides and nitrides (MXenes) for energy storage. *Nature Reviews Materials*, 2017. **2**(2): p. 16098.
21. Anasori, B., M.R. Lukatskaya, and Y. Gogotsi, 2D metal carbides and nitrides (MXenes) for energy storage. *Nature Reviews Materials*, 2017. **2**(2): pp. 1–17.

22. Dhakal, C., S. Aryal, R. Sakidja, and W.-Y. Ching, Approximate lattice thermal conductivity of MAX phases at high temperature. *Journal of the European Ceramic Society*, 2015. **35**(12): pp. 3203–3212.

23. Alhabeb, M., K. Maleski, B. Anasori, P. Lelyukh, L. Clark, S. Sin, and Y. Gogotsi, Guidelines for synthesis and processing of two-dimensional titanium carbide ($Ti_3C_2T_x$ MXene). *Chemistry of Materials*, 2017. **29**(18): pp. 7633–7644.

24. Jun, B.-M., S. Kim, J. Heo, C.M. Park, N. Her, M. Jang, Y. Huang, J. Han, and Y. Yoon, Review of MXenes as new nanomaterials for energy storage/delivery and selected environmental applications. *Nano Research*, 2019. **12**(3): pp. 471–487.

25. Anasori, B., Y. Xie, M. Beidaghi, J. Lu, B.C. Hosler, L. Hultman, P.R. Kent, Y. Gogotsi, and M.W. Barsoum, Two-dimensional, ordered, double transition metals carbides (MXenes). *ACS Nano*, 2015. **9**(10): pp. 9507–9516.

26. Zheng, Y., L. Lin, B. Wang, and X. Wang, Graphitic carbon nitride polymers toward sustainable photoredox catalysis. *Angewandte Chemie International Edition*, 2015. **54**(44): pp. 12868–12884.

27. Kroke, E., M. Schwarz, E. Horath-Bordon, P. Kroll, B. Noll, and A.D. Norman, Tri-s-triazine derivatives. Part I. From trichloro-tri-s-triazine to graphitic C_3N_4 structures. *New Journal of Chemistry*, 2002. **26**(5): pp. 508–512.

28. Cao, S. and J. Yu, g-C_3N_4-based photocatalysts for hydrogen generation. *The Journal of Physical Chemistry Letters*, 2014. **5**(12): pp. 2101–2107.

29. Ong, W.-J., L.-L. Tan, Y.H. Ng, S.-T. Yong, and S.-P. Chai, Graphitic carbon nitride (g-C_3N_4)-based photocatalysts for artificial photosynthesis and environmental remediation: Are we a step closer to achieving sustainability? *Chemical Reviews*, 2016. **116**(12): pp. 7159–7329.

30. Choi, W., N. Choudhary, G.H. Han, J. Park, D. Akinwande, and Y.H. Lee, Recent development of two-dimensional transition metal dichalcogenides and their applications. *Materials Today*, 2017. **20**(3): pp. 116–130.

31. Fu, J., J. Yu, C. Jiang, and B. Cheng, g-C_3N_4-based heterostructured photocatalysts. *Advanced Energy Materials*, 2018. **8**(3): p. 1701503.

32. Manzeli, S., D. Ovchinnikov, D. Pasquier, O.V. Yazyev, and A. Kis, 2D transition metal dichalcogenides. *Nature Reviews Materials*, 2017. **2**(8): pp. 1–15.

33. Zhang, J., Z. Peng, A. Soni, Y. Zhao, Y. Xiong, B. Peng, J. Wang, M.S. Dresselhaus, and Q. Xiong, Raman spectroscopy of few-quintuple layer topological insulator Bi_2Se_3 nanoplatelets. *Nano Letters*, 2011. **11**(6): pp. 2407–2414.

34. Zhou, X., J. Cheng, Y. Zhou, T. Cao, H. Hong, Z. Liao, S. Wu, H. Peng, K. Liu, and D. Yu, Strong second-harmonic generation in atomic layered GaSe. *Journal of the American Chemical Society*, 2015. **137**(25): pp. 7994–7997.

35. Shi, Y., H. Li, and L.-J. Li, Recent advances in controlled synthesis of two-dimensional transition metal dichalcogenides via vapour deposition techniques. *Chemical Society Reviews*, 2015. **44**(9): pp. 2744–2756.

36. Akinwande, D., N. Petrone, and J. Hone, Two-dimensional flexible nanoelectronics. *Nature Communications*, 2014. **5**(1): pp. 5678.

37. Fuhrer, M.S. and J. Hone, Measurement of mobility in dual-gated MoS_2 transistors. *Nature Nanotechnology*, 2013. **8**(3): pp. 146–147.

38. Kalantar-zadeh, K., J.Z. Ou, T. Daeneke, A. Mitchell, T. Sasaki, and M.S. Fuhrer, Two dimensional and layered transition metal oxides. *Applied Materials Today*, 2016. **5**: pp. 73–89.

39. Augustyn, V., Tuning the interlayer of transition metal oxides for electrochemical energy storage. *Journal of Materials Research*, 2017. **32**(1): pp. 2–15.

40. Yang, T., T.T. Song, M. Callsen, J. Zhou, J.W. Chai, Y.P. Feng, S.J. Wang, and M. Yang, Atomically thin 2D transition metal oxides: Structural reconstruction, interaction with substrates, and potential applications. *Advanced Materials Interfaces*, 2019. **6**(1): p. 1801160.

41. Rui, X., Z. Lu, H. Yu, D. Yang, H.H. Hng, T.M. Lim, and Q. Yan, Ultrathin V_2O_5 nanosheet cathodes: Realizing ultrafast reversible lithium storage. *Nanoscale*, 2013. **5**(2): pp. 556–560.

42. Sasaki, T. and M. Watanabe, Semiconductor nanosheet crystallites of quasi-TiO_2 and their optical properties. *The Journal of Physical Chemistry B*, 1997. **101**(49): pp. 10159–10161.

43. Carrasco, J., N. Lopez, F. Illas, and H.-J. Freund, Bulk and surface oxygen vacancy formation and diffusion in single crystals, ultrathin films, and metal grown oxide structures. *The Journal of Chemical Physics*, 2006. **125**(7): p. 074711.

44. Meyer, J., S. Hamwi, M. Kröger, W. Kowalsky, T. Riedl, and A. Kahn, Transition metal oxides for organic electronics: Energetics, device physics and applications. *Advanced Materials*, 2012. **24**(40): pp. 5408–5427.

2 Graphene-Based 2D Nanomaterials for Fuel Cells

Manahil Bakhtiar and Farman Ali
Hazara University

Nisar Ali
Huaiyin Institute of Technology

Syed Sakhwat Shah
Quaid-i-Azam University Islamabad

Muhammad Bilal
Huaiyin Institute of Technology

CONTENTS

DOI: 10.1201/9781003178422-2

2.1 INTRODUCTION

Because of the fast expansion as well as evolution of industrialization, demand for energy is rapidly growing. Current renewable and non-renewable energy supplies, such as natural gas, coal, and minerals, are diminishing and have a number of disadvantages, including expensive costs, poor efficiency, global climate change but also contamination [1]. To solve the aforementioned disadvantages, research is underway to investigate low-cost, environmentally beneficial durable, and extremely effective energy transmission and storage technologies. Electrochemical fuel cell innovation is providing a key role in delivering clean as well as renewable energy conversion systems for such continuous supply of electricity [2]. Low-temperature fuel cells are many of the maximum promising applicants for the dependable and green conversion of hydrogen or alcohol into electric powered electricity in automotive, allotted electricity generation, and transportable digital packages on a huge scale [3]. Graphene sheets are a top-notch catalytic supporter in low-temperature gasoline or fuel cells due to its big surface area, excessive conductivity, precise graphitized basal aircraft structure, plus in all likelihood reasonably priced fabrication price.

2.2 FUEL CELLS

The hunt for affordable and healthy alternative resource for depleting fossil energy supplies has intensified dramatically throughout the period of nanotechnology. Fuel cells, which can produce fuel effectively utilizing alternative fuels, are needed to address the expanding worldwide environmental problems generated by the combustion of fossil fuels. The maintenance of our upcoming surroundings and the development of large renewable energy infrastructures are two of the most pressing concerns that mankind currently faces in the twenty-first generation. When it comes to energy production technologies, the expressions "sustainable energy" and "renewable energy" could be altered. Fuel cells have accomplished good advances in commercialization following its beneficial use in the Apollo lunar expedition as a result; fuel cells seem to be the foremost essential approach for producing renewable energy, which might meet projected energy demands [4]. A traditional fuel cell consists of three basic modules, which include electrolyte membrane, which is packed in between bipolar plates and two porous electrodes are sandwich between these bipolar plates [5]. Bipolar plates are a critical feature of fuel cells because they ensure an equal allocation of fuel/air. Bipolar plates create electrical conduction within the cell, restrict additional gases and coolant from leaking, and eliminate unwanted heat out from catalyst surface, among other benefits. Non-porous graphitic carbon and conducting polymers are perhaps the most common materials employed for bipolar plates [6]. The electrodes must be electronic conductors, while the electrolyte must be ionic. Polystyrene, polyaniline, poly-methyl methacrylate, and polyethylene-di-oxy-thio-phene are all common conducting electrolytes [7]. On the surface of the anode, the fuel is oxidized, and the liberated electrons pass through an outer circuit that reduces O_2 at the cathode. To complete the circuit, the mobile free electrons migrate via electrolytes at the same time. The fuels are classed as phosphoric acid fuel cell (PAFC), polymer electrolyte membrane fuel cell (PEMFC), alkaline fuel cell (AFC), molten carbonate fuel cell (MCFC), and solid-oxide fuel cell (SOFC) based upon the varieties of electrolytes used. PAFC, PEMFC, and AFC are appropriate to automobiles and versatile applications because they function at low temperatures (300°C), but MCFC and SOFC may use a variety of fuels owing to high working temperatures (500°C) and are interesting to stationary applications. Electrochemical, conductivity effectiveness, and endurance are all factors to consider when choosing materials for fuel cell assembly. Graphene as well as its substitutes have exceptional chemical, electrical plus mechanical properties, making these viable resources in fuel cell applications. In recent times, a lot of research has gone into maximizing the possible usage of 2D Graphene-based materials in fuel cells applications [8]. Figure 2.1 represents fuel cell with graphene-based nanomaterials in each component.

2.3 GRAPHENE

Carbon, a past, present, and future foundation of chemistry, has acquired importance in recent years winning the Nobel Prize in Physics in 2010. Electrode innovations, lubricants, carbon-fiber-reinforced polymers, pencil lead, and nanotechnology have

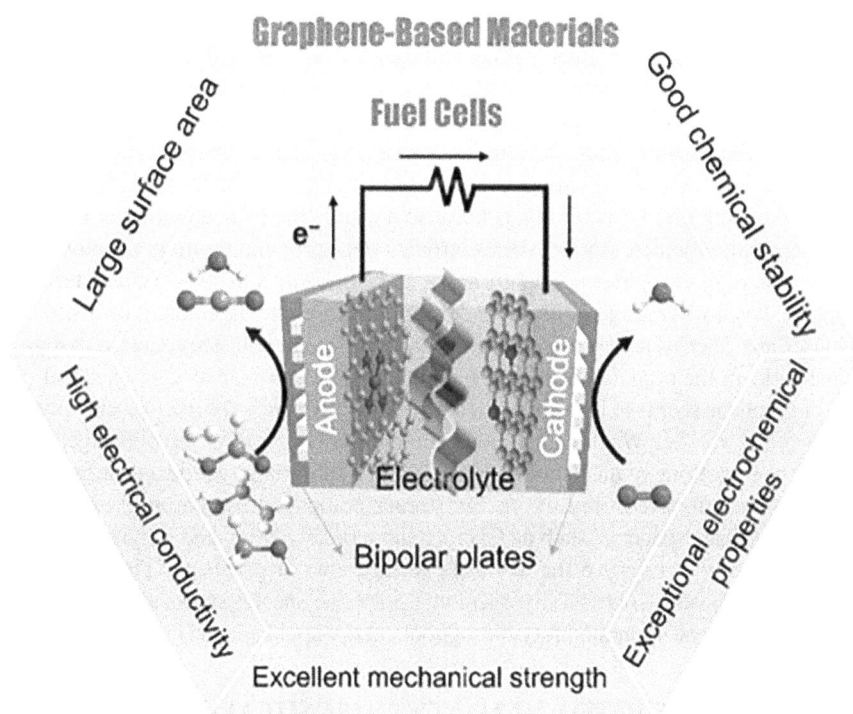

FIGURE 2.1 Schematic representation of fuel cell having Graphene-based nanomaterials in each component. (Adapted with permission from [8]. Copyright © 2020 The Authors. Society of Chemical Industry and John Wiley & Sons Ltd. This is an open access article under the terms of the Creative Commons Attribution License.)

all utilized wide variety of carbon developed at distinct intervals and in varied formats which including C_{12}, coal, and even gasoline [9]. But after the discovery of graphene around 2004 by two prominent physicists representing Manchester University, *K. S. Novoselov* and *Andre Geim* was indeed a huge advance within the discipline of material research. Its breakthrough gave birth to a revolutionary age, and graphene has quickly risen to prominence. Furthermore, P.R. Wallace investigated single-layer graphene theoretically around 1947 and in 1962, *Boehm* along with his companions reported the results of their graphite flakes research. In 1961, he used transmission electron microscopy (TEM) and X-ray diffraction (XRD) to extract and identify single graphene sheets. The name "graphene" was primarily introduced to denote single sheets of graphite during 1987, as well as it was also adopted to represent carbon nanotubes (CNTs) in the early 1990s. Chemists discovered ways to put carbon in graphene monolayers onto other materials in the early 1970s. The two-dimensional (2D) graphene nanomaterial is a monolayer graphite sheet that is millimeters thick as well as it has each single carbon atom sp^2-hybridized and organized in some kind of a honeycomb-structure or hexagonal lattice. Such type of compact packed organization gives it exceptional strength and enables it the world's best material [10]. Ever since successful discovery in 2004, graphene has attracted considerable attention in a variety of domains, including fundamental physics, chemistry, materials research, includes mobile applications. The unusual one-atomic thick but also 2D properties of graphene as well as its great mechanical toughness along with the chemical inertness bring up plenty of possible material uses, notably used as separation membranes mostly in fuel cells applications [11]. We know that in graphene sp_2 hybridization exists between carbon bonds, in these the bonds which are in-plane having $\sigma_{c\text{-}c}$ bonding is considered to be the strongest bonds and as compared to these weakest bonds are those having out of plan π bonds because it causes the delocalization of electrons and as a result of which it offers the electronic conduction in graphene structure [12].

Because of its extraordinary composition and extremely anisotropic characteristics, graphene has a wide range of uses. To develop macroscopic parameters its distinctive qualities are determined by the alignment and displacement of sheets in an ensemble framework. In the domains of power generation with sensors, every geometrical pattern of graphene seems to have its particular practical benefits. Owing to such powerful stacking and van der Waals interactions among graphene sheets, graphene generally utilized as something of an electrode material has a substantially reduced surface area, limiting their effective utility for energy storage equipment. Research has been conducted to utilize "spacers," such as CNTs, nanoparticles (NPs), and polymers, among graphene sheets to improve the available surface area of graphene. These forms of graphene composites successfully overcome graphene sheet restacking concerns and improved graphene's functionality throughout energy applications [13].

2.4   GRAPHENE-BASED 2D NANOCOMPOSITE MATERIALS

Because of inherent exceptional qualities, composites are now receiving a lot of emphasis. Composite materials have stimulated the interests of the 2D material scientific fraternity in the latter years of the 20 eras. Once composited, the high surface-to-volume proportion of 2D material is indeed a major benefit. The 2D-material-based nanocomposite had higher mechanical strengths but also flexibility, increased electrical conductance, excellent optical transparency, resistive to chemical as well as flame endurance, maximum tensile ratios, fatigue stiffness, and overall resistance to corrosive environment are some of the exceptional features of this material [14]. Composite materials have been created by integrating two or more materials having diverse features to create a finished product containing distinct functionalities. Composite materials have been made up of two or more materials, which have been combined to take leverage of every material's finest traits and attributes [15]. Composites are frequently constructed to provide a diverse set of features and properties [16], including the following:

• Stiffness and toughness
• Low expansion coefficient
• Tolerance to exhaustion
• Constructing complicated forms is simple.
• Destroyed structures can be easily repaired.

Nanocomposites had already gotten a lot of publicity recently. Any nanocomposite is a composite in which at least one of its parts, often filler, has proportions of 1–100 nm. Nanomaterials typically introduced into the matrix for specific functions such as toughness, resistivity, electrical conductivity, and magnetic characteristics, among others [17]. Composite as well as nanocomposite classification is divided into numerous groups. The nature of matrix material would be one of those criteria. Here seem to be a variety of composites based on just this foundation, such as:

• **Polymer matrix composites (PMCs)**
  Such composites have quite a superior grade for commercial applications when matched to different varieties of composites. In this form of composite, a variety of filler materials might be incorporated. The thermoplastic as well as thermoset polymers are being used to create this matrix.
• **Metal matrix-composites (MMCs)**
  Non-metal fillers in such a metal matrix make up metal matrix composites (MMCs), which are regarded innovative construction materials. MMCs are mostly employed for engineering disciplines when the working temperature is between 250°C and 750°C. Such composites are commonly made of copper, aluminum, titanium, and super alloys.
• **Ceramic matrix composites (CMCs)**
  This is indeed a type of innovative construction material that consists of metal and non-metallic fillers in a ceramic matrix. CMCs are utilized in engineering fields when the working temperature is between 800°C and 1,650°C [18].
  In most recent time's graphene has also been employed to construct graphene-based noble metal nanocomposites, which is a revolutionary yet prospective technology. Graphene has also been synthesized using a variety of processes up to this point. Chemical exfoliation has indeed been recognized being one of the most promising approaches for producing graphene oxide

(GO) and reduced graphene oxide (rGO) sheets on a massive level even at a reasonable expense. Remarkably, the GO and rGO generated by that kind of method include a large number of reactive oxygen-containing groups, allowing for extensive remodeling but also functionalization. Utilizing GO or rGO nanosheets, a variety of products have been created, including organic crystals, metal-organic frameworks, inorganic nanostructures, CNTs, and biomaterials. Photo-voltaic devices, batteries, super-capacitors, surface-enhanced Raman scattering, and fuel cells are just a few of the possible advantages for most of these composites [19].

2.5   SYNTHESIS OF GRAPHENE AND GRAPHENE-BASED 2D NANOCOMPOSITES

There can be presently a variety of processes for producing 2D materials, the picking of which will be based on the particular materials but also purpose requirements. Bottom-up procedures, such as chemical vapor transport, chemical vapor deposition (CVD), or other physical deposition methods like magnetron sputtering, physical vapor deposition, or molecular beam epitaxy, including colloidal growth, are perhaps the main popular synthesizing procedures. Top-down techniques apart from mechanical exfoliating are indeed extensively used, including exfoliation often taking place in a liquid media as occasionally aided by chemical processes. Surprisingly, so much of the approaches outlined above may be used to make nanostructures with graphene. [20]. Geim and colleagues successfully devised a micromechanical cleavage approach that effectively by using scotch tape exfoliated graphene from pristine graphite [21]. In numerous investigations, GO is generally used as a precursor to create graphene as well as its associated composites, and this can be manufactured from graphite through the enhanced Hummers process [22]. With the discovery of graphene in 2004 through mechanical exfoliation with Scotch-tape, a variety of methods for producing high-quality, large-area graphene are being developed. It's critical to know how and why the graphene was made considering the characteristics of graphene are heavily influenced by the procedures used to make it. The published methodologies can be divided into two main categories: "top-down" and "bottom-up" procedures [23]. Figure 2.2 represents the techniques through which graphene can be synthesized. Top-down include techniques such as sonication, reduction of graphene oxide, electrochemical exfoliation, unzipping CNTs, micromechanical exfoliation, arc discharge, and thermal exfoliation could be used to make graphene. Pyrolysis, CVD, and epitaxial growth on silicon carbide are the additional graphene generation processes that fall under the bottom-up approach. High-quality review papers on the various graphene production procedures and, in specifically, the advantages but also drawbacks of various fabrication approaches were published [24].

2.5.1   MICROMECHANICAL EXFOLIATION USING SCOTCH TAPE

Micromechanical exfoliation can produce atomically skinny, however, numerous layers thick sheets of 2D layered inorganic materials. This process is similar to the one

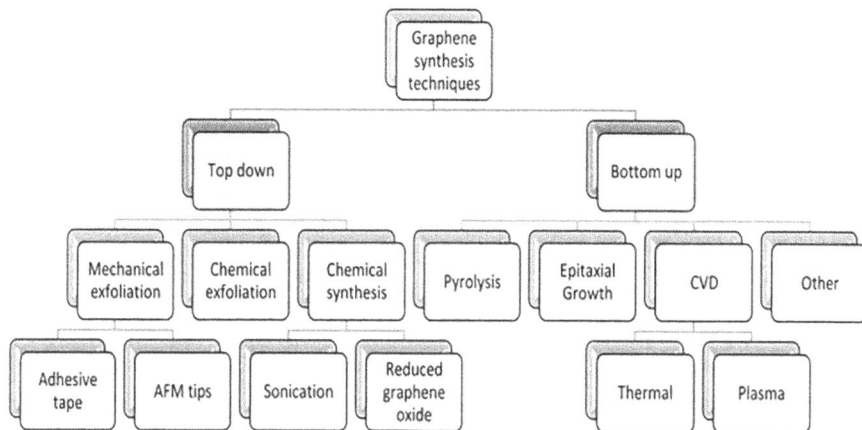

FIGURE 2.2 Approaches used in the synthesis of graphene. (Adapted with permission from [28]. Copyright © 2016, The Author(s). Springer Nature; an open-access article distributed under the terms of the Creative Commons Attribution 4.0 International License (http://creativecommons.org/licenses/by/4.0/.))

Grim and his colleagues developed to extract single-layered graphene from highly oriented pyrolytic graphite. On the photoresist surface, economically on hand inorganic layered substance is rubbed/peeled-off against scotch tape. Because of the dissociation of susceptible van der Waal forces among layers, the entire process generates some of flakes totally on photoresist surface, containing monolayers. Nevertheless, even as this method is easy, quick, yet cost-effective, it does have inherent drawbacks. This technique's monolayer productivity is relatively poor; as a result, it's only appropriate for laboratory studies that can't be used for massive manufacturing for high technical applications [25].

2.5.2 CHEMICAL REDUCTION METHOD

Solvothermal reduction, electrochemical reduction, photo-catalyst reduction, and chemical reagent reduction are all examples of GO chemical reduction that can be done at room temperature. It is the various maximum used strategies for generating graphene and rGO in massive amounts at a minimal price. Chemical exfoliation of graphite flakes creates a weak van der Waals force attraction amongst the layers, according to theory. Chemical exfoliation methods including the modified Hummer's technique's, Enhanced Hummer's technique's but also the conventional Hummer's method had been advanced to create GO having relatively massive size or aspect ratio, excellent productivity, however, additionally low dangerous gases [26].

2.5.3 CHEMICAL VAPOR DEPOSITION

CVD is a few of the extraordinarily powerful procedures closer to generating great monolayer graphene for applications in numerous devices. Through exposing metal to numerous hydrocarbon precursors under elevated temperatures, full-size vicinity samples may be created. There are pretty varieties of CVD techniques, which include

- Plasma-superior CVD
- Thermal CVD
- Hot/cold wall CVD

The specific procedure of graphene creation is dependent upon growth substrate, although it usually begins with the growth of carbon atoms that nucleate upon metal just after hydrocarbons are decomposed, as well as the nuclei grow forming huge domains. Owing to graphene's chemical durability transferring this from the growing substrate to a substrate of concern could be challenging because this can cause flaws and wrinkling throughout the material, whereas thermal variations could also impact the material's durability. Furthermore, the CVD process's complication as well as the technique's significant energy requirements contributes to the task's difficulty, although CVD remained among the most effective strategies toward large-area graphene manufacturing. With considerations of upscaling, CVD is a potential contender in producing high-quality graphene in vast areas, which is why various researcher organizations have concentrated their efforts in this manner. Bae et al. made one of several earliest efforts using a roll-to-roll (RTR) procedure to produce 30-in. graphene sheets, and the RTR technology was later efficiently employed for the ongoing manufacture of graphene. Graphene was yet again synthesized by an RTR technique employing a concentric tube CVD in the latest publication by Polsen et al. Lin et al. devised a surface engineering approach and were successful in creating single-crystalline graphene with a diameter of centimeter [27].

2.5.4   EPITAXIAL GROWTH

One of the really acknowledged ways of graphene synthesis is epitaxial thermal growth on a single crystalline silicon carbide (SiC) surface. The term "epitaxy" originated from the Greek word epi, which indicates "over" or "upon," and "taxis," which signifies "order" or "arrangement." Epitaxial growth occurs when a single crystalline layer is deposited on a single crystalline substrate, resulting in an epitaxial film. It uses single-crystalline SiC substrates to make high-crystalline graphene. Depending on the substrate, there are two different types of epitaxial growth:

- Homo-epitaxial
- Hetero-epitaxial

A homo-epitaxial layer is formed if the film and substrate are composed of the identical component, whereas a hetero-epitaxial layer is formed whenever the film and substrate are made of distinct components. Bommel et al. were the first to describe graphite production on both the 6H–Sic (0001) and 6H–Sic (0002) surfaces around 1975. In 2004, de Heer's et al. examined the electronic characteristics of ultrathin graphite made up of one to three single-mode graphene sheets on the Si completed (0001) face of single-crystal 6H-SiC. Epitaxial graphene growing on SiC is often seen as a feasible technology toward large-scale graphene manufacturing as well as deployment into digital devices. The quantum Hall effect has been used to create graphene on SiC as a new resistance standard. This procedure, however, is extremely costly [28].

2.6 RECENT DEVELOPMENTS IN THE FIELD OF GRAPHENE 2D NANOCOMPOSITES

Multiple scientific institutes have created polymer nanocomposites depending on a number of nano-fillers including CNTs, expanded graphite, and carbon nanofiber. This chapter discusses the application of graphene as well as GO as fillers in a variety of polymeric matrices, including polystyrene (PS), polyurethane, polyaniline (PANI), polycarbonate (PC), poly (3,4-ethyldioxythiophene) (PEDOT), and others. Regarding students concerned in fabricating innovative polymer/graphene and GO-based nanocomposites for diverse purposes, the current debate on polymer/graphene and GO-based nanocomposites might be quite useful.

Applying the melt blending approach, Han et al. created PS/graphene and PS/GO nanocomposites. Cone-calorimetry and thermo-gravimetric analyses were used to assess the thermal properties of the nanocomposites (TGA). Graphene plus GOs was exfoliated inside the PS composites, according to XRD data. GOs but also graphene were found uniformly disseminated across the matrix, with no visible clumps, as seen on scanning electron microscopy pictures [29]. Zhao et al. used in situ polymerization to create a polyaniline/graphene nanosheet/carbon nanotube (PANI/GNS/CNT) composite. Employing an in situ technique, Lee et al. developed waterborne polyurethane/functionalized graphene sheet (FGS) nanocomposites. Using in situ polymerization, Xu et al. created a PEDOT/sulfonated graphene composite. The innovative composites product had outstanding transparencies, high electrical conductivity, thermal stability, plus flexibility, as well as could be treated in both organic and aqueous mediums with convenience. Applying a solution mixing approach, Liang et al. created different categories of nanocomposites. In a thermoplastic polyurethane (TPU) matrix, researchers tried several fillers such as Sulphonated graphene; isocyanate modified graphene, and reduced graphene. Using an in-situ polymerization process, Mohamadi et al. created a polymethyl methacrylate (PMMA)/graphene nanocomposite. Zhao et al. used an in-situ polymerization approach to create a graphene nanosheet (GN)/polythiophene (PTH) hybrid. Liang et al. created poly (vinyl alcohol) (PVA)/graphene nanocomposites by mixing GO into yet another PVA matrix with water mostly as a processing solvent. In an aqueous solution, Zhao et al. generated PVA/fully exfoliated graphene sheets nanocomposites. Ansari and Giannelis used solution processing and compression moulding to make poly (vinylidene fluoride) (PVDF)/FGS from GO plus graphite [30]. To enhance the pyroelectric characteristics of PVDF, Hu et al. exploited the sol–gel method to synthesize PVDF-doped GO. A sol–gel approach is used to make the PVDF layer containing graphene-oxide doping. A thorough investigation of the GO doping method in PVDF has been completed. The permeability of the resultant coatings is the fundamental shortcoming of the sol–gel process; however, this could be substantially reduced by using a 10% PVDF mixture into DMF and solvent-baking under vacuum [31]. To create metal-encapsulated graphene nanocomposites, the researchers combined GO films in a silica mixture. However, the inclusion of metal nanoparticles inside the matrix material together with graphene has captivated the interest of many scholars owing to such benefit of increased inter-particle interaction (i.e., contact between the metal particle and the host matrix). Researchers highlighted many processes connected with graphene and nanocomposites materials, as well as its applicability in the domain of digital memory equipment spanning from electrochemical sensors to instrumentation. By using a straightforward sol–gel, spin-coating, chemical reduction, and thermal-curing approach, a single-pot hydrothermal technique to make carbon-coated SnO-graphene film composites in a green way was developed. Graphene composite has been developed as an anode material for a Li-ion battery and demonstrated great internal storage as well as better cycling performance [32].

2.7 GRAPHENE BASED 2D NANOCOMPOSITES APPLICATIONS IN FUEL CELLS

"For pioneering research referring the two-dimensional substance graphene" Geim and Novoselov were collectively presented the Nobel Prize in Physics in 2010. Graphene is presently one of the foremost well low-dimensional materials, with unique electrical, optoelectronic, physical, including chemical characteristics that have discovered uses in a variety of scientific as well as technical fields. The graphene species has stimulated the interest of a growing variety of scientists researching in a variety of fields, including composite materials and molecular electronics. As a result, numerous interesting graphene-based uses are now being explored [33]. Graphene-based nanomaterials get a wide range of potential applications in the energy and biomedical sectors (Figure 2.3). Here are some present-day excellent instances Activated graphene provides exceptional super capacitors as battery storage graphene electrodes might contribute to a potential strategy toward creating affordable, lightweight, as well as flexible solar cells and fuel cells because multifunctional graphene sheets are attractive foundations for catalytic activity [34].

A fuel cell is indeed an energy storage system, which converts chemical energy to electrical energy using oxygen plus methanol mostly as reaction mixture, the decrease of oxygen plus methanol oxidation within fuel cells might produce green energy. Owing to its outstanding energy conversion effectiveness relatively low temperature, close to negligible emissions, plus energy density, fuel cells offer a huge amount of opportunity mostly as cleaner but sustainable power source in electric vehicles [35]. Beginning from hydrogen/oxygen fuel cells, a variety of alternative fuel cell manufacturing pairings have been tested thus recently platinum is among the foremost common yet commonly utilized components in low-temperature fuel cells. However, the challenge regarding Pt-based catalysts is that Pt is highly expensive, thus its consumption must be limited; but in the other way, lowering Pt proportion reduces fuel cell functionality. Graphene's relatively massive area makes it an appropriate contender for fuel cell purposes. In methanol oxidation and oxygen reduction cells, a Pt/graphene composite has been employed. Because graphene has a 2D composition, all these faces of the materials remain accessible toward the environment, increasing the active surface area and thus increasing cost-effectiveness. RGO's residual oxygen activity (however, minor) eliminates carbon-based ions, allowing it to become more

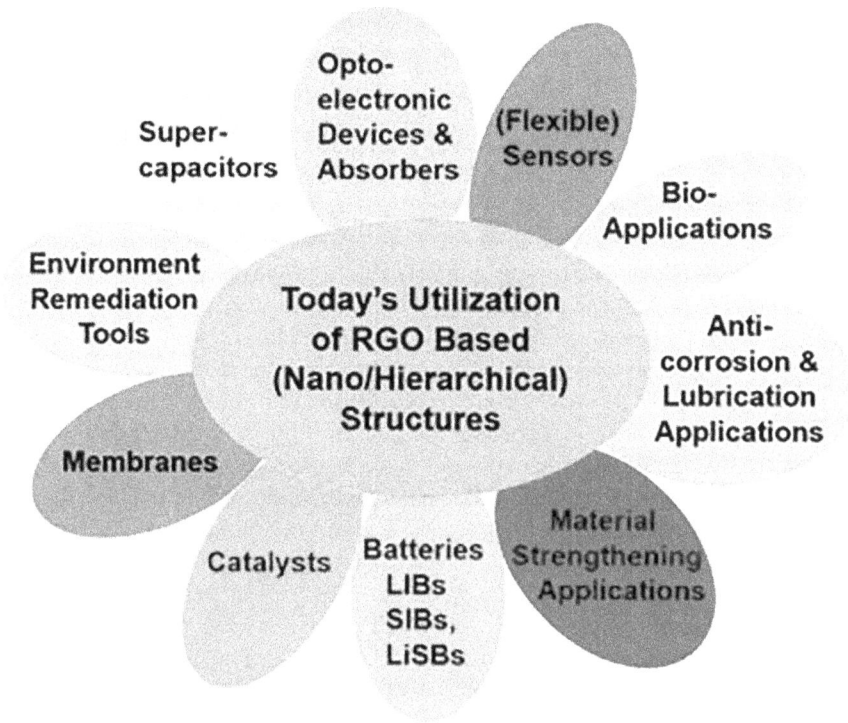

FIGURE 2.3 Schematic representation of graphene-based 2D nanocomposite applications.

resistant to carbon monoxides produced throughout methanol oxidation. Regarding fuel cell applications, nitrogen-doped graphene exhibiting greater permeability has also been employed [36].

2.8 APPLICATIONS OF GRAPHENE-BASED 2D NANOCOMPOSITES

2.8.1 IN MICROBIAL-FUEL CELLS

Potter was the first to establish the conception of MFC in 1910. Previously, it was discovered that the proportion of power generated by living cultures (such as *Escherichia coli* and *Saccharomyces*) was inadequate. Meanwhile, this is recently revealed that by adding electron mediators, the present density as well as power generation may be efficiently increased [37]. MFCs, on either side, are limited in many uses because to their limited energy production in relation to the other aspects that influence MFC effectiveness, like cell design, microbial inoculum, substrate, proton membrane, as well as electrode surface area and material, the cathode could be presumed to become a major constraint for MFC effectiveness due to poor kinetics of oxygen reduction inside the medium. As a result, it's critical to explore toward catalyst materials that have a huge surface area, excellent conductivity, strong catalytic activity, superior sustainability, plus relatively inexpensive in order to assist the cathode process in MFCs [38]. The electrode material selected has a big impact on the MFC's efficiency, and it's crucial for a sustainable implementation of such technique for optimum energy production. Countless unique materials were being investigated as anodes in MFCs over the last two centuries. Graphite rod, graphene felt, carbon cloth, flexible graphite sheets, graphite granules, plus activated carbon were employed in the majority of previous investigations [39]. Owing to their exceptional features, graphene is now considered a promising contender for just an extremely competitive yet cost-effective component in MFCs. The use of graphene as an electrode in microbial fuel cell is important to strengthen performance because to improve anode and cathode properties. The effectiveness of extracellular electron transport has also been enhanced. Additional characteristics that make graphene an ideal option in MFCs are increased surface area plus enhanced contact amongst the substrate and microorganisms. Apparently redesigned electrodes can also be made from molecules that have undergone radical polymerization. In microbial fuel cells, the graphene catalyst at the cathode area has indeed been studied. These were created with the goal of lowering the cathode's expense simultaneously enhancing the kinetics of the oxygen reduction reaction (ORR). And for functionalization of the graphene noble metals are used such as titanium dioxide, nickel foam, plus manganese dioxide are just a few of the popular catalysts indicated as ORR catalysts in microbial fuel cells. Such components seem to be great since they are inexpensive, non-toxic, eco-friendly, as well as chemical stability. An adhesive, such as Nafion or polytetrafluoroethylene PTFE, could be used to make iron oxide/Gr composites but also manganese oxide/Gr. But for manufacturing of Ni/rGO, and TiO_2/Gr SnO_2/rGO TiO_2/Gr composite materials, a hydrothermal approach should be used. Graphene can be chemically doped using heteroatoms, pertaining to the research. That technique is good for improving graphene's physicochemical characteristics. Gr is modified with nitrogen Gr-N to increase conductivity. The ORR electro-catalytic performance of Gr-N in alkaline solutions has been improved, making this the perfect catalyst in FC purposes [40].

2.8.2 In Proton Exchange Membrane Fuel Cells (PEMFC)

Membrane extraction has evolved over the last two centuries mostly as a method for addressing humanity's major concerns, including insufficient commodities as well as ecological issues. Membrane separation, in comparison to traditional separation technologies, is a more energy-efficient but also ecologically friendly approach that takes up minimal time but might run continuously. Having controllable pore size but also shape an optimal membrane offers increased permeation flux, enhanced selectivity plus superior durability Furthermore, to maximize permeability, obtain greater productivity, plus improve membrane functionality, the thickness of the membrane must be reduced. The latest emergence of 2D graphene-based materials in terms of manufacturing as well as fabrication presents an interesting possibility toward building a distinct variety of membranes featuring exceptional separation capabilities. Graphene, GO, and chemically converted graphene are all examples of graphene-based compounds [41]. Anode, cathode, and perhaps a separation membrane are the three essential components of a fuel cell. Pt, Au, Ru, along with their respective alloys constitute the more often utilized cathodic elements; however, they seem to be also very costly. Pt is perhaps the best extensively utilized cathodic material in fuel cells out of all of those elements, but it is also the more costly besides having precious. Because of CO poisoning, the Pt-based catalyst generally has a low resistance, the development of metal-free ORR catalysts having sufficient charge density plus robust materials may replace but rather minimize Pt is extremely difficult. A fabrication of Pt/graphene alloys becomes the major approach. Yoo et al. created PtNP/GNS composite materials with a strength four times that of conventionally produced Pt/C. PtNP/GNS has a CO absorption rate 40 times lower than Pt/C, indicating that it is a suitable candidate in proton exchange membrane fuel cell applications [42]. The Nafion membrane is now the foremost widely employed solid polymer electrolyte for acidic PEMFCs; however, it seems to be insufficient in regards to manufacturing prices, ion conductivity, mechanical strength, and methanol permeation. To increase productivity PEMFCs featuring distinctive frameworks can be implemented by blending with Nafion to create a nanocomposite membrane. A Nafion/GO nanocomposite (Nafion/GO) for a PEMFC membrane was developed, which really showed 40% reduced methanol crossover than a Nafion membrane despite keeping constant ionic conductivity, leading to a significant improvement in PEMFC efficiency [14]. PGO-doped Nafion nanocomposites membranes were constructed with phosphoric acid functionalization. Overall proton conductivity of the nanocomposite's membrane containing 2 wt.% is 6.6% larger than those of the pure Nafion membrane. A sandwich-shaped GO/Nafion composite membrane-based PEMFC with remarkable methanol endurance at 60°C is reportedly shown [43].

2.8.3 Direct-Methanol-Fuel-Cells

Fuel cells have long been widely regarded as being among the purest yet more economical substitutes to classical heat engines in providing electricity having reduced chemical emissions as well as great productivity allowing for continuous startup including in harsh environments. For their excellent energy density green emissions, as well as ambient operation settings, direct methanol fuel cells (DMFCs) employing liquefied plus regenerative methanol have just been regarded as an attractive alternative. Expensive metals have long been widely used as excellent electrocatalysts, however, their expensive price lack of availability combined poor stability prevents massive industrial manufacturing as well as use in DMFCs. Owing to its interesting features, graphene recently prompted a lot of interest in DMFCs. Wang et al. explored a convenient method for reducing then spreading Pt. nanoparticles onto the surfaces of GO nanosheets having particle sizes ranging from 1 to 5 nm. The electrocatalytic activity of manufactured Pt/G composites was stronger than that of Pt./MWCNT. Chen et al. suggested a simple hydrothermal technique for preparing RGO but also improving their reduction capabilities. No extra reductants or surfactants was used for the first place when a spontaneous redox reaction between $PtCl_4^{2-}$ with RGO was observed [44]. Ongoing temperature-related concerns, as well as significant methanol permeability in DMFC applications, have fueled interest in creating alternative forms of proton-conducting membranes. Sulfonate poly (ether sulfone) (SPES) is a particularly appealing material, one of the best potentially synthetic membranes along with its good mechanical properties, resilience to elevated temperatures of up to 300°C, and acidic durability. SPES could be manufactured with such a high degree of sulfonation (DS) that gives these enormous proton conductivity capabilities. When incorporating GO into a Nafion mixture, it demonstrated good DMFC efficiency. Surfactant adsorption shows enhanced proton conductivity as well as lower methanol permeation. The sulfonated poly (ether ether ketone) (SPEEK) has often been studied with GO. Whenever used correctly, GO will greatly strengthen the physical features of a composite even at low loading levels. Because of the large, surface area chemical tenability, plus barrier effect of GO functional groups, polymer/GO nanocomposites in ionic conductors are predicted to deliver outstanding efficiency in DMFC [45]. For DMFC applications, the PVA-grafted GO (PVA/GO) composite has been integrated into sulfonated PVA. These findings appeared satisfactory, so they simply just added to demonstrate the possible benefits of graphene into membrane production in DMFC applications. For the first time, Yusuf et al. used GO paper as an electrolyte [35]. For DMFC applications, graphene was also employed as a proton exchange membrane. The membrane was made by combining graphene oxide nanoplatelets (GO). The greater proton conductance plus constant decline in methanol penetration rate was associated with a size growth of nanoparticles after researching membrane transport properties, the increased transport features could be due to the surface mobility of methanol as well as proton molecules, or even the altered conduction channels formed using graphene. By comparison to commercial Nafion membranes, the GO membrane in DMFCs has exhibited better energy density but also no reported reduction in open circuit potential. The significance of graphene-based membranes in DMFCs is demonstrated by this [46].

2.9 LIMITATIONS

Without a question, these disciplines are continuously progressing mostly through the development but also discovery of miraculous materials. Fundamental electrochemical conceptions are mostly unexplored, but they are critical for grasping mechanical features but also potential drawbacks of various graphene nanomaterials. Mostly by quick advancement made from the last two decades on the artificial factors of graphene-related materials, as well as advancements in in situ spectroscopic methodologies as well as strong theoretical modeling, the upcoming 5 years are quite often constructive for graphene-based electrochemical technological innovations [47]. Multiple scientific organizations have made significant progress in exploiting graphene as a power storage material in recent times, while there are yet certain obstacles to overcome. The increased electrochemical double layer capacitor (EDLC) of graphene, for example, demonstrates its capability to cope with several other power storage devices. Unfortunately, the measured capacitance values are significantly weaker than anticipated due to a variety of factors such as graphitic sheet restacking, insufficient ion transport, hydrophobicity, and so forth. As a result, the restrictions indicated above should be overcome before all graphene materials may be employed in electrical and chemical power storing systems [48]. Noble-metal nanocomposites founded on graphene have shown promise as electro-catalysts toward fuel purposes. Despite significant advances, there are yet hurdles in this domain of study. First, perfect management of graphene nanosheet flaws, doping, voids, but also surface characteristics remains a challenge. The surface engineering of graphene is important for the patterned growth and binding of new metal nanoparticles, which has an impact on electro-catalytic efficiency. The Hummers technique for preparing GO and also the hydrothermal approach for introducing N or S doping really aren't capable of functionalizing graphene nanoparticles having regulated atomic locations [49]. Distributing nanocomposites evenly in polymer matrix remains a technological challenge the cytotoxic consequences of nanocomposites are one of the primary concerns that could have a significant influence on emerging graphene/polymer nanocomposite material applications particularly in the material world of technologies [50].

2.10 CONCLUSION AND FUTURE PROSPECTS

Graphene and its derivatives are strong materials that hold a lot of potential in terms of fixing ecological but also energy issues. GN and GN-based nanoparticles are much more feasible for technologies such as batteries, super-capacitors, including fuel cells. To address the increased demand for GN for film manufacturing composite incorporation, but additionally device integration, chemical as well as physical approaches also have been applied. The extraordinary properties of GN make it a viable candidate for implementation as an energy storage material. Insufficient power production with expensive manufacturing price has hampered the application of large-scale FC technology to present. As a result, more exploration is necessary to find reduced long-lasting, high-performance components that can be employed to design beneficial FC systems. The utilization of graphene as supporting frameworks to regulate the formation of operational nano-size (usually 5 nm) crystals using liquid precursors is the first impressive finding. The utility of such a method is demonstrated in a variety of areas, including fuel cell electrodes for Pt catalyst decorating. Unanticipated capabilities, such as well-dispersed nanometer-size Pt catalyst nanoparticles for increased catalytic activity in fuel cells, could be added to graphene. Even though graphene is frequently regarded as the excellent candidate to substitute transparent yet conductive oxides, researchers affirm that its incorporation in PV systems will necessitate additional efforts in order to achieve the stringent PV parameters. Owing to the fairly substantial early contaminants of metal-catalyzed CVD graphene sheets, material as well as fundamental sciences are also required to develop and appraise a preferable technique to dope CVD graphene but also to determine to what degree such doping may be regulated. Regarding development in the future, an appropriate methodology involving preparation characterization, conceptualization, device manufacturing, plus scale-up is required.

REFERENCES

1. Zhang, X., Cheng, X., & Zhang, Q. (2016). "Nanostructured energy materials for electrochemical energy conversion and storage: A review." *Journal of Energy Chemistry, 25*(6), 967–984.
2. Iqbal, M.Z., Khan, A., Numan, A., Haider, S.S., & Iqbal, J. (2019). "Ultrasonication-assisted synthesis of novel strontium based mixed phase structures for supercapattery devices." *Ultrasonics Sonochemistry, 59*, 104736.
3. Martínez-Huerta, M.V., & M.J. Lázaro. (2017). "Electrocatalysts for low temperature fuel cells." *Catalysis Today, 285*, 3–12.

4. Khan, K., Tareen, A.K., Aslam, M., Zhang, Y., Wang, R., Ouyang, Z., Gou, Z., & Zhang, H. (2019). "Recent advances in two-dimensional materials and their nanocomposites in sustainable energy conversion applications." *Nanoscale, 11*(45), 21622–21678.

5. Steele, B.C. & Heinzel, A. (2011). Materials for fuel-cell technologies. In: Dusastre, V. (Ed.) *Materials for Sustainable Energy: A Collection of Peer-Reviewed Research and Review Articles from Nature Publishing Group* (pp. 224–231). Singapore: World Scientific.

6. Iqbal, M.Z., Siddique, S., Khan, A., Haider, S.S., & Khalid, M. (2020). Recent developments in graphene based novel structures for efficient and durable fuel cells. *Materials Research Bulletin, 122*, 110674.

7. Azadmanjiri, J., Srivastava, V.K., Kumar, P., Nikzad, M., Wang, J., & Yu, A. (2018). "Two-and three-dimensional graphene-based hybrid composites for advanced energy storage and conversion devices." *Journal of Materials Chemistry A, 6*(3), 702–734.

8. Su, H. & Hu, Y.H. (2021). "Recent advances in graphene-based materials for fuel cell applications." *Energy Science & Engineering, 9*(7), 958–983.

9. Ali Tahir, A., Ullah, H., Sudhagar, P., Teridi, M.A.M., Devadoss, A., & Sundaram, S. (2016). "The application of graphene and its derivatives to energy conversion, storage, and environmental and biosensing devices." *The Chemical Record, 16*(3), 1591–1634.

10. Bera, M. & Maji, P.K. (2017). "Graphene-based polymer nanocomposites: Materials for future revolution." *MOJ Polymer Science, 1*(3), 00013.

11. Liu, G., Jin, W., & Xu, N. (2015). "Graphene-based membranes." *Chemical Society Reviews, 44*(15), 5016–5030.

12. Jin, J., Wen, Z., Ma, G., Lu, Y., Cui, Y., Wu, M., Liang, X., & Wu, X. (2013). "Flexible self-supporting graphene–sulfur paper for lithium sulfur batteries." *RSC Advances, 3*(8), 2558–2560.

13. Mao, S., Lu, G., & Chen, J. (2015). "Three-dimensional graphene-based composites for energy applications." *Nanoscale, 7*(16), 6924–6943.

14. Qadir, A., Le, T.K., Malik, M., Min-Dianey, K.A.A., Saeed, I., Yu, Y., Choi, J.R., & Pham, P.V. (2021). "Representative 2D-material-based nanocomposites and their emerging applications: A review." *RSC Advances, 11*(39), 23860–23880.

15. Lawal, A.T. (2019). Graphene-based nano composites and their applications: A review. *Biosensors and Bioelectronics, 141*, 111384.

16. Sen, M. (2020). Nanocomposite materials. In: *Nanotechnology and the Environment* (pp. 1–12). London: IntechOpen.

17. Rafiei-Sarmazdeh, Z. & Ahmadi, S.J. (2019). Graphene-like nanocomposites." In: *Nanorods and Nanocomposites* (p. 141). London: IntechOpen.

18. Beaumont, P.W., Zweben, C.H., Gdutos, E., Talreja, R., Poursartip, A., Clyne, T.W., Ruggles-Wrenn, M.B., Peijs, T., Thostenson, E.T., Crane, R. & Johnson, A. eds. (2018). *Comprehensive Composite Materials II.* Amsterdam, The Netherlands: Elsevier.

19. Tan, C., Huang, X., & Zhang, H. (2013). Synthesis and applications of graphene-based noble metal nanostructures. *Materials Today, 16*(1–2), 29–36.

20. Solís-Fernández, P., Bissett, M., & Ago, H. (2017). "Synthesis, structure and applications of graphene-based 2D heterostructures." *Chemical Society Reviews, 46*(15): 4572–4613.

21. Tan, C., Cao, X., Wu, X.J., He, Q., Yang, J., Zhang, X., & Zhang, H. (2017). Recent advances in ultrathin two-dimensional nanomaterials. *Chemical Reviews, 117*(9): 6225–6331.

22. Yu, F., Wang, C., & Ma, J. (2016). Applications of graphene-modified electrodes in microbial fuel cells. *Materials, 9*(10), 807.

23. Choi, H.J., Jung, S.M., Seo, J.M., Chang, D.W., Dai, L., & Baek, J.B. (2012). Graphene for energy conversion and storage in fuel cells and supercapacitors. *Nano Energy, 1*(4), 534–551.

24. Sengupta, J. (2019). "Different synthesis routes of graphene-based metal nanocomposites." arXiv preprint arXiv:1911.01720.
25. Gupta, A., Sakthivel, T., & Seal, S. (2015). Recent development in 2D materials beyond graphene. *Progress in Materials Science*, *73*, 44–126.
26. Mousavi, S.M., Low, F.W., Hashemi, S.A., Lai, C.W., Ghasemi, Y., Soroshnia, S.,... & Tiong, S.K. (2020). Development of graphene based nanocomposites towards medical and biological applications. *Artificial Cells, Nanomedicine, and Biotechnology*, *48*(1), 1189–1205.
27. Papageorgiou, D.G., Kinloch, I.A., & Young, R.J. (2017). Mechanical properties of graphene and graphene-based nanocomposites. *Progress in Materials Science*, *90*, 75–127.
28. Bhuyan, M.S.A., Uddin, M.N., Islam, M.M., Bipasha, F.A., & Hossain, S.S. (2016). Synthesis of graphene. *International Nano Letters*, *6*(2), 65–83.
29. Han, Y., Wu, Y., Shen, M., Huang, X., Zhu, J., & Zhang, X. (2013). Preparation and properties of polystyrene nanocomposites with graphite oxide and graphene as flame retardants. *Journal of Materials Science*, *48*(12), 4214–4222.
30. Shah, R., Kausar, A., Muhammad, B., & Shah, S. (2015). Progression from graphene and graphene oxide to high performance polymer-based nanocomposite: A review. *Polymer-Plastics Technology and Engineering*, *54*(2), 173–183.
31. Hu, Y.C., Hsu, W.L., Wang, Y.T., Ho, C.T., & Chang, P.Z. (2014). Enhance the pyroelectricity of polyvinylidene fluoride by graphene-oxide doping. *Sensors*, *14*(4), 6877–6890.
32. Dhand, V., Rhee, K.Y., Ju Kim, H., & Ho Jung D. (2013). A comprehensive review of graphene nanocomposites: Research status and trends. *Journal of Nanomaterials*, *2013*, 763953, 1–14.
33. Torres, T. (2017). Graphene chemistry. *Chemical Society Reviews*, *46*(15), 4385–4386.
34. Berger, M. (2019). *Nanoengineering: The Skills and Tools Making Technology Invisible*. London: Royal Society of Chemistry.
35. Yusuf, M., Kumar, M., Khan, M.A., Sillanpää, M., & Arafat, H. (2019). A review on exfoliation, characterization, environmental and energy applications of graphene and graphene-based composites. *Advances in Colloid and Interface Science*, *273*, 102036.
36. Mitra, S., Banerjee, S., Datta, A., & Chakravorty, D. (2016). A brief review on graphene/inorganic nanostructure composites: Materials for the future. *Indian Journal of Physics*, *90*(9), 1019–1032.
37. Kaur, R., Marwaha, A., Chhabra, V.A., Kim, K.H., & Tripathi, S.K. (2020). Recent developments on functional nanomaterial-based electrodes for microbial fuel cells. *Renewable and Sustainable Energy Reviews*, *119*, 109551.
38. Valipour, A., Ayyaru, S., & Ahn, Y. (2016). Application of graphene-based nanomaterials as novel cathode catalysts for improving power generation in single chamber microbial fuel cells. *Journal of Power Sources*, *327*, 548–556.
39. Sonawane, J.M., Yadav, A., Ghosh, P.C., & Adeloju, S.B. (2017). Recent advances in the development and utilization of modern anode materials for high performance microbial fuel cells. *Biosensors and Bioelectronics*, *90*, 558–576.
40. Olabi, A.G., Wilberforce, T., Sayed, E.T., Elsaid, K., Rezk, H., & Abdelkareem, M.A. (2020). Recent progress of graphene based nanomaterials in bioelectrochemical systems. *Science of the Total Environment*, *749*, 141225.
41. Liu, G., Jin, W., & Xu, N. (2015). Graphene-based membranes. *Chemical Society Reviews*, *44*(15), 5016–5030.
42. Dasari, B.L., Nouri, J.M., Brabazon, D., & Naher, S. (2017). Graphene and derivatives– synthesis techniques, properties and their energy applications. *Energy*, *140*, 766–778.
43. He, D., Tang, H., Kou, Z., Pan, M., Sun, X., Zhang, J., & Mu, S. (2017). Engineered graphene materials: Synthesis and applications for polymer electrolyte membrane fuel cells. *Advanced Materials*, *29*(20), 1601741.

44. Liu, Q., Xu, Q.J., Fan, J.C., Zhou, Y., & Wang, L.L. (2015). A review of graphene supported electrocatalysts for direct methanol fuel cells. In: *Advanced Materials Research* (vol. 1070, pp. 492–496). Switzerland: Trans Tech Publications Ltd.
45. Muthumeenal, A., Saraswathi, M.S.A., Rana, D., & Nagendran, A. (2017). Fabrication and electrochemical properties of highly selective SPES/GO composite membranes for direct methanol fuel cells. *Journal of Environmental Chemical Engineering, 5*(4), 3828–3833.
46. Farooqui, U.R., Ahmad, A.L., & Hamid, N.A. (2018). Graphene oxide: a promising membrane material for fuel cells. *Renewable and Sustainable Energy Reviews, 82,* 714–733.
47. Szunerits, S. & Boukherroub, R. (2018). Graphene-based nanomaterials in innovative electrochemistry. *Current Opinion in Electrochemistry, 10,* 24–30.
48. Palaniselvam, T., & Baek, J.B. (2015). Graphene based 2D-materials for supercapacitors. *2D Materials, 2*(3), 032002.
49. Liu, J., Ma, Q., Huang, Z., Liu, G., & Zhang, H. (2019). Recent progress in graphene-based noble-metal nanocomposites for electrocatalytic applications. *Advanced Materials, 31*(9), 1800696.
50. Lawal, A.T. (2020). Recent progress in graphene based polymer nanocomposites. *Cogent Chemistry, 6*(1), 1833476.

3 MXene-Based 2D Nanomaterials for Fuel Cells

P. Dhanasekaran
CSIR-Central Electrochemical Research
Institute-Madras Unit

S. Vinod Selvaganesh
Indian Institute of Technology-Madras

CONTENTS

3.1 INTRODUCTION

The recurrence rate of technology expansion has strengthened more than the leading centuries and will feasibly rise to a great extent in the present century. Also, expanding global eco-friendly concerns began via burning fossil fuels demand green and sustainable energy, like fuel cells producing higher energy competently from green fuels like hydrogen and methanol. Recent progress, significantly decreasing fossil fuel reserves level economically worldwide. The rapid world population and development of larger industrialization have augmented the power demand worldwide, reaching more than 60,000 TWh per year beneath by 2040, and its equivalent carbon-di-oxide emission may be expected between 38 and 45 Gt per year up to

DOI: 10.1201/9781003178422-3

2040 [1]. Hence, the major concerns are energy conversion or storage systems, specifically required CO_2 free; it may replace the existing fossil fuel-based power system. Protecting our environment and develop large-scale energy generation are the main challenges for the 21st century [2]. Also, sustainable fuel production and human safety can be more ensured by making new and additional inventions and modifying the existing energy conversion technologies.

Electrochemical Science and technology systems used to produce fuels are the maximum efficient energy conversion methods, storage, and controlled environmental pollution. Electrochemical technologies are the most admirable way to generate sustainable energy from various sources that can be stored or utilized for portable infrastructure applications. In this regard, chemical bond breaking or construction is an essential step for electrochemistry to produce storage or conversion devices to minimize fossil fuels requirement and convert to increase renewable energy, specifically to the fuel cell system. However, considering other sustainable energy sources like wind power, hydro, solar, and geothermal power, there are a few limitations. Fuel cell technology has received greater attention extensively around the world. Fuel cells system instigates pollution-free and promising power phase/cycle technology that directly converts chemical energy to electrical energy through an electrochemical reaction.

3.2 UNDERSTANDING THE MATERIAL CHALLENGES FOR FUEL CELL APPLICATION

The commercialization of fuel cells is hindered by catalyst stability, slow kinetic, high cost, external component accessibility, and real-time reliability. However, there is a certain issue for catalyst concerns, such as long-life cycle, materials cost, and efficiency, that needs to be addressed before commercialization. Generally, the fuel cell's operating temperature (T) is controlled over the electrolytes and cell performance. Efficiency depends on operating temperature; while an increased operating temperature, the internal resistance, and cell performance are decreased, which equalizes the operating temperature voltage decrease. Dynamic behavior such as an operating voltage variation leads to changes in cell temperature and changes/affects the overall stack component and material (mechanical stress), which may decrease the stack's lifetime. Start-up-time: Required time to reach optimal working temperature, enhancing cell performance with gradually increasing operating temperature [3]. The above parameters provide the first sign to the start-up of possible fuel cell technologies. From an electrochemistry point of view, H_2 gas is the most and best reacting fuel, which reacts directly and gives high power density; however, the supply and production of H_2 gas lead to logistic challenges. Considering liquid fuels is preferable; however, liquid fuel (methanol) delivers less power density than direct H_2 gas.

As far as the various types of fuel cells are considered, like low-temperature polymer electrolyte fuel cell (LT-PEFC), direct methanol fuel cell, air-breathing fuel cell, high-temperature polymer electrolyte fuel cell (HT-PEFC), alkaline polymer electrolyte fuel cell (APEFC), and phosphoric acid fuel cell, still rely on platinum (Pt) supported on high surface area carbon as the state of the art electrocatalyst. Besides, Pt has a relatively high work function, corrosion resistance than other materials and thus is the most desired/preferred catalyst for fuel cell application.

However, a cell operated at variable load condition (0.6–1 V) and vigorous operation will gradually decrease the Pt active area due to Pt agglomeration followed by dissolution, structural, and morphology changes on the carbon layer leading to decreased fuel cell performance [4]. Thus, a proper strategy has to be adopted to address the issues related to the oxidation of carbon followed by Pt dissolution is the need of the hour.

To fulfill a catalyst's prerequisites, it is vital to explore innovative catalysts with well-ordered nanostructures for drastic improvement in catalytic activity, power density, durability, and efficiency [5]. Generally, fuel cell catalyst is also influenced by the electrode nanomaterials, desorption/adsorption at the cathode, kinetic reaction, and chemical energy change to electrical energy. According to various literature reports, Pt impregnated on 1D (1-Dimensional) nanomaterials shows better fuel cell stability [6–8]. Selvaganesh et al. reported that Pt particles impregnated on functionalized multi-walled carbon nanotubes (f-MWCNT) via the polyol method. Pt/f-MWCNT delivers 750 mW cm^{-2} of power density and retains 90% of cell performance after 100 hours at 1.2 V [9,10]. Likewise, Dhanasekaran et al. prepared amorphous carbon semi-coated 1-dimensional TiO_2 nanorod (CCT) on Pt via a colloidal method. The designated Pt/CCT catalyst exhibited initial fuel cell performance of 1.5 mA cm^{-2} at 0.6 V. Besides, Pt/CCT retains 82% of the initial fuel activity after 100 hours potential hold at 1.2 V [11,12]. However, Pt distribution and dispersion on the surface are major drawbacks for 1D nanoparticles toward the commercial aspect.

2D nanomaterials like metal nitrides, carbides, graphene, and modified metal oxides are the most sought-after supports/composites for noble metal for actual fuel cell application. These 2D nanostructures can act as both cathode and anode nanocatalysts in fuel cell application, which can reduce a major portion of issues related to reaction kinetic as well as significantly cut down the cost. Interestingly, 2D materials have extraordinary mechanical/breaking strength, more than 200 times higher than steel, and unpredictable, flexible nature. Those properties can further turn surface structure (formation of more defect states and functional groups), making it lightweight and stronger than other dimensional materials [13]. In view of this, 2D nanomaterials are being perceived as the most promising and cost-effective alternative owing to their larger surface area and active sites for quick electron or inter-facial charge transfer, which is more desired material characteristics for fuel cell catalyst and membrane application. Many researchers are exploring 2D nanomaterials as catalysts or support in LT-PEFC and APEFC applications. Dhanasekaran et al. reported an optimum graphene 2D nanostructure hybrid with carbon-supported Pt (Pt/CG 2) prepared by the versatile chemical route method. Pt/CG 2 nanohybrid materials delivered 450 mW cm^{-2} of power density with a flexible amount of 50 µg cm^{-2} of Pt at the cathode in LT-PEFC application. In addition, the as-prepared catalysts were also shown excellent fuel cell stability and were able to retain 60% of initial activity after 6,000 cycles from 1 to 1.5 V [14]. Selvaganesh et al. explored that Pt supported on 3,4-ethylene-dioxythiophene-reduced graphene oxide prepared via a simple chemical mixture. It has shown superior polymer electrolyte fuel cell (PEFC) performance of 750 mW cm^{-2} with metal loading of 200 µg cm^{-2} and retained more than 80% of PEFC performance after 10,000 cycles between 1 and 1.5 V. The above result indicates that 2D materials have higher mechanical integrity, improve Pt dispersion on its surface, and cell performance and stability [15].

Detailed studies of carbon materials, modified metal oxide system, chalcogenide, and phosphide-based 2D nanomaterials with Pt or Pt-free catalysts were explored in-depth in the previous chapter. In the present chapter, we explain the evaluation and improvement of fuel cell performance in terms of stability by depositing Pt or Pt-free on modified forms of MXene-based 2D nanostructure and a clear understanding of the degradation mechanism anode–cathode catalyst layer.

3.3 THE BASIC STRUCTURE OF MXene 2D NANOMATERIALS AND THEIR PROPERTIES

Since the 1970s, several nanostructures of carbides and nitrides materials have taken initial steps to improve the surface area and activity, which have been proposed for energy application. In the 20th century, the group from four to six of transition metal nitrides and carbides showed excellent chemical stability, hard, high-temperature, and superior wear-resistant materials for various applications, especially for energy applications [16,17]. Especially, MXene two-dimensional group has drastically gained attention as promising materials, attributed to its outstanding electronic properties, larger surface area, and a larger group of 2D nanomaterials [18,19]. MXene is a new class of 2D nanomaterial with the structural formula of $M_{n+1}X_n$ or $M_{n+1}X_nT_x$, which is derivative from the MAX structure. Here, M represents a group from four to six (Sc, Ti, Y, Ta, Hf, Zr, Cr, Ta, Nb, and W), X represents only for carbon or nitrogen, $n = 1$–3, and T is designate the surface ending groups that are have usually OH, O, and F or Cl and x in T_x attributed the availability of surface functionalities [20,21].

Generally, 2D nanomaterials have a sheet structure with a thickness of 1 nm. The thickness of 2D nanomaterials can precisely be controlled by tuning the n in MXenes from M_2X, M_3X_2, and M_4X_3. MXene has been prepared via selective etching of A layers from MAX phases [22]. Generally, the group from 8 to 16 (Periodic table) are used for A elements in the MAX phases. However, the metallic bonding of M-A layers is much weaker than M-X bonds, which is selectively etching of A layers from the MAX phase resulting in the formation of MXene phase. To date, most of MXene 2D nanomaterials are prepared via a top-down approach. The advantage of a top-down approach is the preparation of surface-termination-free MXene structure in relation to the bottom-up approach. Various nanosheet or layered MXene 2D structure synthesis from MAX phases via a top-down approach. To synthesize MXene 2D structure from MAX, the fluoride-based acidic solution is usually used for the etching of the A group.

Moreover, a top-down selective chemical etching is the key approach to MXene preparation. MXene properties mainly depend on the synthesis and termination groups. The number of sheets mainly depends on the synthesis method. While altering the etching parameter, the number of nanosheets also varied from M_4X_3, M_3X_2, and M_2X. Similarly, the surface terminations (T_x) is varied with a different group (OH, O, and F) and different ratio during the etching process resulting in changes in MXene properties. Usually, the etching process has resulted in tuned surface morphology. A more ambitious etching may increase the amount of defects state in MXene phases.

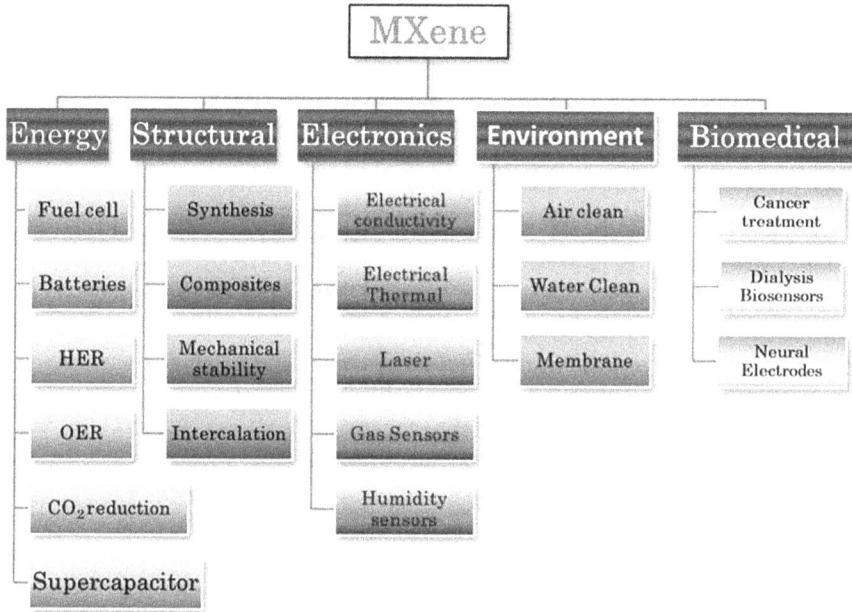

FIGURE 3.1 MXene 2D nanomaterials toward various applications explored and properties till date.

MXene nanosheets have excellent unique physical properties and promising nanomaterials for different applications, which is represented in Figure 3.1. The as-prepared MXene is hydrophilic in nature and can easily be discredited in water and different solvents like propylene, ethanol, *N, N*-dimethylformamide. Besides, the zeta potential of MXene shows a negative range of −30 to 80 mV. These high negative Zeta potential values make MXene easily preparable in the form of thin films or prepare a colloidal solution under stable conditions [22]. Also, MXene colloids can easily be mixed with a polymer solution and different nanomaterials to prepare hybrid and effective composites for energy applications. Similarly, the chemical oxidation stability also needs to be addressed before going to energy application. MXene-based nanomaterial oxidation stability depends highly on chemical structure, composition, and morphology.

3.4 IMPORTANCE OF MXene 2D NANOMATERIALS FOR FUEL CELL APPLICATION

MXenes have fascinated worldwide attention, especially for energy conversion and storage application, due to their structural resemblance to graphene adjustable electronic configuration and rich surface abundant groups. Since Ti_3C_2 is the first MXene materials, which were reported in 2011, MXenes have seen major developments in current years; more than 20 classes of MXenes has been reported with remarkable thermal stability and conductivity, such as Ti_2C, Nb_2C, V_2C, Mo_2C, $(Ti, V)_2C$, Ti_3C_2, $(Ti, Nb)_2C$, $Ti_3(C, N)_2$, $(Mo_2Ti_2)C_3$, $(Ti, V)_3C_2$, Zr_3C_2, $(Cr, V)_3C_2$,

$(Cr_2Ti)C_2$, $(Mo_2Ti)C_2$, Nb_4C_3, Ti_4N_3, $(Ti, Nb)_4C_3$, Ta_4C_3, and $(Nb, Zr)_4C_3$. In addition, the major consequence and implication of synthesis techniques were used to open the door for the formation of a new type of MXene structure, especially HF etching techniques. The etching techniques using HF could result in the formation of a new structure of MXene materials such as Ti_2CT_x, $Ti_3CN_xT_x$, $TiNbCT_x$, $Ta_4C_3T_x$, V_2CT_x, Nb_2CT_x, Mo_2CT_x, $Nb_4C_3T_x$, $(Nb_{0.8}Zr_{0.2})_4C_3T_x$, $(Nb_{0.8}Ti_{0.2})_4C_3T_x$, $Hf_3C_2T_x$, $Zr_3C_2T_x$, Mo_2C, and Zr_3C_2 [21–23].

MXene has excellent electrochemical stability and has been widely studied in energy-related fuel cell applications. MXene is especially attractive in terms of outstanding chemical and thermal stability and excellent electron inter-conducting behavior compared to various carbon and metal oxide systems. Especially, a two-layer MXene-based system has essential characteristics that can utilize for energy conversion PEFC application. MXene-based 2D nanomaterials also act as a catalyst as well as support materials for fuel cell application. The strong interaction between MXene-based 2D nanomaterials and Pt metal might be ameliorating the catalytic activity and stability.

According to Gao et al., as per density functional theory (DFT) calculation, Pt d-electron adsorption center is more negative than transition metal atoms, Pt metal nanoparticles show excellent adsorption energies, particularly with $Ti_3C_2T_x$ as support. Pt nanoparticles uniformly deposited on the optimum layered $Ti_3C_2T_x$ nanosheet, which formed higher intermetallic compounds, resulting in higher selectivity. 2D layer-structured transition metal nitrides and carbides (MXene) have great interest in process ultrafast charge transfer kinetics, larger active surface, and strong interfacial coupling, excellent surface tunable physicochemical properties [24].

3.5 MXene STRUCTURE AND THEORETICAL STUDIES

MXene nanosheet, in particular, the monolayer has a hexagonal structure with six-fold symmetry, almost similar to graphene nanostructure. The top view reveals the rhombus unit cell. Whereas from the side view, 2D MXene, consists of a tri-layer nanosheet with an X-layer sandwich between two transition metal groups (M) or layers. 2D MXene nanosheets are organized via six bonds with almost the closest X atoms. The above distinguishing character with six co-ordination shows that transition metal ions lead to the formation of different X groups like O_2, F_2, and OH. Generally, MXene nanostructure has the tendency to bond/interact with d-blocks heavy transition metal groups (4d and 5d). Hence, their electronic structure arrangement is greatly persuaded via relativistic spin–orbital combination and improves the band structure and surface properties.

According to theoretical studies, MXene and MXene-based composites ameliorate the fuel cell activity due to the nature of electronic interaction remarkably improving electrochemical performance. Ti_2CO_2 has the lowest oxygen reduction overpotential of 0.10 eV in the theoretical studies, which also suggested it as the best catalyst for Li-O_2 batteries. However, Ti_2CO_2 nanostructure shows a higher oxygen reduction potential of 0.69 V compared to TiC bulk materials [25]. Besides, Liu et al. theoretically performed simulations for the Pt/v-$Ti_{n+1}C_nT_x$ ($n = 1$–3, T = O and/or F) heterostructures via DFT calculation. The DFT simulations indicated that F-terminal

MXene nanosheets were envisaged to showing higher oxygen reduction activity than O-terminal MXene. The terminal group with F may exhibit lower long-term stability due to weaker chemical bonding [26].

Hence, V_2C hybridized with graphene showed an excellent kinetic barrier of 0.2 eV and a much lower overpotential of 0.36 V compared to 1.24 V for N-doped graphene and even could be a cost-effective alternative to Pt (0.45 V) catalyst [27]. Cheng et al. systematically studied the oxygen reduction activity using Ag, Pd, Cu, Au, and Pt thin layers impregnated on MXene Mo_2C nanostructure via DFT. He concluded that Mo_2C supported Au thin layers show higher oxygen reduction activity, which was comparable to Pt surface-supported carbon. The free-energy diagram (Figure 3.2a) of oxygen reduction pathway of Au thin layer supported on Mo_2C at zero electrode potential in acidic medium showed notable improvement in electrochemical activity and oxygen 4-electron pathway, which was ascribed to the shielding layer of Au thin layer on Mo_2C surface, which might be retained metal-support solid interaction (Figure 3.2b) [28].

The theoretical simulation results of 2D MXene PDDA-$Ti_3C_2(OH)_x$ nanosheet supported Pt nano worms nanoparticles with higher adsorption energy levels such as -3.09 eV atom^{-1} compared to -1.75 eV atom^{-1} for Pt nanoparticles supported $Ti_3C_2(OH)_x$ nanosheet. The above adsorption energy reveals that Pt nano worms nanoparticles interact more strongly with $Ti_3C_2(OH)_x$ surface than Pt nanoparticles. Moreover, the strong interface interaction direct to large d-band centers is shifted to down concerning Fermi energy and weaker Co adsorption [29]. Similarly, another report, for Pt-free Fe-N-C structure (Zeta potential +30 mV) has arranged with MXene (Zeta potential -39.7 mV) to produce ultra superlattice characteristics with a Brunauer–Emmett–Teller (BET) area of 30 m^{2}g^{-1} and the repeated dimension of 0.4 and 2.1 nm. The superlattice-like hetero-characteristic structure shows higher oxygen reduction activity with a more positive onset value of 0.92 V with the direct four-electron transfer [30].

3.6　PT FREE MXene NANOSHEET, AND IT IS COMPOSITE FOR FUEL CELL APPLICATION

Yang et al. fabricated a Pt-free multiwall carbon nanotube decorated with an optimum amount of MoS_2 quantum dots and MXene $Ti_3C_2T_x$ structure for the alkaline fuel cell. The as-prepared composite catalyst with an optimum composition of $MoS_2QDs@Ti_3C_2T_xQDs$ with enriched surface functional group uniformly coated on MWCNTs resulted in good electrochemical oxygen reduction reaction (ORR) and methanol oxidation activity [31]. In addition, in energy conversion applications, the exploitation of MXene has an additional advantage to improve volumetric energy density. The fabrication of $MoS_2QDs@Ti_3C_2T_xQDs@MWCNTs$ composite has good ORR and oxygen evolution activity (Figure 3.3). MXene composite has better stability in an alkaline medium and retains 93% initial activity than Pt/C (63%) at 0.6 V even after 6,000 cycles.

Wen et al. experimentally demonstrated that FeNC hybrid with MXene nanosheets shows better electrochemical activity and 25 mV higher half-wave potential than commercial Pt/C. MXene with FeNC hybrid composition improves the electronic

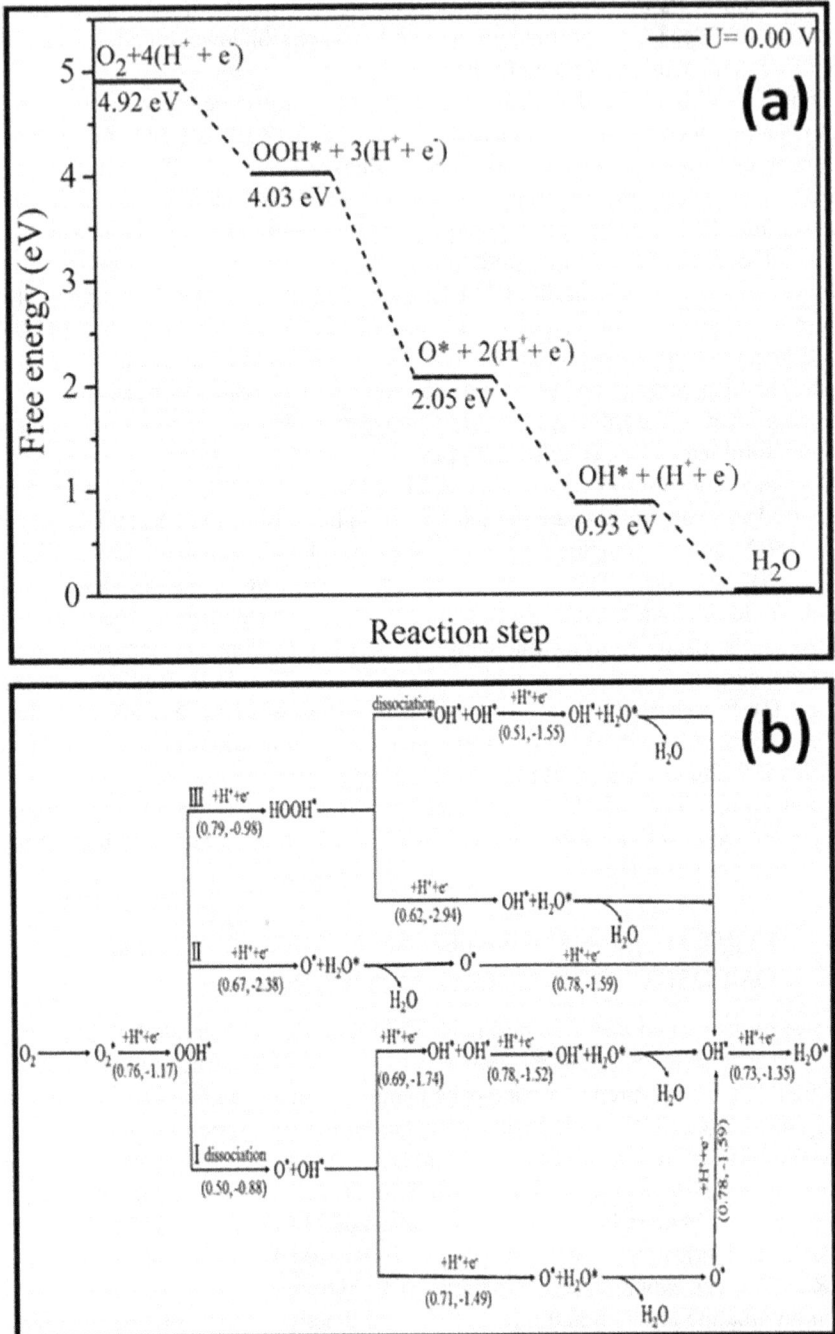

FIGURE 3.2 (a) Shows a free-energy diagram of Au thin layer supported Mo$_2$C, (b) proposed electrochemical oxygen reduction mechanism of Au thin layer on Mo$_2$C MXene support. (Adapted with permission from Ref. [28]. Copyright (2018) IOP Publishing Ltd.)

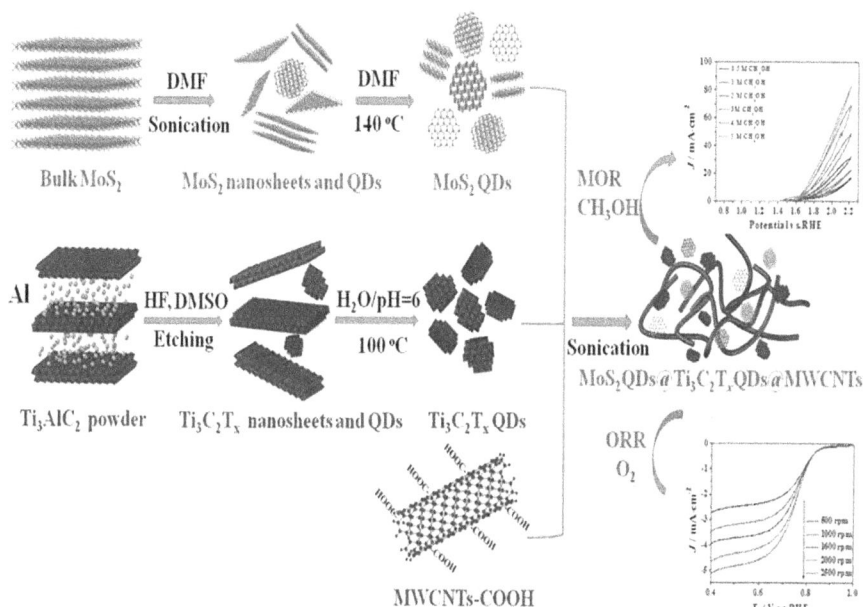

FIGURE 3.3 Formation of $MoS_2QDs@Ti_3C_2T_xQDs@MWCNTs$ and its energy conversion application. (Adapted with permission from Ref. [31]. Copyright (2018) Elsevier publishing Ltd.)

pathway and stabilizes FeNC active sites during the durability test. The stability test result reveals that a 2.6% reduction occurs even after the 2000s during continuous tests [32]. Similarly, cobalt was incorporated in CNT nanostructure with Ti_3C_2 thin layer nanosheets shows a remarkable half-wave potential of 0.82 V. An optimum level of Co-CNT 1D nanostructure uniformly composited with MXene Ti_3C_2 nanosheets was found to improve the diffusion limiting current density under mass transfer region, which even surpasses the commercial carbon-supported Pt [33].

According to Lin et al., the ultrathin MXene 2D Ti_3C_2 sheets with a thickness of 0.5–2.0 nm showed superior oxygen reduction activity in an alkaline medium. Ti_3C_2 nanosheets have a more positive onset potential of 0.85 V and a current density of 2.3 mA cm^{-2}. It is noteworthy that a Tafel slope value of 64 mV dec^{-1} was recorded for Ti_3C_2 in 0.1 mol L^{-1} KOH [34]. Similarly, Zhang et al. experimentally studied an optimum level of Ag nanoparticles deposited over the MXene system, which exhibited better oxygen reduction activity with the four-electron process. Similarly, $Ag_{0.9}Ti_{0.1}$ nanowires supported MXene composite exhibited higher electronic conductivity and catalytic efficiency with a recorded half-wave potential of 0.78 V, which has a 0.22 V more positive shift as compared to 0.56 V for Ag/C catalyst [35].

MXene nanosheets can support rapid electron transport by/through varying the electrophilicity of available catalyst or composite catalysts and thus enhance their electrochemical activity. Tahir et al. prepared an optimum level of $NiFe_2O_4$ composited with $(Ti_3C_2T_x)$ MXene nanosheet supported carbon felt synthesis via facile dip and dry hydrothermal method. The modified methodology and optimum composition of $NiFe_2O_4$-MXene@CF used as anode catalyst for enhancing microbial fuel cell considerably and show a higher peak power density of 1,385 mW m^{-2}, which is much higher than NiFe$_2$O$_4$@CF (400 mW m^{-2}) and MXene@CF (900 mW m^{-2}) as represented in Figure 3.4. The above superior microbial fuel cell performance may be ascribed to high electron conductivity and low charge transfer resistance, and the presence of numerous catalytic active sites may be formed in NiFe$_2$O$_4$-MXene@CF composition [36].

Zilan Li et al. recently reported a thin layer of MXene supported iron phthalocyanine (FePc) macrocycles enhancing twofold higher specific ORR activity. FePc on Ti$_3$C$_2$T$_x$ MXene electrocatalyst was prepared by homogeneous dispersion in dimethylformamide (DMF) followed by an ultrasonic condition and ORR studies of the as-prepared FePc/Ti$_3$C$_2$T$_x$ catalyst performed in 0.1 M KOH as electrolyte. The interaction between MXene and FeN$_4$ structure led to more comfortable oxygen adsorption and desorption on FeN$_4$ sites. Furthermore, the coupling of Ti$_3$C$_2$T$_x$ MXene with FePc leads to significant changes in electron Fe 3d delocalization and spin configuration, resulting in strong interaction cause active FeN$_4$ sites that more freely absorb the oxygen species resulting in improved ORR activity [37].

Figure 3.5 represents the ORR activity of Pt/C, FePc, and FePc/Ti$_3$C$_2$T$_x$ used as cathode electrocatalysts. FePc catalyst shows a higher half-wave potential of 0.86 V as compared to 0.84 V for Pt/C. Introducing Ti$_3$C$_2$T$_x$ into FePc structure resulted in remarkably improved ORR activity and a more positive shift of 25 mV. In addition, FePc/Ti$_3$C$_2$T$_x$ catalyst shows superior kinetic current density at 0.9 V of 15.5 mA cm^{-2}, which is much higher than Pt/C and FePc catalysts of 2.6 and 6.9 mA cm^{-2} (Figure 3.5b and c). Besides, the turn of frequency (TOF) of active FeN$_4$ sites is observed to be 15 e s^{-1} for FePc/Ti$_3$C$_2$T$_x$, which is more than fourfold higher as compared to pure FePc (3.9 e s^{-1}) as represented in Figure 3.5d.

The corresponding K-L plots exhibited linearity and reflected a four electron transfer (Figure 3.6e). Besides, FePc/Ti$_3$C$_2$T$_x$ catalyst also clearly indicated <1% of H$_2$O$_2$ formation during ORR activity, as shown in Figure 3.5f. Ti$_3$C$_2$T$_x$ MXene

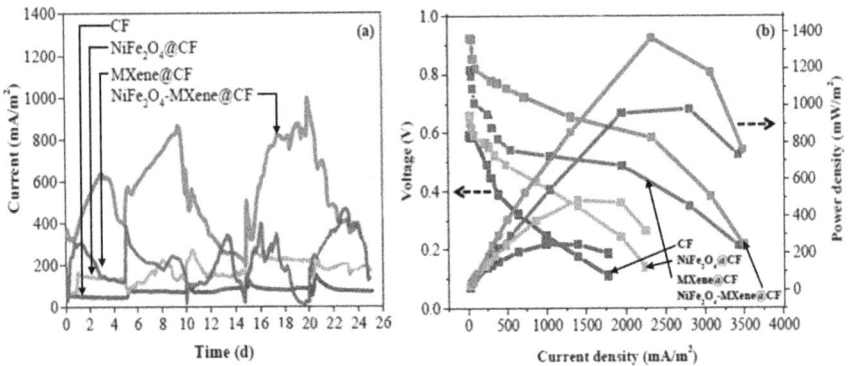

FIGURE 3.4 (a) Shows the current density vs MFC five cycles (25 days) at 0.2 V, (b) steady-state performance for MXene@CF, CF, NiFe$_2$O$_4$-MXene@CF and NiFe$_2$O$_4$@CF electrode. (Adapted with permission from Ref. [36]. Copyright (2018) Elsevier publishing Ltd.)

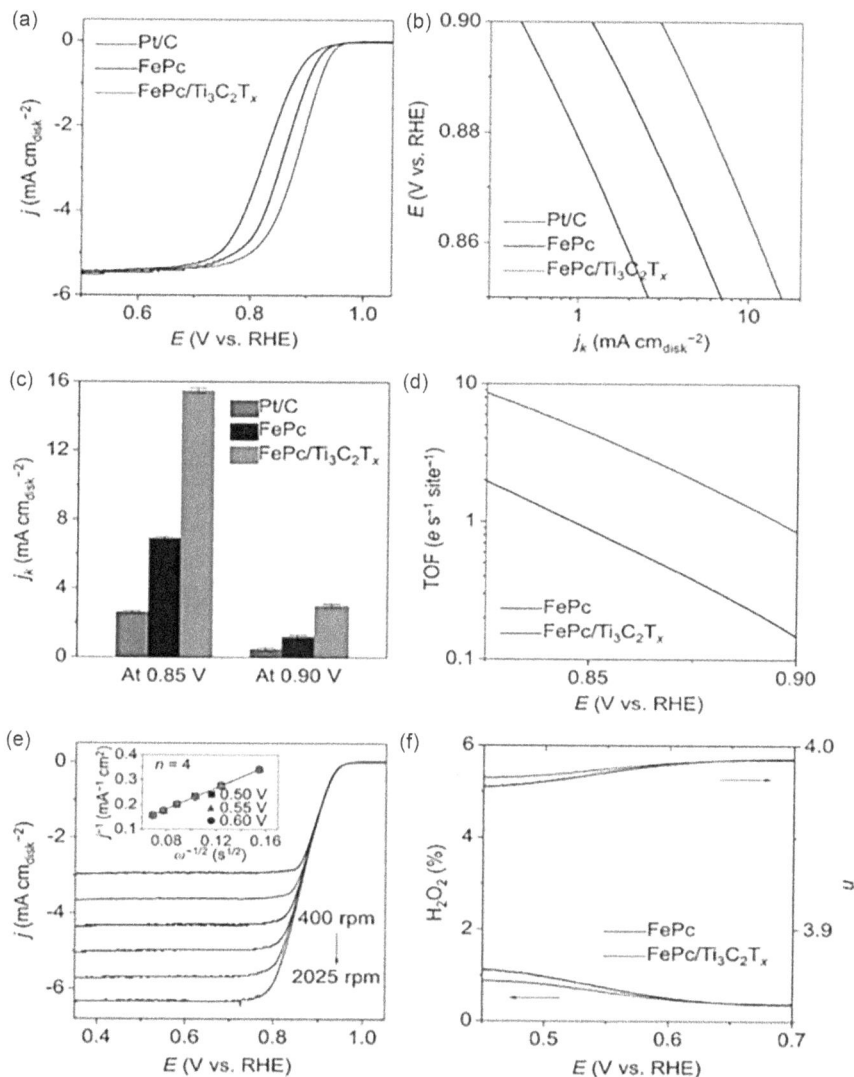

FIGURE 3.5 (a) LSV of oxygen reduction reaction curve, (b) Tafel plots, (c) comparison of current density at 0.9 and 0.85 V, (d) TOF of FeN$_4$ active sites in pure FePc and FePc/Ti$_3$C$_2$T$_x$, (e) K-L plot a different potential, and (f) Percentage of per-oxide formation and electron transfer number. (Adapted with permission from Ref. [37]. Copyright (2018) Wiley-VCH Publishing Ltd.)

supported FePc can retain more than 74% of ORR activity during continuous experiments compared to only 49% for Pt/C and 65% for FePc. The improvement of ORR activity and stability is mainly due to decreased isomer shift of FePc/Ti$_3$C$_2$T$_x$ matrix than pure FePc and an increase in the density of s electron around Fe (II) centers. The isomer shift initiates from the weakened shielding effect of Fe

3d electrons that can also expose a Fe 3d delocalization results in the strong inter-action between $Ti_3C_2T_x$ and FeN_4 moieties, which can induce spin-state transition the electron density re-distribution via van der Waals forces.

Wen et al. synthesized hybrid nanosheets of $FeNC/Ti_3C_2$ MXene for the oxygen reduction reaction. The Ti_3C_2 MXene 2D nanomaterials were prepared from MAX phase Ti_3AlC_2 by HF etching method. Ferrous acetate and 1,10-phenanthro-line monohydrochloride mixed in the ethanol to form a metal-ligand matrix [32]. According to the report by Wen et al., Ti_3C_2 MXene sheet was added slowly into a metal-ligand matrix and was further annealed in an inert atmosphere. The electro-chemical kinetic studies revealed remarkable higher oxygen reduction performance of FeNC supported MXene sheet with more than 24 mV higher half-wave potential in relation to Pt/C. In addition, FeNC supported MXene also exhibited excellent stabil-ity, with 97% of activity being retained after a 20,000 s of a continuous test, which is markedly higher than 84% for Pt/C electrocatalyst. The above results indicated that the improvement of ORR activity is mainly due to the contribution of MXene in the FeNC matrix, which can extract a 2D platform with outstanding electronic pathway as well as significant exposure of FeN active centers for ORR activity.

3.7 PT IMPREGNATED ON MXene NANOSHEET AND ITS COMPOSITE FOR FUEL CELL APPLICATION

Zhang et al. investigated the fabrication of $Ti_3C_2T_x$ (MXenes) via a modified mild and efficient HCl/LiF etching method. The obtained $Ti_3C_2T_x$ nanosheets were found to possess better conductivity, hydration properties, and abundant hydroxyl group on layer surface. According to Zhang et al., the Pt metal nanoparticles uniformly deposited on as-prepared $Ti_3C_2T_x$ nanosheets by sodium borohydride method showed excellent oxygen reduction activity in both alkalines and acidic mediums [38]. However, a multilayer thickness shows poor solubility resulting in difficulty in forming a uniform homogeneous solution. At the same time, mild etching methods produced flakes with higher lateral dimensions, which exhibited a more negative charge on $Ti_3C_2T_x$ nanosheet surface and excellent catalytic performance [39]. The oxygen reduction reaction was evaluated via rotating disk electrode (RDE) in both acid and alkaline. The OCV, as shown $Pt/Ti_3C_2T_x$, is 0.95 V, which is much higher than Pt/C (0.91 V). Similarly, a current density of 1.32 mA cm^{-2} was recorded for $Pt/Ti_3C_2T_x$ in relation to 1.1 mA cm^{-2} for Pt/C. The superior activity is mainly due to the modified electronic d-band center of Pt. Similarly, $Pt/Ti_3C_2T_x$ electrocatalyst was able to retain 99.5% of its initial ORR activity under various potential regimes in relation to 91% for Pt/C.

Wang et al. proposed Pt nanowires incorporated over a novel $Ti_3C_2T_x$ and CNT hybrid framework for high-temperature fuel cells. According to Wang et al., an opti-mum level of MXene could facilitate and create more active or anchoring sites for Pt nanowire's strong adsorption, making MXene ideal catalyst support for fuel cell application. However, the poor charge transfer process and irreversible agglomeration (strong van der Waals) associated with MXene were considered major drawbacks for further use. To prevent these drawbacks, the author tried with 1D CNT grafted with 2D MXene, preventing the Pt agglomeration and improved uniform distribution [40].

The nanosheet framework of MXene and the 1D framework of CNT can both afford the electronic pathway toward enhancing high-temperature fuel cell (HT-FC) performance. The Pt NWs/$Ti_3C_2T_x$-CNT hybrid framework shows better HT-FC activity of 182 mW cm^{-2} and stability even after 15,000 s and retains more than 90% of initial HT-FC performance (Figure 3.6a and b).

Xie et al. reported that Pt metal nanoparticles impregnated on layer nanosheet $Ti_3C_2T_2$ MXene improved the ORR performance and stability in relation to Pt/C. The Pt/$Ti_3C_2T_2$ hybrid nanosheet exhibited a superior half-wave potential of 0.87 V in O_2-saturated 0.1 M $HClO_4$ in comparison to 0.834 V for Pt/C [41]. Similarly, Zhang et al. fabricated a bi-functional electrocatalyst of Pd and $Ti_3C_2T_x$ nanoparticles uniformly deposited over CNT via the ethylene glycol reduction method. The optimum composition of Pd/$Ti_3C_2T_x$-supported CNT nanotube (1:2) shows excellent specific activity and mass activity of 3.3 and 4.4 times higher as compared to Pd/C for oxygen reduction activity in alkaline medium. The optimum composition of 1:2 Pd/$Ti_3C_2T_x$-CNT showed an open-circuit potential and half-wave potential of 1.085 and 0.925 V, respectively, which is much higher than Pd/C [42]. In addition, the authors also performed an actual fuel cell experiment under alkaline conditions at two different temperatures (25°C and 60°C) and observed an excellent cell performance of more than 40 mW cm^{-2} of peak power density with metal loading of cathode side about 0.4 mg cm^{-2} (Figure 3.6c and d).

Similarly, Xu et al. developed a hybrid composite catalyst of MXene ($Ti_3C_2T_x$) nanosheet and CNT, demonstrating higher corrosion stability and electrical conductivity. The hybrid composite supported Pt nanoparticle was found to show higher cell performance. The optimum composition of CNT-$Ti_3C_2T_x$ (1:1) supported Pt cathode electrocatalyst showed a higher performance of 180 mW cm^{-2} [43]. Besides, 22 cell stacks with an active area of 1 cm^2 were assembled, and the stack test was performed at room temperature. The portable stack delivered a maximum stack power of 138 W at 11 V (Figure 3.6e–f). The lower overpotential loss and higher oxygen reduction ability might be ascribed to faster interfacial oxygen kinetics and strong metal, and support interaction of Pt supported the CNT-$Ti_3C_2T_x$ (1:1) framework. Similarly, single-cell durability reveals more than 94% of initial performance observed after 360 minutes in relation to only 75% for Pt/C.

Yang et al. prepared Pt nano-worms directly grown on poly(diallyl dimethyl ammonium chloride) surface-functionalized MXene $Ti_3C_2T_x$ nanosheets. According to Yang et al., they were introducing PDDA on $Ti_3C_2T_x$ nanosheets, which significantly improved their surface charge properties to reinforce an electrostatic interaction between MXene and metal nanoparticles. In addition, PDDA on $Ti_3C_2T_x$ nanosheets had a tuning effect on the surface properties resulting in the stereo assembly of worms-shaped Pt metal nanoparticles. Pt nano-worms on the PDDA-$Ti_3C_2T_x$ surface show superior electrocatalytic activity, stability, and strong antipoisoning methanol oxidation. The methanol oxidation reactions were performed in 1 M methanol as an electrolyte. Pt nano-worms/PDDA-$Ti_3C_2T_x$ showed a higher anodic peak of 17.2 mA cm^{-2} in relation to 3.3 mA cm^{-2} for Pt nanoparticles supported on carbon. The high performance is mainly due to the higher distribution of Pt nano-worms and the high active PDDA-$Ti_3C_2T_x$ matrix, promoting a larger reaction population in the triple-phase interface [29].

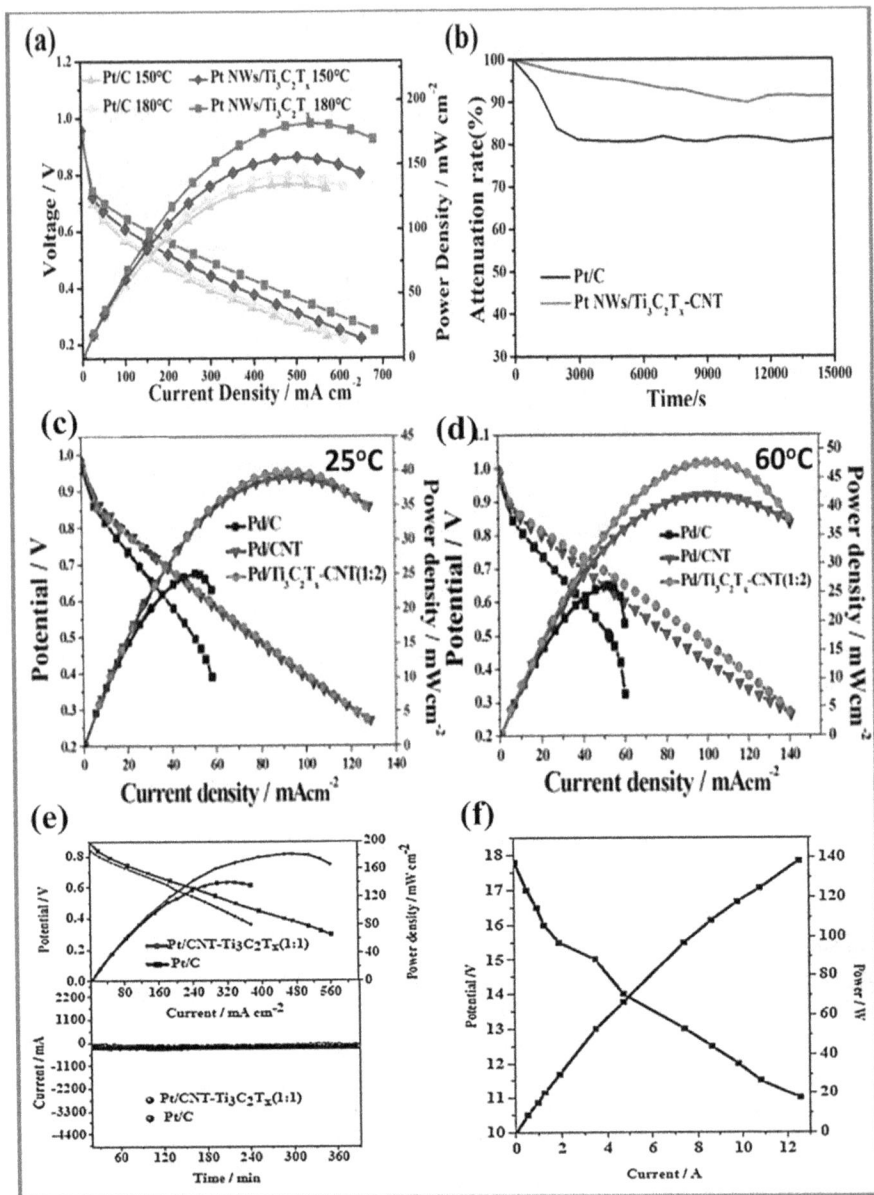

FIGURE 3.6 (a) Steady-state performance with different temperatures, (b) stability test carried out in H_2/O_2 configuration at 150°C. (Adapted with permission from Ref. [40]. Copyright (2018) Elsevier publishing Ltd.) (c and d) Shows the steady-state polarization is carried out at 25°C and 60°C. (Adapted with permission from Ref. [42]. Copyright (2020) Wiley-VCH Publishing Ltd.) (e) Steady-state single-cell polarization for Pt/CNT-$Ti_3C_2T_x$ and long-term stability, (f) fuel cell stack performance data. (Adapted with permission from Ref. [43]. Copyright (2018) ACS publishing Ltd.)

3.8 MXene/POLYMER MEMBRANE FOR FUEL CELL APPLICATION

Thin MXene nanosheet was effectively used as universal 2D fillers for inorganic–organic composite membrane, especially for PEFC application. MXene-based filler for polymer composite membrane effectively improved proton transfer and physical properties than base membrane. Fei et al. reported an optimum level of MXene-based filler in polybenzimidazole (PBI) membrane for intermediated polymer electrolyte fuel cell. 3 wt.% of $Ti_3C_2T_x$-MXene 2D nanosheet incorporated into polybenzimidazole membrane has increased the tensile strength (18 MPa) and Young's modulus (609 MPa) compared to PA-doped PBI (12 and 352 MPa) at 150°C. Furthermore, 3 wt.% of $Ti_3C_2T_x$-MXene-PBI membrane showed superior fuel cell performance (200 mW cm^{-2}) than PBI-based membrane (130 mW cm^{-2}) [44].

MXene 2D material is also used for solid oxide fuel cell membrane for effectively improving oxygen ionic conductivity. For example, 5 wt.% $Ti_3C_2T_x$-MXene filler incorporated into $Sm_{0.2}Ce_{0.8}O_{1.9}$ resulted in enhanced ionic conductivity and two-fold higher fuel cell performance in relation to $Sm_{0.2}Ce_{0.8}O_{1.9}$ based membrane at 500°C [45]. Similarly, the $Ti_3C_2T_x$-MXene nanosheet was functionalized with quaternary ammonium, an effective filler for the quaterized polysulfone polymer membrane. $Ti_3C_2T_x$-MXene-quaterized polysulfone composite membrane is used for anion exchange polymer electrolyte fuel cell. $Ti_3C_2T_x$-MXene-a quaterized polysulfone membrane has almost two times higher hydroxide ionic conductivity and fuel cell performance (100 mW cm^{-2}) than a quaterized polysulfone-based membrane (50 mW cm^{-2}). The combination of $Ti_3C_2T_x$-MXene-quaterized polysulfone composite shows better temperature tolerance, water retention properties, and superior hydroxide ionic conductivity than the quaterized polysulfone membrane [46].

The MXene-based material like $Ti_3C_2T_x$ nanosheets was dispersed in Nafion polymer via simple solution casting techniques. The recast membrane shows a glass transition temperature (thermal stability) of 126°C, where the glass transition temperature increased up to 132°C while incorporating $Ti_3C_2T_x$ MXene into the Nafion membrane. Similarly, the presence of 2D MXene filler within the polymer framework can reduce the elongation properties. In addition, by improving the amount of $Ti_3C_2T_x$ nanosheet into the Nafion membrane, the water uptake properties also increased. The optimum composition of Nafion-$Ti_3C_2T_x$-10 membrane delivers 200 mW cm^{-2} of power density than recast Nafion (150 mW cm^{-2}) [47]. Similarly, 1 wt.% of $Ti_3C_2T_x$ MXene nanosheet incorporated into SPEEK-based membrane showed improved proton conductivity of 0.066 S cm^{-1} than SPEEK membrane (0.056 S cm^{-1}). The optimum composition of MXene filler provides a new vision into the potential application toward the MXene family for proton conduction and membrane formation.

3.9 FUTURE SCOPE FOR 2D MXene FRAMEWORK IN ENDOWING PORTABLE FUEL CELL APPLICATION

In the future, MXene 2D nanostructured materials with tunable flexible structure and morphology could be explored toward improving chemical and thermal stability. Especially, lightweight MXene coated engineering components like a current collector and bi-polar plate could be expected as an alternative component toward the engineering aspect to reduce cost and improve materials durability in fuel cell application in the future. MXene 2D nanostructure materials are comparatively easy to fabricate via the etching process from MAX, and as a cost-effective technique, it can be used to fabricate the different types of MXene with various functional groups on the top of the surface in the future. Hence, modified MXene 2D nanostructure could be concurring the reliability and actual fuel cell performance toward automobile application. Moreover, considering the electrocatalyst aspect and expecting, MXene 2D nanostructure fabrication with ultra-thin monolayer structure, better conductivity, and higher surface area can help improve overall fuel cell activity.

In addition, tuning the morphology, varying composition with different carbon materials or metal oxide systems, and controlling the surface defect may further improve fuel cell performance. Besides, understanding the underlying surface mechanism with Pt-supported MXene 2D nanostructure or Pt-free MXene nanostructure is vital to further improve fuel cell performance. Furthermore, thin metal layer utilization on the MXene surface is more challenging and essential to reduce costs and improve stack durability. MXene 2D nanosheet framework via $Ti_3C_2T_x$ carbide-based electrocatalyst or support materials or undercoating layer in carbon electrode can permit fuel cell operation in ambient atmosphere, thereby improving water retention properties and helping to reduce auxiliary components further and making them more convenient for portable application in future.

An optimum level of MXene 2D nanostructure incorporated with electrolyte membranes such as Nafion and SPEEK is more challenging and may improve the water retention properties, proton conductivity, and mechanical strength. In addition, it might further improve the membrane performance and stability. We expect modified MXene 2D nanostructure might change the catalyst utilization and should be explored to meet technology ordination requirements in the future.

ACKNOWLEDGMENT

P. Dhanasekaran and S. Vinod Selvaganesh thank CSIR-for Senior Research Associateship (Scientist's Pool Scheme-9123-A and 9178-A, respectively).

REFERENCES

1. Z. W. She, J. Kibsgaard, C. F. Dickens, I. Chorkendorff, J. K. Nørskov, and T. F. Jaramillo, "Combining theory and experiment in electrocatalysis: Insights into materials design," *Science*, 2017, 355, 1–12.
2. J. M. Campos-Martin, G. Blanco-Brieva, and J. L. G. Fierro, "Hydrogen peroxide synthesis: An outlook beyond the anthraquinone process," *Angewandte Chemie: International Edition*, 2006, 45(42), 6962–6984.
3. J. Garche and L. Jörissen, "Applications of fuel cell technology: Status and perspectives," *Electrochemical Society Interface*, 2015, 24, 39–43.
4. P. Dhanasekaran, S. Vinod Selvaganesh, and S. D. Bhat, "Nitrogen and carbon doped titanium oxide as an alternative and durable electrocatalyst support in polymer electrolyte fuel cells," *Journal of Power Sources*, 2016, 304, 360–372.

5. K. Khan, A. K. Tareen, L. Jia, U. Khan, A. Nairan, Y. Yuan, X. Zhang, M. Yang, and Z. Ouyang, "Facile synthesis of tin-doped mayenite electride composite as a non-noble metal durable electrocatalyst for oxygen reduction reaction (ORR)," *Dalton Transactions*, 2018, 47, 13498–13506.

6. F. Kong, M. Banis, L. Du, L. Zhang, L. Zhang, J. Li, K. Doyle-Davis, J. Liang, Q. Liu, X. Yang, R. Li, C. Du, G. Yin, and X. Sun, "Highly stable one-dimensional Pt nanowires with modulated structural disorder towards the oxygen reduction reaction," *Journal of Materials Chemistry A*, 2019, 7, 24830–24836.

7. P. Dhanasekaran, S. V. Selvaganesh, and S. D. Bhat, "Enhanced catalytic activity and stability of copper and nitrogen doped titania nanorod supported Pt electrocatalyst for oxygen reduction reaction in polymer electrolyte fuel cells," *New Journal of Chemistry*, 2017, 41, 13012–13026.

8. P. Dhanasekaran, S. Vinod Selvaganesh, and S. D. Bhat, "Preparation of TiO_2: TiN composite nanowires as a support with improved long-term durability in acidic medium for polymer electrolyte fuel cells," *New Journal of Chemistry*, 2017, 41, 2987–2996.

9. S. V. Selvaganesh, P. Sridhar, S. Pitchumani, and A. K. Shukla, "Pristine and graphitized-MWCNTs as durable cathode-catalyst supports for PEFCs," *Journal of Solid State Electrochemistry*, 2014, 18, 1291–1305.

10. S. Vinod Selvaganesh, P. Dhanasekaran, and S. D. Bhat, "Nanocomposite TiO_2-f-MWCNTs as durable support for Pt in polymer electrolyte fuel cells," *Journal of Solid State Electrochemistry*, 2017, 21, 2997–3009.

11. P. Dhanasekaran, S. Vinod Selvaganesh, A. Shukla, N. Nagaraju, and S. D. Bhat, "Boosting Pt oxygen reduction reaction activity and durability by carbon semi-coated titania nanorods for proton exchange membrane fuel cells," *Electrochimica Acta*, 2018, 263, 596–609.

12. P. Dhanasekaran, B. Saravanan, S. Vinod Selvaganesh, and S. D. Bhat, "Addressing LT-PEFC 15 cell stack durability using carbon semi-coated titania nanorods-Pt electrocatalyst," *International Journal of Hydrogen Energy*, 2019, 44, 1940–1952.

13. J. C. Meyer, A. K. Geim, M. I. Katsnelson, K. S. Novoselov, T. J. Booth, and S. Roth, "The structure of suspended graphene sheets," *Nature*, 2007, 446, 60–63.

14. P. Dhanasekaran, S. Vinod Selvaganesh, A. Shukla, and S. D. Bhat, "Synergistic interaction of graphene-amorphous carbon nanohybrid with thin metal loading for enhanced polymer electrolyte fuel cell performance and durability," *Materials Letters*, 2021, 282, 2–6.

15. S. Vinod Selvaganesh, P. Dhanasekaran, R. Chetty, and S. D. Bhat, "Microwave assisted poly(3,4-ethylenedioxythiophene)-reduced graphene oxide nanocomposite supported Pt as durable electrocatalyst for polymer electrolyte fuel cells," *New Journal of Chemistry*, 2018, 42, 10724–10732.

16. S. Ted Oyama, J. C. Schlatter, J. E. Metcalfe, and J. M. Lambert, "Preparation and characterization of early transition-metal carbides and nitrides," *Industrial and Engineering Chemistry Research*, 1988, 27, 1639–1648.

17. R. B. Levy and M. Boudart, "Platinum-like behavior of tungsten carbide in surface catalysis," *Science*, 1973, 181, 547–549.

18. M. Naguib, O. Mashtalir, J. Carle, V. Presser, J. Lu, L. Hultman, Y. Gogotsi, M. W. Barsoum, "Two-dimensional transition metal carbides," *ACS Nano*, 2012, 6, 1322–1331.

19. M. Naguib, M. Kurtoglu, V. Presser, J. Lu, J. Niu, M. Heon, L. Hultman, Y. Gogotsi, M. W. Barsoum "Two-dimensional nanocrystals produced by exfoliation of Ti_3AlC_2," *Advanced Materials*, 2011, 23, 4248–4253.

20. W. Sun, S. A. Shah, Y. Chen, Z. Tan, H. Gao, T. Habib, M. Radovic, M. J. Green, "Electrochemical etching of Ti_2AlC to Ti_2CT_X (MXene) in low-concentration hydrochloric acid solution," *Journal of Materials Chemistry A*, 2017, 5, 21663–21668.

21. B. Anasori, M. R. Lukatskaya, and Y. Gogotsi, "2D metal carbides and nitrides (MXenes) for energy storage," *Nature Reviews Materials*, 2017, 2(16098), 1–17.
22. S. Kuchida, T. Muranaka, K. Kawashima, K. Inoue, M. Yoshikawa, and J. Akimitsu, "Superconductivity in Lu_2SnC," *Physica C: Superconductivity and Its Applications*, 2013, 494, 77–79.
23. K. Maleski, V. N. Mochalin, and Y. Gogotsi, "Dispersions of two-dimensional titanium carbide MXene in organic solvents," *Chemistry of Materials*, 2017, 29, 1632–1640.
24. Y. Gao, Y. Y. Vao, Y. B. Gu, H. Zhuo, G. L. Zhuang, S. W. Deng, X. Zhong, Z. Z. Chen, J. H. Chen, X. Pan, J. G. Wang, "Functionalization Ti_3C_2 MXene by the adsorption or substitution of single metal atom," *Applied Surface Science*, 2019, 465, 911–918.
25. Y. Yang, Y. Qin, X. Xue, X. Wang, M. Yao, and H. Huang, "Intrinsic properties affecting the catalytic activity of 3d transition-metal carbides in $Li-O_2$ battery," *Journal of Physical Chemistry C*, 2018, 122, 17812–17819.
26. C. Y. Liu and E. Y. Li, "Termination effects of $Pt_{/v.}Ti_{n+1}C_nT_2$ MXene surfaces for oxygen reduction reaction catalysis," *ACS Applied Materials and Interfaces*, 2019, 11, 1638–1644.
27. J. K. Nørskov, J. Rossmeisl, A. Logadottir, L. Lindqvist, J. Kitchin, T. Bligaard, H. Josson, "Origin of the overpotential for oxygen reduction at a fuel-cell cathode," *Journal of Physical Chemistry B*, 2004, 108, 17886–17892.
28. C. Cheng, X. Zhang, Z. Fu, and Z. Yang, "Strong metal-support interactions impart activity in the oxygen reduction reaction: Au monolayer on Mo_2C (MXene)," *Journal of Physics Condensed Matter*, 2018, 30, 475201–475212.
29. C. Yang, Q. Jiang, H. Huang, H. He, L. Yang, and W. Li, "Polyelectrolyte-induced stereoassembly of grain boundary-enriched platinum nanoworms on $Ti_3C_2T_x$ MXene nanosheets for efficient methanol oxidation," *ACS Applied Materials and Interfaces*, 2020, 12, 23822–23830.
30. L. Jiang, J. Duan, J. Zhu, S. Chen, and M. Antonietti, "Iron-cluster-directed synthesis of 2D/2D Fe-N-C/MXene superlattice-like heterostructure with enhanced oxygen reduction electrocatalysis," *ACS Nano*, 2020, 14, 2436–2444.
31. X. Yang, Q. Jia, F. Duan, B. Hu, M. Wang, L. He, Y. Song, and Z. Zhang, "Multiwall carbon nanotubes loaded with MoS_2 quantum dots and MXene quantum dots: Non–Pt bifunctional catalyst for the methanol oxidation and oxygen reduction reactions in alkaline solution," *Applied Surface Science*, 2019, 464, 78–87.
32. Y. Wen, C. Ma, Z. Wei, X. Zhu, and Z. Li, "FeNC/MXene hybrid nanosheet as an efficient electrocatalyst for oxygen reduction reaction," *RSC Advances*, 2019, 9, 13424–13430.
33. J. Chen, X. Yuan, F. Lyu, Q. Zhong, H. Hu, Q. Pan, and Q. Zhang, "Integrating MXene nanosheets with cobalt-tipped carbon nanotubes for an efficient oxygen reduction reaction," *Journal of Materials Chemistry A*, 2019, 7, 1281–1286.
34. H. Lin, L. Chen, X. Lu, H. Yao, Y. Chen, and J. Shi, "Two-dimensional titanium carbide MXenes as efficient non-noble metal electrocatalysts for oxygen reduction reaction," *Science China Materials*, 2019, 62, 662–670.
35. Z. Zhang, H. Li, G. Zou, C. Fermandez, B. Liu, Q. Zhang, J. Hu, and Q. Peng, "Self-reduction synthesis of new MXene/Ag composites with unexpected electrocatalytic activity," *ACS Sustainable Chemistry and Engineering*, 2016, 4, 6763–6771.
36. K. Tahir, W. Miran, J. Jang, N. Maile, A. Shahzad, M. Morztahida, A. Ghani, B. Kim, H. Jeon, S. Lim, and D. S. Lee, "Nickel ferrite/MXene-coated carbon felt anodes for enhanced microbial fuel cell performance," *Chemosphere*, 2021, 268, 128784.
37. Z. Li, Z. Zhuang, H. Zhu, L. Zhou, M. Luo, J. Zhu, A. Lang, S. Feng, W. Chen, L. Mai, and S. Guo, "The marriage of the FeN_4 moiety and MXene boosts oxygen reduction catalysis: Fe 3d electron delocalization matters," *Advanced Materials*, 2018, 30, 1–8.

38. C. Zhang, B. Ma, Y. Zhou, and C. Wang, "Highly active and durable Pt/MXene nano-catalysts for ORR in both alkaline and acidic conditions," *Journal of Electroanalytical Chemistry*, 2020, 865, 114142.

39. V. Gajdosova, L. Lorencova, M. Prochazka, M. Omastova, M. Micusik, S. Prochazkova, F. Kveton, M. Jerigova, D. Velic, P. Kasak, and J. Tkac, "Remarkable differences in the voltammetric response towards hydrogen peroxide, oxygen and Ru(NH3)6^{3+} of elec-trode interfaces modified with HF or LiF-HCl etched $Ti_3C_2T_x$ MXene," *Microchimica Acta*, 2020, 187, 52.

40. R. Wang, Z. Chang, Z. Fang, T. Xiao, Z. Zhu, B. Ye, C. Xu, J. Chen, "Pt nanowire/$Ti_3C_2T_x$-CNT hybrids catalysts for the high performance oxygen reduction reaction for high temperature PEMFC," *International Journal of Hydrogen Energy*, 2020, 45, 28190–28195.

41. X. Xie, S. Chen, W. Ding, Y. Nie, and Z. Wei, "An extraordinarily stable catalyst: Pt NPs supported on two-dimensional $Ti_3C_2X_2$ (X=OH, F) nanosheets for oxygen reduction reaction," *Chemical Communications*, 2013, 49, 10112–10114.

42. P. Zhang, R. Wang, T. Xiao, Z. Chang, Z. Fang, Z. Zhu, C. Xu, L. Wang, J. Cheng, "The high-performance bifunctional catalyst Pd/$Ti_3C_2T_x$–carbon nanotube for oxygen reduc-tion reaction and hydrogen evolution reaction in alkaline medium," *Energy Technology*, 2020, 8, 7.

43. C. Xu, X. Zhang, C. Fan, H. Chen, X. Liu, Z. Fu, R. Wang, T. Hong, J. Cheng, "MXene ($Ti_3C_2T_x$) and carbon nanotube hybrid-supported platinum catalysts for the high-performance oxygen reduction reaction in PEMFC," *ACS Applied Materials and Interfaces*, 2020, 12, 19539–19546.

44. M. Fei, R. Lin, Y. Deng, H. Xian, R. Bian, X. Zhang, J. Cheng, C. Xu, D. Cai, "Polybenzimidazole/MXene composite membranes for intermediate temperature poly-mer electrolyte membrane fuel cells," *Nanotechnology*, 2018, 29, 3.

45. H. Xian, C. Fan, P. Zhang, R. Wang, C. Xu, H. Zhai, T. Hong, J. Cheng, "Effect of MXene on oxygen ion conductivity of $Sm_{0.2}Ce_{0.8}O_{1.9}$ as electrolyte for low temperature SOFC," *International Journal of Electrochemical Science*, 2019, 14, 7729–7736.

46. X. Zhang, C. Fan, N. Yao, P. Zhang, T. Hong, C. Xu, J. Cheng "Quaternary $Ti_3C_2T_x$ enhanced ionic conduction in quaternized polysulfone membrane for alkaline anion exchange membrane fuel cells," *Journal of Membrane Science*, 2018, 563, 882–887.

47. Y. Liu, J. Zhang, X. Zhang, Y. Li, and J. Wang, "$Ti_3C_2T_x$ filler effect on the proton conduction property of polymer electrolyte membrane," *ACS Applied Materials and Interfaces*, 2016, 8, 20352–20363.

4 2D Nanomaterials in Flexible Fuel Cells

Tauqir A. Sherazi and Misbah Jabeen
COMSATS University Islamabad

Tahir Rasheed
King Fahd University of Petroleum and Minerals (KFUPM)

Syed Ali Raza Naqvi
Government College University

CONTENTS

DOI: 10.1201/9781003178422-4

59

4.1 INTRODUCTION OF 2D NANOMATERIALS

The materials that have at least one dimension in the range of 1–100 nm are classi-fied as nanomaterials. The 2D nanomaterials are those which have two dimensions greater than 100 nm, while the third must be less than 100 nm that is normally thick-ness. The 2D nanomaterials exist in the form of sheets. Some common examples of 2D nanomaterials are Graphene, MXene, transition metal dichalcogenides (TMDs), metal organic frameworks (MOFs), covalent organic frameworks (COFs), hexagonal boron nitride, and layered double hydroxides [1] (Figure 4.1).

4.2 CHARACTERISTICS OF 2D NANOMATERIALS

2D nanomaterials have a lot of optical and electronic properties due to their electron confinement pattern. They can act as electrocatalysts, photo electrocatalysts, elec-trochemical sensors, and colorimetric sensors. TMDs, MOFs, and COFs are used in energy cleaning process, catalysis and also in water remediation techniques [2].

The smaller size offers high surface-to-volume ratio, while the dangling bonds present on the surface of nanomaterials are responsible for their high surface energy. The 2D nanomaterials possess high mechanical and thermal stability and at the same time due to their thin sheet-like structures they experience elasticity and can be molded or bent in any form. Graphene, now a day, is used in portable electronics and energy storage devices. These materials also show magnetic properties such as ferromagnetism with a variety of applications such as cellular phones and electric motors. Optical properties of 2D nanomaterials are responsible for the development of light emitting diode. These materials due to their thin layered structures can also show additives properties and can act as solid lubricants [3]. Hence, 2D nanomateri-als, due to their extraordinary characteristics, are useful for alternate and sustainable solutions of environment and energy challenges.

4.3 APPLICATIONS OF 2D NANOMATERIALS IN ENERGY DEVICES

The 2D nanomaterials exhibit unique features as compared to other geometries of nanomaterials like zero-dimensional (0D) nanomaterials such as quantum dots, one-dimensional (1D) nanomaterials, i.e., nanorods or nanowires, and three-dimensional (3D) networks or even their bulk counterparts. In 2D nanomaterials, charge carriers are confined along the thickness but are allowed to move along the plane of these

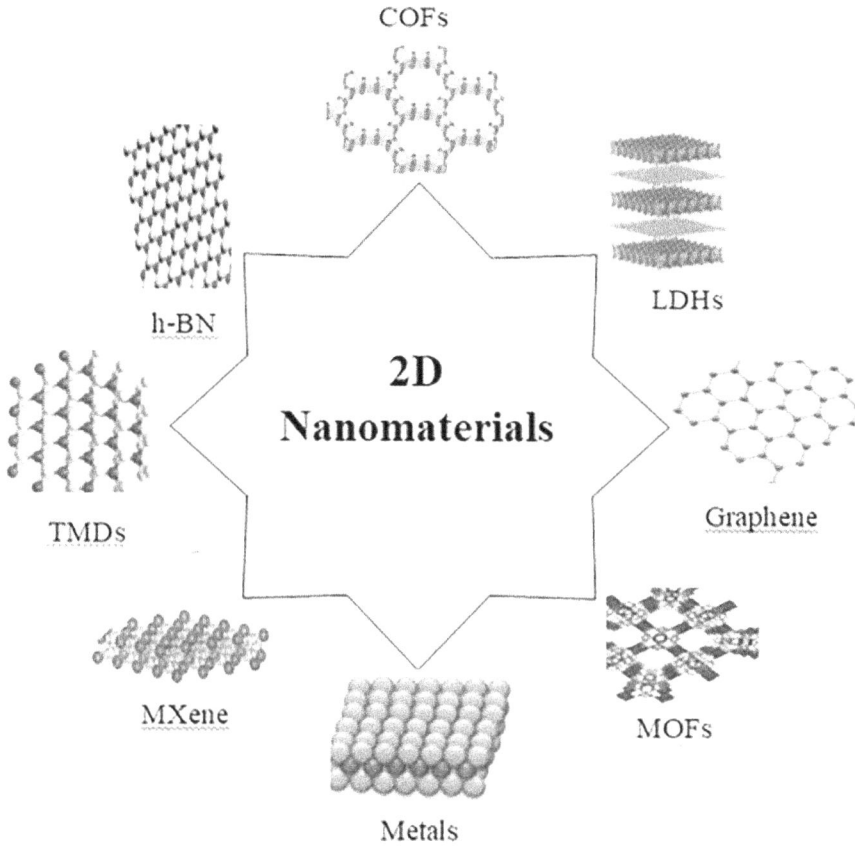

FIGURE 4.1 Various types of 2D nanomaterials.

nanomaterials. This unique characteristic makes them responsible for vast applications in electrical and electronic devices.

Electronic properties of 2D nanomaterials can be modified by tuning their thickness, which is not possible in 0D, 1D, and 3D nanomaterials. 2D nanomaterials are highly sensitive to external stimulus, such as chemical modification, chemical doping, adsorption of other molecules or materials, and mechanical deformation due to their larger surface areas. Owing to these outstanding properties and morphological advantages, 2D nanomaterials are promising materials for a wide range of applications such as in electrochemistry, optoelectronics, energy storage and conversion devices, biomedicines, catalysis, sensors, and many others.

4.3.1   2D Nanomaterials for Energy Conversion and Storage

The carbon-supported platinum (Pt/C) catalyst in fuel cells has many deficiencies such as slow oxygen reduction reaction (ORR) rate, higher cost of platinum, thermal instability, and loss of catalytic efficiency due to carbon monoxide (CO) poisoning. In order to achieve maximum use of metal catalysts, Pt nanomaterials that are supported with a low-cost porous material have been formed that have reduced expenses of Pt material consumption. Graphene and its derivatives such as graphene oxide or reduced graphene oxide (rGO) are important materials that can be used as carbon supports for electrocatalysts to improve electrochemical performance during fuel oxidation in fuel cells [4]. Similarly, 2D metal oxides thin sheet is also used in energy conversion and storage process due to its flexibility, stability, and higher conductivity. Transition metals dichalcogenides, i.e., MoS_2 show superior performance in hydrogen evolution reaction during water splitting, while transition metal oxides such as nickel oxide (NiO) are useful for oxygen evolution reaction [5].

4.3.2 2D NANOMATERIALS IN SUPERCAPACITORS

Owing to their high specific surface area, well-confined charge transport and tunable surface properties, 2D nanomaterials are very promising for electrode materials in supercapacitors.

Graphene and graphene-based composite materials are widely used as electrode materials for supercapacitors. Graphene is a 2D nanomaterial with thickness of one atomic layer, hexagonally arranged network of sp^2-hybridized carbon atoms. Due to this structure, graphene has unique physical properties like an ultra-high specific surface area (having a theoretical value of $2,600\,m^2g^{-1}$), high light transmittance (for a single-layered sheet, it is 97%), and high breaking strength ($42\,N\,m^{-1}$). Hence, these unique features have widened graphene's applications in the fields that require high electrical conductivity and surface area, and supercapacitor is one of them [6].

In graphene-based supercapacitors, energy is stored via electrical double-layer capacitance. Due to its high specific surface area and high electrical conductivity, graphene enhances the charge storage capacity in supercapacitors [7].

4.3.3 2D NANOMATERIALS IN FUEL CELL AND BATTERIES

Durability of 2D nanomaterials, their surface functionalization and metal doping capability enhance their performances and applications in electronic and portable devices. The development of advanced, lightweight, inexpensive, and highly efficient fuel cells and lithium-ion batteries electrode materials, electrocatalysts and electrolytes involve use of 2D nanomaterials, e.g., graphene sheets or metal oxides nanosheets are used nowadays in proton exchange membrane fuel cell (PEMFC), and others [8].

4.4 FUEL CELL

Fuel cell is an electrochemical device that converts chemical energy of fuel into electrical energy. Mostly the fuel used in these fuel cells is hydrogen, thus it does not lead to carbonaceous emissions. The chemical reaction is aided by oxidant gases, while water produces as by-product. Fuel cells have a variety of applications ranging from stationary power production to portable devices, and in transportation [9]. Fuel cell produces electric energy similar to batteries, but it differs from batteries such that fuel cells are provided with a continuous supply of fuel contrary to recharging phenomenon in batteries. They are highly efficient and are environmentally benign. Major components of a simple hydrogen fuel cell are polymer electrolyte membrane (PEM), cathodic catalyst layer, cathode, anodic catalyst layer, anode, and current collectors [10]. Different catalysts are used in fuel cell depend on cell type such as nickel is used for high temperature and platinum is used in low-temperature fuel cell [11] (Figure 4.2).

4.5 TYPES OF FUEL CELL

Fuel cells are classified into different types on the basis of electrolyte present. Fuel cell having solid electrolyte are PEMFC, solid oxide fuel cell, and direct methanol fuel cell, while those possessing liquid electrolyte membrane are phosphoric acid fuel cell, alkaline fuel cell, and molten carbonate fuel cell [12]. The brief description of major types of fuel cell is given below:

4.5.1 Proton Exchange Membrane Fuel Cells (PEMFC)

PEMFC is also known as PEM fuel cell. In 1960, General Electric was first who invented the PEMFC in Gemini spacecraft for NASA. It consists of solid polymer membrane as an electrolyte acting as ionic (proton) conductor and electronic insulator. This fuel cell is developed for the stationary and transport applications and also for portable applications because of its low working temperature range (80°C–200°C) [13]. These cells are recognized as clean power production technology because of its almost pollution-free operation and easy handling. Despite of its characters its commercialization is still limited because of many aspects like low power density at low temperature (40°C–45°C), low electrocatalyst tolerance toward carbon monoxide poisoning, high cost, and need to supply excess fuel to achieve better performance [14].

4.5.2 Phosphoric Acid Fuel Cell (PAFC)

PAFC is categorized into sub-grouped of low-temperature fuel cell (150°C–200°C). It uses phosphoric acid as an electrolyte. Conduction of protons occurs through electrolyte and electrons pass through external circuit to generate current. PAFC is considered as clean technology because water is produced as by-product [15].

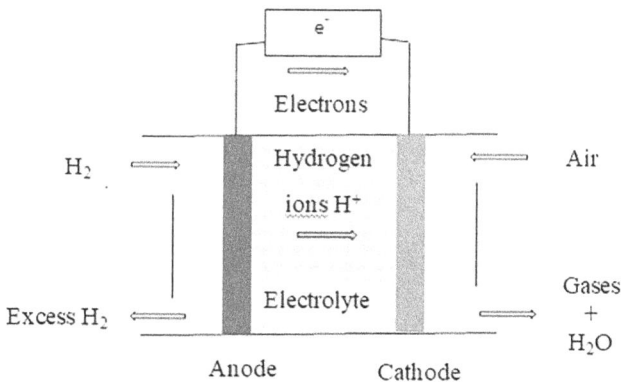

FIGURE 4.2 Schematic of hydrogen fuel cell.

4.5.3 Solid Oxide Fuel Cell (SOFC)

SOFCs are classified as high-temperature fuel cell working at a temperature range of 600°C–1,000°C. It was found by an engineer George Westinghouse. Electrolyte used in SOFC is ceramic (yttria-stabilized zirconia). SOFC exhibits high efficiency up to 70% and has internal hydrogen extraction capability. Extra electricity can be generated by SOFC by recycling the waste heat. However, the commercialization of SOFCs is restrained due to their slow start-up. They are suitable for medium- to large-sized power applications. Solid electrolyte makes SOFC stable even at higher temperatures [16].

4.5.4 Alkaline Fuel Cell (AFC)

AFC is also called Bacon fuel cell as it was invented by Francis Bacon for NASA's space mission cell. Due to its efficiency of 60%–70% AFC is also called low-temperature fuel cell. Aqueous solution of potassium hydroxide works as an electrolyte to transport anions from anode to cathode [17]. AFC has ability of quick start-up but it has great sensitivity to carbon dioxide, which restricts its use for commercial application. Applications of this type of fuel cell are present in space shuttle and in transport [18].

4.5.5 Direct Methanol Fuel Cell (DMFC)

In DMFC, liquid methanol is directly used as an anode fuel for oxidation and polymer membrane as an electrolyte. DMFCs have an internal reforming mechanism such as at anode liquid methanol and water react to produce hydrogen and subsequently release of electrons and protons. Protons move across the electrolyte membrane toward cathode where they get reduced in the presence of oxygen, while electrons travel through external circuit offering electricity. Due to the presence of internal reformer in their constructive design DMFC has low cost for operation [19]. Unfortunately, these cells have limited economic values because of methanol crossover and slow reaction kinetics ensuing low efficiency. DMFCs have the potential to replace batteries for computers, cameras, and other electronic applications [20].

4.5.6 Molten Carbonate Fuel Cell (MCFC)

These fuel cells used the mixture of alkali metal carbonates and operate at high-temperature range (600°C–700°C) that is above the melting point of carbonate and carbonate ion. Thus, transportation of ions takes place through molten electrolyte [21]. When MCFCs are coupled with turbine their efficiency increases and reaches up to 65%. The catalysts for electrochemical reactions used in MCFCs are cheaper, which make them cost-effective. These fuel cells are feasible for power generation plants based on coal that uses natural gas and anaerobic digester. Basic principle of MCFC is the same as in PEMFC with the difference of ionic conduction through molten carbonate [22].

4.5.7   Direct Ethanol Fuel Cell (DEFC)

DEFC is the type of PEMFC because it produces electric power by the same electrochemical reaction using polymer membranes such as Nafion as an electrolyte. It utilized ethanol as fuel instead of hydrogen gas. The extraction of ethanol is easy from fermentation of biomass like sugar cane and wheat corn. That's why it is cheaper fuel and non-toxic [23].

4.5.8   Solid Alkaline Fuel Cell (SAFC)

In SAFC also known as anion exchange membrane fuel cell, solid electrolyte membrane is used contrary to the liquid alkali (aqueous KOH) electrolyte used in alkaline fuel cell. Instead of pure noble metal catalyst, SAFC can utilize non-noble metal base catalysts like Sn, C, and Cu/Ni-modified. Due to fascinating features, the SAFCs have consideration an economical, ecofriendly, efficient alternate energy source, which can operate even at low temperature [24,25].

4.5.9   Direct Formic Acid Fuel Cell (DFAFC)

DFAFC generates electric current by the oxidation of formic acid as an anodic fuel. DFAFC is characterized by efficiency of 20%, low fuel crossover and safe in use but high over potential and high cost limit its commercialization. In this fuel cell, carbon monoxide is produced in electrocatalytic reaction as an intermediate product. Thus to render the production of CO, Pt or Pd-based electrocatalysts are used in DFAFC. These fuel cells are used in portable electronics [26,27].

4.5.10   Zinc Air Fuel Cell (ZAFC)

In 1968, Leclanche introduced a metal air battery containing a negatively charged anode, which is made of metal and positively charged oxygen permeable cathode. Zinc is replaced with the metal that acts as anodic material in metal air battery to design ZAFC [28]. ZAFCs are divided into two types depending on charging process, non-rechargeable fuel cells are called as primary ZAFCs and rechargeable fuel cells are called secondary ZAFCs. ZAFC with capacity of 20 W is suitable source of power generation in large vehicles [29].

4.5.11   Regenerative Fuel Cell (RFC)

RFC works in reversible manner that it has ability to regenerate chemical reactants electrochemically using photovoltaic cells and produce electricity. These fuel cells are equipped with lightweight pressure vessel to contain the gases and energy storage system. Energy storage for long term in RFCs and have high energy density make them superior over the secondary batteries. They yield energy up to 400 Wh kg^{-1}. The RFCs are designed to use in space, high-altitude airships, aircrafts, long-term missions and as well as for military use because of their 100% charging and discharging cycle [30].

4.5.12 Microbial Fuel Cell (MFC)

MFCs are green source for the generation of electricity by using domestic wastewater or biological wastewater and different fuels like glucose and lactate, and it operates at low temperature (20°C–40°C). These fuel cells have two chambers having bacteria at anode side and separated from cathode by PEM. In MFCs, the bacteria acts as biocatalyst for the anaerobic oxidation of substrate and generates electrons that combine with the oxygen and protons to generate electricity and form water as only byproduct. The efficiency of MFC is approximately up to 50%. Some factors like power output and cost limit the commercialization of MFCs. These fuel cells are used in small electronic devices such as biosensors [31]. All the types of fuel cells are summarized with reference to their operating temperature, electrolyte used, efficiency, and major applications as presented in Table 4.1.

4.6 ENERGY APPLICATIONS OF FUEL CELLS

Fuels cells are ecofriendly, highly efficient, easy to handle, and are feasible for the production of energy. They have a wide range of applications like portable, stationary, backup and automotive but few are available on a commercial scale. The use of these fuel cells in different fields depends upon the parameters such as working efficiency, size, weight, power output, and fuel type [32]. Some of the main categories in which fuel cells are used are following:

- Portable applications
- Stationary applications
- Mobile applications

4.6.1 Portable Applications

Portable fuel cells are lightweight, long lasting, and have long operation time and used for portable application such as laptops, computers, cell phones, power tools, battery chargers, sensors, military equipment, etc. These are generally low-temperature range fuel cells such as PER fuel cells, DMFC is used due to having superior start-up performance [33]. Power range for portable application is from 5 to 50 W. For portable applications, fuel cells are the best option for substitution of batteries due to extended operation time, nonstop production of power, and lightweight [34].

4.6.2 Stationary Applications

Fuel cells are widely used in stationary applications for the production of heat and power. Both the low and high-temperature range fuel cell types are suitable for these applications. However, quick startup is only achieved by the low-temperature range fuel cells like PEMFCs. High-temperature range fuel cells suitable for stationary application are SOFCs, AFCs, and MCFCs. They can be attached with other equipment like batteries, photovoltaics, wind turbines, capacitors, and generators to generate primary or secondary power. Fuels such as compressed hydrogen, propane,

TABLE 4.1

Summary of Fuel Cell Types and Their Characteristics

Fuel Cell Type	Operating Temperature (°C)	Electrolyte Used	Efficiency (%)	Applications
Proton exchange membrane fuel cell (PEMFC)	80–200	Solid membrane	40–50	Vehicles, mobiles, and low power combined heat and power systems
Phosphoric acid fuel cell (PAFC)	150–200	Phosphoric acid	40–50	Combined heat and power systems of 200 kW
Solid oxide fuel cell (SOFC)	600–1,000	Ceramics	60–70	All sizes of combined heat and power system
Alkaline fuel cell (AFC)	50–200	KOH	60–70	Space vehicles
Direct methanol fuel cell (DMFC)	20–90	Solid membrane	20–50	In portable electronics of low power
Molten carbonate fuel cell (MCFC)	600–700	Carbonated metal	65	Medium and large scale combined heat and power system
Direct ethanol fuel cell (DEFC)	30–60	Nafion	80	Portable applications
Solid alkaline fuel cell (SAFC)	80–100	Solid electrolyte membrane in alkaline medium	20	In electronic devices
Direct formic acid fuel cell (DFAFC)	60	Nafion membrane	20	Portable electronics
Zinc air fuel cell (ZAFC)	50–70	Alkaline electrolyte	96	Power generation in large vehicles
Regenerative fuel cell (RFC)	700–1,000	Solid oxide electrolyte	50–80	In long-term energy storage and in space applications
Microbial fuel cell (MFC)	20–40	Phosphate buffer	50	In small electronic devices and in biosensors

methanol, bio-gas, oil-based fuels, synthesis gas, town gas, digester gas, and land fill gas are more commonly used for stationary applications. Fuel cells power for stationary applications is in the range of 1–50 MW for large-scale applications; however, small-scale applications like telecommunication range is 1–100 kW [35].

4.6.3 MOBILE APPLICATIONS

Fuel cells are also used in different types of transport like buses, bicycles, scooters, and vehicles. In transport, internal combustion engines are replaced by the fuel cell because no air pollution is caused by the fuel cell. Automobile uses PEMFC and DMFC fuel cells type because of the fuels (hydrogen and methanol) used in these types of fuel cells are easily available. Power range for transport applications is about 20–250 kW. PEMFCs, ZAFCs, PAFCs, and DMFCs fuel cell types are in use as power source for Fuel Cell buses. Among all types of fuel cells, PEMFCs are leading in such applications. Major issue in commercialization of fuel cell in automobiles is storage and distribution of hydrogen gas. To resolve this issue, bio-fuels like ethanol and methanol are used, they have low-cost, easily available, non-toxic, and renewable [36].

4.7 INCORPORATION OF 2D NANOMATERIALS IN FUEL CELL

2D nanomaterials are being used extensively in developing fuel cells components to reduce size and enhanced efficiency. 2D nanomaterials have excellent properties such as high surface area, lightweight, higher conductivity, higher stability, surface functionalization, and overall excellent performance.

The best electrocatalyst that was used first for the electrolytic reaction in fuel cells was Pt based. In the start, a large amount of Pt was used, which cause the addition of extraordinary cost, making the fuel cell extremely expensive. However, the nanotechnology and the use of Pt in nanoscale reduces the use of Pt several hundred times without compromising the fuel cell efficiency thus exponentially reducing the fuel cell cost. Another approach is replacement of noble metal catalyst to non-noble metal catalyst. For this purpose Pd metal-based electrocatalysts containing 2D nanomaterials proved the best alternatives. These Pd-based 2D nanomaterials show greater efficiency against ORR at cathode and fuel oxidation reaction at anode in fuel cell because of high electron mobility and improved surface functionalization [37]. Similarly, graphene and its derivatives such as graphene doped by nitrogen (N-graphene) are used as good metal free electrode in AFC due to its greater electrocatalytic activity and long-term stability [38]. Nanomaterials based on graphene also perform best as metal-free electrocatalysts and catalyst support in PEMFC [39] (Figure 4.3).

Hence 2D nanomaterials are of great importance and show efficient performance when incorporated in fuel cells such as graphene or its derivatives. Other nanomaterials such as transition metal oxides and TMDs also perform with greater efficiency in electrocatalytic reaction such as ORR in fuel cell.

4.8 FLEXIBLE FUEL CELL

Recently flexible energy devices are getting more attention due to their beneficial properties such as lightweight, good mechanical strength, flexible nature, shape diversity, and portability. For this purpose, there is a need to replace the high weight traditional energy sources like solar cells, supercapacitors, batteries, and fuel cells to produce lightweight flexible energy sources such as flexible solar cells, flexible fuel cells, and flexible supercapacitors to fit in portable flexible energy devices. These flexible fuel cells such as flexible PEMFC are used in portable electronics for mobile applications, i.e., in mobile phone.

A flexible fuel cell is low cost and low weight bendable fuel cell that can be molded in any form to fit in an energy device. It has a thin chip-like structure composed of nanocomposites. The energy conversion property of these flexible fuel cells is never disturbed on bending and molding their shape [41]. An example of flexible fuel cell is

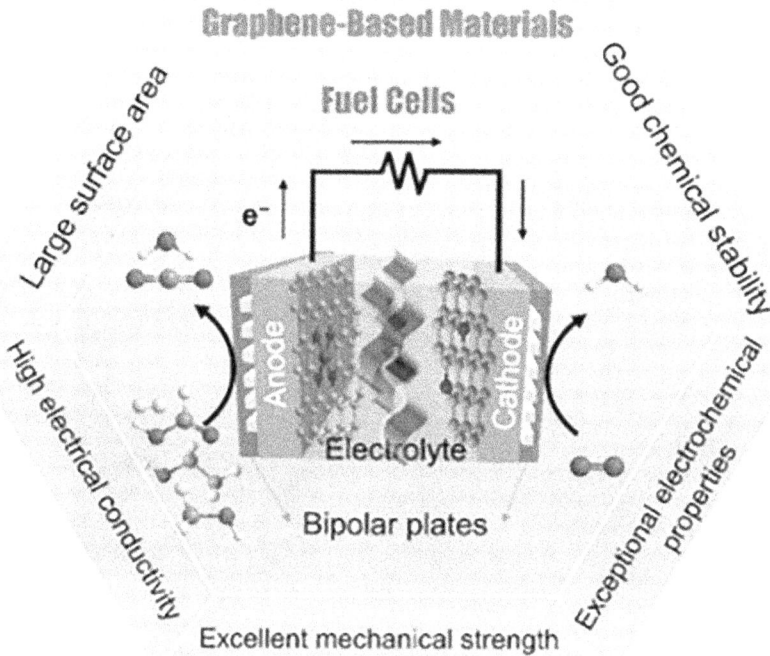

FIGURE 4.3 Advantages of using graphene-based materials in fuel cells. (Adapted with permission from Ref. [40]. Copyright (2020) Wiley Online Library.)

DEFC. The flexibly fabricated fuel cell shows high performance of 21.48 mW cm^{-2}. Due to its unique flexibility it is used in electronic clock and smart phone to power them [42] (Figure 4.4).

4.8.1 FLEXIBLE PROTON EXCHANGE MEMBRANE FUEL CELL

The traditional PEMFC consists of graphitic or metallic plates for current collector and flow channels are too much rigid and heavy that they are unable to fit in flexible devices. In PEMFC, oxygen in air needs to be pumped into the cell for electrochemical reaction. In order to make these cells favorable for portable electronic devices. The air breathing system inside the PEMFC was invented through which air naturally diffuses into the cell without pump. However, they still contained bulky and expensive metallic or graphitic plates. Moreover, the specific area power density is usually between 100 and 200 mW cm^{-2}, which causes a much lower specific weight power densities (1.761 W kg^{-1}). So far, there is an urgent need for a lightweight and flexible PEMFC satisfying the demands of flexible electronic devices. For this purpose, the preparation of flexible electrode is challenging in flexible PEMFC. In traditional PEMFCs, carbon paper is mostly used as the electrode due to its high stability but it is not suitable to directly use carbon paper in a flexible PEMFC due to its brittleness and relatively low electrical conductivity. In recent flexible devices

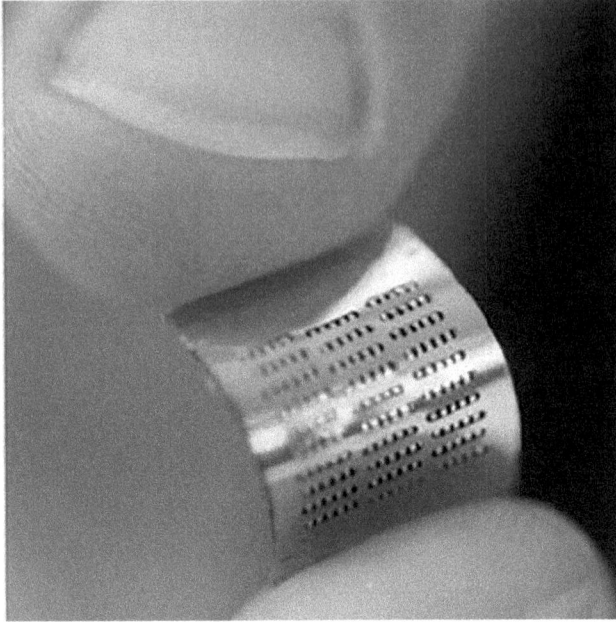

FIGURE 4.4   Flexible micro fuel cell demonstrator: size 1 cm², thickness 200 μm, 80 mW cm⁻². (Adapted with permission from Ref. [43]. Copyright (2004) De Gruyter.)

1D nanomaterial like carbon nanotubes (CNTs), 2D nanomaterials like graphene, conducting polymers and nanowires of different transition metal atoms are used as flexible electrode due to their higher surface area to volume ratio, higher stability, higher air permeability, fine flexibility, and higher electrical conductivity.

4.8.2   Fabrication of Flexible PEMFC

In the case for fabrication of flexible, lightweight, and compact PEMFC, the huge change in design of traditional PEMFC is required. For this purpose, rigid, bulky, and heavy graphitic or metallic plates are replaced with plates made up of nano-composites. As we know the main two functions of graphitic or metallic plates that they act as flow feed and current collector. In the air breathing flexible PEMFCs, highly conductive and flexible electrode made up of nanocomposites works as current collector, also responsible for air diffusion. Moreover, the flow field anodic plate is replaced with lightweight plastic plate while cathodic plate is completely removed.

These modifications in traditional PEMFC result in reduced volume and weight from 25 cm³ and 68.6 g for traditional 1–0.028 cm³ and 0.065 g for flexible fuel cell, respectively, with the same working area (1 cm²) [44].

It is interesting that the flexible PEMFC has even higher performance (145.2 mW cm⁻²) as compared to traditional one (90.3 mW cm⁻²). The higher performance of flexible PEMFC is attributed to the removal of cathodic plate because it can block the air transfer due to its bulky nature [44].

4.8.3 Fabrication of Flexible Electrode

Fabrication of flexible electrode in PEMFC is a key challenge because in such a case the reactants and products require low resistance for transfer through electrode. Therefore, flexible electrode should have higher stability, higher flexibility, and gas permeability. Flexible electrode should also have higher electrical conductivity and can act as an excellent support surface for catalyst. An example of flexible electrode is composite electrode having porous CNT membrane and carbon paper for a flexible PEMFC. The CNT membrane is so dense and needs higher permeability. For this purpose, a uniform circular micrometer-sized array from laser marking array is passed through this membrane to increase its permeability. Still micrometer-sized pores are too large to support electrocatalyst. Thus to solve this problem a flexible composite electrode is prepared by attaching carbon paper with CNTs-based membrane with a paste of active carbon and polytetraflouroethylene (PTFE). Finally flexible and highly permeable electrode is used in PEMFC to make it flexible and efficient [44].

4.9 2D NANOMATERIALS BASED FLEXIBLE FUEL CELL

The flexible fuel cells containing 2D nanomaterials such as graphene, MXene, and their derivatives are promising devices for clean power generation. In PEMFC, the membrane used as an electrolyte for transport of ions is made up of polymer. To improve the performance of the polymeric membrane, a small portion of graphene or its derivative is added because of good permeability and stability of graphene. As graphene is flexible in nature due to its thin sheet-like structure, it can be molded in any form without affecting its surface functionalization [45].

Similarly, for small-size electronic and portable devices, flexible electrodes are required that can be achieved through 2D nanomaterials. For this purpose, a thin graphene paper or rGO nanosheets are used as support for electrode material. These nanosheets have good mechanical strength, high conductivity, and small degree of bending property that cannot affect its mechanical strength. Besides these flexible electrodes, flexible electrolytes and bipolar plates based upon 2D nanomaterials can also be used to make low cost and lightweight energy conversion devices such as fuel cell that occupies little space due to flexible nature. Enzymatic biofuel cells (EBFCs) are also used in biomedical devices like implantable or wearable devices that require flexible and biocompatible electrode. EBFCs are the type of bioelectrochemical devices in which chemical energy is converted into electrical energy in presence of a biological catalyst (enzyme) [46].

Flexible electronics is a major technology for wearable devices in future. Still flexible lithium ion batteries and solar panels are under research. Lithium-ion batteries have the highest energy density among different secondary batteries, but low energy density compared to fuel cell. For flexible fuel cell high energy density and higher flexibility both are needed. A flexible fuel cell with great performance under its torsion based on polydimethylsiloxane (PDMS) that was coated with porous Ag nanowires network as shown in Figure 4.5.

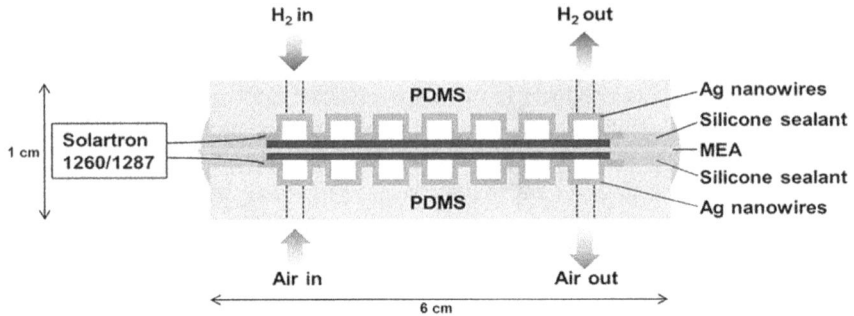

FIGURE 4.5 Cross-sectional view of a flexible fuel cell with Ag nanowires-based current collector and PDMS-based flow channel structure. (Adapted with permission from Ref. [47]. Copyright (2014) Copyright the Authors, some rights reserved; exclusive license [ECS Transactions].)

4.10 APPLICATIONS OF 2D NANOMATERIALS-BASED FLEXIBLE FUEL CELLS IN ENERGY DEVICES

Recently 2D nanomaterials-based flexible fuel cells have a variety of applications due to their tremendous properties. They have replaced expensive, less efficient, and large-sized fuel cells that were difficult to fit in portable electronics. Flexible fuel cells are clean and efficient source of energy that can be used in all size electronic devices. They can be molded in any shape. Their activity or energy storage and conversion property remains the same, no matter how much turns it takes during bending. When they combine with 2D nanomaterials, these flexible cells become advantageous for emerging technological applications. They can be used in electrical devices for power generation and backup. Besides the use of flexible fuel cells just for energy production, these can also be used in implantable and wearable biomedical devices [48].

4.10.1 FLEXIBLE PAPER-BASED FUEL CELL FOR POWER APPLICATIONS AT SMALL SCALE

A flexible paper-based hydrogen fuel cell is successfully developed recently for small power applications. To prevent hydrogen leakage and avoid bulky hydrogen storage, an innovative hydrogen fuel cell structure is designed having hydrogen source containing Al foil into the paper substrate, paper capsule here acts as the hydrogen chamber. Other components of this fuel cell, such as current collectors and electrodes are deposited directly on the surface of paper leading to a compact, lightweight, and flexible hydrogen fuel cell.

This paper-based fuel cell is intrinsically a flexible power source due to the presence of its all flexible components along with paper substrate, Al foil for hydrogen supply, ink-based electrode, and silver-based grid current collector. Thus, it is suitable for powering different flexible electronic devices including biosensors and wearable electronics. In order to increase the working and lifetime of this paper-based flexible fuel cell for real applications, both thin-film cell package outside the fuel cell and higher Al foil loading inside the fuel cell can be implemented to reduce hydrogen leakage and enhance hydrogen storage, respectively.

A single hydrogen fuel cell in comparison to stacked fuel cells can provide only a voltage output of less than 1 V, which is insufficient for a variety of applications. Thus, to obtain higher voltage, fuel cell stacking is done in a series manner. Instead of the conventional method of multi-layer stacking of fuel cells, a planar paper-based fuel cell stack is designed that contains 4 single cells in the same part of paper substrate [49].

4.10.2 Flexible Fuel Cell Pack as Bifunctional Catalyst

Recently, flexible energy sources with long-term stability and low weight have attracted great attention due to the increasing demand of flexible electronics, i.e., flexible and folding smartphones, wearable electronic, flexible displays, and implantable medical devices. For this purpose, flexible composite electrodes and flexible fuel cells are being designed but it is still challenging to fabricate entire fuel cell pack with flexible nature due to the involvement of its several components, complicate arrangement, and requirement of onsite hydrogen generator or reformer.

It is possible to fabricate a flexible hydrogen generator by using formic acid because formic acid has shape variability and fine fluidity. In addition, some other properties of formic acid are high energy density (1,725 Wh kg^{-1}) and high-quality hydrogen production with a catalyst. Still formic acid usage for fabrication of flexible hydrogen generator faces problem as it is prone to leakage during adaptability and flexibility. Moreover, formic acid-based hydrogen generator needs two components, the container for formic acid storage and the reactor for hydrogen production. To overcome these obstacles a unique bifunctional catalyst containing aerogel has abilities of liquid formic acid storage into the pores of silica aerogel to prevent it from leakage during flexibility and then it catalyzes the stored formic acid to hydrogen. A flexible hydrogen generator with high compatibility, adaptability, and flexibility can be designed by using aerogel containing bifunctional catalyst.

By combining this flexible hydrogen generator with flexible fuel cell stack can make an adaptable and flexible fuel cell pack with bifunctional catalytic property having high energy density (135.9 Wh kg^{-1}). Such a flexible fuel cell pack is highly promising to meet the increasing demand of flexible electronics [50].

4.11 FUTURE PERSPECTIVE

The studies reveal that due to emerging applications and amazing properties of 2D nanomaterials they are playing important role in different technologies especially in electrical and electronic devices. Most of the 2D nanomaterials are investigated and utilized for improved performance of an individual component of the devices, which does not guarantee the fabrication of flexible device on the whole. However, due to flexible and sheet-like structure of 2D nanomaterials, these have excellent potential to fabricate the overall flexible machines along with enhanced performance. The flexible fuel cells that can be molded and bent without the concern of its breakage have taken charge in the field of portable electronics. The merit of such flexible fuel cells must include mechanical flexibility, low cost, high performance, and facile engineering. The peak power output and stability are broadly comparable with existing state-of-the-art fuel cells to meet the demands for practical application. The 2D nanomaterials need to explore much to build up the individual flexible components of fuel cells with the capability of fuel cell fabrication and even stacking of these fuel cells with retention of flexibility. The material selection and design should be explored such that the bending phenomenon could not affect the electronic properties of the electrodes, leakage of gases or reactants, heat control, and release of byproducts. Such fuel cells offer perspectives for advancement in wearable and portable applications. The extreme diversity of 2D materials provides huge opportunities in developing flexible fuel cells. One of the major concerns of using nanomaterials is its environmental hazards, thus the 2D nanomaterials developed and used in flexible fuel cell fabrication must be highly stable and compatible, otherwise, these could be a serious issue for the environment.

REFERENCES

1. Azadmanjiri, J., Berndt, C., Wang, J., Kapoor, A., & Srivastava, V., Nanolaminated composite materials: Structure, interface role and applications. *RSC Advances*, 2016. **6**: pp. 109354–109360.
2. Zhang, H., Chhowalla, M., & Liu, Z., 2D nanomaterials: Graphene and transition metal dichalcogenides. *Chemical Society Reviews*, 2018. **47**(9): pp. 3015–3017.
3. Singh, N.B. & Shukla, S.K., Chapter 3: Properties of two-dimensional nanomaterials. In: R. Khan and S. Barua (Eds.), *Two-Dimensional Nanostructures for Biomedical Technology* (pp. 73–100). 2020, Amsterdam, Netherlands: Elsevier.
4. Khan, A.H., Ghosh, S., Pradhan, B., Dalui, A., Shrestha, L.K., Acharya, S., & Ariga, K., Two-dimensional (2D) nanomaterials towards electrochemical nanoarchitectonics in energy-related applications. *Bulletin of the Chemical Society of Japan*, 2017. **90**(6): pp. 627–648.
5. Moraes, L.P.R., Mei, J., Fonseca, F.C., & Sun, Z., Two-dimensional metal oxide nanomaterials for sustainable energy applications. In: S. Zafeiratos (Ed.), *2D Nanomaterials for Energy Applications* (pp. 39–72). 2020, Amsterdam, Netherlands: Elsevier.
6. Down, M.P. & Banks, C.E., 2D materials as the basis of supercapacitor devices. In: S. Zafeiratos (Ed.), *2D Nanomaterials for Energy Applications* (pp. 97–130). 2020, Amsterdam, Netherlands: Elsevier.
7. Geng, J., Chen, S., & Chen, X., State-of-the-art applications of 2D nanomaterials in energy storage. In: L. Singh & D.M. Mahapatra (Eds.), *Adapting 2D Nanomaterials for Advanced Applications* (pp. 253–293). 2020, American Chemical Society.
8. Hou, J., Shao, Y., Ellis, M.W., Moore, R.B., & Yi, B., Graphene-based electrochemical energy conversion and storage: Fuel cells, supercapacitors and lithium ion batteries. *Physical Chemistry Chemical Physics*, 2011. **13**(34): pp. 15384–15402.
9. Öztürk, A., Akay, R.G., Erkan, S., & Yurtcan, A.B., Chapter 1: Introduction to fuel cells. In: R.G. Akay & A.B. Yurtcan (Eds.), *Direct Liquid Fuel Cells* (pp. 1–47). 2021, London: Academic Press.
10. Winter, M. & Brodd, R.J., What are batteries, fuel cells, and supercapacitors? *Chemical Reviews*, 2004. **104**(10): pp. 4245–4270.
11. Haile, S.M., Fuel cell materials and component, *Journal of Acta Materialia*, 2003. **51**(19): pp. 5981–6000.
12. Hemmat Esfe, M. & Afrand, M., A review on fuel cell types and the application of nanofluid in their cooling. *Journal of Thermal Analysis and Calorimetry*, 2020. **140**(4): pp. 1633–1654.

13. Alaswad, A., Palumbo, A., Dassisti, M., & Olabi, A.G., Fuel cell technologies, applications, and state of the art: A reference guide. In: *Reference Module in Materials Science and Materials Engineering*. 2016, Amsterdam, Netherlands: Elsevier. https://www.sciencedirect.com/science/article/pii/B9780128035818040091?via%3Dihub

14. Costamagna, P. & Srinivasan, S., Quantum jumps in the PEMFC science and technology from the 1960s to the year 2000. Part II. Engineering, technology development and application aspects. *Journal of Power Sources*, 2001. **102**: pp. 253–269.

15. Sundén, B., Chapter 8: Fuel cell types – overview. In: B. Sundén (Ed.), *Hydrogen, Batteries and Fuel Cells* (pp. 123–144). 2019, Amsterdam, Netherlands: Elsiever.

16. Choudhury, A., Chandra, H., & Arora, A., Application of solid oxide fuel cell technology for power generation: A review. *Renewable and Sustainable Energy Reviews*, 2013. **20**: pp. 430–442.

17. Ferriday, T.B. & Middleton, P.H., Alkaline fuel cell technology: A review. *International Journal of Hydrogen Energy*, 2021. **46**(35): pp. 18489–18510.

18. Gülzow, E., Alkaline fuel cells. *Fuel Cells*, 2004. **4**(4): pp. 251–255.

19. Samimi, F. & Rahimpour, M.R., Chapter 14: Direct methanol fuel cell, In: A. Basile and F. Dalena (Eds.), *Methanol* (pp. 381–397). 2018, Amsterdam, Netherlands: Elsevier.

20. Kamarudin, S.K., Achmad, F., & Daud, W.R.W., Overview on the application of direct methanol fuel cell (DMFC) for portable electronic devices. *International Journal of Hydrogen Energy*, 2009. **34**(16): pp. 6902–6916.

21. Yuh, C., Hilmi, A., Farooque, M., Leo, T., & Xu, G., Direct fuel cell materials experience. *ECS Trans*, 2009. **17**(1):637.

22. Antolini, E., The stability of molten carbonate fuel cell electrodes: A review of recent improvements. *Applied Energy*, 2011. **88**(12): pp. 4274–4293.

23. Zakaria, Z., Kamarudin, S.K., and Timmiati, S.N., Membranes for direct ethanol fuel cells: An overview. *Applied Energy*, 2016. **163**: pp. 334–342.

24. Park, J.S., Park, S.H., Yim, S.D., Yoon, Y.G., Lee, W.Y., & Kim, C.S., Performance of solid alkaline fuel cells employing anion-exchange membranes. *Journal of Power Sources*, 2008. **178**(2): pp. 620–626.

25. Dekel, D.R., Review of cell performance in anion exchange membrane fuel cells. *Journal of Power Sources*, 2018. **375**: pp. 158–169.

26. Aslam, N.M., Masdar, M.S., Kamarudin, S.K., & Daud, W.R.W., Overview on direct formic acid fuel cells (DFAFCs) as an energy sources. *APCBEE Procedia*, 2012. **3**: pp. 33–39.

27. El-Nagar, G.A., Hassan, M.A., Lauermann, I., & Roth, C., Efficient direct formic acid fuel cells (DFAFCs) anode derived from seafood waste: Migration mechanism. *Scientific Reports*, 2017. **7**(1): pp. 17818.

28. Neburchilov, V., Wang, H., Martin, J.J., & Qu, W., A review on air cathodes for zinc–air fuel cells. *Journal of Power Sources*, 2010. **195**(5): pp. 1271–1291.

29. Sapkota, P. & Kim, H., Zinc–air fuel cell, a potential candidate for alternative energy. *Journal of Industrial and Engineering Chemistry*, 2009. **15**(4): pp. 445–450.

30. Pu, Z., Zhang, G., Hassanpour, A., Zheng, D., Wang, S., Liao, S., & Sun, S., Regenerative fuel cells: Recent progress, challenges, perspectives and their applications for space energy system. *Applied Energy*, 2021. **283**: pp. 116376.

31. Rahimnejad, M., Adhami, A., Darvari, S., Zirepour, A., & Oh, S.E., Microbial fuel cell as new technology for bioelectricity generation: A review. *Alexandria Engineering Journal*, 2015. **54**(3): pp. 745–756.

32. Badwal, S.P.S., Giddey, S., Kulkarni, A., Goel, J., & Basu, S., Direct ethanol fuel cells for transport and stationary applications: A comprehensive review. *Applied Energy*, 2015. **145**: pp. 80–103.

33. Dyer, C.K., Fuel cells for portable applications. *Journal of Power Sources*, 2002. **106**: pp. 31–34.

34. Wang, Y., Chen, K.S., Mishler, J., Cho, S.C., & Adroher, X.C., A review of polymer electrolyte membrane fuel cells: Technology, applications, and needs on fundamental research. *Applied Energy*, 2011. **88**(4): pp. 981–1007.
35. Volkart, K., Densing, M., De Miglio, R., Priem, T., Pye, S., & Cox, B., Chapter 23: The role of fuel cells and hydrogen in stationary applications. In: M. Welsch, P.D. Keles, A. Faure-Schuyer, A. Dobbins, A. Shivakumar, P. Deane, & M. Howells (Eds.), *Europe's Energy Transition* (pp. 189–205). 2017. Amsterdam, Netherlands: Elsevier.
36. Belmonte, N., Luetto, C., Staulo, S., Rizzi, P., & Baricco, M., Case studies of energy storage with fuel cells and batteries for stationary and mobile applications. *Challenges*, 2017. **8**(1): 9.
37. Xu, H., Shang, H., Wang, C., & Du, Y., Recent progress of ultrathin 2D Pd-based nanomaterials for fuel cell electrocatalysis. *Small*, 2021. **17**(5): p. 2005092.
38. Qu, L., Liu, Y., Baek, J.-B., & Dai, L., Nitrogen-doped graphene as efficient metal-free electrocatalyst for oxygen reduction in fuel cells. *ACS Nano*, 2010. **4**(3): pp. 1321–1326.
39. Zhou, X., Qiao, J., Yang, L., & Zhang, J., A review of graphene-based nanostructural materials for both catalyst supports and metal-free catalysts in PEM fuel cell oxygen reduction reactions. *Advanced Energy Materials*, 2014. **4**(8): p. 1301523.
40. Su, H. & Hu, Y.H., Recent advances in graphene-based materials for fuel cell applications. *Energy Science & Engineering*, 2020. **9**: pp. 958–983.
41. Tominaka, S., Nishizeko, H., Mizuno, J., & Osaka, T., Bendable fuel cells: On-chip fuel cell on a flexible polymer substrate. *Energy & Environmental Science*, 2009. **2**(10): pp. 1074–1077.
42. Wang, J., Pei, Z., Liu, J., Hu, M., Feng, Y., Wang, P., & Huang, Y., A high-performance flexible direct ethanol fuel cell with drop-and-play function. *Nano Energy*, 2019. **65**: p. 104052.
43. Hahn, R., Energieversorgung von energieautarken Sensoren (autonomous energy supply for wireless sensors). *TM-Technisches Messen*, 2007. **74**: pp. 456–465.
44. Ning, F., He, X., Shen, Y., Jin, H., Li, Q., Li, D., & Zhou, X., Flexible and lightweight fuel cell with high specific power density. *ACS Nano*, 2017. **11**(6): pp. 5982–5991.
45. Alekseeva, O.K., Pushkareva, I.V., Pushkarev, A.S., & Fateev, V.N., Graphene and graphene-like materials for hydrogen energy. *Nanotechnologies in Russia*, 2020. **15**(3): pp. 273–300.
46. Shen, F., Pankratov, D., Halder, A., Xiao, X., Toscano, M.D., Zhang, J., & Chi, Q., Two-dimensional graphene paper supported flexible enzymatic fuel cells. *Nanoscale Advances*, 2019. **1**(7): pp. 2562–2570.
47. Park, T., Chang, I., Lee, J., Ko, S.H., & Cha, S.-W., Performance variation of flexible polymer electrolyte fuel cell with Ag nanowire current collector under torsion. *ECS Transactions*, 2014. **64**: pp. 927–934.
48. Li, Z., Zheng, Z., Xu, L., & Lu, X., A review of the applications of fuel cells in microgrids: Opportunities and challenges. *BMC Energy*, 2019. **1**(1): pp. 8.
49. Wang, Y.F., Kwok, H.Y.H., Zhang, Y.G., Pan, W.D., Zhang, H.M., Lu, X., & Leung, D.Y.C., A flexible paper-based hydrogen fuel cell for small power applications. *International Journal of Hydrogen Energy*, 2019. **44**(56): pp. 29680–29691.
50. Wang, H., Bai, C., Zhang, T., Wei, J., Li, Y., Ning, F., Shen, Y., Wang, J., Zhang, X., Yang, H., Li, Q., & Zhou, X., Flexible and adaptable fuel cell pack with high energy density realized by a bifunctional catalyst. *ACS Applied Materials & Interfaces*, 2020. **12**(4): pp. 4473–4481.

5 2D Nanomaterials for Portable and Flexible Fuel Cells

S. Vinod Selvaganesh
Indian Institute of Technology-Madras

P. Dhanasekaran
CSIR-Central Electrochemical Research Institute-Madras

Bincy George Abraham and Raghuram Chetty
Indian Institute of Technology-Madras

CONTENTS

DOI: 10.1201/9781003178422-5

5.1 INTRODUCTION

Among the various alternative energy technologies explored to replace conventional fossil fuel-based energy systems, fuel cells offer impressive features, including high fuel utilization and environmentally benign operations. Intensified research and development toward fuel cell technology have paid dividends for being employed in automotive and stationary applications. As a prime example, Japan demonstrates the potential of hydrogen economy in the Tokyo 2020 Olympics. Hydrogen-powered fuel cells were employed to power the athlete village as well as to operate vehicles for transporting people. Such initiatives and other government-industry partnerships on a global scale toward fuel cell technology can help bring down the barriers to fuel cell commercialization.

Apart from automotive and stationary applications, the fuel cell is also being extensively explored as a flexible and portable power source for handheld gadgets due to its inherent modular nature. Fuel cells are capable of providing higher and reliable power for a longer time compared to batteries. This characteristic property of fuel cells makes them attractive over batteries toward application in electronic devices. The architecture of fuel cells is modified based on the required power output and nature of the application. To reduce complexity and improve compactness, passive systems utilizing natural capillary forces, diffusion, evaporation, etc., are used as fuel delivery mechanisms. Flexible and portable fuel cells are often operated in an air-breathing mode wherein the natural diffusion of ambient air provides the required oxidant for the cathode reaction. Significant advancements have also been made to enhance the performance and durability of the cells by suitably modifying flow field patterns, membrane electrode assembly (MEA) design, cell configuration, and stacking, depending on the application. Nevertheless, the costs of fuel cell systems for portable and flexible applications are still higher to be competitive with present state-of-the-art energy systems such as batteries and supercapacitors. Hence, to target significant reduction of costs for fuel cell systems, it is indispensable to explore suitable and reliable fuel cell materials.

5.2 MATERIAL CHALLENGES: PRE-REQUISITES FOR A RELIABLE MATERIAL FOR FUEL CELL APPLICATIONS

Cost reduction can be attained by optimizing the architecture for the end-use, enhancing catalyst utilization, and minimizing the material costs by employing efficient materials for fuel cell components such as catalysts, catalyst supports, and gas diffusion layer. Modifications are necessary to be made on conventionally rigid fuel cell components to make the fuel cells more suitable for portable and flexible devices. For example, the conventionally rigid gas diffusion media has higher tendency for fatigue failure under constant bending conditions making them less suitable for application in flexible power sources. Hence, several nanostructured materials are explored as flexible gas diffusion media. Similarly, it is crucial to optimize the various fuel cell components for flexibility without compromising on performance toward flexible applications.

The choice of material for the flexible and portable fuel cell is also decided by the cell architecture and its end-use. Portable fuel cell systems of various architectures viz. planar (Figure 5.1a) [1], tubular (Figure 5.1b) [2], cylindrical (Figure 5.1c) [3], etc., are reported and mostly employ perforated metal/metal mesh as current collectors/electrodes and polymers/metals for the exterior casing of the cell. Flexible fuel cells systems for bendable (Figure 5.1d) [4] and wearable applications (Figure 5.1e) are also reported wherein polymer substrates like polydimethylsiloxane [5], polycarbonate, etc. are often used to provide the necessary structural degrees of freedom while conductive inks/ultrathin metal nanowires/mesh are used as current collectors/electrodes.

5.3 THE FASCINATING CONCEPT OF 2D MATERIALS

Strategic use of material science and engineering can help to prepare tunable materials with beneficial structural and electrochemical properties, thereby opening new possibilities to design performance materials. In this regard, 2D materials offer remarkable chemical, electronic, optical, and mechanical properties. In addition, they are known for their flexible structure as well as their high thermal and electrical conducting nature.

The properties of the 2D materials can be diversified due to their tunable architectures. The catalytic properties of pristine 2D materials can be tuned by introducing defects, the addition of dopants, compositing with various support materials, hybridizing with transition metals, etc. Additionally, 2D materials as supports can

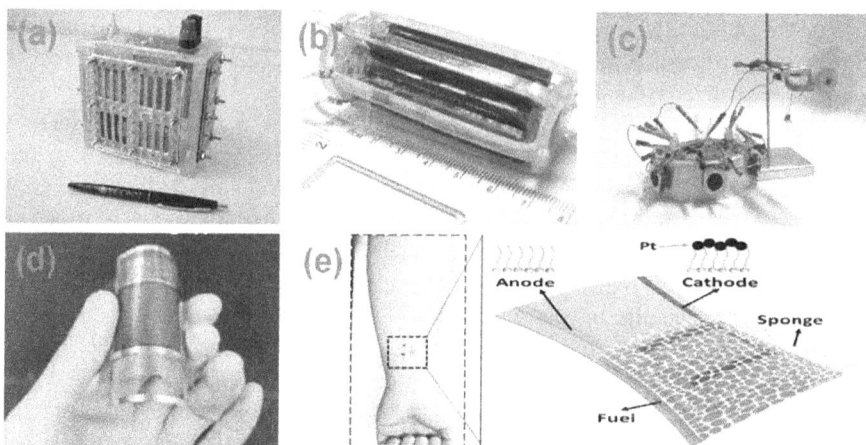

FIGURE 5.1 Photograph of portable stack prototype with (a) planar architecture. (Reproduced with permission from Ref. [1] Copyright © 2007, Elsevier.) (b) Tubular architecture. (Reproduced with permission from Ref. [2] Copyright © 2008, The American Society of Mechanical Engineers.) (c) Cylindrical architecture. (Reproduced with permission from Ref. [3] Copyright © 2020, Elsevier.) (d) Photograph of a flexible air-breathing fuel cell. (Reproduced with permission from Ref. [4] Copyright © 2017, The Author(s). The article was printed under a CC-BY license.) (e) Photograph of stretchable fuel cell for wearable applications with the right-side panel indicating the various components of the cell. (Reproduced with permission from Ref. [5] Copyright © 2019, Wiley-VCH Verlag GmbH & Co. KGaA, Weinheim.)

offer a high surface area to disperse catalyst nanoparticles and increase stability, among other beneficial properties. Pristine as well as composite 2D materials are investigated for applications as membranes because of their high chemical stability and their potential to selectively control the passage of ions. Furthermore, intra-plane covalent bonding of the layered pristine 2D materials offers strong in-plane mechanical properties, making them suitable candidates for preparing durable materials. The conductivity of several reported 2D materials was found to be close to their bulk counterparts, clearly indicating their promising functionality toward thinner conductive materials for future electronic devices.

Hence, the various beneficial properties and unique characteristics offered by 2D nanostructured materials, especially the high surface area, tunable functionalities, and remarkable anisotropic physical and electronic properties, can be exploited for materials in fuel cells.

The family members of 2D materials have increased significantly and have extended from graphene to graphene derivatives (such as graphene, graphene oxide, fluoro-graphene, and chloro-graphene) and graphene analogs (such as boron nitride, metal mono- and di-chalcogenides, metal oxides, metal halides, and MXenes). Among the various 2D materials, some of the most explored materials for fuel cell applications include graphene, hexagonal boron nitride (h-BN), metal di-chalcogenides, and MXenes.

Graphene consists of carbon atoms positioned in an order similar to honeycomb lattice. h-BN has a structure analogous to graphene, with the original carbon atom replaced with boron and nitrogen atoms. h-BN has polar B-N bonds as opposed to non-polar C–C bonds of graphene. Hence, h-BN can be stacked into various modes in the bulk form, consequently making h-BN very stable.

Metal dichalcogenides are of the type MX_2 where the metal atom M (such as Mo, Ti, and W) is sandwiched between chalcogen atoms (such as S, Se, and Te). The highly crystalline and almost defect-free nature of MX_2 offers them excellent mechanical properties, making them highly suitable to be employed as reinforcing components, especially in flexible electronics. The group of 2D materials comprising carbides, nitrides, and carbo-nitrides of transition metals are commonly referred to as MXenes. Intriguingly, MXenes exhibit hydrophilicity and excellent electronic conductivity, which is an attractive combination of properties for 2D materials.

Hence, with this preface of fascinating properties of 2D materials, further discussion is focused on reviewing reports that have strategically employed 2D nanostructures-based materials in various fuel cell components toward augmenting fuel cell performance.

5.4   CURRENT STATUS OF 2D MATERIALS IN FUEL CELLS

5.4.1   2D MATERIALS AS CATALYSTS AND CO-CATALYSTS

A significant barrier to the commercialization prospects of fuel cells is the cost of noble metal-based catalysts as well as its are susceptibility to degradation and poisoning from reaction intermediates and products, resulting in a limited operational life.

Hence, considerable effort has been expended in identifying suitable alternative materials for catalysts. In this regard, several research avenues have considerably helped to identify and develop extremely effective 2D material-based catalysts [6].

5.4.1.1 Catalysis of Oxygen Reduction Reaction (ORR)

Low-cost and abundant 2D materials with enhanced oxygen reduction reaction (ORR) activity are being investigated as alternatives to noble metals as catalysts. Several reports have suggested that doped 2D MoS_2 has the potential to replace noble metals like Pt as an ORR catalyst. Zhang and colleagues used Density Functional Theory (DFT) calculations to understand the effect of heterodoping of MoS_2 using N and P toward ORR catalysis in an acidic system. The authors reported that N or P doping increased spin density in the MoS_2 basal plane, shifting the reaction steps toward a four-electron pathway for ORR. N-doped MoS_2 monolayers do not interact strongly with the intermediate species after O_2 activation, resulting in better catalytic activity compared to P-doped MoS_2, which interacts strongly with O* and OOH* species, resulting in lower catalytic activity. The overpotential of the N-doped MoS_2 monolayer was 0.67 V, which is comparable to the overpotential of Pt-based electrocatalyst [7]. Apart from doping, MoS_2 is also composited with various materials to enhance catalytic properties. Mao et al. synthesized MoS_2 composited with Ni_3S_2, which offered abundant heterointerfaces. As shown in Figure 5.2a, the Ni_3S_2/MoS_2 demonstrated improved alkaline ORR performance and showed a higher half-wave potential of 0.885 V vs. RHE. The improved performance was attributed to the presence of several Mo-Ni-S sites and Mo edges, which offered a large number of reactive sites for ORR along with providing suitable chemisorptions properties for oxygen to adsorb on these reactive sites for the ORR [8].

Graphene-based catalysts are also being pursued as promising alternative non-Pt, cost-effective catalysts toward ORR. Sheng and colleagues synthesized boron-doped graphene by thermally annealing graphite oxide in the presence of boron oxide. The boron atoms act as active sites for adsorbing and activating oxygen, resulting in improved ORR activity (as shown in Figure 5.2b), and selectivity compared to pristine graphene [9]. Hou and co-workers synthesized 2D porous N-doped carbon on graphene in which the pore structures were tailored using interfacial self-assembly engineering. The fabrication of porous N-doped carbon on graphene with spherical (mNC-rGO-S) or cylindrical pores (mNC-rGO-C) was carried out by modifying pluronic block copolymers in solution. The cylindrical pores of mNC-rGO-C allowed for faster in-plane ion diffusion, resulting in increased ORR activity in 0.1 M KOH with a half-wave-potential of 0.74 V, which is comparable to commercial Pt/C catalysts, as shown in Figure 5.2c [10].

MXene is another attractive 2D material that has been investigated for ORR catalysis due to its excellent electrochemical stability. Chen and co-workers fabricated iron-cobalt (FeCo) catalysts hybridized with Ti_3C_2 MXene and investigated the performance of ORR in an alkaline medium. MXene offered enhanced charge transfer as well as the specific surface area to help the hybrid catalyst exhibit outstanding ORR activity as evident from a higher onset and half-wave potential as compared to conventional 20 wt.% Pt/C catalyst in 0.1 M KOH solution [11].

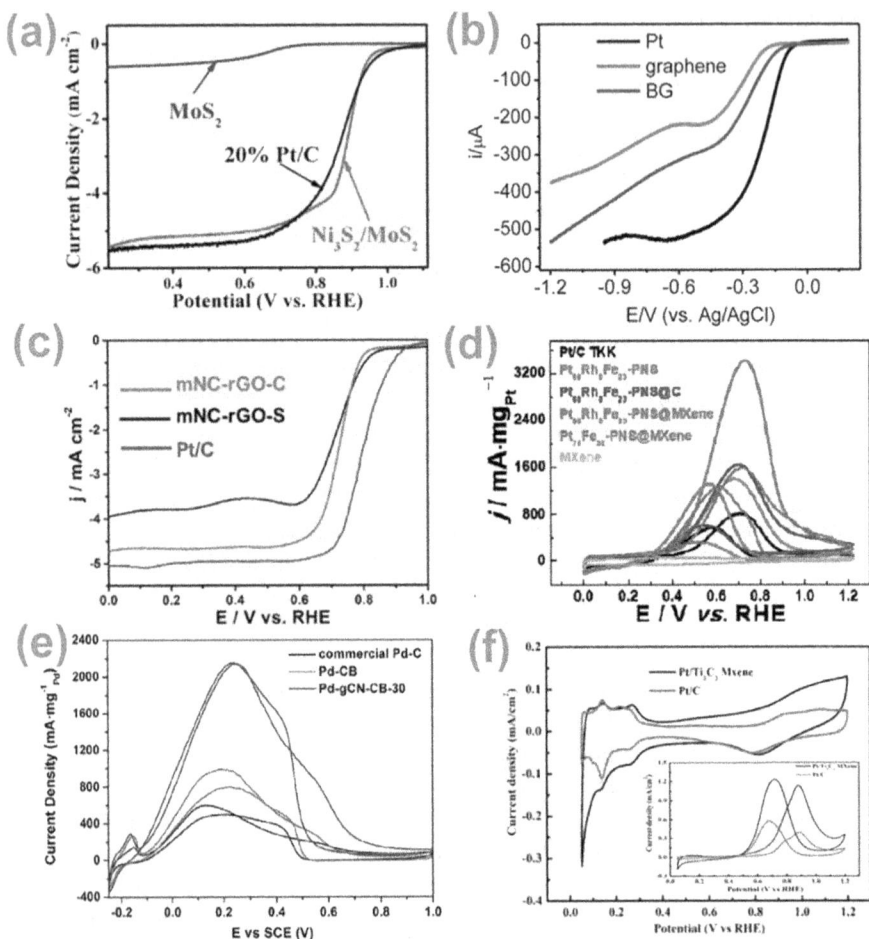

FIGURE 5.2 (a) Plot comparing linear sweep voltammetric (LSV) curves of Ni_3S_2/MoS_2 composite sheets with pristine MoS_2 nanosheets , and commercial 20% Pt/C at 1,600 rpm in O_2-saturated 0.1 M KOH (scan rate: 10 mV s[-1]). (Reproduced with permission from Ref. [8] Copyright © 2019, The Royal Society of Chemistry.) (b) LSV curves of graphene, boron-doped graphene and bulk Pt disk electrode at 1,200 rpm in an O_2-saturated 0.1 M KOH (scan rate: 10 mV s[-1]). (Reproduced with permission from Ref. [9] Copyright © 2012, The Royal Society of Chemistry.) (c) LSV curves of mNC-rGO-C (red), mNC-rGO-S and commercial Pt/C catalyst at 1,600 rpm in O_2-saturated 0.1 M KOH (scan rate: 10 mV s[-1]). (Reproduced with permission from Ref. [10] Copyright © 2019, Wiley-VCH Verlag GmbH & Co. KGaA, Weinheim.) (d) Cyclic voltammogram (CV) curves of various compositions of $Ti_3C_2T_x$ MXene nanosheets supported PtRhFe nanospheres compared with pristine $Ti_3C_2T_x$ MXene and commercial Pt/C in 1 M KOH + 1 M ethanol solution at the scan rate of 50 mV s[-1]. (Reproduced with permission from Ref. [12] Copyright © 2019, Elsevier.); (e) CV curves of hybrid catalysts of Pd and graphitic carbon nitride on carbon black (Pd-gCN-CB) compared to Pd supported on carbon black (Pd-CB) and commercial Pd-C in 0.5 M HCOOH + 0.5 M H_2SO_4 solution at the scan rate of 50 mV s[-1]. (Reproduced with permission from Ref. [14] Copyright © 2014, Elsevier.) (f) CV curves of Pt/Ti_3C_2 MXene compared to commercial Pt/C in 0.5 M H_2SO_4 with inset showing corresponding CV curves in 0.5 M H_2SO_4 + 0.5 M CH_3OH. (Reproduced with permission from Ref. [15] Copyright © 2018, Elsevier.)

5.4.1.2 Catalysis for Alcohol Oxidation Reaction

2D materials are extensively investigated to overcome the slow kinetics of electro-chemical oxidation of alcohols for developing direct liquid fuel cells. Wang et al. pre-pared porous ternary PtRhFe nanospheres combined with $Ti_3C_2T_x$ MXene nanosheets to explore its activity toward ethanol oxidation reaction (EOR). The porous nano-spheres presented several active sites and channels, consequently promoting charge and mass transfer. At the same time, MXene improved the catalytic performance toward EOR. Figure 5.2d shows the cyclic voltammogram comparing $Ti_3C_2T_x$ MXene nanosheets supported PtRhFe nanospheres with pristine $Ti_3C_2T_x$ MXene and com-mercial Pt/C. The improved performance for MXene nanosheets supported PtRhFe nanospheres could be attributed to the rich termination groups such as $-O$, $-OH$, and $-F$ on the surfaces and the uniform interface with the trimetallic alloys [12].

Lin et al. synthesized Mo_2C hybridized Pt, which was loaded on activated car-bon as catalysts for the EOR. The extent of binding interaction of Pt and Mo_2C was controlled by optimizing the process parameters for the high-temperature carburiza-tion. The authors observed that Mo_2C promoted the EOR activity, with the mass-specific activity of the catalyst being nearly three times higher than the commercial PtRu/C catalyst. Additionally, hybridization with Mo_2C particles results in a 100 mV negative shift of the peak potential for CO oxidation compared to Pt/C, indicating the Pt-Mo_2C catalyst's tolerance to poisoning [13]. Qian et al. synthesized hybrid catalysts of Pd and graphitic carbon nitride (g-C_3N_4) on carbon black to investigate the activity and stability toward alcohol oxidation. The authors observed significant synergistic effects between the components attributed to the hybrid catalyst exhibit-ing nearly four times higher activity toward formic acid oxidation in acidic media (as shown in Figure 5.2e), as well as toward methanol oxidation reaction (MOR) in alka-line media compared to commercial Pd/C. Additionally, durability tests indicated improved corrosion resistance due to the modification with g-C_3N_4 [14].

5.4.2 2D MATERIALS AS CATALYST SUPPORT

Catalysts are significantly prone to degradation in fuel cell operating conditions. Hence, it is subservient to employ corrosion-resistant catalyst supports, which can also offer high surface area and excellent conductivity. In this regard, various 2D materials are being extensively explored due to their various beneficial electrochemi-cal properties. Wang et al. explored the suitability of Ti_3C_2-MXene as a support for Pt toward MOR and evaluated the performance and activity with respect to Pt/C. The layered structure of Ti_3C_2-MXene assisted in achieving evenly distributed Pt nanoparticles. Pt loaded on Ti_3C_2-MXene offered higher electrochemical surface area and three times enhanced MOR activity than Pt/C as observed in Figure 5.2f and its inset, respectively. Furthermore, the designated electrocatalyst also showed enhanced cycling stability by retaining more than 85% of the initial surface area compared to nearly 53% for Pt/C [15].

Zhang et al. synthesized ultrathin WO_3 nanosheets with oxygen vacancies through physical exfoliation by Ar plasma etching of bulk WO_3. The authors observed that the surface electronic and physical structure of nanosheets could be effectively tuned by modifying the synthesis parameters, which had a significant impact on the conse-quent catalytic activity toward MOR and stability of Pt when loaded on the prepared ultrathin WO₃ nanosheets [16]. Woo et al. utilized Vulcan carbon as a spacer and obstruct the horizontal stacking preference of graphene layers. PtRu supported on graphene-Vulcan carbon composite exhibited high electrochemical activity and stability for the MOR compared with PtRu/G and PtRu/C. The enhanced performance can be attributed to the modified graphene sheets, which provides increased mass transport and a greater number of active sites for electrochemical reactions [17].

5.4.3 2D MATERIAL-BASED GAS DIFFUSION MEDIA

Gas diffusion media often consists of a microporous layer (MPL) coated over the macroporous gas diffusion layer (GDL), which minimizes interfacial contact resistance (ICR), cushions the membrane from protruding GDL fibers, and promotes water/gas transport. Yuan et al. fabricated an anodic MPL layer comprising of graphene-carbon nanotubes (G-CNTs) composite material. Exfoliation of graphite material was performed to yield 2D nanosheets of reduced graphene oxide (rGO), which offered hydroxyl, carbonyl, and carboxyl functional groups on the surface, consequently enabling rGO to be easily dispersed in solvents. Fuel cell studies showed the hybrid MPL to outperform the conventional MPL (as shown in Figure 5.3a), which is ascribed to the even surface exposure of CNTs, higher conductivity of both the constituents and its porous nature resulting in improved reactant transport at the catalyst sites [18].

Leeuwner et al. synthesized graphene foam MPL and compared its performance with MPL made with other morphological structures such as commercial carbon paper, perforated graphitic layer, and perforated stainless steel to determine the relationship between material properties as well as interfacial characteristics of MPL with catalysts. The authors reported that interfacial effects had a significant impact on fuel cell performance, with the graphene foam MPL found to favor water retention than the commercial carbon MPL. Furthermore, graphene foam MPL exhibited better durability after continuous polarization cycles and repeated stress tests, compared to other prepared MPLs [19]. Najafabadi and co-workers prepared electrochemically exfoliated graphene (EGN) micro-sheets, which improved the electronic conductivity as well as electrodes' interfacial contact with the catalyst layer due to the unique planar structure of EGN sheets. Further, EGN was hybridized with CB to help improve water removal properties as well as help in overcoming the mass-transfer limitations arising due to the densely stacked EGN sheets. The polarization curves, represented in Figure 5.3b, indicate the enhanced performance for MPL with EGN-CB composite compared to MPL with EGN only and MPL with Vulcan carbon, which was ascribed to the effective water management and reduced ohmic losses [20].

5.4.4 2D MATERIAL-BASED MEMBRANES

Membranes play an essential role in providing the electrodes with ionic conductivity as well as electronic separation. They also prevent the mixing of reactants during fuel cell operation. However, methanol crossover is relatively a severe issue in direct methanol fuel cell (DMFC) due to the crossing of methanol through the membrane, resulting in degradation of performance. Hence, membranes are altered

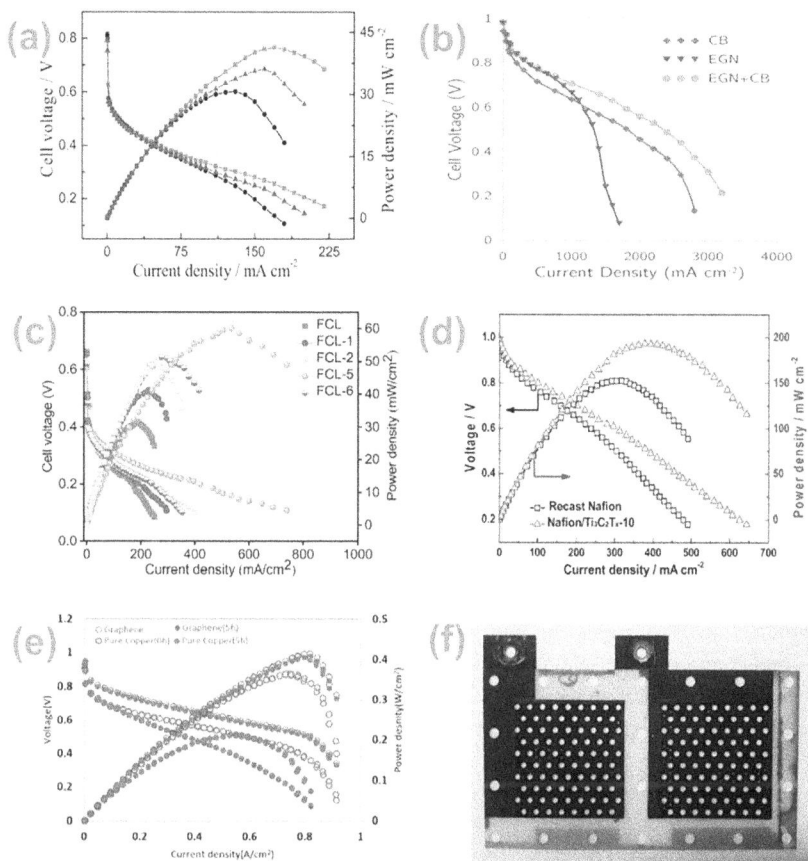

FIGURE 5.3 (a) Polarization curves comparing the performance of passive DMFCs employing anodic MPL comprising of XC-72R carbon, graphene-carbon nanotubes, and a 1:1 mass ratio composite of graphene-carbon nanotubes and XC-72R carbon using a 4M methanol solution. (Reproduced with permission from Ref. [18] Copyright © 2014, Elsevier Ltd.) (b) Polarization curves comparing the performance of fuel cells with MPLs comprising of Vulcan XC72R (CB), exfoliated graphene microsheets (EGN), and a composite EGN and CB (EGN+CB) with a 1:1 mass ratio. (Reproduced with permission from Ref. [20] Copyright © 2016, Wiley-VCH Verlag GmbH & Co. KGaA, Weinheim.) (c) Polarization curves comparing the performance of DMFCs operating on 2M methanol employing pristine sPES and s-MoS$_2$/SPES composite of different s-MoS$_2$ content from 1 to 6 wt.% (Reproduced with permission from Ref. [21] Copyright © 2020, The Author(s). The article was printed under a CC-BY license.) (d) Polarization curves of PEM fuel operating at 90°C, 80% RH employing recast Nafion membrane compared to Nafion-Ti$_3$C$_2$T$_x$ composite membrane. (Reproduced with permission from Ref. [26] Copyright © 2016, American Chemical Society.) (e) Polarization curves of the PEM fuel cell indicating the durability of graphene-coated copper bipolar plate compared to pure copper bipolar plate after 5 hours of operation at 0.6 V and 40°C. Hollow circles and filled circles indicate polarization values before and after 5 hours of operation, respectively. (Reproduced with permission from Ref. [28] Copyright © 2017, Elsevier.) (f) Photograph of current collector comprising of graphene layer applied on copper-coated FR-4 substrate. (Reproduced with permission from Ref. [32] Copyright © 2020, The Author(s). The article was printed under a CC-BY license.)

with suitable materials to minimize fuel crossover. Yadav and co-workers explored sulfonated polyether sulfone (sPES) composited with sulfonated molybdenum sulfide (s-MoS$_2$) as membranes for application in DMFC. Incorporation of 5 wt.% of s-MoS$_2$ as a composite to sPES membrane resulted in a significant reduction (up to 91%) in methanol permeation, which can be ascribed to the methanol permeation barrier created by s-MoS$_2$. Additionally, better proton transport offered by s-MoS$_2$ resulted in enhanced DMFC performance for s-MoS$_2$-sPES composite membrane as compared to pristine sPES membrane. The 5 wt.% of s-MoS$_2$ into the composite membrane delivered a peak power performance of 60 mW cm^{-2} compared to 30 mW cm^{-2} for pristine sPES membrane (as shown in Figure 5.3c) operating with 2 M methanol in a DMFC with PtRu/C of 2 mg cm^{-2} and Pt/C of 2 mg cm^{-2} as anode and cathode catalyst, respectively [21].

The strengthening properties of exfoliated MoS$_2$ nanosheets in polymers are discussed in several reports. Khan et al. evaluated the mechanical properties of MoS$_2$ nanosheets polystyrene composite membranes, which were prepared using a simple blending and casting technique. The authors reported that three to four layers of MoS$_2$ enhanced Young's modulus by around four times compared to that for pristine polystyrene, which was mainly ascribed to the layered structure of exfoliated MoS$_2$ nanosheets [22].

Ye et al. prepared a nanocomposite membrane by blending exfoliated graphene nanosheets and polyvinyl alcohol. The graphene nanosheets in the composite membrane formed a continuous, well connected, and tortuous ionic channel, thereby endowing the membrane with improved ionic transport. Additionally, the incorporation of graphene nanosheets resulted in decreased methanol permeation and increased mechanical properties of membranes [23].

Budak and Devrim prepared polybenzimidazole/graphene oxide (PBI/GO) composite membranes to be used for high-temperature Proton-exchange membrane (PEM) fuel cells in a micro-cogeneration application. The 12-cell HT-PEFC stack with 150 cm^2 active area employing PBI/GO membrane offered a maximum power of 546 W, which was higher than 468 W provided by the same stack utilizing the PBI membrane. The incorporation of GO was able to tackle the acid leaching in the membrane due to its hydrophilic structure, consequently improving the proton conductivity and ultimately enhancing the stack performance [24].

Fei et al. explored MXenes as filler for polybenzimidazole (PBI) membrane toward application in intermediate temperature polymer electrolyte fuel cells. The conductivity of 3 wt.% Ti$_3$C$_2$T$_x$-PBI membrane was almost twice that of pristine PBI membranes, which is due to hydrogen bonding of the oxygen-containing groups on Ti$_3$C$_2$T$_x$ with PBI, resulting in efficient hopping sites and formation of suitable proton transfer channel in the membrane. At intermediate temperatures, the proton transfer may follow hopping jump rather than vehicle carriage due to the lack of water in the membrane. The incorporation of Ti$_3$C$_2$T$_x$ sheets improved proton conductivity, mechanical properties, and thermal stability of PBI membranes for operations up to 150°C [25]. In another study, Liu and co-workers investigated the influence of Ti$_3$C$_2$T$_x$ nanosheets on various matrices viz. Nafion, sulfonated polyether ether ketone (SPEEK), and chitosan polymers. Enhancement of conductivity was observed for all polymer matrices when Ti$_3$C$_2$T$_x$ nanosheets were incorporated. Polarization curves shown in Figure 5.3d, clearly show nearly 30% improvement in performance when employing Nafion-Ti$_3$C$_2$T$_x$ composite membrane compared to recast Nafion membrane. The enhancement was attributed to the fact that in addition to inducing hydrogen-bonding interaction and thereby providing efficient hopping sites, the Ti$_3$C$_2$T$_x$ nanosheets also offered high surface area and aspect ratio to construct suitable interfacial pathways within the membrane [26].

Akel and co-workers investigated the effect of nano hexagonal boron nitride (nano h-BN) particles in the properties of nano hexagonal BN nanosheet-Nafion composite membrane. Various amount of nano h-BN was composited with Nafion polymer and results indicated increasing hydrophobic nano h-BN content in the blend decreased the swelling property and methanol retention of the membranes. The composite membranes were thermally stable up to approximately 270°C. The proton transport was through the sulfonic acid groups of Nafion and the amine and −OH groups on the BN surface [27].

5.4.5 2D Materials as Bipolar Plates and Current Collectors

Metals are favored as materials for bipolar plates over graphite due to significantly better mechanical strength and good electrical conductivity. However, the oxide layers that may form on the metallic surfaces can decrease their conductivity. In this regard, a facile method using 2D nanomaterials effectively coated on a metallic bipolar plate or current collector plate could improve ICR and other electrochemical properties. A thin coating layer of corrosion-resistant 2D materials is often employed over the metallic bipolar plate to suppress the oxide layer formation.

Lee et al. prepared a graphene layer as a coating for copper bipolar plates. The graphene coating was prepared using chemical vapor deposition on copper plates, which offers good quality graphene layer with enhanced thermal and electronic properties. Studies for a 9 cm^2 active area cell in H$_2$-air configuration, as shown in Figure 5.3e, clearly presented enhancement in performance for graphene-coated Cu bipolar plates, which delivered 400 mW cm^{-2} compared to pristine Cu bipolar plates, which offered 350 mW cm^{-2} of power. Durability studies showed that graphene was able to prevent the dissolution and migration of Cu from the Cu bipolar plates [28]. Another role being explored for 2D materials in bipolar plates is to improve the mechanical properties of graphitic bipolar plates by compositing with materials to enhance mechanical strength and durability without compromising on conductivity. Kakati et al. developed composite bipolar plates comprising graphite, carbon black, and carbon fiber, hybridized with 1% 2D monolayer graphene. The components are bound into a unit using phenol-formaldehyde resin. The incorporation of carbon fiber enhanced the mechanical strength, while the electrical conductivity was improved by the addition of graphene. Ten percent improvement was observed in the cell performance for the graphene incorporated composite bipolar plate compared to the graphene-less composite bipolar plate. The authors concluded that enhanced through-plane and in-plane electrical conductivities arising due to incorporation of flake shape 2D graphene were the reason for the improvement in performance [29].

Sim and co-workers prepared highly crystalline multilayer graphene through radiative heating of polymethyl methacrylate (PMMA). The graphene was then used as a protective coating on Ni foam. The graphene coating provided a hydrophobic surface for water management while resisting corrosion. Results demonstrated that graphene coating was able to maintain low corrosion current density and retain a better ICR of 9.3 mΩ cm^2 at 10 kg cm^{-2} after stability tests compared to bare Ni foam. PEM fuel cell employing graphene-coated Ni foam showed superior fuel cell performance and delivered a maximum power density of 967 mW cm^{-2} at 0.5 V in H$_2$:air configuration [30].

Jiang and Drzal explored a facile method and optimum composition of exfoliated graphene nanoplatelets (GNP), carbon black, and polyphenylene sulfide (PPS) mixed matrix to be used for bipolar plate. Polyphenylene sulfide has excellent chemical and thermal tolerance. PPS/GNP was prepared via ball milling followed by compression molding. The mechanically flexible GNP electrode was made via vacuum filtration paper. The optimum composite of PPS and graphene showed superior gas barrier and mechanical properties. Both PPS/GNP and PPS/CB/GNP nanocomposites demonstrated an increase in flexure strength with an increase in GNP composition [31].

During fuel cell operating conditions, current collectors are also prone to corrosion. Researchers are exploring alternate materials, among which 2D materials are quite attractive. Kuan et al. prepared graphene thin film as a coating for copper-based thin current collector for PEM fuel cell as can be seen in Figure 5.3f. FR-4 was taken as a substrate onto which a 0.5 mm thick Cu film was coated using thermal evaporation. The coating of graphene was attempted for corrosion resistance through three strategies viz. using graphene ink, graphene dispersion, and graphene suspension. A thin 2D graphene layer (~3 nm) is effectively coated via spin coating method followed by vacuum oven process. Graphene ink and graphene suspension methods showed a higher corrosion rate behavior compared to graphene dispersion on Cu current collector plate, which showed the lowest corrosion rate [32].

5.5 MANIFESTATIONS OF 2D MATERIALS IN FLEXIBLE AND PORTABLE FUEL CELL SYSTEMS

Successful implementation of the fuel cell technology for flexible and portable applications would require the perfect choice of materials and architecture. The application of 2D materials in various components of fuel cells, as discussed in the previous section, clearly highlights the potential of 2D materials for supporting the specific requirements of flexible and portable systems. Various research groups have tried to exploit the different beneficial properties of 2D material-based nanostructures and use them to demonstrate in portable and flexible fuel cell systems. In this regard, various approaches that are implemented are briefly discussed.

Shen et al. explored 2D graphene paper for use as a carrier for enzyme immobilization in a flexible enzymatic biofuel cell (EBFC). The thin and flexible graphene paper, as shown in Figure 5.4a, with a 2D active surface was fabricated by assembling graphene oxide into a paper-like architecture and subsequently reducing it to form a layered and cross-linked matrix, which can be easily visualized in Figure 5.4b. The prepared, flexible graphene paper exhibited good thermal and mechanical properties in addition to its excellent conductivity. The graphene paper served multiple roles as it functioned as a current collector in addition to operating as anode and cathode electrodes hosting the corresponding enzymes, which can be visualized from the schematic shown in Figure 5.4c. The fabricated glucose/oxygen EBFCs

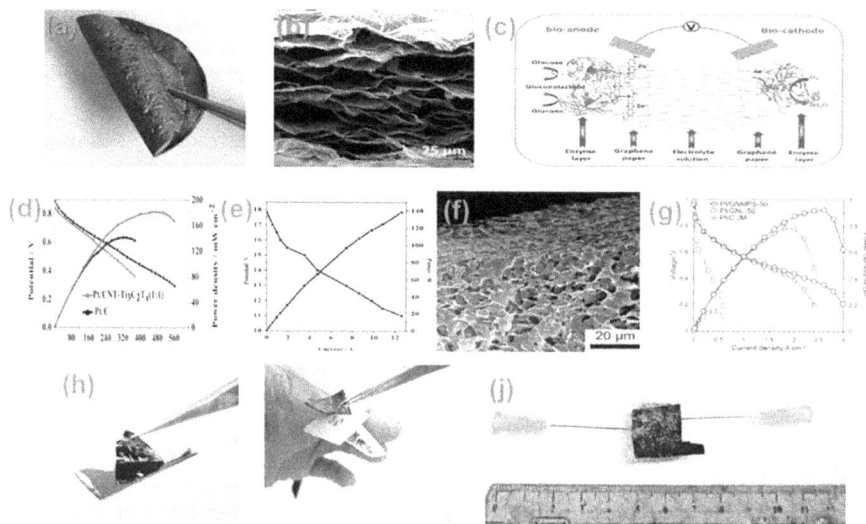

FIGURE 5.4 (a) Photograph and (b) SEM image of layered assembly of graphene paper. (c) Schematic of enzymatic biofuel cell with the enzymes immobilized onto the graphene paper. (Reproduced with permission from Ref. [33] Copyright © 2019, The Author(s). The article was printed under a CC-BY license.) (d) Polarization curves comparing Pt/CNT–Ti$_3$C$_2$T$_x$ (1:1) and Pt/C catalysts employed in H$_2$/air PEM fuel cell operated at 60°C and (e) performance of 22 cell-stack with Pt/CNT-Ti$_3$C$_2$T$_x$ (1:1) electrocatalyst operated at 25°C. (Reproduced with permission from Ref. [35] Copyright ©2020, American Chemical Society.) (f) SEM image of graphene-Nafion mesoporous composite scaffold (GNMPS) and (g) comparison of polarization curves for Pt supported on GNMPS and Pt supported on lamellar-structured graphene-Nafion composite (GNL) to commercial Pt/C. (Reproduced with permission from Ref. [37] Copyright © 2014, The Royal Society of Chemistry.) Photographs of (h) graphene nanosheets deposited on a nickel substrate, (i) assembling of flexible graphene into a membrane electrode assembly and (j) the microbial fuel cell prepared using the flexible graphene nanosheet anode, with syringe tips incorporate to facilitate feeding of wastewater as fuel. (Reproduced with permission from Ref. [38] Copyright © 2013, Wiley-VCH Verlag GmbH & Co. KGaA, Weinheim.)

exhibited remarkable mechanical flexibility along with a relatively high open-circuit voltage (OCV) of 0.66 V and a power density of 4 µW cm^{-2} [33]. Dhanasekaran et al. developed a 2D graphene sheet-carbon hybrid support for Pt catalysts, which exhibited excellent fuel cell performance even with a low metal loading of 50 µg cm^{-2}. The prepared 2D support demonstrated long-term stability with retention of more than 50% of initial activity even after 5,000 potential cycles under stringent operating conditions. The authors concluded that an optimum graphene content, when interconnected with amorphous carbon, can expand the synergistic interactions resulting in increasing cell performance and long-term stability. Furthermore, with such ultra-low loading, the PEM fuel cell performance clearly shows the potential of graphene supports for enhanced catalysts utilization, which is essential for portable and flexible systems [34]. Xu et al. fabricated a portable fuel cell stack comprising Pt supported on a Ti$_3$C$_2$T$_x$ MXene nanosheet incorporated with carbon nanotubes. The optimum composition of MXene and CNT 2D nanomaterials supported Pt nanoparticles showed superior PEM fuel cell performance and 3.4 fold higher mass activity than Pt/C, as observed from Figure 5.4d. The improved performance and stability are attributed to the fact that the MXene 2D nanosheet had higher corrosion resistance and better electrical conductivity when hybridized with CNT materials. During stack studies, Pt/CNT-Ti$_3$C$_2$T$_x$ (1:1) electrocatalyst was mixed with Pt/C and used as cathode catalysts. The 22 cell-stack with Pt/CNT-Ti$_3$C$_2$T$_x$ (1:1) electrocatalyst showed an OCV of ~18 V and delivered a remarkable power of 138 W at 11 V, as shown in Figure 5.4e. It is noteworthy that the cell stack was operated with a low anode and cathode metal loading of 200 μg cm^{-2} [35].

Liu et al. investigated the effect of reducing agents used to prepare rGO catalysts on the performance of micro-DMFC when employed as support composited with Nafion for Pt-Ru catalyst. The authors observed that Pt-Ru/rGO-Nafion electrode fabricated via sodium borohydride reduction method performed better when compared to the same with graphene oxide fabricated via ethylene glycol reduction. The micro-DMFC was fabricated with a traditional structure consisting of two endplates (polymethyl methacrylate), current collectors, gaskets, and MEA. The micro-DMFC was able to deliver a maximum power density of nearly 20 mW cm^{-2} with an optimum composition of 30% Pt content, 1:1.5 Pt-Ru molar ratio, and Nafion (10%) over the rGO [36].

Xia and co-workers explored a macroporous scaffold of graphene-Nafion composite as effective support for Pt. As shown in Figure 5.4f, the scaffold structure offered an effective porous structure for efficient channels for gas/liquid transport. The graphene component of the scaffold provided electron conductivity while the Nafion ionomers performed the proton transport. Figure 5.4g clearly shows that Pt/ Graphene-Nafion composite exhibited 20% higher fuel cell performance than the same loading of commercial Pt/C, indicating that the graphene-Nafion composite scaffold provided suitable support to achieve ultra-high Pt utilization [37]. Mink et al. employed graphene nanosheet-based anode for a flexible micrometer-sized microbial fuel cell (MFC). The 2D graphene nanosheets were deposited on a nickel substrate (as shown in Figure 5.4h) and assembled to form a membrane electrode assemble (as shown in Figure 5.4i). Syringe tips were incorporated to facilitate the feeding of wastewater as fuel to the MFC prepared using the flexible graphene nanosheet anode (as shown in Figure 5.4j). The MFC was able to offer 1 nW of power directly from wastewater liquid. The authors attributed the relatively high performance to the contribution of graphene in lowering the internal resistance [38].

5.6   FUTURE SCOPE FOR 2D MATERIAL-ENDOWED FLEXIBLE AND PORTABLE FUEL CELL APPLICATIONS

With advances in synthesis and fabrication strategies of useful 2D materials, much effort is directed toward employing 2D nanostructured materials as cost-effective and durable alternates to existing fuel cell materials. Future developments should focus on employing 2D materials and composites to ultimately replace noble metal catalysts without compromising activity or durability, consequently reducing the fuel cell system costs. Besides, tuning 2D materials to support monolayer, atomic layer, and optimum composites of catalyst can enhance the performance and stability as well as the effective utilization of catalysts.

2D materials-based membranes have demonstrated exceptional performance in fuel cell applications. In the future, 2D material-based membranes may also be possible to show state-of-the-art performance as fuel cell electrolytes in a wide variety of applications by modifying the architecture (interlayer spacing and pore size), chemical structure, composition, filler ratio of the molecular transport route, etc., which make them a potential family of superior alternative to traditional membranes. Further deployment of these hybrid membranes emphasizes two key issues: (i) membrane stability over long periods of operation and (ii) facile techniques to facilitate large-scale production of reliable and defect-free membranes. Further efforts are directed toward facile techniques for scalable manufacturing. 2D nanomaterials on commercially viable cell architecture and materials for flexible fuel cells should be suitably chosen. Further efforts are also required to suitably engage thin 2D material-based layer composites as a current collector and flow-field channels.

The recent advances and developmental strategies employed toward flexible and portable fuel cells have significantly attracted global attention in regards to a convenient and handy power source to serve various consumer needs. In this regard, 2D materials are expected to play a significant role to thrust further the prospects of flexible and portable fuel cells toward commercialization.

ACKNOWLEDGMENTS

Dr. S. Vinod Selvaganesh (Scientist's Pool Scheme-9178-A) and Dr. P. Dhanasekaran (Scientist's Pool Scheme-9123-A) would like to thank CSIR for the Senior Research Associateship funding under Scientist Pool Scheme. Prof. Raghuram Chetty and Dr. Bincy George Abraham would like to acknowledge the Ministry of New and Renewable Energy (MNRE Grant No: 102/61/2009-NT), Government of India.

REFERENCES

1. Guo Z, Faghri A. Development of a 1 W passive DMFC. *International Communications in Heat and Mass Transfer* **2008**, 35, 225–239.
2. Lee M-S, Chen L-J, Hung M-F, Lo M-Y, Sue S-J, Lo C-H, Wang Y-P. A novel design of a cylindrical portable direct methanol fuel cell. *Journal of Fuel Cell Science and Technology* **2008**, 5, 031004.
3. Abraham BG, Chetty R. Design and fabrication of a quick-fit architecture air breathing direct methanol fuel cell. *International Journal of Hydrogen Energy* **2021**, 46, 6845–6856.
4. Park T, Kang YS, Jang S, Cha SW, Choi M, Yoo SJ. A rollable ultra-light polymer electrolyte membrane fuel cell. *NPG Asia Materials* **2017**, 9, e384–e384.
5. Zhai Q, Liu Y, Wang R, Wang Y, Lyu Q, Gong S, Wang J, Simon GP, Cheng W. Intrinsically stretchable fuel cell based on enokitake-like standing gold nanowires. *Advanced Energy Materials* **2020**, 10, 1903512.
6. Khan K, Tareen AK, Aslam M, Zhang Y, Wang R, Ouyang Z, Gou Z, Zhang H. Recent advances in two-dimensional materials and their nanocomposites in sustainable energy conversion applications. *Nanoscale* **2019**, 11, 21622–21678.
7. Zhang H, Tian Y, Zhao J, Cai Q, Chen Z. Small dopants make big differences: Enhanced electrocatalytic performance of MoS_2 monolayer for oxygen reduction reaction (ORR) by N– and P–doping. *Electrochimica Acta* **2017**, 225, 543–550.

8. Mao J, Liu P, Du C, Liang D, Yan J, Song W. Tailoring 2D MoS$_2$ heterointerfaces for promising oxygen reduction reaction electrocatalysis. *Journal of Materials Chemistry A* **2019**, 7, 8785–8789.

9. Sheng Z-H, Gao H-L, Bao W-J, Wang F-B, Xia X-H. Synthesis of boron doped graphene for oxygen reduction reaction in fuel cells. *Journal of Materials Chemistry* **2012**, 22, 390–395.

10. Hou D, Zhang J, Tian H, Li Q, Li C, Mai Y. Pore engineering of 2D mesoporous nitrogen-doped carbon on graphene through block copolymer self-assembly. *Advanced Materials Interfaces* **2019**, 6, 1901476.

11. Chen L, Lin Y, Fu J, Xie J, Chen R, Zhang H. Hybridization of binary non-precious-metal nanoparticles with d-Ti$_3$C$_2$ MXene for catalyzing the oxygen reduction reaction. *ChemElectroChem* **2018**, 5, 3307–3314.

12. Wang P, Cui H, Wang C. In situ formation of porous trimetallic PtRhFe nanospheres decorated on ultrathin MXene nanosheets as highly efficient catalysts for ethanol oxidation. *Nano Energy* **2019**, 66, 104196.

13. Lin L, Sheng W, Yao S, Ma D, Chen JG. Pt/Mo$_2$C/C-cp as a highly active and stable catalyst for ethanol electrooxidation. *Journal of Power Sources* **2017**, 345, 182–189.

14. Qian H, Huang H, Wang X. Design and synthesis of palladium/graphitic carbon nitride/carbon black hybrids as high-performance catalysts for formic acid and methanol electrooxidation. *Journal of Power Sources* **2015**, 275, 734–741.

15. Wang Y, Wang J, Han G, Du C, Deng Q, Gao Y, Yin G, Song Y. Pt decorated Ti$_3$C$_2$ MXene for enhanced methanol oxidation reaction. *Ceramics International* **2019**, 45, 2411–2417.

16. Zhang Y, Shi Y, Chen R, Tao L, Xie C, Liu D, Yan D, Wang S. Enriched nucleation sites for Pt deposition on ultrathin WO$_3$ nanosheets with unique interactions for methanol oxidation. *Journal of Materials Chemistry A* **2018**, 6, 23028–23033.

17. Woo S, Lee J, Park SK, Kim H, Chung TD, Piao Y. Enhanced electrocatalysis of PtRu onto graphene separated by vulcan carbon spacer. *Journal of Power Sources* **2013**, 222, 261–266.

18. Yuan T, Yang J, Wang Y, Ding H, Li X, Liu L, Yang H. Anodic diffusion layer with graphene-carbon nanotubes composite material for passive direct methanol fuel cell. *Electrochimica Acta* **2014**, 147, 265–270.

19. Leeuwner MJ, Wilkinson DP, Gyenge EL. Novel graphene foam microporous layers for PEM fuel cells: Interfacial characteristics and comparative performance. *Fuel Cells* **2015**, 15, 790–801.

20. Najafabadi AT, Leeuwner MJ, Wilkinson DP, Gyenge EL. Electrochemically produced graphene for microporous layers in fuel cells. *ChemSusChem* **2016**, 9, 1689–1697.

21. Yadav V, Niluroutu N, Bhat SD, Kulshrestha V. Sulfonated poly(ether sulfone) based sulfonated molybdenum sulfide composite membranes: Proton transport properties and direct methanol fuel cell performance. *Materials Advances* **2020**, 1, 820–829.

22. Khan MB, Jan R, Habib A, Khan AN. Evaluating mechanical properties of few layers MoS$_2$ nanosheets-polymer composites. *Advances in Materials Science and Engineering* **2017**, 1–7, 25.

23. Ye YS, Cheng MY, Xie XL, Rick J, Huang YJ, Chang FC, Hwang BJ. Alkali doped polyvinyl alcohol/graphene electrolyte for direct methanol alkaline fuel cells. *Journal of Power Sources* **2013**, 239, 424–432.

24. Budak Y, Devrim Y. Micro-cogeneration application of a high-temperature PEM fuel cell stack operated with polybenzimidazole based membranes. *International Journal of Hydrogen Energy* **2020**, 45, 35198–35207.

25. Fei M, Lin R, Deng Y, Xian H, Bian R, Zhang X, Cheng J, Xu C, Cai D. Polybenzimidazole/Mxene composite membranes for intermediate temperature polymer electrolyte membrane fuel cells. *Nanotechnology* **2018**, 29, 035403.

26. Liu Y, Zhang J, Zhang X, Li Y, Wang J. $Ti_3C_2T_x$ filler effect on the proton conduction property of polymer electrolyte membrane. *ACS Applied Materials & Interfaces* **2016**, 8, 20352–20363.

27. Akel M, Ünügür Çelik S, Bozkurt A, Ata A. Nano hexagonal boron nitride-nafion composite membranes for proton exchange membrane fuel cells. *Polymer Composites* **2016**, 37, 422–428.

28. Lee YH, Noh S, Lee J-H, Chun S-H, Cha SW, Chang I. Durable graphene-coated bipolar plates for polymer electrolyte fuel cells. *International Journal of Hydrogen Energy* **2017**, 42, 27350–27353.

29. Kakati BK, Ghosh A, Verma A. Efficient composite bipolar plate reinforced with carbon fiber and graphene for proton exchange membrane fuel cell. *International Journal of Hydrogen Energy* **2013**, 38, 9362–9369.

30. Sim Y, Kwak J, Kim SY, Jo Y, Kim S, Kim SY, Kim JH, Lee CS, Jo JH, Kwon SY. Formation of 3D graphene-Ni foam heterostructures with enhanced performance and durability for bipolar plates in a polymer electrolyte membrane fuel cell. *Journal of Materials Chemistry A* **2018**, 6, 1504–1512.

31. Jiang X, Drzal LT. Exploring the potential of exfoliated graphene nanoplatelets as the conductive filler in polymeric nanocomposites for bipolar plates. *Journal of Power Sources* **2012**, 218, 297–306.

32. Kuan Y-D, Ke T-R, Lyu J-L, Sung M-F, Do J-S. Development of a current collector with a graphene thin film for a proton exchange membrane fuel cell module. *Molecules* **2020**, 25, 955.

33. Shen F, Pankratov D, Halder A, Xiao X, Toscano MD, Zhang J, Ulstrup J, Gorton L, Chi Q. Two-dimensional graphene paper supported flexible enzymatic fuel cells. *Nanoscale Advances* **2019**, 1, 2562–2570.

34. Dhanasekaran P, Vinod Selvaganesh S, Shukla A, Bhat SD. Synergistic interaction of graphene-amorphous carbon nanohybrid with thin metal loading for enhanced polymer electrolyte fuel cell performance and durability. *Materials Letters* **2021**, 282, 128837.

35. Xu C, Fan C, Zhang X, Chen H, Liu X, Fu Z, Wang R, Hong T, Cheng J. MXene ($Ti_3C_2T_x$) and carbon nanotube hybrid-supported platinum catalysts for the high-performance oxygen reduction reaction in PEMFC. *ACS Applied Materials & Interfaces* **2020**, 12, 19539–19546.

36. Liu C, Hu S, Yin L, Yang W, Yu J, Xu Y, Li L, Wang G, Wang L. Micro direct methanol fuel cell based on reduced graphene oxide composite electrode. *Micromachines* **2021**, 12, 1–12.

37. Xia Z, Wang S, Jiang L, Sun H, Qi F, Jin J, Sun G. Rational design of a highly efficient Pt/graphene–Nafion® composite fuel cell electrode architecture. *Journal of Materials Chemistry A* **2015**, 3, 1641–1648.

38. Mink JE, Qaisi RM, Hussain MM. Graphene-based flexible micrometer-sized microbial fuel cell. *Energy Technology* **2013**, 1, 648–652.

6 Role of 2D Materials in Organic and Perovskite Photovoltaics

Jazib Ali
University of Rome Tor Vergata

Sehrish Gull
Shenzhen University

Ghulam Abbas Ashraf
Zhejiang Normal University

Muhammad Bilal
Huaiyin Institute of Technology

CONTENTS

6.1   INTRODUCTION

Two-dimensional (2D) materials have gained enormous attraction among the researchers since 2004, after successfully achieving single-layered graphene (Gr) through micromechanical cleaving from graphite. Typically, these materials are composed of

DOI: 10.1201/9781003178422-6

atomic thin sheets with out-of-plane week van der Waals interactions and strong in-plane covalent bonding, which are gently exfoliated into cantilevered monolayer or few layer thin flakes with unique and distinguishing optoelectronic properties. Here, we focus on the development of some most commonly investigated 2D materials along with the synthesizing techniques.

6.2 VARIOUS CATEGORIES OF 2D MATERIALS

6.2.1 HEXAGONAL BORON NITRIDE

Hexagonal boron nitride (h-BN) is a monolayer with a 6eV bandgap that shares a tight lattice structure with Gr and is composed of alternate boron and nitrogen atoms arranged in a honeycomb crystal lattice [1]. This insulating material exhibited an extraordinary chemical stable nature and atomically flat surface, which makes it an appealing substrate or dielectric layer for the production of various electronic systems [2].

6.2.2 TRANSITION METAL DICHALCOGENIDES (TMDS)

TMDs are a significant category of nanostructured materials having a generic composition MX_2, wherein M denotes a transition metal such as tungsten (W) and molybdenum (Mo) while X denotes a chalcogen namely selenium (Se), silicon (S), or tellurium (Te) [3]. TMDs are typically made up of monolayer metal atoms wedged between bilayers of chalcogen atoms by covalent M-X bonding, with different layer thicknesses of TMDs (in bulk forms) tied by feeble van der Waals forces, allowing exfoliation into a single layer system. In contrast to their bulk counterparts, which have an indirect bandgap energy, 2D monolayers have a direct bandgap energy. As a result, TMDs like WS_2 and MoS_2 have configurable bandgaps with layer-dependent characteristics. TMDs have achieved significant advances in a variety of applications attributable to their customizable electronic structures and solution processability [4].

6.2.3 GRAPHENE AND ITS DERIVATIVES

Graphene, a 2D honeycomb hexagonal lattice of carbon with atomically thin carbon layered structures, was the first 2D material to be studied extensively and is still assumed the rudimentary allotrope of all other types of carbon materials, like carbon nanotubes, graphite, and fullerenes. Gr exhibits many rare characteristics including more carrier mobility, excellent electrical conductivity (zero-bandgap semimetal), better mechanical strength (the strongest material constantly measured), admirable optical transparency, extensive light response array, outstanding thermal conductivity, bigger specific surface area, cheap, strong chemical stability, and tuneable energy bandgap, all of them make it highly tempting for numerous applications.

Gr is hydrophobic, but following oxidation using Hummer's process, which involves treating it with strong acids and oxidizing chemicals, it transforms into graphene oxide (GO). Because of the connection of various functional groups during the synthesis mechanism, GO is regarded as a vital derivative of Gr with a hydrophobic property. So, the connected functional groups on the edges and surfaces, such as epoxide, hydroxyl, and carbonyl groups, are hydrophilic, large surface area homogeneous films with higher production are feasible, making them desirable in industrial applications. In contrast, Gr, as generated GO, has an insulating behavior and weak electrical conductivity, then after a chemical reduction procedure, it can be converted to reduced graphene oxide (rGO), which has a more conductive tendency and transparency [5]. Due to persistent flaws, the structure of graphene in rGO cannot be restored, even though the solution processability of rGO allows for the easy manufacture of conductive films or junctions at low temperatures. Graphene quantum dots and Gr-nanoribbons are two further types of Gr that have been studied in addition to GO and rGO.

6.3 SYNTHESIS OF 2D MATERIALS

Exfoliation and CVD are two of the most used processes for preparing 2D materials (CVD) [6]. Mechanical exfoliation with scotch tape was initially employed in 2004 to get Gr-flakes from bulk graphite crystals [7]. Bulk crystals (with layered structure) were turned into single or few layers nano flakes by repeatedly attaching and peeling them using scotch tape. The flakes generated using this method are often crystalline in character, with a low density of traps, which is necessary for fundamental research. Furthermore, for large-scale consumption, it is tough to accomplish error-free control over the size and thickness of the flakes. For large-scale production, solution-based exfoliation processes are typically used, which involve disseminating 2D materials into various solvents using chemical or ultrasonic procedures [8]. Solvent molecules usually permeate the layers of 2D materials via chemical reaction and ultrasonic irradiation operations, enabling film production easy by deposition of the resulting dispersion on diverse substrates. Solution-based exfoliation techniques, in contrast to mechanical exfoliation, allow for commercial-scale synthesis of 2D materials.

To satisfy the demands of mass manufacturing, remarkably crystalline 2D properties can be utilized by mechanical exfoliation and a variety of solution-based strategies, but it is harder to establish higher quality, wide area, and consistent monolayer or few layered films of 2D materials unless the named CVD procedure is used. For illustration, using a surface catalyzed process, big area or few layer Gr-sheets can be grown on a copper substrate using the CVD process [9]. Generally, in a CVD method, metal substrates (such as Cu, Ni, and Pt) are subjected to precursor vapors, that either react or disintegrate at the substrate's surface to generate the desired coatings. Materials formed using the CVD process have superior morphological and optoelectronic properties than solution-based 2Ds. Wet transfer techniques were used to transfer the generated films of 2D materials to other substrates for various applications [10].

6.4 2D MATERIALS IN ORGANIC SOLAR CELLS (OSCs)

OSCs, which are made up of available and simple modifiable polymeric or tiny organic molecules, are considered as a fantastic photovoltaic technology that will provide excellent energy conversion efficiency at a low cost thanks to simple production procedures. Similarly, inorganic solar cell is limited due to expensive fabrication techniques, but OSCs are spurred to minimal prices by cheap active materials [11]. OSCs include a device construction that includes a charge transporting layers (CTL), electrodes, and photoactive layer [12]. In both planar and bulk hetero junction-based OSCs, the photoactive layer is constructed of organic materials (fullerene or non-fullerene) or conjugated polymer as p-type (donor) and n-type (acceptor), respectively. Planar hetero junctions had a double-layer structure comprising of acceptor and donor molecules created by successive layer deposition, whereas bulk hetero junctions had acceptor and donor areas mixed.

6.4.1 2D Materials as Electrodes for OSCs

Although indium tin oxide (ITO) is a popular material for transport of conductive electrodes in OSCs and perovskite solar cells (PSCs), it has a number of drawbacks. Furthermore, radio frequency sputtering, which includes plasma or ion bombardment, is used to create transparent conductive oxide layers, which causes multiple damages to the underlying layers [13]. The need for high temperatures for post-deposition heating of certain oxide coatings causes deterioration of the underneath organic substance [14]. As the rear electrode is placed on the organic CTL, OSCs with flipped orientation are susceptible to the concerns mentioned above [15]. Gr could potentially replace ITO, delivering similar results in terms of conductivity, transparency, material availability, flexibility, and operational costs [16]. The majority of papers are on portable devices; however, economical deposition procedures for bulky-Gr films have been investigated and may be recommended for future applications [17].

The increased conductivity, appropriate surface shape, and desirable work function (WF) of Gr as an electrode in OSCs can be ascribed to it [18]. In general, three methods have been used to improve the electrical conductivity of Gr-based films: (i) "Multilayer Stack," which involves the upright piling of Gr-films; (ii) "Chemical Doping," that involves the use of wet chemical dopants, e.g., HNO_3, $AuCl$, and others; and (iii) "Electrostatic Doping" of Gr-based ferroelectric polymers. Incorporating the first and second approach dramatically enhanced the conductivity of Gr sheets and make them a viable candidate for ITO. Furthermore, Gr doping aids in corresponding its WF with the surrounding CTL; when WF of pristine-Gr is associated with HOMO levels of various organic semiconductor donors, considerable hole clumping occurs (Gr-anode). While WF of pristine-Gr aligns with the LUMO levels of widely applied as organic acceptors, the accumulation of electrons causes an ohmic contact, rendering Gr-cathode. Scientists explored the impact of the number of layers of a layer-by-layer mounted Gr-anode on the cell characteristics of OSCs [19].

HNO_3-doped P-doped Gr-material aids in reducing sheet resistance and, as a result, the output device's series resistance. Without any further chemical procedures, the sheet resistance of monolayered Gr (1LG) and seven-layered Gr (7LG) was reduced from 512.5 to 150.6 Ω sq^{-1}, correspondingly. After adding HNO_3, the value drops to 36.6 Ω sq^{-1}, which is lower than that of ITO. Three layers of CVD-grown Gr sheets were used in OSCs on PET substrates, yielding a 4.33% efficiency [18]. Chemical doping is indeed a very successful approach; however, sustaining sheet resistance after doping is a key feature. Moisture and oxygen adsorption in Gr-multi-layered stacking lowers the electrical conductivity of doped Gr-sheets over time, affecting the stability efficiency of solar cell. To solve this issue, researchers used electrostatic doping of few layers of Gr with a multilayer ferroelectric polymer covering, poly(vinylidene fluoride-cotrifluoroethylene), because the dipoles keep their alignment for lengthy periods of time in ambient circumstances, the doping impact is extremely stable [19].

Gr is utilized as an anode material in solar cells in several reports, but there have been few investigations on its use as a cathode material. When comparing the Gr layer's high sheet resistance to traditional cathode materials, there is a significant concern. The cathode material for OSCs has been investigated as a single stacked Gr layer [20]. Contact doping has been found to improve system output by aligning WFs, as well as increasing power conversion efficiency (PCE) owning to the transparency of Gr. Furthermore, it has been discovered that using composites such as aluminum titanium oxide (Al-TiO$_2$) enhances the effectiveness of Gr-cathode [21].

It was discovered that solution-based TiO$_2$ improves the yield of Gr-cathode due to its excellent electron-transport capacity, whereas evaporating aluminum nanoclusters in composite dropped Gr WF for better energy level alignment. The OSCs exhibited a PCE that increased from 6.1% to 7.1%, when Gr is used as anode or cathode with a BHJ that is composed of fullerene derivative [6,6]-phenyl C71 butyric acid methyl ester (PCBM) and a polymer thieno [43]-thiophene-alt-benzodithiophene (PTB7 [22]). Furthermore, scientists have examined the feasibility of using Gr as either an anode or cathode in a similar device. For clearly transparent OSCs, Gr electrodes have been produced on flexible substrates. Figure 6.1a shows the normalized PCE of Gr/Gr (cathode/anode) devices on paper, with a critical radius of about 0.7 mm and twisted cyclically to different radii. After 100 folding cycles, the device's properties remained the same, indicating that the folding radius is ≈50% more than the critical radius. The critical radius varied between Gr-based electrodes and ITO electrodes, ranging from 0.7 mm for Gr to 2.0 mm for ITO (Figure 6.1b), indicating that Gr is more flexible than ITO, with the ability to bend to lesser radii comparable to ITO [23].

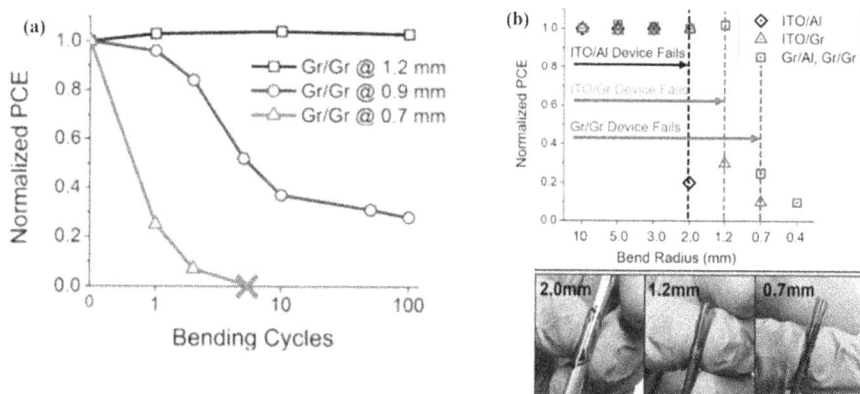

FIGURE 6.1 (a) Normalized device PCE of Gr/Gr-based flexible OSCs at different bending cycles. (b) Normalized PCE of Gr/Gr-based devices and photograph at different bending radii of curvature. (Adapted with permission from [23]. Copyright © 2016 Wiley-VCH Verlag GmbH & Co. KGaA, Weinheim. John Wiley & Sons.)

PCE values of 2.8%–3.8% were found for clearly transparent devices having all Gr electrodes (anode and cathode) spanning the visible regime, with optical transmittance of 54%–61%. This research shows that Gr may be used in place of ITO electrodes to fabricate cost-effective, flexible, and energy-harvesting devices [24]. Furthermore, the manufacturing of adaptable OSCs with a Gr electrode doped with the ferroelectric polymer layer (also serving as a flexible substrate) has been established, with a performance of 2.07% and the ability to be spun around onto a 7 mm diameter cylinder. It's worth mentioning that OSCs fabricated on glass or other polymeric flexible substrates, such as polyethylene terephthalate (PET) with CVD-based Gr or ITO electrodes, exhibited almost identical PCEs of 1.18% (Gr-electrodes) and 1.27% (ITO electrode) [25]. The ITO film produced significant cracks rather than the Gr layer that had no micro cracks or fractures. Increased film sheet resistance is caused by mechanical stresses in ITO caused by microfractures, resulting in a decrease in fill factor (FF) and, as a result, a decrease in PCE. A polyimide (PI) substrates with Gr-electrode-based package-free flexible OSCs were demonstrated, which exhibited superior flexibility due to Gr. Importantly, it exhibited an excellent bending stability after 1,000 bending cycles and its PCE drop to 8% [26].

6.4.2 2D MATERIAL AS CTL FOR OSCs

The most frequently employed interfacial layer for charge extraction in organic and PSCs is PEDOT:PSS, which has a very acidic (pH ≈ 1) and hygroscopic nature, demanding the effective and efficient replacement material for this layer [27]. Inorganic materials, e.g., NiO, MoO_3, and V_2O_5, which are viable substitutes, necessitate the use of a costly high-vacuum plasma-improved deposition techniques [28]. The use of cost-effective roll-to-roll Gr and derivatives as a substitute for these metallic oxides appears to be a very advantageous option [29]. To allow the intrinsic electric field to pass the active layer and the holes to travel in the direction of the anode, the hole extraction layer involves a high WF. On the other hand, for successful electron transportation toward the cathode, a low WF is required for the extraction of the electronic layer. Because of their ambipolar nature, Gr and its derivatives allow for WF adjustment, making them appropriate for both HTL and ETL. GO, a one-atom-thick Gr-derivative, has hydroxyl (OH), epoxy, and carboxyl (COOH) groups on its base and COOH group at its corners. The occurrence mixed sp^2 and sp^3 hybridizations GO has a unique electrical structure, when inserted between active layer and ITO, GO (2 nm thickness) performed as an efficient HTL in OSCs, with an average PCE of $3.5 \pm 0.3\%$ [30]. The output of a solar cell can be improved by adjusting the concentration of GO.

Spin coating at varied amounts (1, 2, and 4 mg mL^{-1}) of GO was used to deposit the manufactured $PCDTBT$: $PC_{71}BM$ OSCs on the mosaic ITO-coated glass substrate. For 1 and 2 mg mL^{-1} dosages, the optical transmittance held steady, while for 4 mg mL^{-1}, the transmittance was 15% inferior to pristine ITO. Due to homogenous morphology, decreased series resistance (6.89 Ω cm^{-2}) was achieved for a GO HTL of 1 mg mL^{-1} in contrast to dosages of 2 and 4 mg mL^{-1}, which revealed resistance values of 9.54 and 11.51 Ω cm^{-2}, accordingly. As a result, the performance achieved 2.73% at the lowermost dosage (1 mg mL^{-1}) with a width of 1–3 nm, which is comparable to dosages of 2 mg mL^{-1} (competence is 0.67%) and 4 mg mL^{-1} (competence is

0.22%) [31]. GO/rGO with oxygen comprising functional groups is accomplished with exceptional solubility in both water and polar organic solvents, making it a possible solution-processable interface material for CTL over a wide temperature range [32]. Sulfated GO with OSO_3H groups produces an HTL with WF that is ideally suited to the organic layer, resulting in improved device performance [33]. When functionalized with a sulfonic acid group, rGO acts as a substantial interfacial layer, resulting in improved photovoltaic performance [34]. Better efficiency has been accomplished due to better corresponding of sulfonic acid groups with the HOMO of photoactive materials and used in a variety of donor-based system, together with polymers, e.g., poly-1-(6-4,8-bis(2-ethylhexyl)oxy-6-methylbenzo1,2-b: 4,5-b′dithiophen-2- (TQ1) and polythieno-3,4-b thiophene/benzodithiophene (PTB7). Moreover, appropriate results for PBDTTT-CF and PTB7 have been published, with a remarkable output of 7.18%, analogous to PEDOT: PSS [34].

In contrasted to chemically altered Gr, which has WF of ≈4.66 eV, functionalized GO has a higher WF of 5.04 eV [34]. It was discovered that fluorinated and chlorinated GO were also used as HTL, substituting PEDOT: PSS, with efficiencies of 7.59% for chlorinated GO [35] and 8.6% for fluorine-functionalized rGO [32]. Different organic blends, such as PCDTBT: PC70BM, allow SC performance to increase by ≈5% by employing GO as an efficient HTL, compared to PEDOT: PSS across a dynamic area of $0.64 cm^2$ [36]. Murray et al. have established a single layer of electronically adjusted GO with an efficiency of 7.4% [37] to replace PEDOT: PSS with an organic blend of PTB7:PC71BM. GO has the disadvantage of generating an excellent ohmic contact and also delivering full insurance by coating [38]. As a result, a hybrid of GO and PEDOT: PSS has also been stated as HTL by various groups in OSC.

The incorporation of solution treated GO/rGO, which has the potential to be used as an ETL, has also been investigated. PTB7:PC71BM has been successfully treated using GO-grafted polyethylenimine ethoxylate, which has an efficiency of 8.21% [39]. In addition, when compared to solely metal oxide devices, hybrid metal oxide/rGO ETL produces higher output. It's attributable to the circumstance that rGO has a long enough mean free path length for charge transfer as associated with the typical metal oxides utilized in ETLs [40].

In inverted OSCs topologies, the most typically employed ETLs are ZnO and TiO_x [41]. 2D GO can be incorporated into ETL to create a composite that improves solar cell production. When used in the TiO_x ETL, rGO can help to prevent charge recombination and raise the PCE value to 2.7% [42]. When compared to a device that does not have rGO, the efficiency increase is 8%. Hence, non-reduced GO produces an excellent ETL with ZnO, with the maximum PCE value of 1.109%, which is 34% greater than ZnO ETL devices' 0.824% PCE [43].

6.5 2D MATERIALS IN PEROVSKITE SOLAR CELLS (PSCs)

Perovskite is another promising absorbing material having ABX_3 configuration, where A is organic cation (CH_3NH_3, $CH_3CH_2NH_3$, and NH_2CHNH_2), B is divalent metal cation (Pb, Sn, etc.), and X is halide (Cl, Br, and I). This material exhibited extraordinary properties such as small binding energy, tuneable bandgap, large absorption coefficient, bipolar charge carrier transportations, and large diffusion length. PSCs stated efficiency has increased from 3.9% to 25.5%. This perovskite material is sandwiched between the ETL and HTL that extract the carriers on both electrode sides. The PSCs structure is comprised of fluorine or tin-doped tin oxide (FTO or ITO) layered glass, while a thermally evaporated metal (gold, aluminum, silver, and copper) electrode used on the backside.

6.5.1 2D MATERIAL AS ETL IN PSCs

Higher carrier mobility and suitable energy level alignment are the key consideration of an efficient ETL in PSCs [44]. Black phosphorous (BP) is an astonishing material due to its optoelectronics properties such as decent carrier mobility and suitable bandgap [45,46]. The bulk BP has a bandgap of 0.3 eV, which can be modified by changing the number of layers. The ultra-small BP (dubbed BP quantum dots (BPQDs)) has successfully substituted the ETL in PSCs [44].

The diameter of liquid-phase exfoliated BPQDs ranging from 3 to 10 nm has dramatically improved the efficiency as an ETL, as compared to ETL-free PSCs. The highest efficiency with BPQDs of 11.26% and 14.6% was achieved after optimization of layer thickness for the PSCs fabricated on polymeric and rigid substrate, respectively. A compact and homogeneous perovskite films with great improvement in the electron extraction, resulting in a suitable energy level alignment between perovskite and BPQDs, which helped to the efficient electron injection and transport. The BPQD-based flexible device showed a small minor modification after 200 bending cycles in current density (J_{sc}) and open circuit voltage (V_{oc}), while a rapid reduction has been seen in FF and PCE that might be a reason of increase in sheet resistance.

Moreover, substantial effort has been utilized in the synthesis procedure of solution processed few-layer BP (FL-BP) nanosheets, but a higher-quality, more stable, and extremely reliable FL-BP has proven to be difficult [47]. To use a vortex fluidic device and pulsed laser illumination, Batmunkh et al. [48] established a simple, effective, and rapid synthesizing process to generate stable and excellent performance of FL-BP. Phosphorene nanosheets that had been solution-processed as developed were then utilized in an optimized ETM for cold temperature TiO_2-based planar PSCs. The phosphorene's rapid charge movement and proper band energy compatibility assisted significantly to the improvement in PCEs. As a consequence, PSCs depending on TiO_2 with phosphorene nanosheets included producing an estimated PCE of 16.53% that was higher than the control device's (14.32%) (Figure 6.2).

In recent years, the PCE of p-i-n PSCs has improved significantly, exceeding 21% [49]. The substantial mismatch of the energy levels of the device components demonstrated a significant barrier, leading in a restricted V_{oc} value. The standard HTM (PEDOT:PSS) has a WF of $\approx$5 eV [50], signifying non-optimal energy level compatibility with $MAPbI_3$, which has the HOMO level at -5.4 eV. Furthermore, the PCBM ETL has a LUMO level of roughly 4.0 eV, which is significantly different from the WF including both Au (5.1 eV) and Ag (4.7 eV). Wang et al. [51] claimed MoS_2 and GO as interfacial layers between PEDOT: PSS and $MAPbI_3$ and between PCBM and Ag, accordingly, to improve energy level matching. In contrast, the PCEs improved from 14.15% to 16.89% after introducing 3 nm thin sheet of MoS_2, and equal to 19.14% after introducing GO film (1 nm, verified by AFM analysis). A huge boost was seen due to

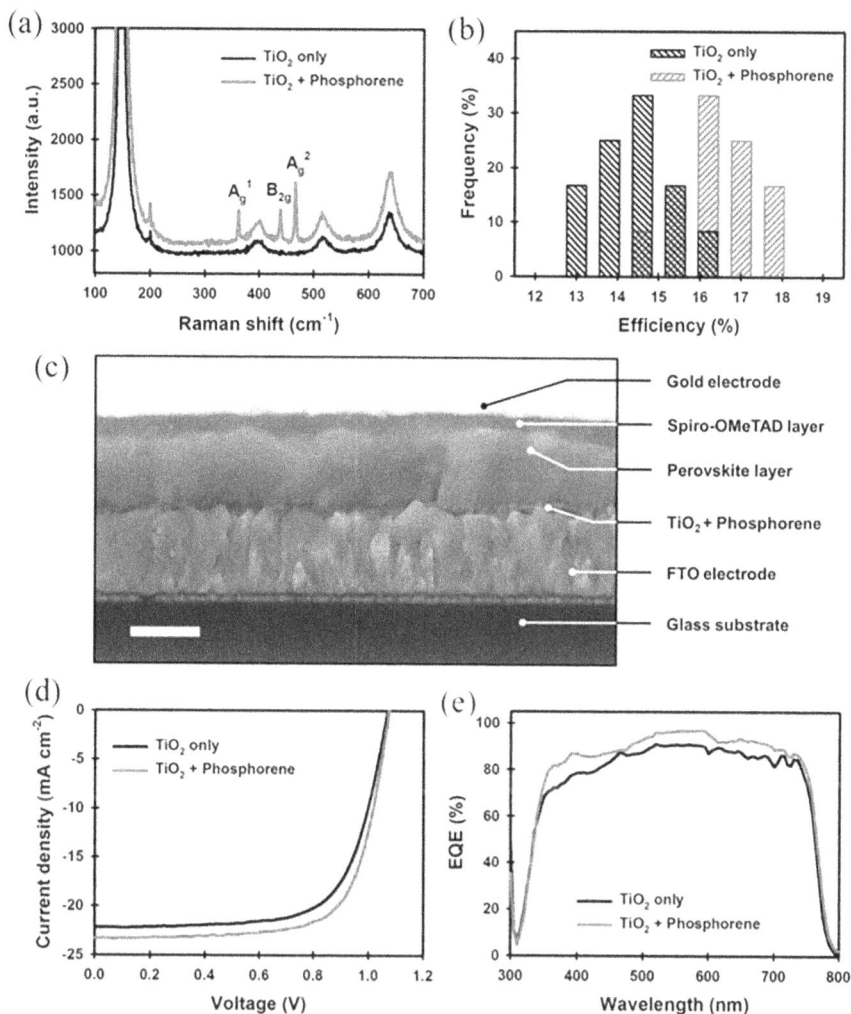

FIGURE 6.2 (a) Raman spectroscopy, (b) histogram of efficiency of TiO_2 and TiO_2+ phosphorene ETL based devices, (c) complete cross–sectional image of PSCs, (d) current density curve, and (e) EQE spectra of PSCs based on TiO_2 and TiO_2+ phosphorene ETM. (Adopted with permission from [48]. Copyright © 2019 Wiley-VCH Verlag GmbH & Co. KGaA, Weinheim. John Wiley & Sons.)

improved energy level alignment, which reduced nonradiative losses of charge carriers, as evidenced by an improved V_{oc} (0.962–1.135 V).

6.5.2 2D MATERIAL AS HTL IN PSCS

With the application of Spiro OMeTAD HTM, the efficiency of PSCs has attained an amazing benchmark of over 20%, with instability posing a severe problem [52]. The unsatisfactory impacts of the dopants (e.g., lithium bis(trifluoromethanesulfonyl) imide and tert-butylpyridine) employed to dope the Spiro OMeTAD to counteract the minor electrical conductivity are mainly associated with the problem of durability [53]. However, multiple publications have shown that the iodide ion-migration from perovskite layer to electrode through the Spiro-OMeTAD has a detrimental effect on the device efficiency as well as stability [54]. Capasso et al. [55] inserted a buffer layer of liquid-phase exfoliated MoS$_2$ nanosheets, which played two roles: (i) stopping the formation of shunt paths at perovskite/electrode interface, and (ii) efficiently improve the charge extraction and injection process. The N-methyl-2-pyrrolidone solvent is the best choice for effective MoS$_2$ exfoliation [56], but its incompatibility with perovskite is still a big challenge for PSCs. Thus, a solvent-exchange method is necessary to introduce a suitable and perovskite-friendly solvent, which indicated that the content of oxygen and nitrogen was minimum with associated flake in the perovskite friendly purified solvent IPA. To investigate the capability of MoS$_2$ as a hole collector, a device having a mesoporous structure with HTL free and MoS$_2$ containing layer exhibited a PCE of 1.5% and 4.5%, respectively, signifying the effectiveness of MoS$_2$ buffer layer as a good hole collector. Similarly, MoS$_2$ used as atomic buffer layer with Spiro-OMeTAD based HTL in PSCs, it exhibited an efficiency of 13.3% and 11.5%, while without this buffer layer an efficiency of 14.2% and 11.4% achieved with an active device area of 0.1 and 1.05 cm^2, respectively (Figure 6.3). This MoS$_2$ layer containing devices attained around 93% of their initial efficiency as compared to ≈66% of Spiro-based PSCs after 550 hours stability test.

Kim et al. employed WS$_2$, MoS$_2$, and GO HTLs substituted with traditional PEDOT:PSS in planar p-i-n PSCs [57]. The CVD approach was used to synthesize WS$_2$ and MoS$_2$, whereas Hummer's method was employed to prepare the GO. The WFs of WS$_2$, GO, MoS$_2$, and PEDOT:PSS were measured to be 4.95, 5.0, 5.1, and 5.25 eV, and its surface root mean roughness were calculated to be 1.778, 1.391, 1.108, and 0.841 nm, respectively, but these HTLs based PSCs have almost similar efficiency, i.e., 9.93 (PEDOT:PSS), 9.62 (GO), 9.53 (MoS$_2$), and 8.02% (WS$_2$), respectively. It is clear that the efficiency attained for these HTMs are influenced by their

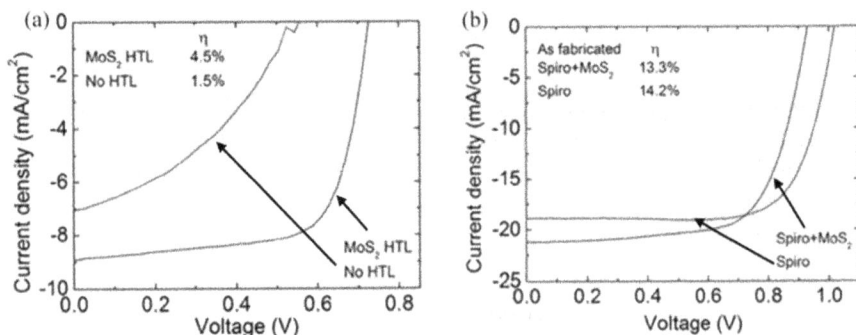

FIGURE 6.3   (a) Current–density curve of PSCs with and without MoS$_2$ HTL with a device area of 0.1 cm^2, (b) current–density curve of PSCs-Spiro based HTL MoS$_2$ HTL with and without MoS$_2$ buffer layer with active device area of 0.1 cm^2. (Reproduced with permission from [55]. Copyright © 2016 Wiley-VCH Verlag GmbH & Co. KGaA, Weinheim. John Wiley & Sons.)

WF values because it has a direct impact on the V_{oc} of the devices. As a result, the WS$_2$ based PSCs efficiency slightly declined in contrast to other devices, owing to the lower V_{oc} value (0.82 V) that come from its lower WF as compared to other materials.

Huang and colleagues [58] used lithium intercalation–based exfoliation approach to prepare water-soluble WS$_2$ and MoS$_2$ as HTMs. AFM was used to measure the thickness of the MoS$_2$ and WS$_2$ flakes, which were found to be 3.6–8.0 nm and 2.5–4.0 nm, respectively. It should be well-known that MoS$_2$ and WS$_2$ layers had an average of six and three layers, respectively. The impact of two crystalline phases of MoS$_2$ and WS$_2$ (semiconducting 2H and metallic 1T phases) on PSCs performance was explored in this research. The composition of the 1T phase in TMDs HTLs was found to have a significant impact on device performance. The phase transformation generated by the exfoliation procedure and post heating of TMDs at 90°C and 200°C was investigated, and it was discovered that 1T % improved after they were exfoliated from their majority counterparts, but then decreased after heating. The devices were built and compared using 11 nm thick MoS$_2$ and WS$_2$ HTL with more 1T content (>50%, 1T rich) and less 1T content (20%, 1T poor).

With decreasing 1T concentration, the PV performance of these devices substantially declined. A PCE of 15%, 14.35%, and 12.44% has been achieved with WS$_2$, 1T rich, and PEDOT: PSS HTLs in PSCs, respectively. The MoS$_2$ and WS$_2$ HTL-based devices exhibited a good V_{oc} and J_{sc} that might be due to suitable energy level alignment with perovskite and thought to be the primary contributor to the improvement in the PCE. Meanwhile, these devices exhibited promising stability in the nitrogen-filled glove box and maintained 78% (MoS$_2$) and 72% (WS$_2$) of their initial PCEs after 56 days, while the PEDOT: PSS-based PSCs completely decomposed after 35 days.

6.5.3   2D MATERIAL IN ACTIVE LAYER IN PSCs

The interfacial and additive engineering approach is thought to be an effective technique to grow the crystallization and surface treatment of perovskite films [59]. Guo et al. [60] proposed adding MXene (Ti$_3$C$_2$T$_x$) to the perovskite precursor that improve crystal size, minimize defect density, and improve charge transfer performance. They discovered a less amount of MXene (0.03 wt.%) that effectively declines the crystallization rate, resulting in bigger perovskite crystals (usual grain size like as 150–358 nm). Ti$_3$C$_2$T$_x$ has been observed to be critical in delaying the perovskite nucleation process. Consequently, a maximal PCE of 17.41% achieved that was superior to the control (15.54%) device. The electron transfer mechanism was aided by the decent conductivity of MXene. Furthermore, the MXene addition was shown to passivate the charge traps and grain boundaries, resulting in a negligible hysteresis. Wang et al. [61] demonstrated a small amount of 2D BP to the perovskite precursor solution, which efficiently improved the efficiency and photostability of PSCs. The average size and thickness of the fabricated BP flakes are ≈0.5–2 μm and ≈8 nm, respectively.

The existence and homogeneous distribution of BP in perovskite have been confirmed by elemental mapping of the perovskite precursor. By applying BP on MAPbI$_3$ in PSCs with regular planar device structure, the PCE boosted to 20.65% with negligible hysteresis. An exceptional performance with better charge transport properties has been achieved by passivating the carrier traps on the perovskite layer in the presence of BP. Interestingly, PSCs based on the MAPbI$_3$/BP displayed exceptional photostability in a N$_2$ glovebox when exposed to a white light emitting diode for illumination (100 mW cm^{-2}) for 1,000 hours, retaining 94% of the initial performance. The existence of BP with their excellent self-healing property has been linked to the passivation of lead defect generation under irradiation [62]. The potential of BP is to enhance the stability of PSCs, which is a key requirement of such cutting-edge technology. The role of 2D materials as additive in perovskite layer is still not widely used and extensive research is needed in this way. These materials have the ability (phosphorene) to absorb near-infrared, strong UV and entire visible wavelength region, making them useful for harnessing the entire sun spectrum. Efficaciously optimizing the materials and devices will also necessitate a significant amount of effort.

6.6 CONCLUSIONS

In this chapter, we have summarized the current development of graphene and other 2D materials in third generation of photovoltaic technology. The benefits of these 2D materials in the organic and perovskite photovoltaic have been explored owing to their astonishing properties such as decent charge transport properties, highly stable nature, and cost effectiveness. 2D materials not only improve the homogeneous film formation ability but also effectively play an important role in the interface engineering and device optimization in organic and perovskite photovoltaics. Despite these extraordinary properties, various approaches have been successfully demonstrated for the mono and few layer 2D materials preparations such as liquid-phase exfoliation, mechanical exfoliation, and CVD, but the long processing time, expensive, low yield, and lower reproducibility are the major obstacles and must need to be solved for their efficient use in optoelectronics devices.

REFERENCES

1. K. Watanabe, T. Taniguchi, H. Kanda, Direct-bandgap properties and evidence for ultraviolet lasing of hexagonal boron nitride single crystal, *Nature Materials*, 3 (2004) 404–409.
2. C.R. Dean, A.F. Young, I. Meric, C. Lee, L. Wang, S. Sorgenfrei, K. Watanabe, T. Taniguchi, P. Kim, K.L. Shepard, Boron nitride substrates for high-quality graphene electronics, *Nature Nanotechnology*, 5 (2010) 722–726.
3. M. Chhowalla, H.S. Shin, G. Eda, L.-J. Li, K.P. Loh, H. Zhang, The chemistry of two-dimensional layered transition metal dichalcogenide nanosheets, *Nature Chemistry*, 5 (2013) 263–275.
4. D. Jariwala, V.K. Sangwan, L.J. Lauhon, T.J. Marks, M.C. Hersam, Emerging device applications for semiconducting two-dimensional transition metal dichalcogenides, *ACS Nano*, 8 (2014) 1102–1120.
5. D.R. Dreyer, S. Park, C.W. Bielawski, R.S. Ruoff, The chemistry of graphene oxide, *Chemical Society Reviews*, 39 (2010) 228–240.
6. Z. Liu, S.P. Lau, F. Yan, Functionalized graphene and other two-dimensional materials for photovoltaic devices: Device design and processing, *Chemical Society Reviews*, 44 (2015) 5638–5679.
7. K.S. Novoselov, A.K. Geim, S.V. Morozov, D.-E. Jiang, Y. Zhang, S.V. Dubonos, I.V. Grigorieva, A.A. Firsov, Electric field effect in atomically thin carbon films, *Science*, 306 (2004) 666–669.

8. X. Cai, Y. Luo, B. Liu, H.-M. Cheng, Preparation of 2D material dispersions and their applications, *Chemical Society Reviews*, 47 (2018) 6224–6266.

9. X. Li, W. Cai, J. An, S. Kim, J. Nah, D. Yang, R. Piner, A. Velamakanni, I. Jung, E. Tutuc, Large-area synthesis of high-quality and uniform graphene films on copper foils, *Science*, 324 (2009) 1312–1314.

10. C.E.N.E.R. Rao, A.E.K. Sood, K.E.S. Subrahmanyam, A. Govindaraj, Graphene: The new two-dimensional nanomaterial, *Angewandte Chemie International Edition*, 48 (2009) 7752–7777.

11. L. Zhu, M. Zhang, G. Zhou, T. Hao, J. Xu, J. Wang, C. Qiu, N. Prine, J. Ali, W. Feng, Efficient organic solar cell with 16.88% efficiency enabled by refined acceptor crystallization and morphology with improved charge transfer and transport properties, *Advanced Energy Materials*, 10 (2020) 1904234.

12. L. Zhu, W. Zhong, C. Qiu, B. Lyu, Z. Zhou, M. Zhang, J. Song, J. Xu, J. Wang, J. Ali, Aggregation-induced multilength scaled morphology enabling 11.76% efficiency in all-polymer solar cells using printing fabrication, *Advanced Materials*, 31 (2019) 1902899.

13. H. Kanda, A. Uzum, A.K. Baranwal, T.N. Peiris, T. Umeyama, H. Imahori, H. Segawa, T. Miyasaka, S. Ito, Analysis of sputtering damage on I–V curves for perovskite solar cells and simulation with reversed diode model, *The Journal of Physical Chemistry C*, 120 (2016) 28441–28447.

14. R. Kang, S.-H. Oh, S.-I. Na, T.-S. Kim, D.-Y. Kim, Investigation into the effect of post-annealing on inverted polymer solar cells, *Solar Energy Materials and Solar Cells*, 120 (2014) 131–135.

15. F. Li, C. Chen, F. Tan, C. Li, G. Yue, L. Shen, W. Zhang, Semitransparent inverted polymer solar cells employing a sol-gel-derived TiO_2 electron-selective layer on FTO and $MoO_3/Ag/MoO_3$ transparent electrode, *Nanoscale Research Letters*, 9 (2014) 1–5.

16. R. Garg, S. Elmas, T. Nann, M.R. Andersson, Deposition methods of graphene as electrode material for organic solar cells, *Advanced Energy Materials*, 7 (2017) 1601393.

17. H. Park, P.R. Brown, V. Bulović, J. Kong, Graphene as transparent conducting electrodes in organic photovoltaics: Studies in graphene morphology, hole transporting layers, and counter electrodes, *Nano Letters,* 12 (2012) 133–140.

18. H. Kim, S.-H. Bae, T.-H. Han, K.-G. Lim, J.-H. Ahn, T.-W. Lee, Organic solar cells using CVD-grown graphene electrodes, *Nanotechnology*, 25 (2013) 014012.

19. K. Kim, S.-H. Bae, C.T. Toh, H. Kim, J.H. Cho, D. Whang, T.-W. Lee, B. Özyilmaz, J.-H. Ahn, Ultrathin organic solar cells with graphene doped by ferroelectric polarization, *ACS Applied Materials & Interfaces*, 6 (2014) 3299–3304.

20. P. Misra, *Applied Spectroscopy and the Science of Nanomaterials*. Berlin/Heidelberg: Springer, 2014.

21. D. Zhang, F. Xie, P. Lin, W.C. Choy, Al-TiO_2 composite-modified single-layer graphene as an efficient transparent cathode for organic solar cells, *ACS Nano*, 7 (2013) 1740–1747.

22. H. Park, S. Chang, X. Zhou, J. Kong, T.s. Palacios, S. Gradečak, Flexible graphene electrode-based organic photovoltaics with record-high efficiency, *Nano Letters*, 14 (2014) 5148–5154.

23. Y. Song, S. Chang, S. Gradecak, J. Kong, Visibly-transparent organic solar cells on flexible substrates with all-graphene electrodes, *Advanced Energy Materials*, 6 (2016) 1600847.

24. Q. Wang, Q. Dong, T. Li, A. Gruverman, J. Huang, Thin insulating tunneling contacts for efficient and water-resistant perovskite solar cells, *Advanced Materials*, 28 (2016) 6734–6739.

25. L. Gomez De Arco, Y. Zhang, C.W. Schlenker, K. Ryu, M.E. Thompson, C. Zhou, Continuous, highly flexible, and transparent graphene films by chemical vapor deposition for organic photovoltaics, *ACS Nano*, 4 (2010) 2865–2873.

26. Z. Liu, J. Li, F. Yan, Package-free flexible organic solar cells with graphene top electrodes, *Advanced Materials*, 25 (2013) 4296–4301.
27. B. Xu, S.-A. Gopalan, A.-I. Gopalan, N. Muthuchamy, K.-P. Lee, J.-S. Lee, Y. Jiang, S.-W. Lee, S.-W. Kim, J.-S. Kim, H.-M. Jeong, J.-B. Kwon, J.-H. Bae, S.-W. Kang, Corrigendum: Functional solid additive modified PEDOT: PSS as an anode buffer layer for enhanced photovoltaic performance and stability in polymer solar cells, *Scientific Reports*, 7 (2017) 46779–46779.
28. L. Barkat, M. Hssein, Z. El Jouad, L. Cattin, G. Louarn, N. Stephant, A. Khelil, M. Ghamnia, M. Addou, M. Morsli, Efficient hole-transporting layer MoO_3: CuI deposited by co-evaporation in organic photovoltaic cells, *Physica Status Solidi (a)*, 214 (2017) 1600433.
29. J. Yun, J. Yeo, J. K, H.G. Jeong, D.Y. Kim, Y.J. Noh, S.S. Kim, B.C. Ku, S.I. Na, Solution-processable reduced graphene oxide as a novel alternative to PEDOT: PSS hole transport layers for highly efficient and stable polymer solar cells, *Advanced Materials*, 23 (2011) 4923.
30. S.-S. Li, K.-H. Tu, C.-C. Lin, C.-W. Chen, M. Chhowalla, Solution-processable graphene oxide as an efficient hole transport layer in polymer solar cells, *ACS Nano*, 4 (2010) 3169–3174.
31. S. Rafique, S.M. Abdullah, H. Alhummiany, M.S. Abdel-Wahab, J. Iqbal, K. Sulaiman, Bulk heterojunction organic solar cells with graphene oxide hole transport layer: Effect of varied concentration on photovoltaic performance, *The Journal of Physical Chemistry C*, 121 (2017) 140–146.
32. X. Cheng, J. Long, R. Wu, L. Huang, L. Tan, L. Chen, Y. Chen, Fluorinated reduced graphene oxide as an efficient hole-transport layer for efficient and stable polymer solar cells, *ACS Omega*, 2 (2017) 2010–2016.
33. J. Liu, Y. Xue, L. Dai, Sulfated graphene oxide as a hole-extraction layer in high-performance polymer solar cells, *The Journal of Physical Chemistry Letters*, 3 (2012) 1928–1933.
34. J.-S. Yeo, J.-M. Yun, Y.-S. Jung, D.-Y. Kim, Y.-J. Noh, S.-S. Kim, S.-I. Na, Sulfonic acid-functionalized, reduced graphene oxide as an advanced interfacial material leading to donor polymer-independent high-performance polymer solar cells, *Journal of Materials Chemistry A*, 2 (2014) 292–298.
35. Z. Yin, J. Zhu, Q. He, X. Cao, C. Tan, H. Chen, Q. Yan, H. Zhang, Graphene-based materials for solar cell applications, *Advanced Energy Materials*, 4 (2014) 1300574.
36. S.P. Koenig, R.A. Doganov, H. Schmidt, A. Castro Neto, B. Özyilmaz, Electric field effect in ultrathin black phosphorus, *Applied Physics Letters*, 104 (2014) 103106.
37. I.P. Murray, S.J. Lou, L.J. Cote, S. Loser, C.J. Kadleck, T. Xu, J.M. Szarko, B.S. Rolczynski, J.E. Johns, J. Huang, Graphene oxide interlayers for robust, high-efficiency organic photovoltaics, *The Journal of Physical Chemistry Letters*, 2 (2011) 3006–3012.
38. J.C. Yu, J.I. Jang, B.R. Lee, G.-W. Lee, J.T. Han, M.H. Song, Highly efficient polymer-based optoelectronic devices using PEDOT: PSS and a GO composite layer as a hole transport layer, *ACS Applied Materials & Interfaces*, 6 (2014) 2067–2073.
39. J. Kim, H. Lee, S.J. Lee, W.J. Da Silva, A.R. Bin Mohd Yusoff, J. Jang, Graphene oxide grafted polyethylenimine electron transport materials for highly efficient organic devices, *Journal of Materials Chemistry A*, 3 (2015) 22035–22042.
40. K.I. Jayawardena, R. Rhodes, K.K. Gandhi, M.R. Prabhath, G.D.M. Dabera, M.J. Beliatis, L.J. Rozanski, S.J. Henley, S.R.P. Silva, Solution processed reduced graphene oxide/metal oxide hybrid electron transport layers for highly efficient polymer solar cells, *Journal of Materials Chemistry A*, 1 (2013) 9922–9927.
41. Z. Lin, C. Jiang, C. Zhu, J. Zhang, Development of inverted organic solar cells with TiO_2 interface layer by using low-temperature atomic layer deposition, *ACS Applied Materials & Interfaces*, 5 (2013) 713–718.

42. Y. Zhang, S. Yuan, Y. Li, W. Zhang, Enhanced electron collection in inverted organic solar cells using titanium oxide/reduced graphene oxide composite films as electron collecting layers, *Electrochimica Acta*, 117 (2014) 438–442.

43. J.-S. Ahn, J.-W. Ahn, Y.-C. Park, S.-J. Kim, E.-M. Han, Characterization of ZnO-Graphene electron transfer layer co-deposited by cyclic voltammetry for inverted organic solar cells, *Molecular Crystals and Liquid Crystals*, 654 (2017) 22–26.

44. N. Fu, C. Huang, P. Lin, M. Zhu, T. Li, M. Ye, S. Lin, G. Zhang, J. Du, C. Liu, Black phosphorus quantum dots as dual-functional electron-selective materials for efficient plastic perovskite solar cells, *Journal of Materials Chemistry A*, 6 (2018) 8886–8894.

45. L. Li, Y. Yu, G.J. Ye, Q. Ge, X. Ou, H. Wu, D. Feng, X.H. Chen, Y. Zhang, Black phosphorus field-effect transistors, *Nature Nanotechnology*, 9 (2014) 372–377.

46. A.N. Rudenko, M.I. Katsnelson, Quasiparticle band structure and tight-binding model for single-and bilayer black phosphorus, *Physical Review B*, 89 (2014) 201408.

47. M. Batmunkh, M. Myekhlai, A.S. Bati, S. Sahlos, A.D. Slattery, T.M. Benedetti, V.R. Gonçales, C.T. Gibson, J.J. Gooding, R.D. Tilley, Microwave-assisted synthesis of black phosphorus quantum dots: Efficient electrocatalyst for oxygen evolution reaction, *Journal of Materials Chemistry A*, 7 (2019) 12974–12978.

48. M. Batmunkh, K. Vimalanathan, C. Wu, A.S. Bati, L. Yu, S.A. Tawfik, M.J. Ford, T.J. Macdonald, C.L. Raston, S. Priya, Efficient production of phosphorene nanosheets via shear stress mediated exfoliation for low-temperature perovskite solar cells, *Small Methods*, 3 (2019) 1800521.

49. D. Luo, W. Yang, Z. Wang, A. Sadhanala, Q. Hu, R. Su, R. Shivanna, G.F. Trindade, J.F. Watts, Z. Xu, Enhanced photovoltage for inverted planar heterojunction perovskite solar cells, *Science*, 360 (2018) 1442–1446.

50. C. Winder, N.S. Sariciftci, Low bandgap polymers for photon harvesting in bulk heterojunction solar cells, *Journal of Materials Chemistry*, 14 (2004) 1077–1086.

51. Y. Wang, S. Wang, X. Chen, Z. Li, J. Wang, T. Li, X. Deng, Largely enhanced V_{oc} and stability in perovskite solar cells with modified energy match by coupled 2D interlayers, *Journal of Materials Chemistry A*, 6 (2018) 4860–4867.

52. D. Bi, W. Tress, M.I. Dar, P. Gao, J. Luo, C. Renevier, K. Schenk, A. Abate, F. Giordano, J.-P.C. Baena, Efficient luminescent solar cells based on tailored mixed-cation perovskites, *Science advances*, 2 (2016) e1501170.

53. J. Liu, Y. Wu, C. Qin, X. Yang, T. Yasuda, A. Islam, K. Zhang, W. Peng, W. Chen, L. Han, A dopant-free hole-transporting material for efficient and stable perovskite solar cells, *Energy & Environmental Science*, 7 (2014) 2963–2967.

54. S. Wang, Y. Jiang, E.J. Juarez-Perez, L.K. Ono, Y. Qi, Accelerated degradation of methylammonium lead iodide perovskites induced by exposure to iodine vapour, *Nature Energy*, 2 (2016) 1–8.

55. A. Capasso, F. Matteocci, L. Najafi, M. Prato, J. Buha, L. Cinà, V. Pellegrini, A.D. Carlo, F. Bonaccorso, Few-layer MoS_2 flakes as active buffer layer for stable perovskite solar cells, *Advanced Energy Materials*, 6 (2016) 1600920.

56. A. Jawaid, D. Nepal, K. Park, M. Jespersen, A. Qualley, P. Mirau, L.F. Drummy, R.A. Vaia, Mechanism for liquid phase exfoliation of MoS_2, *Chemistry of Materials*, 28 (2016) 337–348.

57. Y.G. Kim, K.C. Kwon, Q. Van Le, K. Hong, H.W. Jang, S.Y. Kim, Atomically thin two-dimensional materials as hole extraction layers in organolead halide perovskite photovoltaic cells, *Journal of Power Sources*, 319 (2016) 1–8.

58. P. Huang, Z. Wang, Y. Liu, K. Zhang, L. Yuan, Y. Zhou, B. Song, Y. Li, Water-soluble 2D transition metal dichalcogenides as the hole-transport layer for highly efficient and stable p–i–n perovskite solar cells, *ACS Applied Materials & Interfaces*, 9 (2017) 25323–25331.

59. J. Ali, Y. Li, P. Gao, T. Hao, J. Song, Q. Zhang, L. Zhu, J. Wang, W. Feng, H. Hu, Interfacial and structural modifications in perovskite solar cells, *Nanoscale*, 12 (2020) 5719–5745.
60. Z. Guo, L. Gao, Z. Xu, S. Teo, C. Zhang, Y. Kamata, S. Hayase, T. Ma, High electrical conductivity 2D MXene serves as additive of perovskite for efficient solar cells, *Small*, 14 (2018) 1802738.
61. Y. Wang, H. Zhang, T. Zhang, W. Shi, M. Kan, J. Chen, Y. Zhao, Photostability of MAPbI$_3$ perovskite solar cells by incorporating black phosphorus, *Solar RRL*, 3 (2019) 1900197.
62. R. Long, W. Fang, A.V. Akimov, Nonradiative electron–hole recombination rate is greatly reduced by defects in monolayer black phosphorus: ab initio time domain study, *The Journal of Physical Chemistry Letters*, 7 (2016) 653–659.

7 Graphene-Based 2D Nanomaterials for Solar Cells

Nur Ezyanie Safie and Mohd Asyadi Azam
Universiti Teknikal Malaysia Melaka

CONTENTS

7.1 INTRODUCTION

Graphene, a semi-metal possessing zero bandgap and ambipolar electrical properties, offers outstanding mechanical characteristics, including a large specific surface area [1], good thermal conductivity [2], and a high Young's modulus [3]. Because of its exciting and new material properties, graphene, a carbon allotrope with the arrangement of carbon atoms in a honeycomb lattice, has piqued the attention of the material engineering world. Graphene has an excellent transmittance characteristic in the range of 97.7%–98% in the visible spectrum and high electrical conductivity. The remarkable characteristics of graphene allow it to be employed in a diverse field of applications, including electronics [4]; photonic [5] and optoelectronic circuits [6]; spintronics [7]; and energy storage [8,9] and conversion [10]. Along with its unique 2D characteristics, it can be used as a thin, flexible, yet robust display screen, particularly in different biomedical devices [11] and developing areas such as graphene-based intelligent materials. Graphene has often been employed in energy-efficiency applications to acquire a high-performance and efficient device, along with low-cost, lightweight, and flexible energy devices to fulfill the world's constrained energy resources.

DOI: 10.1201/9781003178422-7

These graphene 2D materials, which contain an inherent bandgap and diverse crystal structures, have emerged as potential prospects for various applications in the next-generation emerging field of electronics and optoelectronics, especially in various types of solar cells, including the silicon (Si)-based and emerging thin film generations. For instance, graphene has been utilized as different operating layers in solar cells to produce high-performance PV devices; as electrodes in supercapacitors [8] and lithium-ion batteries [12] to provide optimized working devices and low-cost energy storage; as well as catalysts and proton-exchange membranes in fuel cells. Graphene has piqued the interest of researchers not only because of its semiconducting nature and ultralightweight but also because of the difference in their mechanical characteristics depending on the in- and out-of-plane structure. Figure 7.1 depicted the chemical structure of the 2D graphene and its derivatives, such as graphene quantum dots (GQDs), graphene oxide (GO), and reduced graphene oxide (RGO). This chapter aims to provide information on graphene and its derivatives to their suitable multi-function in PV devices.

7.2 GRAPHENE IN SILICON SOLAR CELLS

Due to its narrow bandgap, affordability, nontoxicity, and well-established manufacturing processes, Si has become the most common semiconductor employed in photovoltaics (PVs). The conversion efficiency of the photon into electricity and the cost per watt of power generated are indeed the two most important variables in the viability of Si-based solar devices as PV technology. Si solar cells have occupied the PV industry for the past 50 years, with remarkable power conversion efficiencies (PCEs). In spite of the dominance of solar modules with bulk crystalline Si solar cells owing to their high PCE, the rapid growth industry has increased the requirement of thinner Si solar cells to lower the costs, where the cross-section thickness obtained for low-temperature-fabricated Si solar cells is ~100 μm [14]. Reducing the thickness of the active absorber in a Si solar cell is essential to lower the cost per watt and increase the watt per gram usage. Soon, the Si solar PV technology roadmap predicts a substrate thickness of fewer than 25 μm. Figure 7.2 below depicted the illustration of conventional Si solar cells and the futuristic thin Si solar cells.

Researchers and the industrial sector have recently targeted achieving the Shockley–Queisser limit by utilizing thin monocrystalline Si substrates. Absorbers that are thinner and lighter needed lesser Si but still possess a lower efficiency of photon-absorption. As a result, the thickness of Si absorber has effect on solar-cell performance which has been identified as possibly the most significant issue to solve. Thus, to boost solar-cell PCE, several light-trapping techniques and extrinsic of the bulk and surface recombination processes are utilized in thinner absorber materials. Also, with requirements that are developing solar cells be manufactured on flexible substrates and susceptible to elevated roll-to-roll fabrication, Si progressively advances toward a flexible application, reaching flexibility at thicknesses <50 μm. Such flexible Si substrates might result in more efficient Si utilization, however, changed encapsulation processes, decreased kerf loss, sawing technologies, novel handling ideas, and wafer fabrication would need to be improved. As a result, the Si PV sector is anticipating a paradigm shift soon.

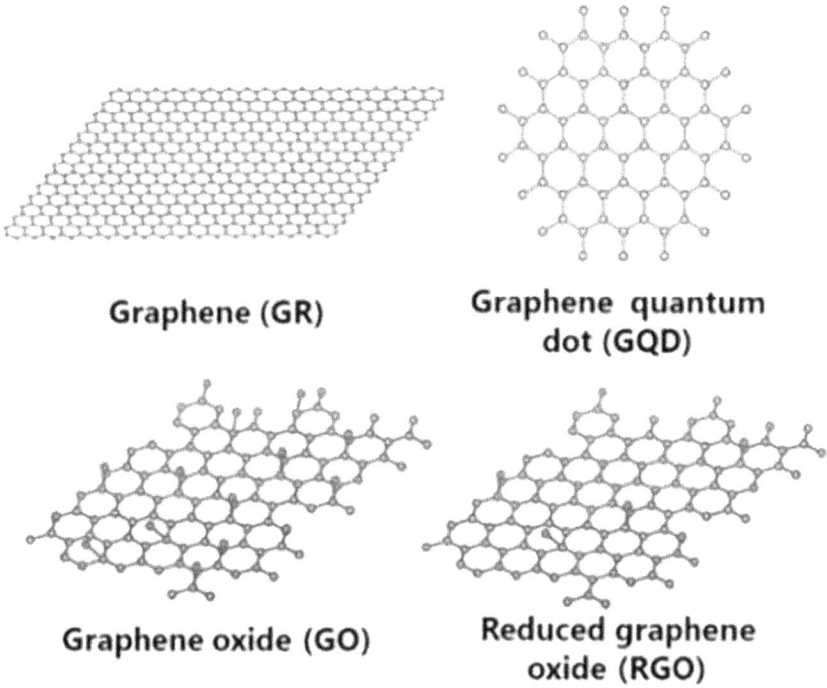

FIGURE 7.1 Chemical structure of 2D graphene and its commonly studied derivatives. (Adapted with permission from Ref. [13]. Copyright (2020) Copyright The Authors, some rights reserved; exclusive licensee [Multidisciplinary Digital Publishing Institute (MDPI)].)

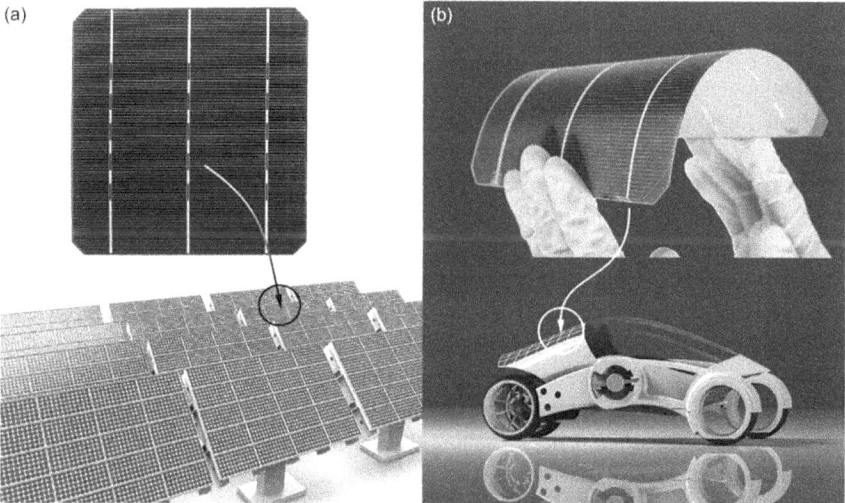

FIGURE 7.2 Illustration of (a) conventional Si solar cells and (b) futuristic thinner Si solar cells. (Adapted with permission from Ref. [14]. Copyright (2021) Elsevier.)

7.2.1 GRAPHENE AS TRANSPARENT CONDUCTING ELECTRODES

Because of its outstanding physical and electrical characteristics, graphene has piqued the interest of numerous researchers since its first production in 2004. As a viable material for implementation as the transparent electrodes, graphene has outstanding optical and electronic properties, including excellent electronic properties, high optical transmittance, and high thermal conductivity, making graphene an excellent replacement to indium tin oxide (ITO) or other noble metals in PV applications. Moreover, graphene sheets produced by chemical vapor deposition (CVD) [15] can be loaded and attached to a substrate without the need for complex procedures. As graphene's layered structure offers excellent transparency, conductivity, and thermal and chemical resilience, it is contemplated as a promising 2D nanomaterial for PV devices. Graphene's work function varies depending on its thickness and is reported in the range of 4.89–5.16 eV [16], which is in the same range for silicon and silicon dioxide, between 4.05 and 4.5 eV. Therefore, graphene loaded on Si creates a metal–semiconductor contact known as Schottky contact and functions as a transparent active material in the device. For instance, Singh et al. [17] employed a low cost with an effective technique for transferring graphene sheets on a Si wafer and evaluating their characteristics via AFM and Raman spectroscopy. In this work, the device performance parameters of Schottky solar cells are obtained as 0.65 V, 0.11 mA cm^{-2}, 62%, and 0.04% reported for open-circuit voltage (V_{oc}), short-circuit current density (J_{sc}), fill factor (FF), and PCE, respectively. The results remark the higher PV parameters obtained than the other reported devices developed based on graphene/p-Si solar cells.

Furthermore, the effectiveness of these graphene/Si Schottky junction cells could be enhanced by utilizing nanostructured semiconductors appropriately. For instance, a considerably lower reflectance value is achievable by the Si nanowire arrays without the need for specific designed antireflection coatings throughout a broad-spectrum range compared to the thin films. Taking into account the potential application for the Si nanowire, a further improvement in graphene/Si device performance has been studied employing nanowires rather than the planar Si. Arefinia and Asgari [18] provide scalable analysis of heterojunction solar cells based on p-type Si quantum wires and graphene (denoted as p-SiQWs/G). A correlated optical and electrical model is used to study scaling difficulties in the performance of p-SiQWs/G solar cells, emphasizing SiQWs diameter, the distance within SiQWs, and graphene layers. For SiQWs of average diameter, the minimum PCE of p-SiQWs/G solar cells is observed. The p-SiQWs/G solar cell with a narrow diameter indicates great PCE owing to the broad bandgap of SiQWs with a narrow diameter. A p-SiQWs/G solar cell with a wide diameter, on the other hand, exhibits excellent PCE due to the solid light absorption of SiQWs caused by the large surface area. The findings also demonstrate that tighter SiQWs and monolayer graphene packing increase the PCE of p-SiQWs/G solar cells. As means to optimize the device performance, light reflection from the solar cells' surface should be reduced since it is one of the main primary factors restricting their performance.

Moreover, the refractive index value also needs to be considered because it is linked with the reflectance value. For example, Si has a high refractive index that inhibits the efficient electron-hole movement in the device, restricting Si solar cells' performance. However, Si is recognized as a key material utilized for solar cell commercialization because of its compatibility with conventional semiconductor device processes. One way to reduce the reflection caused by the Si is by introducing antireflection (AR) coating. AR coating can prolong the carrier lifespan and enhance the electrophysical characteristics of solar cells, hence improving the device performance. The AR coating is presently being researched primarily through organic materials due to its changeable characteristics. Therefore, it is vital to identify an AR material that is coherent with Si solar cells. Porous silicon (PSi) is one of the numerous potential materials that are appealing for solar cell applications because of its excellent AR effect and other advantages, including surface passivation effects, bandgap widening, and broad optical absorption as well as the transmission range. Shin et al. [19] reported doped silver nanowires (Ag NWs) to the graphene transparent conductive electrode (TCE) to develop PSi Schottky-type heterojunction solar cells. The PSi is created via a metal-assisted chemical etching method and regulating its porosity via changing the Ag nanoparticles etching deposition time. The greater PCE of the Ag NWs-doped graphene/PSi solar cells reported in this study is 4.03% at deposition time $(t_d) = 3$ s/concentration (n_A) of Ag NWs $= 0.1$ wt. percent (wt.%). It is found that the diode ideality factor and light absorption rise as t_d escalates. As n_A rises, V_{oc} and transmittance drop, but absorption of light and sheet resistance increase. Because of this exchange, the PCE is optimized at $t_d = 3$ s/$n_A = 0.1$ wt.%.

However, graphene's electronic and physical characteristics are still factoring which inhibit the pristine G/n-Si solar cells' performance. Hence, many approaches have been explored to encounter the limitation. Doping graphene improves G/n-Si Schottky barrier solar cells (SBSC) performance by minimizing its sheet and series resistance. Also, graphene helps in increasing the built-in potential due to its varied work function. Boron, HNO_3, $SOCl_2$, auric chloride ($AuCl_3$), 1-pyrenecarboxylic acid (PCA), and bis(trifluoromethanesulfonyl) amide are among widely used materials in a variety of chemical and electrochemical doping techniques (TFSA). The investigation of doping treatments must consider the varied reactivity of graphene's most inert basal planes compared to its edge planes. Lancellotti et al. [20] studied the reaction of two additives, including thionyl chloride ($SOCl_2$) and nitric acid (HNO_3) coated on a multilayer G/n-Si SBSC using a double antireflection coating (DARC) technique. Following the DARC layer production, a second acid exposure is required to adjust for a partial de-doping consequence and restore graphene loading. Moreover, the electrical analysis reveals that the DARC/G/n-Si finite solar cell acts differently with the two dopants. Involvement of DARC does not interfere with eventual re-doping with HNO_3, permitting the production of device with PCE of 8.5%. However, different from $SOCl_2$, this efficacy is influenced by the chemical interactions between the coated DARC, graphene, and additive compounds.

7.2.2 Graphene as a Junction Layer

Increasing the photo-generated carrier's lifetime by inhibiting the recombination at the surface of bulk materials is possible for improving Si solar cells' PCE while still maintaining low cost. As a result, research into novel materials that may decrease surface recombination in bulk materials is critical because it could enhance minority carrier lifetime in commercial PV devices. Vaqueiro-Contreras et al. [21] investigated the efficacy of GO for Si solar cells as a surface passivation layer. Here, the surface recombination velocity as low as $14.4\,cm\,s^{-1}$ is obtained. The mechanism for the passivation is completed through the negative fixed charge originating from the oxygen-containing groups, particularly hydroxyl groups surrounding the GO's structure. Thus, prior predictions of dominating chemical passivation are disregarded. Because of the feasibility in which GO may be deposited and removed without causing surface damage, this method is a viable option for short-term surface passivation in bulk lifetime measurements. In addition, using an applicable encapsulation technique or capping layer deposition, the GO can be potentially used due to its low-cost procedure compared to surface passivation materials presently utilized for Si solar cells.

Furthermore, Si solar cells show a less effective PCE in the ultraviolet (UV)-blue region due to their capability to demonstrate substantial penetration depth in the visible and near-infrared irradiation compared to UV-blue region and high optical reflectance. Hence, improving Si solar cells' PCE by strengthening the UV-blue responsivity for employing more high-energy photons excitation is another challenge that needs to be addressed. Sabetghadam et al. [22] synthesized GQDs using the hydrothermal technique and spin-coated GQDs that acted as luminescent down-shifting (LDS) layers on commercial Si solar cells. Correlative improvement is obtained, resulting in J_{sc} of mc-Si and c-Si solar cells up to 5.14% and 6.82%, respectively. Besides, J_{sc} is not affected by the Si solar cells' reflection caused by the GQD-filled LDS layers. Luminescent GQDs are wide-bandgap semiconductors that have gained much interest lately due to their unusual chemical, electrical, and optical characteristics. Meanwhile, the photon down-shifting characteristic of GQDs is an exciting aspect for PV spectrum conversion applications.

On the other hand, growing interest has been developed to replace bulky Si-based p-n and Schottky junction devices with device that have atomically thin architectures. However, the high refractive index of Si yields high reflection on the surface, limiting light absorption in the device. An approach to overcome this restriction is to cover the Si with a new material possessing significant transparency. Rehman et al. [23] exhibit vertical graphene nano hills (VGNH) based solar cells without a catalyst. The electrical characteristics of a PV device based on VGNH grown on top of an interfacial layer Al_2O_3 are compared to bare Si. It is discovered that the interfacial layer's function is to reduce surface recombination and maximize the built-in potential. A simple critical approach is used to build large-area devices to prevent an unstable transfer procedure. With the aim to trade-off within transmittance and conductivity, graphene–Si junction solar cells have excellent transparency to inject more photons while maintaining a low sheet resistance value.

7.3 GRAPHENE IN EMERGING THIN-FILM SOLAR CELLS

Perovskites have recently gained popularity as a PV material due to their longer charges diffusion length than Si, which results in higher PCE and the utilization of less expensive low-temperature manufacturing conditions for a light-harvesting active layer. Alternatively, dye-sensitized solar cells (DSSCs) have demonstrated outstanding potential PV technology offerings for producing high PCE at a low-cost fabrication method. It is important to note that high PCE perovskite solar cells (PSCs) have previously been achieved, and progress is being made to produce large-area cells with consistent PCE, stability, and reproducibility. Although these new technologies are unlikely to replace Si solar cells, researchers are confident that they will discover usage application areas where solar cells must be manufactured on flexible substrates using low-cost processes.

Semi-transparent flexible solar cells, for example, are getting more attention because of their appealing application in building integrated PVs. High-performance DSSCs and PSCs with high transparency throughout the visible spectral region are excellent for these applications and could be produced via simple, low-temperature methods, which standard Si technology does not allow. Highly transparent, conductive, and flexible materials such as graphene hold great promise for tremendously escalating the performance of these devices. Hence, graphene finds functions in DSSCs and PSCs as electrodes, charges extraction layers, and interfacial layers that are important to passivate, encapsulate, and buffer the reaction between each layer. However, the short exciton diffusion length in DSSCs, which has a comparably smaller light absorption range and the environmental volatility of PSCs, is a restraining aspect that must be overcome to improve performance further. The short diffusion length restricts the thickness of the dye absorber layer, limiting effective optical absorption within the material's specified thickness.

Although many dye materials have naturally high absorption coefficients, inadequate photon harvesting occurs because of the restricted absorption proportion of the dye absorber materials. As a result, practical tactics and procedures must use solar energy over its whole spectral range. Besides, PSCs have performed excellently with PCE exceeding 20%, but their lack of stability has proven to be a hurdle to commercialization. In this circumstance, graphene 2D materials have been used as a third component in DSSCs, employing materials with synergistic absorption ranges as the alternative donor or acceptor to generate high-energy and low-energy photons to boost optical absorption in the thin active absorber layers. Besides, graphene materials could also act as a stabilizer material for PSCs to enhance the perovskite material stability. Figure 7.3 shows the historical evolution of DSSCs and PSCs.

7.3.1 Graphene as Electrodes

In DSSCs, counter electrodes (CEs) are made up of two components: the current collector and a catalytic film made of platinum (Pt) particles or nanocarbons. Due to its low sheet resistance (R_{sheet}), high electrical conductivity, and optical transparency of more than 80% in the visible wavelength range, fluorine-doped tin oxide (FTO) is frequently employed as the electron collector. In general, platinized-FTO (FTO/Pt) has accelerated the electrolyte species restoration in the conventional triiodide redox couple. Platinized-FTO (FTO/Pt) and graphene nanoplatelets-coated FTO is introduced to decrease cobalt (Co)-based redox mediators. Notably, polypyridine complexes of Co^{2+}/Co^{3+} have a higher redox potential (>0.4 V vs. NHE) than the triiodide redox couple (0.35 V vs. RHE), prompting PCE above 12% to be reached closer to 14.3%. Nevertheless, similar findings are acquired employing FTO, which are fragile and require high-temperature procedures to fabricate. These undesirable characteristics create significant problems for FTO coating on plastic substrates, which are

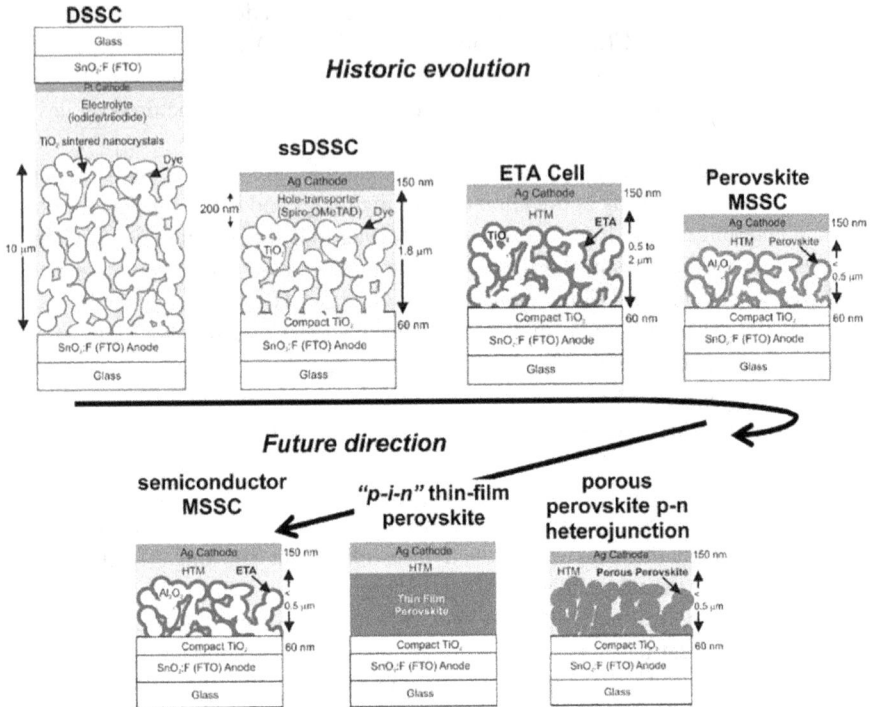

FIGURE 7.3 Evolution of DSSCs and PSCs toward the emerging thin-film solar cells. (Adapted with permission from Ref. [24]. Copyright (2013) American Chemical Society.)

required for flexible solar cells. Besides, whenever exposed to liquid electrolytes, Pt-based CEs deteriorate, reducing the lifespan stability of the cells.

Carbon-based materials have been effectively proven in this context as a viable substitute catalyst replacing Pt in CE. Graphite, graphene flakes, graphene nanoplatelets, GO, activated carbons, carbon nanotubes (CNTs), hard carbon spherules, and hybrid carbon nanocomposites are among the materials in the category. Graphene nanoplatelets coated on FTO, in particular, have been studied as an effective catalyst for DSSC based on the polypyridine redox pair, surpassing FTO/Pt CE and yielding PCE of up to 13%. However, the creation of semi or fully transparent CEs based on nanocarbons wholly behaving as catalysts and current collectors has received less attention to date. Nevertheless, the loss of transparency induced by the "bulky" CNTs fibers restricts the practicality of illuminating the DSSCs from either side in a standard architecture and the manufacturing of DSSCs with on-demand color embedded in buildings. In this specific situation, graphene has seemed to be the right fit to perform numerous roles as it combines high charge carrier mobility, optical transparency, good thermal conductivity, and outstanding mechanical properties. These characteristics apply to single-layer graphene (SLG) flakes, and reproducing them in large-area samples remains challenging.

Current studies have shown that graphene-based electrodes are employed in various devices, including PV, energy storage devices, light-emitting diodes, sensors, touch screens, and transistors. The effectiveness of such graphene-based electrodes, particularly when contrasted with other carbon allotropes, is primarily due to efficient production and processing processes. Moreover, CVD can be used to achieve large-area development of pristine SLG, which could be applied onto arbitrary surfaces using both wet and dry techniques. These transfer methods allow for the fabrication of flexible electrodes with the transmittance of more than 85% and sheet resistance <30 Ω sq^{-1}, comparable to conventional transparent conductive oxide (TCO), e.g., ITO and FTO. Capasso et al. [25] applied the merge of a graphene film generated by CVD (CVD-graphene) film that acts as a charge collector, with few-layer graphene (FLG) flakes produced through a liquid-phase exfoliation that acts as a catalyst for reducing the electrolyte redox couple. Under 1 SUN illumination, these as-prepared CVD-graphene/FLG CEs are evaluated in both triiodide redox electrolyte- and cobalt redox electrolyte-based semi-transparent DSSCs, yielding PCE of 2.1% and 5.09%, respectively. In addition, cobalt redox electrolyte-based DSSCs attain a PCE of up to 6.87% at 0.1 SUN. According to these studies, the optical, electrical, chemical, and catalytic characteristics of graphene-based dual films created through mixing FLG flakes and CVD-graphene are viable substitutes to FTO/Pt CEs.

Besides, RGO is a typical two-dimensional (2D) nanomaterial synthesized from three-dimensional (3D) graphite made up of sp^2 carbon atoms arranged hexagonally to construct a honeycomb lattice structure. The study explored by Tang et al. [9] employed the CE that is made using the three-dimensional graphene networks (3DGNs) synthesis through the CVD technique, and the CE based on graphene material provided the resultant DSSC with good PV performance characteristics. The 3DGNs' high quality and uniform structure offered a route for efficient electron transport. Meanwhile, the RGO allowed close contact at the interface within the redox electrolyte and graphene basal plane. The PCE achieved is directly connected to the utilized RGO's degree of reduction and mass fraction. The subsequent optimization gave a PCE of 9.79% for the DSSCs based on the 3DGN–RGO CE, equivalent to a Pt CE-based device, suggesting excellent candidates for the as-prepared CE. RGO has exhibited promise in catalysis, energy storage, and conversion applications because of its high conductivity, enhanced hydrophilic characteristics, feasible processing, and surface functionalization.

Significantly, many studies have concentrated on RGO-based materials as extremely effective triiodide reduction catalysts, intending to replace the noble metal Pt CE in DSSCs. RGO, on the other hand, has a restricted active site, often linked with the structural distorted and oxygen-containing functional groups for triiodide redox electrocatalysis. Besides, when extracted from suspensions, RGO films tend to re-aggregate or re-stack quickly because of the p–p stacking. Eventually, the specific area and electrocatalytic active sites will reduce causing inadequate catalytic performance on triiodide redox reduction. Wang et al. [26] discovered that sulfur-doped RGO/MoS$_2$ (S-rGO/MoS$_2$) composites are effectively produced using a simple one-step annealing approach using a CS$_2$ sulfur source, as shown in Figure 7.4. Electrochemical tests suggested that the as-prepared composite outperformed the Pt electrode in catalytic activity during triiodide reduction. Under optimal circumstances, the N719-based DSSC with an S-rGO/MoS$_2$ CE obtained a PCE of 6.96%, exceptionally near the device with a thermally coated Pt CE (7.35%). The excellent electrocatalytic activity of the composite is ascribed to the high conductivity derived

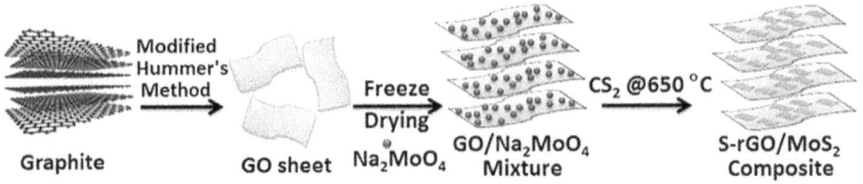

FIGURE 7.4 Schematic procedure to synthesize S-rGO/MoS$_2$ composites. (Adapted with permission from Ref. [26]. Copyright (2018) Elsevier.)

from the 2D conductive property of S-rGO and MoS$_2$, as well as the exposure of many edges giving many active sites. Therefore, this study proved that employing S-rGO/MoS$_2$ composite as a viable substitute for the noble metal Pt in the process of triiodide reduction is a straightforward way to use 2D nanomaterials as multifunctional materials PV technology.

7.3.2 GRAPHENE AS AN ADDITIVE AND INTERFACIAL LAYERS

In composition or interfacial engineering, especially in PSCs, the interaction of molecular additives and dopants with the perovskite materials is a practical approach for producing high-quality perovskite films while minimizing grain boundaries. Besides, their commercialization is hampered by substantial deterioration under outdoor circumstances. Yi et al. [27] studied the encapsulation method to counter the poor stability of PSCs. The structural and optical characteristics of encapsulated perovskite films with a varying number of layers are carefully under well-controlled moisture conditions via optical spectroscopy and grazing-incidence wide-angle X-ray scattering investigations. The graphene device preserved 95% from the original PCE after 100 hours of exposure to 85%–90% humidity and 82.4% after 3,700 hours in ambient condition, as illustrated in Figure 7.5. Consequently, the graphene-encapsulated perovskite films surpassed a novel film concerning stability.

Meanwhile, Li et al. [28] introduced GQD as an additive to poly(3,4-ethylenedioxythiophene) polystyrene sulfonate (PEDOT:PSS) that is employed as a hole transporting layer (HTL) for inverted PSCs device architecture. Based on its excellent transparency, low-temperature procedure, and adequate conductivity, PEDOT:PSS is the frequent hole carrying layer utilized in inverted PSCs. Nevertheless, PSCs' highest PCE fabricated on PEDOT:PSS is revealed to be around 19%, which is lesser than assorted excellent-performance PSCs. The pristine PEDOT:PSS work function is significantly lesser than the valence band of the perovskite active material. The misalignment in energy level causes energy losses throughout the charges regeneration, and a low V_{oc} of <1.0 V is commonly noticed. Simultaneously, PEDOT:PSS conductivity is lesser than the commonly utilized electron transport layer (ETL), [6,6]-phenyl-C$_{61}$-butyric acid methyl ester (PC$_{61}$BM). Inconsistent hole injection in the HTL and the ETL could result in charge carrier concentration, generating a high leakage current and a poor FF. By employing the PEDOT: PSS/GQDs composite, the current density–voltage (J–V) curves of the PEDOT:PSS/GQDs composite indicate that GQDs have integrated into the PEDOT:PSS films and the conductivity of

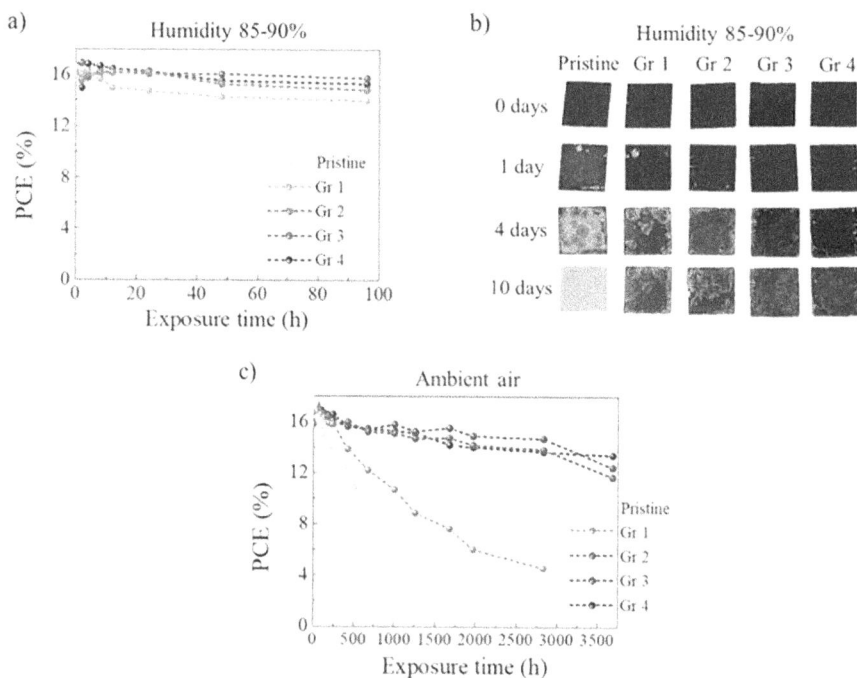

FIGURE 7.5 (a) Aging test under humidity 85%–90%, (b) photograph of all samples from day 0–10, and (c) aging test ambient air. (Adapted with permission from Ref. [27]. Copyright (2020) Elsevier.)

the films rises. The improved conductivity of the PEDOT:PSS/GQDs film results in improved hole–electron regeneration in the device.

On the other hand, Vasanth et al. [29] employed electrophoretic deposition (EPD) to deposit GO on a titanium dioxide (TiO_2) nanoparticle layer for DSSCs application. Due to the enormous surface area, nanocrystalline metal oxides are greatly favored as electron acceptors in DSSCs, allowing for effective dye adsorption. Despite the surface area of nanocrystalline metal oxides being thought to be an essential element in achieving larger photocurrents in DSSCs, it has been shown that defects have a significant impact on charge transport characteristics. Structural disorders in materials cause several defects to be dispersed in their bulk and on their surface. This condition harms the performance of DSSCs; hence, nanostructured functional materials have been incorporated into defect passivation techniques. In this study, the performance inhibits the recombination activity as the TiO_2/dye/electrolyte interfaces of DSSCs are terminated when loaded with a GO passivation layer on the TiO_2 electrode. It is found that the DSSCs that used 5 minutes of EPD loaded of GO on the TiO_2 nanoparticle layer indicated a 5% increase in PCE and a 5% rise in short circuit current density. Effective inhibition of electron–hole recombination activity at the TiO_2/dye/electrolyte interfaces is ascribed to GO coating on TiO_2. The increased thickness of the GO on the TiO_2 nanoparticle layer hampered charge transfer and negatively impacted the performance of different DSSCs. Thus, it is proposed that escalating the thickness of the GO layer reduced the likelihood of photo-excitation electron injection from dye absorption layer to TiO_2. The results show that GO can be a viable alternative for nanostructured solar cells as the surface passivation.

In addition, multi-element co-doping has an impact and could improve semiconductor photocatalytic activity compared to single-element doping. According to atomic orbital theory, nitrogen integration creates a doping level in TiO_2's first energy level N 1s and O 2p orbital, trapping electrons and holes after photoexcitation in the conduction band and the valence band, respectively, thus, inhibiting the electron–hole pairs recombination process. Gao et al. [30] applied electrospinning to synthesize graphene-containing TiO_2 nanofibers, and the nitrogen element is integrated using the hydrothermal technique. Nitrogen doping greatly enhances the specific surface area of the fiber and its capacity for dye absorption. According to X-ray photoelectron spectroscopy, nitrogen is effectively integrated into the fiber, whereas graphene and TiO_2 stay unaltered. Compared to the nitrogen-free solar cell, the J–V curve shows that V_{oc} is 0.71 V, FF is 0.46, and PCE is enhanced by 26% after nitrogen doping.

7.4   GRAPHENE IN FLEXIBLE PEROVSKITE SOLAR CELLS

Flexible and lightweight PSCs have lately received much interest because of the wide range of possible applications, particularly wearable electronics. Under sintering temperatures over 450°C, the high PCE of PSCs is often dependent on mesoporous TiO_2 as an ETL. For large-scale roll-to-roll production, basic planar structure via low-temperature technique is highly favored. However, low-temperature annealing results in poor electron mobility of ETL. Graphene is integrated with metal oxides as hybrid ETLs to improve electron generation to achieve higher performance. For instance, Tin (IV) oxide (SnO_2)/graphene nanosheets (GNs) would boost the transportation of the free electron and flexibility as well. Furthermore, the C_{60}-self-assembled monolayer (C_{60}-SAM) may expedite electron transport from the perovskite active material to metal oxide ETLs and passivate the trap states of the perovskite active material, significantly reducing or even abolishing J–V hysteresis. Liu et al. [31] introduced the GNs that are weakly oxidized and integrated into SnO_2 to improve electron transportation and flexibility of the ETL. The C_{60}-SAM placed within the ETL and the perovskite active material could prevent the electron–hole recombination process. A reverse voltage scan yielded a PCE of 13.36%, a V_{oc} of 1.10 V, a J_{sc} of 18.39 mA/cm², and an FF of 0.66, whereas a forward scan voltage yielded a PCE of 12.81%, a V_{oc} of 1.10 V, a J_{sc} of 18.19 mA/cm², and an FF of 0.64, suggesting negligible hysteresis. The PSCs produced through a low-temperature procedure are suitable for the roll-to-roll manufacturing process.

Besides, organic materials such as phenyl C_{61} butyric acid methyl ester (PCBM), fullerene, and amine-based fullerene [6,6]-phenyl-C_{61}-butyric acid 2-((2-(dimethylamino)-ethyl) (methyl)amino)ethyl ester (PCBDAN) are captivating as ETLs in PSCs due to the fabrication based on the low-temperature and straightforward method. The stability of PSCs in the ultraviolet (UV) region is recently proven by dopants addition to the PCBM ETL. However, the PCBM's intrinsic poor electrical conductivity and electron mobility still hampered the device performance. GQDs, in particular, are efficiently integrated into PCBM to improve the PV parameters and endurance of PSCs on rigid substrates. Shin et al. [32] employed PCBM doped with GQDs that act as an ETL for flexible PSCs. When the concentration of GQDs is increased to 2.5 mg L^{-1}, the PCE of the PSCs with graphene (GR)/(3-Aminopropyl)triethoxysilane (APTES) TCE rises to 16.4% and 15.0% on rigid and flexible substrates, respectively, owing to the reducing charge recombination at the ETL/perovskite interface and enhanced ETL conductivity. Because of the increased flexibility provided by the APTES interlayer, the flexible PSCs demonstrate good bending stability, successfully retaining 80% of the initial PCE, although after 3,000 bending cycles at a curvature radius of 4 mm.

In addition, because traditional TCE based on conductive oxides is stiff and rigid, highly flexible devices cannot be manufactured. An SLG sheet has a transmittance of 97.7% in the visible range which is more significant compared to a standard ITO. Graphene also has excellent mechanical flexibility and bending endurance, making it perfect for flexible TCE in flexible PSCs. Flexible organic solar cells without any TCO, in particular, demonstrated significant PCE when utilizing graphene-based TCEs. Numerous researchers have created incredibly effective p-i-n-type TCO-free PSCs with no hysteresis by utilizing integrated SLG as an anode TCE rather than ITO. Nevertheless, it is acknowledged that the impact of loading on graphene has become less beneficial with time, yielding a loss in device performance owing to a rise in the sheet resistance of graphene TCEs. Kim et al. [33] disclosed ultra-flexible p-i-n-type MAPbI$_3$ PSCs with numerous graphene layers as anode TCEs that are efficient and dependable. The solar cells are demonstrated to be greater at layer number ($L_n = 2$), where the PCE displays 13.35% and 13.94% for forward and reverse scans, respectively, with practically no hysteresis in the current density–voltage curves, by varying the layer number (L_n) of graphene TCEs. It is reported that the PCE is retained at 90% from the initial value after 1,000 bending cycles, demonstrating excellent curving stability against bending deformation, even at a bending radius of 2 mm, as depicted in Figure 7.6.

Nevertheless, large-scale graphene production necessitates processing temperatures of 1,000°C or higher, necessitating a transfer procedure to polymer substrates.

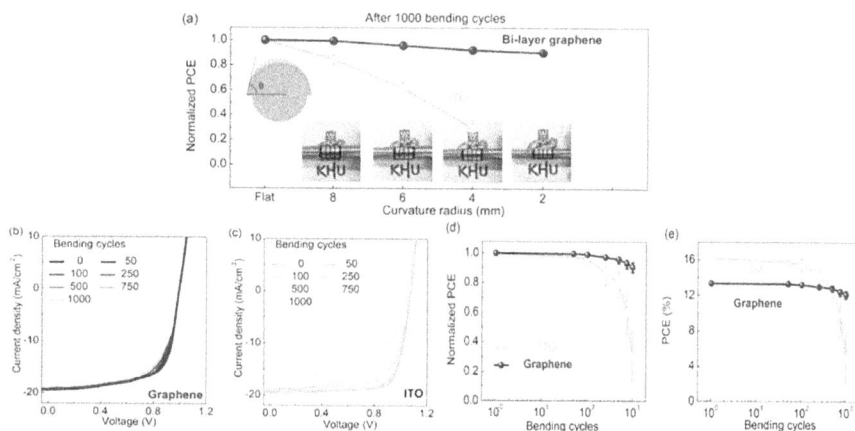

FIGURE 7.6 (a) Normalized PCEs for the PSCs with pristine ITO and graphene ($L_n = 2$), (b and c) J–V curves of graphene ($L_n = 2$) and ITO, respectively, (d) normalized value of PCEs, and (e) absolute value of PCEs for both graphene ($L_n = 2$) and ITO. (Adapted with permission from Ref. [33]. Copyright (2018) Elsevier.)

When transferring a free-standing graphene sheet to a flexible substrate at such high temperatures, structural inhomogeneities within the graphene film, such as wrinkles and ripples, impair active electrodes' physical and electrical characteristics. Transfer-graphene may fulfill the flexibility criteria, but its utility for high-performance applications is restricted owing to high sheet resistance levels caused by deficiencies in wrinkles, crystals, and grain and domain sizes. Tran et al. [34] studied a transfer-free large-scale monolayer graphene bottom electrode that is transparent and flexible. Through plasma-assisted thermal chemical vapor deposition (PATCVD), high-quality graphene is directly generated on a polymer substrate at 150°C. Such a transparent AZO/Ag/AZO (AAA) multilayer is also employed as a top CE to develop semi-transparent PSCs with excellent mechanical flexibility. The 300-nm-thick PSCs employing PATCVD-graphene exhibited a high transmittance of 26% at a wavelength of 700 nm. The graphene-based PSCs retain a PCE level that is more than 90% higher than the value after 1,000 bending cycles at a tensile strain of 1.5%. This increased bending resilience highlights the prospect of forming a hybrid structure for a flexible solar cell utilizing a graphene-based perovskite active material without a transfer method.

However, graphene TCE seems far from workable requirements despite its outstanding qualities due to its poor conductivity. The n- and p-type graphene TCEs with low resistance, long-term stability, as well as excellent transparency are effectively synthesized utilizing bis (trifluoromethanesulfonyl)-amide (TFSA) and triethylenetetramine (TETA), respectively. Moreover, semi-transparent organic solar cells (OSCs) using TFSA/TETA-doped graphene TCEs are successfully fabricated by Jang et al. [35]. They reported flexible PSCs with anode and cathode electrodes prepared by integration of graphene (GR) with TFSA and TETA, respectively. The polyethene terephthalate (PET)/TETA-doped GR (TETA-GR)/ZnO/MAPbI$_3$/poly-triarylamine and the PEDOT:PSS/TFSA-doped GR (TFSA-GR)/PET layers are laminated to form the flexible PSCs. The PCE shows an increment pattern after the addition of Ag reflectors to the device. Moreover, PCE retains 70% or more of its original value, although after 1,000 bending cycles at an 8 mm bending radius, indicating that the flexible PSCs have viable applications as next-generation semi-transparent PV devices.

7.5 CONCLUSION

The multi-functional 2D graphene and its derivatives are widely explored by utilizing a TCE, Schottky junction, catalytic CE, passivation on the active layer material, and additives in the charges transport layer for different PV devices. In Si solar cells, graphene is utilized as a TCE given its outstanding characteristics are low sheet resistance and high transmittance. Besides, graphene and its derivatives' role in DSSCs is commonly found as catalyst CE and charges transport layer to improve the device performance. Meanwhile, the integration of graphene and its derivatives in PSCs generally improves the stability of the ionic reaction between perovskite material and the charges transport layer. Overall, graphene acts as an interfacial layer and is functional as passivation, encapsulation, and additive to yield excellent PCE of PV devices. Moreover, the graphene role in flexible and durable PV devices could help produce high PCE on a large scale, especially for flexible commercial applications.

REFERENCES

1. S. Zhang, H. Wang, J. Liu, and. Bao. Measuring the specific surface area of monolayer graphene oxide in water. 2020. *Materials Letters*, 261, 127098.
2. A. A. Balandin. Raman-based technique for measuring thermal conductivity of graphene and related materials. 2018. *Journal of Raman Spectroscopy*, 49(1), 106–120.
3. T. Zhong, J. Li, and K. Zhang. A molecular dynamics study of Young's modulus of multilayer graphene. 2019. *Journal of Applied Physics*, 125(17), 175110.
4. M. Sang, J. Shin, K. Kim, and K. J. Yu. Electronic and thermal properties of graphene and recent advances in graphene based electronics applications. 2019. *Nanomaterials*, 9, 374.
5. M. Romagnoli, V. Sorianello, M. Midrio, F. H. L. Koppens, C. Huyghebaert, D. Neumaier, P. Galli, W. Templ, A. D'Errico, and A. C. Ferrari. Graphene-based integrated photonics for next-generation datacom and telecom. 2018. *Nature Reviews Materials*, 3(10), 392–414.
6. J. Wang, J. Song, X. Mu, and M. Sun. Optoelectronic and photoelectric properties and applications of graphene-based nanostructures. 2020. *Materials Today Physics*, 13, 100196.
7. I. Choudhuri, P. Bhauriyal, and B. Pathak. Recent advances in graphene-like 2D materials for spintronics applications. 2019. *Chemistry of Materials*, 31(20), 8260–8285.
8. R. N. A. R. Seman, M. A. Azam, and M. H. Anib. Graphene/transition metal dichalcogenides hybrid supercapacitor electrode: Status, challenges, and perspectives. 2018. *Nanotechnology*, 29(50), 502001.
9. B. Tang, H. Yu, W. Huang, Y. Sun, X. Li, S. Li, and T. Ma. Three-dimensional graphene networks and RGO-based counter electrode for DSSCs. *RSC Advances*, 9(28), 15678–15685.
10. S. Choi and C. Kim. Reduced graphene oxide: Based materials for electrochemical energy conversion reactions. 2021. *Carbon Energy*, 1, 85–108.
11. A. N. Banerjee. Graphene and its derivatives as biomedical materials : Future prospects and challenges. 2018. *Interface Focus*, 8, 20170056.
12. R. Kumar, S. Sahoo, E. Joanni, R. K. Singh, W. K. Tan, K. K. Kar, and A. Matsuda. Recent progress in the synthesis of graphene and derived materials for next generation electrodes of high performance lithium ion batteries. 2019. *Progress in Energy and Combustion Science*, 75, 100786.
13. X. Dai, P. Koshy, C. C. Sorrell, J. Lim, and J. S. Yun. Focussed review of utilization of graphene-based materials in electron transport layer in halide. 2020. *Energies*, 13(23), 6335.
14. D. Shin, J. Rok, W. Shin, C. Lee, and G. Kang. Layup-only modulization for low-stress fabrication of a silicon solar module with 100 µm thin silicon solar cells. 2021. *Solar Energy Materials and Solar Cells*, 221, 110903.
15. M. A. Azam, N. E. S. A. A. Mudtalib, and R. N. A. R. Seman. Synthesis of graphene nanoplatelets from palm-based waste chicken frying oil carbon feedstock by using catalytic chemical vapour deposition. 2018. *Materials Today Communications*, 15, 81–87.
16. S. M. Song, J. K. Park, O. J. Sul, and B. J. Cho. Determination of work function of graphene under a metal electrode and its role in contact resistance. 2012. *Nano Letters*, 12, 3887–3892.
17. S. Singh, S. Powar, P. Mahala, S. Majee, T. Eshwar, A. Kumar, and J. Akhtar. Fabrication and characterization of graphene based silicon Schottky solar cell. 2018. *Superlattices and Microstructures*, 120, 637–641.
18. Z. Arefinia and A. Asgari. Scaling issues of Schottky junction solar cells based on graphene and silicon quantum wires in the sub-10-nm regime. 2018. *Optik*, 153, 65–72.
19. D. H. Shin, J. H. Kim, J. H. Kim, C. W. Jang, S. W. Seo, H. S. Lee, S. Kim, and S. H. Choi. Graphene/porous silicon Schottky-junction solar cells. 2017. *Journal of Alloys and Compound*, 715, 291–296.

20. L. Lancellotti, E. Bobeico, A. Castaldo, P. Delli Veneri, E. Lago, and N. Lisi. Effects of different graphene dopants on double antireflection coatings/graphene/n-silicon hetero-junction solar cells. 2018. *Thin Solid Films*, 646, 21–27.
21. M. Vaqueiro-Contreras, C. Bartlam, R. S. Bonilla, V. P. Markevich, M. P. Halsall, A. Vijayaraghavan, and A. R. Peaker. Graphene oxide films for field effect surface passivation of silicon for solar cells. *Solar Energy Materials and Solar Cells*, 187, 189–193.
22. S. A. Sabetghadam, Z. Hosseini, S. Zarei, and T. Ghanbari. Improvement of the current generation in silicon solar cells by utilizing graphene quantum dot as spectral converter. 2020. *Materials Letters*, 279, 128515.
23. M. A. Rehman, S. B. Roy, D. Gwak, I. Akhtar, N. Nasir, S. Kumar, M. F. Khan, K. Heo, S. H. Chun, and Y. Seo. Solar cell based on vertical graphene nano hills directly grown on silicon. 2020. *Carbon*, 164, 235–243.
24. H. J. Snaith. Perovskites : The emergence of a new era for low-cost, high-efficiency solar cells. 2013. *The Journal of Physical Chemistry Letters*, 4, 3623–3630.
25. A. Capasso, S. Bellani, A. L. Palma, L. Najafi, A. E. D. R. Castillo, N. Curreli, L. Cina, V. Miseikis, C. Coletti, G. Calogero, V. Pellegrini, A. D. Carlo, and F. Bonaccorso. CVD-graphene/graphene flakes dual-films as advanced DSSC counter electrodes. 2019. *2D Materials*, 6, 035007.
26. Y. Wang, Y. Guo, W. Chen, Q. Luo, W. Lu, P. Xu, D. Chen, X. Yin, and M. He. Sulfur-doped reduced graphene oxide/MoS$_2$ composite with exposed active sites as efficient Pt-free counter electrode for dye-sensitized solar cell. 2018. *Applied Surface Science*, 452, 232–238.
27. A. Yi, S. Chae, S. Won, H. J. Jung, I. H. Cho, J. H. Kim, and H. J. Kim. Roll-transferred graphene encapsulant for robust perovskite solar cells. 2020. *Nano Energy*, 77, 105182.
28. W. Li, N. Cheng, Y. Cao, Z. Zhao, Z. Xiao, W. Zi, and Z. Sun. Boost the performance of inverted perovskite solar cells with PEDOT: PSS/graphene quantum dots composite hole transporting layer. 2019. *Organic Electronics*, 78, 105575.
29. A. Vasanth, N. S. Powar, D. Krishnan, S. V. Nair, and M. Shanmugam. Electrophoretic graphene oxide surface passivation on titanium dioxide for dye sensitized solar cell application. 2020. *Journal of Science Advanced Materials and Devices*, 5(3), 316–321.
30. N. Gao, T. Wan, Z. Xu, L. Ma, S. Ramakrishna, and Y. Liu. Nitrogen doped TiO$_2$/graphene nanofibers as DSSCs photoanode. 2020. *Materials Chemistry and Physics*, 255, 123542.
31. X. Liu, X. Yang, X. Liu, Y. Zhao, J. Chen, and Y. Gu. High efficiency flexible perovskite solar cells using SnO$_2$/graphene electron selective layer and silver nanowires electrode. 2018. *Applied Physics Letters*, 113(20), 1–6.
32. D. H. Shin, J. M. Kim, S. H. Shin, and S. H. Choi. Highly-flexible graphene transparent conductive electrode/perovskite solar cells with graphene quantum dots-doped PCBM electron transport layer. 2019. *Dye and Pigments*, 170, 107630.
33. S. Kim, H. S. Lee, J. M. Kim, S. W. Seo, J. H. Kim, C. W. Jang, and S. H. Choi. Effect of layer number on flexible perovskite solar cells employing multiple layers of graphene as transparent conductive electrodes. 2018. *Journal of Alloys and Compounds*, 744(12), 404–411.
34. V. D. Tran, S. V. N. Pammi, B. J. Park, Y. Han, C. Jeon, and S. G. Yoon. Transfer-free graphene electrodes for super-flexible and semi-transparent perovskite solar cells fabricated under ambient air. 2019. *Nano Energy*, 65, 104018.
35. C. W. Jang, J. M. Kim, and S. H. Choi. Lamination-produced semi-transparent/flexible perovskite solar cells with doped-graphene anode and cathode. 2019. *Journal of Alloys and Compound*, 775, 905–911.

8 Chalcogenide-Based 2D Nanomaterials for Solar Cells

Shiv Kumar Pal and Neeraj Mehta
Banaras Hindu University

CONTENTS

8.1 INTRODUCTION

The most plentiful supplier of renewable energy for our planet is the sun that meets up with the current and forthcoming worldwide energy mandates. The best possible and existing solution for this problem is the solar cells that provide the direct conversion of sunlight into operational energy deprived of creating any hurtful derivatives. One of the possible entrants for the synthesis of competent solar cells is chalcogenide-based semiconducting thin films (TFs) since they can be employed directly for electricity production. Solar cells prepared by using 2D chalcogenides may provide the better solution for the most capable renewable energy for resolving the problem of energy shortage over the globe. They can satisfy the growing energy demand by collecting sufficient electrical power from the continuous source of energy (i.e., sun) due to the good photosensitivity of chalcogenides [1–3].

In the last few decades, the huge manufacturing prices of the silicon-based photovoltaic industry have forced a barricade for the replacement of natural oil products like gasoline, kerosene, and petrol. Consequently, it is a crucial necessity to grow a cost-effective photovoltaic approach to confirm that the solar-energy-based industry has become cost-competitive. The 2D chalcogenides in TF form have been developed in the last decade for the fabrication of photovoltaic devices [4–9]. Some examples are binary transition metal chalcogenides (FeS_2, CdS, $CuSe$), ternary chalcogenides ($CuInS_2$, $AgInSe_2$, $CuGaS_2$), and quaternary chalcogenides [$Cu(In, Ga)Se_2$, $ZnSn(S, Se)_4$].

DOI: 10.1201/9781003178422-8

These 2D chalcogenides have fascinated much consideration because of their capable photoelectrical features as well as a low financial burden during their fabrication. The 2D chalcogenides also prove a capable device proficiency because they are less affected by the bandgap tailoring, defects in the absorber layer, and modification in the interfacial layers corresponding to the transportation of carriers. Furthermore, the TFs of binary/ternary layered chalcogenides such as antimony sulfide (Sb_2S_3), iron sulfide (FeS_2), and copper tin sulfide (Cu_2SnS_3) are discovered as absorber coatings for utilization in solar cells. Correspondingly, the chalcogenide substances to manufacture the solar cell materials comprising non-hazardous elements are also equally significant for the confirmation of the absence of any harmful effects on human beings. Consequently, it is important to acquire knowledge about the physical growth and quality of coating of solar cell materials and their optical, electrical, and photoelectrical properties (e.g., electroluminescence, photosensitivity, photoconductivity, luminescence, and absorption coefficient). The knowledge of such substance properties helps us to optimize efficiency and robustness of solar cells. In addition, the studies of the optoelectronic characteristics of potential 2D chalcogenides-based substances (e.g., trapping centers, recombination probability, the number density of defects, and charge carrier mobilities) cover the method for enhancements in the performance of 2D chalcogenides-based solar cells. Besides, the layers employed for the movements of charge carriers play a significant role for optimum alignment of band and well-organized transportation of the carriers. Examples of such layers are transparent conductive oxide (TCO) layers, buffer layer, and window layer.

One of the examples of 2D nanomaterials that is recently recognized as potential candidates for solar cells fabrications is the family of TMDCs. We denote TMDCs by the chemical formula MX_2. In this formula, the letter "M" is a transition metal, while the letter "X" denotes a chalcogen element. Two major examples of transition metal are tungsten (W) and molybdenum (Mo). TMDCs are the substance materials with robust co-planar covalent bonds and feeble interlayer bonds of van der Waals types. Usually, they consist of three thick atoms in each layer having a metallic layer packed between two chalcogenide layers.

There are numerous crystal assemblies that might be adopted by TDMCs. The 2H-phase having trigonal symmetry is widely observed in MoS_2, WS_2, and $MoSe_2$ and it contributes to semiconducting features of TMDCs. When we utilized TMDCs in bulk form, they possess bandgaps of indirect-type. As soon as we obtained them in the form of monolayers, the nature of the bandgap is converted to direct-type. Consequently, they become striking materials for making the optoelectronic devices that can operate in the visible region of electromagnetic spectrum.

When we want to synthesize chalcogenide-based 2D nanomaterials for utilization in solar cells, the preparation techniques are designed and other elements are selected by keeping in mind the earth-abundance of absorber materials so that the fabrication cost can be controlled. The earth-abundance of some widely used absorber elements along with chalcogens is shown in Figure 8.1.

From the aforesaid discussion, we see that one of the foremost compounds in TF automation is chalcogenides. They are potential substances for using in immensely competent photovoltaic devices as absorbers. This chapter emphasizes the perceptions, designing, and the foremost developments on the photovoltaic characteristics

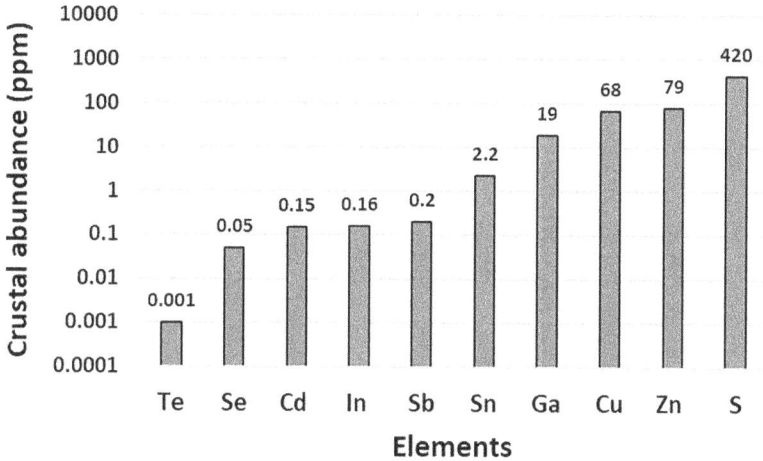

FIGURE 8.1 Earth-abundance of absorber materials along with chalcogen elements.

of two kinds of solar cells: (i) that are directly made of chalcogenides and (ii) those in which chalcogenides played the dominating role. The detailed discussion about past/present perspectives and future scope of solar cells based on binary, ternary, and quaternary chalcogenides is also done. Besides, a transitory appraisal of emergent technologies based on chalcogenides is also reported and the overview of novel approaches for utilization of 2D chalcogenides in photovoltaics is also embedded.

8.2 SYNTHESIS OF CHALCOGENIDE-BASED 2D NANOMATERIALS

Keeping in mind the aim of successfully learning the growing features of the chalcogenides-based 2D materials, the unswerving manufacturing of their TF is necessary. The separation of monolayer flakes was initially achieved through mechanical or solution segmentation [10,11]. But these approaches have serious drawbacks like low manufacturing and poor quality. To overcome these shortcomings, such approaches have been industrialized that can be used for the direct growth of monolayers. The vapor phase methods are the most fruitful approaches for realizing the finest sense of balance between eminence, handling, comfort, and extensibility.

The preparation of TFs or monolayers of transition metal dichalcogenides is extensively attained by employing chemical vapor deposition and oxide precursor. In such methods, the powdered transition metal oxide (TMO) precursors play the role of a source of transition metal source. The powders of the oxide precursor and chalcogen element are located in high- and low-temperature regions inside a furnace of cylindrical geometry. After appropriate heating, both powders are vaporized and they react with the desired sample that is obtained as a coating on a substrate due to the reaction between their vapor phases (Figure 8.2).

Another suitable approach is the physical vapor deposition that is used for the preparation of chalcogenide-based 2D nanomaterials when the physical vapor deposition approach is not presently manageable. In this approach, the chalcogenide compound itself plays the role of precursor. It is heated in powdered form so that an inert

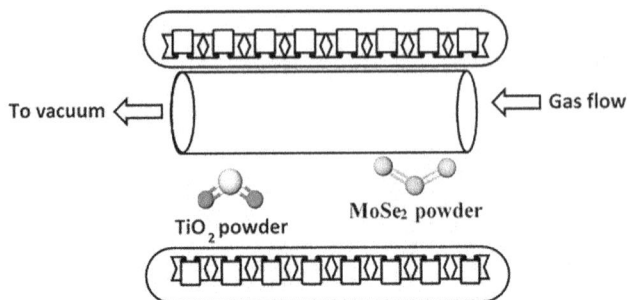

FIGURE 8.2 Chemical vapor transport approach for the preparation of thin films or mono-layers of transition metal dichalcogenides.

gas may carry it after sublimation under a high vacuum so that its coating can be achieved on a receiving substrate through downstream processing.

Generally, a p-n junction is designed in chalcogenide-based TF solar cells by employing an absorber and a window of p-type and n-type, respectively. The n-type window comprises a translucent conductive oxide coating and one or more safeguard coatings. Some widely accepted synthesis techniques are used to prepare chalcogenide-based 2D nanomaterials in TF form. Such TFs are used directly or indirectly in cost-effective photovoltaic devices. Generally, a wide diversity of TF coating approaches has been used to synthesize chalcogenide films. The performance of solar cells differs significantly subject to the procedures used to produce the 2D chalcogenides, e.g., solar cell of copper indium gallium selenide (CIGS). A characteristic TF structure of 2D chalcogenides employed with slight alteration to manufacture the solar cells is shown in Figure 8.3 in which a CIGS absorber has been used.

Figure 8.3 illustrates the classical single-step co-evaporation process (see the direct process in leftward). The rightward indirect process is based on the double-step plan that begins with a precursor synthesis by electrodeposition, sputtering or screen printing, or stacked layers.

8.3   APPLICATIONS OF CHALCOGENIDE-BASED 2D NANOMATERIALS IN SOLAR CELLS

The industrial TF devices of chalcogenide substances (e.g., CdTe, CIGS, and CZTS) having an absorption coefficient larger than 10^4 cm^{-1} and a direct optical gap of about 1.5 eV are possible absorber substances that may be employed for the production of chalcogenide-based solar cells [12].

Now we will discuss the fabrication details of an industrial chalcogenide solar cell, i.e., CIGS that is one of the widely used absorbers. The CIGS cells offer high values of absorption coefficient and thermal durability. These cells are less expansive too.

Figure 8.4 illustrates the outlines for the arrangement of deposition of CIGS by employing the co-evaporation technique. The first outline is the static arrangement that is used to control the fluidities over time (Figure 8.4a). The inline arrangement is the second approach that arranges the sources for delivering the motion of precisely

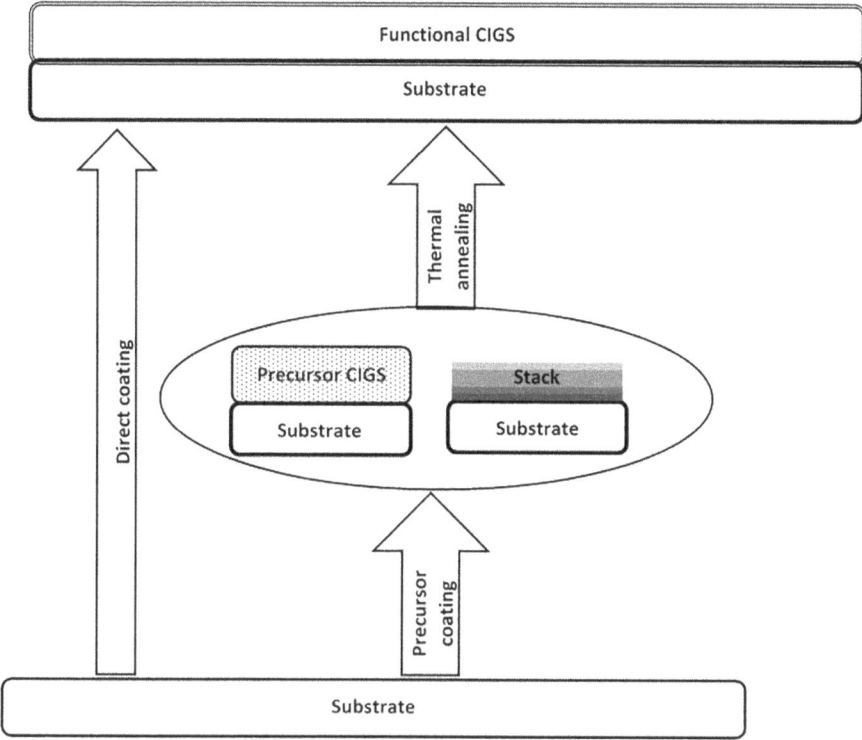

FIGURE 8.3 Illustration of the approaches used in the manufacturing of solar cells of typical chalcogenide-based thin-film.

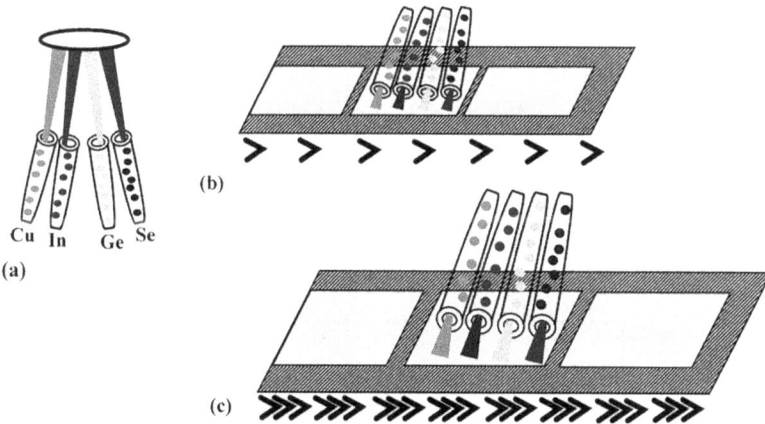

FIGURE 8.4 Illustration of different pre-arrangement for the deposition of CIGS (a) static arrangement, (b) arrangement of the sources for delivering the motion of precisely constant fluidities on the surface of substrate, (c) industrial arrangement.

constant fluidities on the surface of substrate (Figure 8.4). The coating on substrates in large scale and comparatively faster rate is achieved in the industrial arrangement (Figure 8.4c).

There are multiple layers in a typical CIGS solar cell as shown in Figure 8.5. The CIGS layer (i.e., absorber) is packed between the following layers:

 (i) A layer playing the role of the substrate. It is generally made by using a glassy substance or a polymer material for guaranteeing elasticity.
 (ii) The next layer is made of Mo metal that is coated on the substrate by employing sputtering. This layer is also known as the back contact and it helps the solar cell in reflecting the unabsorbed sunlight toward the backward direction so that the CIGS layer may absorb the maximum possible intensity of light. The maximum efficiency is achieved by using a triple-stage co-evaporation.
(iii) On the top of the CIGS layer, the coating of a buffer layer of cadmium sulfide (CdS) is done that plays the role of a window that helps in the transition of the electrical signals from the absorber layer to the top translucent contact of aluminum-doped zinc-oxide contact. The buffer is made of a material that provides a bandgap layer between the absorber and the electrical transmitter.

Thus, this solar cell has a heterojunction consisting of an extremely thin layer of n-type cadmium sulfide and a p-type layer with highly resistive intrinsic zinc oxide and back contact layer of molybdenum, respectively, on other faces.

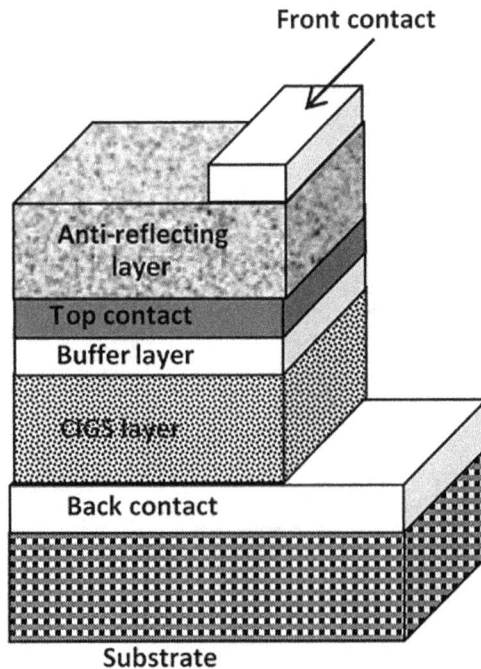

FIGURE 8.5 Design of various multiple layers in a representative CIGS solar cell.

For the desirable ultimate quality of the solar cells, the selection of a suitable substrate is a subject of crucial importance. The substance chosen as a substrate for solar cells made of CIGS absorber has to satisfy numerous necessities as listed below [13]:

(i) It must show good thermal stability up to 823 K.
(ii) There must be no chance of any chemical reaction of substance either with absorber layer or Mo.
(iii) It must possess resistance against harmful environmental situations (e.g., moisture) to preserve the durability and performance of the device.
(iv) Its surface must be smooth enough and the mechanical strength of the surface must be large.
(v) It must possess the coefficient of thermal expansion whose order is comparable to that of absorber layer or Mo.

The usage of TF chalcogenide substances for solar cell applications has increased recently owing to their interesting properties [14]. When we think about window layering in TF-based solar cells, we found that binary chalcogenide compound like CdS is frequently in demand as potential substances for this purpose. However, numerous deposition procedures are accessible for the synthesis of chalcogenide-based 2D nanomaterials, but electrochemical deposition is a favorite method due to its specific benefits (e.g., modest operating environment and effortless scalability). In this method, chalcogenide-based 2D nanomaterials can be deposited having either n-type or p-type conduction mechanisms by appropriately regulating the controlling parameters of deposition. Execution of current methods for the synthesis of solar cells of chalcogenide-based 2D nanomaterials has removed the necessity of annealing at high temperature after the coating steps. Correspondingly, in the case of aqueous electrolytes, the challenge of the fine working potential range has been cracked with the usage of ionic liquid-based electrolytes. The maximum conversion efficiency of solar cells fabricated by using electrodeposited chalcogenide-based 2D layers is nearly 13%.

At present, the consideration of scientists and technologists is fixated on semiconducting substances that possess reasonable efficiency of photon absorption. Some globally famous examples are cadmium telluride (CdTe), copper zinc tin selenide (CZTS), copper indium selenide (CIS), and copper indium gallium selenide (CIGS) because their mass production is inexpensive and performance is fine at odd conditions (e.g., high operating temperatures and dim sunlight) [13].

Some research groups reported their early attempts in noteworthy papers about the first development in rising the efficiency for chalcogenide-based solar cells by employing thermal evaporation techniques to prepare heterojunction of CdTe and CdS in the TF form [13,14]. The combination of cadmium chloride with a heterojunction cell of cadmium sulfide or cadmium telluride TFs is another outstanding advancement to fabricate CdTe solar cells. In this design, the chemical bath deposition is used to synthesize an n-type buffer layer of CdS, while close-spaced sublimation is employed for deposition of a p-type absorber layer of CdTe. For further improvement in the performance of the cell that is coated on an elevated surface of a glass substrate, we use the treatments of cadmium chloride vapor and alterations in window-layer processing.

In addition to coating cadmium telluride over the surface of a substrate of a rigid glass, a coating of CdS/CdTe as a cell was prepared on a thin and elastic sheet of polyimide. For this purpose, the evaporation was done under vacuum at a temperature <723 K. For hardening of the substances, a successive annealing environment was set up without vacuum to first heat them up and then cool down. Noteworthy upgrading in the efficiency of heterojunction CdTe solar cells was reported compared to the previous cells developed by EMPA. Here, the absorber layer of the CdS/CdTe cell was developed by the close space sublimation method. State-of-the-art highest efficient heterojunction CdTe solar cell has been developed by close space sublimation method.

For CIGS solar cells, the first outstanding developments in their efficiency were reported between 1975 and 1985 [13]. In this period, attempts were made to fabricate a photovoltaic cell of CIGS by using n-type cadmium sulfide and p-type copper-indium diselenide. The efficiency of ~8.3 was achieved by depositing the absorber layer on elastic molybdenum foil [15]. This was done by employing an e-beam evaporation with successive selenization. The gallium was doped in the CIS matrix for the construction of solar cells to achieve sufficiently high efficiency (~17%) by depositing CIGS absorber on Mo-coated substrate of soda-lime-silica glass [16]. A breakthrough is reported by Kato et al. [17] when they designed a CIGS solar cell of maximum efficiency (~22.5%).

8.4 PAST AND PRESENT STATUS

Table 8.1 covers the short overview of the attempts done by various research groups to upgrade the performance of solar cells fabricated by using chalcogenides. In some cases, the workers have also used the chalcogenides as a dominant component for upgrading the productivity of solar cells of other materials [18–59].

Kakavelakis et al. demonstrated that few layers of WSe_2 2D crystal synthesized by liquid-phase exfoliation methods improved the power conversion efficiencies (PCEs) from 8.10% to 9.3% (~15%) of PC71BM fullerene domain in bulk heterojunction organic solar cell (OSC) for which they reported the highest ever PCE for 2D material-based OSCs [18]. Wang et al. reported that the WSe_2-modified perovskite solar cell has capability to demonstrate PCE of 10.47%. This value represents the maximum efficiency among $FASnI_3$-based perovskite solar cells [19]. Vikraman et al. reported first time the fabrication of WSe_2/MoS_2 heterostructures on the surface of the substrates of fluorine-doped tin oxide by using them as counter electrodes (CEs) in a dye-sensitized solar cell (DSSC) and achieved the maximum value of PCE (8.44%) as compared to predictable efficiency (8.73%) of high-cost platinum [20,21]. Si et al. [22] have synthesized MoS_2/WSe_2 heterojunction photoanodes by using liquid-phase exfoliation and vacuum filtration techniques and observed that these photoanodes have greater photocurrent intensity and better photo response activity than earlier, which encourages the photogenerated electron–hole pair separation and suppresses their recombination. Yuan et al. [23] observed that the growth of MoS_2 nanosheets on reduced graphene oxide in the vertical direction (MoS_2@RGO) acts as a CE in DSSC and enhanced the PCE (6.82%). This value is high as compared to that of CE of high-cost platinum (6.44%).

TABLE 8.1
Details of the Utilization of Various 2D Chalcogenide Materials in a Solar Cell by Different Research Groups

S. No.	2D Chalcogenide Materials	Role in Solar Cell	References
1	Few layers of WSe_2	To improve the power conversion efficiency of organic/ perovskite solar cell	[18,19]
2	WSe_2/MoS_2 heterostructure on FTO	As a counter electrode in dye-sensitized solar cell	[20–22]
3	MoS_2@RGO	As a counter electrode in dye-sensitized solar cell	[23]
4	Bas/InTe	As tandem solar cell	[24]
5	MoS_2	As counter electrode to enhance the catalytic performances in quantum dot sensitized solar cells	[25]
6	Graphene-MoS_2	As the hole extraction layer in PTB7-Th: PC71BM organic solar cells	[26,27]
7	Honeycomb spherical metallic $1T$-MoS_2	As an efficient counter electrode (CE) for quantum dot sensitized solar cells	[28]
8	MoO_x	As a back contact buffer for thin film n-CdS/p-CdTe solar cell	[29]
9	n-WSe_2	As an excellent transparent electrode in Si heterojunction solar cell	[30]
10	Ultra-thin WS_2	To improve the performance of n-type WS_2/p-type c-Si heterojunction solar cells	[31]
11	GaSe	As very thin buffer layer to improve the efficiency of $CuInGaSe_2$-based solar	[32,33]
12	CdS/CdTe	As an emitter/absorber material in CdTe based solar cell which used in terrestrial application	[34–38]
13	$MoSe_2$	As a counter electrode in dye-sensitized solar cell	[39,40]
14	GaTe	As heterojunction with InGaZnO and formed transparent solar cell	[41]
15	InSe	As an ultra-thin flexible solar cell	[42]
16	Cu-doped ZnTe (ZnTe:Cu)	As a back contact interface layer for the thin film CdS/CdTe solar cell	[43,44]
17	MoS_2/Si	To improve the power conversion efficiency of solar cell	[45–48]
18	$MoSe_2/Cu_2S$	As an efficient counter electrode (CE) for quantum dot sensitized solar cells	[49]
19	MoS_2-P_3HT	To improve the efficiency, moisture stability, and performance of perovskite solar cell	[50,51]
20	CuS/WS_2 and CuS/MoS_2	As counter electrode in dye-sensitized solar cell	[52]
21	TFSA-GR/MoS_2 interlayer	To enhance the performance of graphene/porous Si solar cell	[53,54]
22	$MoS_2/MoTe_2$	As counter electrode in dye-sensitized solar cell	[55]
23	$MoSe_2$ NS/PEDOT:PSS	As counter electrode in dye-sensitized solar cell	[56,57]
24	n-MoS2/p-InP	To enhance the performance and efficiency of solar cell	[58]
25	Mo film	As back electrode in CIGS solar cells	[59]

Xie et al. [24] informed a synthesis of solar cell having conversion efficiency more than 30% in which they used 2D Bas/InTe as base material. To grow the oxygen-containing MoS_2 sheets on 3D graphene vertically, Wu et al. [25] used the hydrothermal techniques and created a CE from such samples for using in quantum dot sensitized solar cells (QDSSCs). They observed that this CE increases the charge and mass transport and enhanced catalytic performances. They observed that the value of PCE of such solar cell was 4.13%, which was ~63% and ~34% greater than that of Pt and pure MoS_2 CE, respectively.

A tandem solar cell is nothing but it is effectively a stack of different solar cells having dissimilar absorption bands on top of each other. It is a variety of multi-junction cells and can either be individual or connected in series. Zheng et al. [26] synthesized the hole extraction layer by using graphene-MoS_2 heterostructures. They prepared this layer by using liquid-phase exfoliation method and utilized it in solar cells of organic PTB7-Th:PC71BM material. They reported the highest value of PCE for this device that was 9.5%. Some workers used different approaches for improving efficiency by using MoS_2 [27–29]. Krishnamoorthy and Prakasam [27] synthesized a photoanode of MoS_2/graphene nanocomposite for using it in DSSCs and obtained the outstanding results (PCE = 8.92%) with this photoanode. This value of PCE is greater than that reported for photoanode of MoS_2 (PCE = 3.36%). Tian et al. [28] fabricated a metallic 1T-MoS_2 in the form of honeycomb spherical geometry through the hydrothermal method and they utilized it as CE for QDSSCs. They reported that the PCE of QDSSCs reached 6.03% by using substrate of Ti-mesh and CEs of MoS_2. A vapor-deposited TMO TF, i.e., molybdenum oxide (MoO_x) was developed by Lin [29], which was utilized as low resistance back contact buffer layer between cadmium telluride of p-type and the back electrode for n-CdS/p-CdTe solar cells. He showed that the efficiency and stability of such cells have also been improved.

Tiwari et al. [30] conveyed n-WSe_2 as a translucent conducting electrode in Si heterojunction solar cell and found that the value of PCE is ~13.09%. This value is reduced and attains a value of 8.9% when the number of several layers of WSe_2 rises from 3 to 40. Finally, when simulation was done by choosing parameters of commercially accessible silicon wafer, they also showed that the maximum value of PCE is 6.8%. Huang et al. [31] chose solar cells made of n-type WS_2/p-type c-Si heterojunction for simulation of structural properties and used 2.5 version of AFORS-HET software at air mass 1.5. They studied the way to improve the functioning of this solar cell by simulation and found an extreme efficiency nearly equal to 20.6%. Using solar cell cpacitance simulator, Boudaoud and Boudaoud [32] simulate the GaSe buffer layer on the p-CuInGaSe$_2$-based solar cells and analyzed the consequence of variation in the thickness of buffer layer on the efficiency of the solar cell. They investigated that when the thickness of GaSe was 10 nm, the efficiency was 30%, and at 20–70 nm, the efficiencies were found stable, that is, 32.02%. Fleutot et al. [33] reported the formation of GaSe at the Cu(In, Ga)Se$_2$/Mo (CIGS)/Mo interface for flexible design of solar cells using a convenient mechanical lift-off process and found an efficiency of 14.3% which was greater than efficiency (13.8%) of earlier solar cells of CIGS. Several authors also investigated different properties of CdTe-based solar cells [34–38].

Vikraman et al. [39] synthesized molybdenum diselenide ($MoSe_2$) atomic layers on the surface of fluorine-doped tin oxide (FTO) substrates via wet chemical approach and employed this combination as a CE in DSSCs directly and reported the PCE of this solar cell 7.28% as compared to 7.40% of CE of expansive platinum. Yuan et al. [40] fabricated $MoSe_2$ nanoflowers (300–400 nm) with nanosphere hierarchical style by using hydrothermal techniques and utilized it as CE in DSSCs. They found a higher remarkable PCE of 7.01% than standard Pt CE (PCE = 6.38%).

Cho et al. [41] demonstrated the first fully transparent solar cell of a p-n heterojunction with InGaZnO with high transparency ~90% in which gallium telluride plays the role of semiconducting 2D layer of p-type. Liu et al. [42] showed that the monolayer of InSe can be utilized as a flexible solar cell. Some research groups developed back-contact made of Cu-doped ZnTe interface layer for utilizing in solar cells of cadmium telluride and cadmium sulfide TFs [43,44]. The MoS_2/Si heterojunction was fabricated by several researchers by using different synthesizing techniques which enhance the efficiency of solar cells by more than 15% [45–48]. Hassan and Sapra [49] constructed an efficient CE using $MoSe_2$–Cu_2S nano heterostructure utilized in QDSSCs and they found that the efficiency of this CE is 3.6%. This value was 17% higher than that of Cu_2S CE. Ray et al. [50] used poly(3-hexylthiophene-2,5-diyl)– 2D molybdenum disulfide (P_3HT-MoS_2) nanohybrid in solar cells of perovskite to improve their performance and moisture steadiness. They found 41% enhancement in efficiency that makes this cell more durable as compared to P_3HT. Sun et al. [51] also demonstrated MoS_2-P_3HT hybrid micro heterostructure and found TiO_2/MoS_2-P_3HT based solar cell exhibiting PCE of 1.28%. This value was 58% greater than that of TiO_2/P_3HT-based solar cell. Hussain et al. fabricated CE by using the heterostructures of CuS/WS_2 and CuS/MoS_2 instead of costly platinum (Pt) in DSSC [52]. They found that the respective values of PCE for CE of CuS/WS_2 and CuS/MoS_2 are 8.2% and 7.1%. These values are approximately similar to the respective values of PCE for the traditional CE of platinum (PCE = 8.7%) and pristine WS_2 (PCE = 6.0%), MoS_2 (PCE = 6.3%), and CuS (PCE = 6.4%) [36].

Jang et al. [53] first pointed out the fabrication of translucent electrode of conducting nature that was made of (trifluoromethanesulfonyl)-amide (TFSA)-doped graphene (TFSA-GR). They used this electrode and molybdenum disulfide interlayer in graphene/ porous Si (GR/PSi) solar cells, in which the inclusion of MoS_2 interlayer enhanced the PCE from 8.83% to 12.09%. Shin et al. [54] pointed out the design of first flexible organic solar cells by using hole transport layer (HTL) translucent conductive electrode made of molybdenum disulfide and TFSA-doped graphene, respectively. When they doped the graphene quantum dots in active layer, they observed the noteworthy rise in PCE. Hussain et al. [55] reported 8.07% value of PCE by using hybrid CE of MoS_2/$MoTe_2$ in DSSCs as compared to 8.33% value of PCE for platinum CE. Huang et al. [56] synthesized a composite film of molybdenum di-selenide nanosheets and poly(3,4-ethylene dioxythiophene): poly (styrene sulfonate) ($MoSe_2$ NS/PEDOT: PSS) that are utilized as the CE in DSSC and achieved 7.58% value of PCE as compared to 7.81% value of PCE in the case of platinum CE.

Wang et al. [57] incorporated the MoS_2 nanoflakes as HTLs in inverted (p-i-n) perovskite solar cells and observed the formation of a hybrid HTL layer due to mixing of these flakes in the layer of PEDOT:PSS. This device displayed remarkable enhancement in PCE and stability simultaneously. Pawar et al. [58] used metal-organic chemical vapor deposition method to obtain coating of molybdenum disulfide on

SiO_2/Si substrate. Later, this coating was moved on a substrate of indium phosphide by employing an easy way of transfer. They used a heterojunction of molybdenum disulfide and indium phosphide of n-type and p-type, respectively, in the place of usually utilized p-Si substrate. Finally, they found a 0.11% value of PCE. The study of microstructure and surface morphology of molybdenum (Mo) TF synthesized by a magnetron sputtering method and used as back contacts for CIGS solar cells have been done by Li et al. [59].

Bandgap energy of various multicomponent (i.e., binary, ternary, and quaternary) chalcogenide-based 2D nanomaterials is shown in Figure 8.6a. The comparison of best research cell efficiency, best module efficiency, theoretical efficiency limit, and energy payback time of a-Si/c-Si and CdTe/CIGS solar cells is shown in Figure 8.6b. More details about citations can be found in Ref. [60]. The efficiency comparison of 1 cm^2 cell and module of chalcogenide-based solar cells with other main solar cells is shown in Figure 8.6c. More details about citations can be found in Ref. [61].

8.5 FUTURE SCOPE

Some chalcogenide-based semiconducting substances having appropriate values of bandgap are CdS, CdSe, PbSe, GeS, CdTe, CIGS, and CZTS. These members of 2D chalcogenides are capable to perform well when they are exposed to visible light. Various researchers have tried to explore them broadly [1]. The utilization of chalcogenide substances designed at nano-scale like semiconducting TFs has been explored in solar cells because of enhancement in the photoelectrical properties. The photocurrent or photoconductivity is increased markedly thanks to simple synthesis approach, large surface area, bandgap alteration, and better kinetics of charge separation. Cadmium telluride is the widely used material whose TFs are globally employed in solar cell industry as a consequence of a project on first solar's utility-scale. In First Solar's event, the TFs of Cd and Te are obtained by using a glass substrate for deposition. It was in the year 2016 a breakthrough came in this field as the solar cell of cadmium telluride having efficiency of 22.1% with average value 17% of efficiency for corresponding modules. It is expected that the Series 6 module should create a power of 420 W, whereas initial version (i.e., series 4 modules) creates a peak value of power nearly 100 W.

The modules of CIS and CIGS are frequently fabricated by employing co-evaporation or co-deposition. The higher concentration of gallium in CIGS is particularly important for an alternative n-type buffer layer. The maximum efficiency values are 0.228 and 0.223, respectively, which can be achieved in CIGS solar cells if we use Zn(O, S) and CdS as buffer layers. The mixing of elements Cu, In, Se, and Ga (occasionally) is done by depositing them on a substrate at different chosen temperatures. The record value of efficiency for CIS cell is 22.9%. For the solar spectrum of 2D chalcogenides having wide band gap values, the low values of optical absorption can be overwhelmed by vigilant alteration of morphology.

Consequently, the present attempts are especially motivated on the alteration of the nanostructuring so that the chalcogenide-based semiconducting substances become more responsive under the exposure of visible light since it has almost 40% contribution in the solar energy coming at earth from the sun.

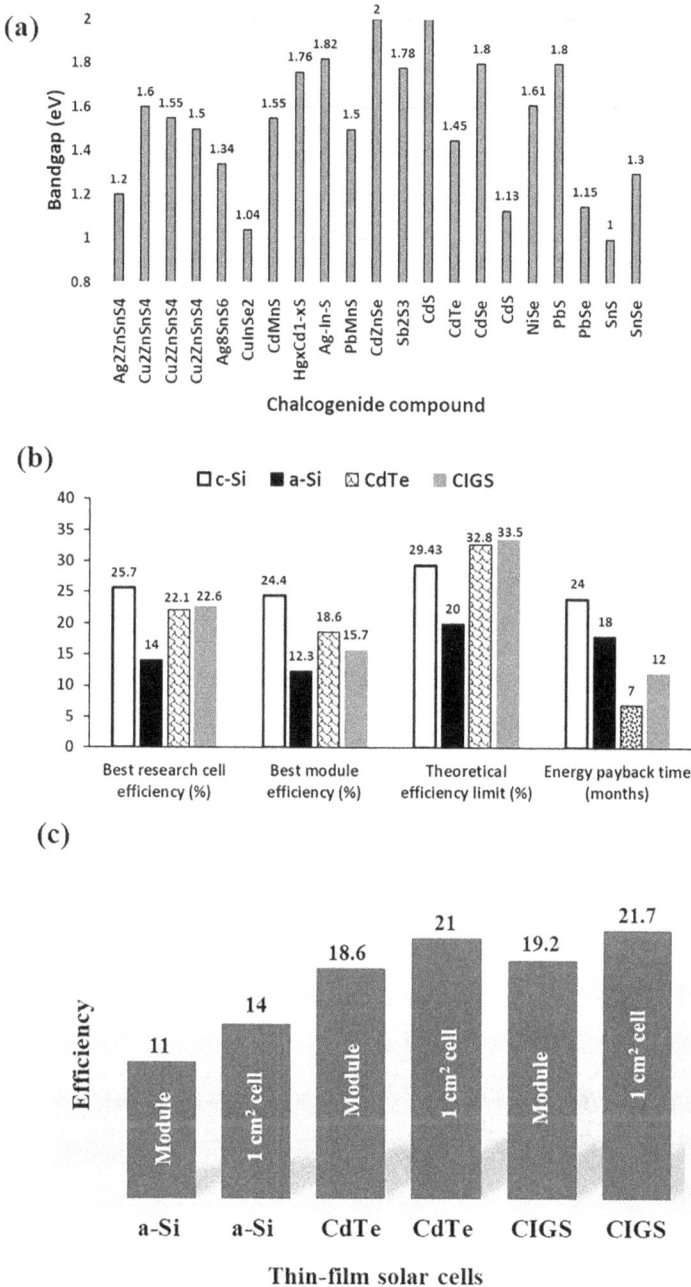

FIGURE 8.6 (a) Band gap energy of various multicomponent (i.e., binary, ternary, and quaternary) chalcogenide-based 2D nanomaterials; (b) comparison of conventional photovoltaic crystalline-silicon (c-Si) sell and TF solar cells of amorphous-silicon (a-Si), cadmium telluride (CdTe), and copper indium gallium selenide (CIGS); and (c) comparison of the performance of cell/module of different industrial solar cells.

8.6 CONCLUSION

Photovoltaic devices like solar cells are considered as an outstanding application of 2D chalcogenide materials or their TFs having wide-gap values. The chalcogenide-based TF belongs to the main categories of solar cells. In the research field of photovoltaics, the members of the CIGS group have a longstanding past since they are more than 20 years old in the time surrounding the first oil crash. At that time, some attempts were made to investigate the semiconducting samples of the metal-one/metal-two/ chalcogenide group as an alternative to energy demands. Since then, a lot of research was done in this direction and it became established that the members of the CIGS family demonstrated realistically for the fulfillment of commitments. The earnest manufacturing began nearby 2005. Till the present time, the initial big factories are successfully running. We have already raised a few issues related to CIGS solar cells in the aforesaid discussion. The solar cells fabricated with CIS possess the operating life comparable to that of those made by silicon. Further, their manufacturing cost is lower than that of silicon-based solar cell as coating deposition technology is used. A world record has been set by demonstrating 22.6% for solar cells of CIGS and they hold this record since 2016. This is the maximum obtainable value of any solar cell in TF form that is more than the record efficiency 21.9% of multicrystalline silicon.

It is worth remembering that it is not sufficient to employ continuously a p-type CIGS coating of 1–2 μm depth. There is a necessity of a p-n junction and good electrical contacts. After the research and development on numerous chalcogenide materials, the achievable approximate values of efficiency for CIGS cells are 20% and 3%, respectively, on the laboratory and industrial scales. Thus, the 2D chalcogenide substances that are used frequently in solar cells applications are restricted to solitarily a limited number of candidates. However, the solar cells of CIGS demonstrate excellent performance, better efficiency (~20%), and low production cost. Due to these features, they are currently viewed as a foremost rival to conventionally available silicon cells.

ACKNOWLEDGMENTS

Shiv Kumar Pal is grateful to UGC, New Delhi, India for providing fellowship under the JRF scheme for National Eligibility Test (NET) qualified scholars. Neeraj Mehta is thankful to his university for providing an incentive grant under the IoE scheme (Dev. Scheme No. 6031).

REFERENCES

1. E. Mikheeva, K. Koshelev, D.-Y. Choi, S. Kruk, J. Lumeau, R. Abdeddaim, I. Voznyuk, S. Enoch, Y. Kivshar, Photosensitive chalcogenide metasurfaces supporting bound states in the continuum, *Opt. Exp.* (2019) 27, 33847–33853.
2. N. Ho, J. M. Laniel, R. Vallee, A. Villeneuve, Photosensitivity of As2S3 chalcogenide thin films at 1.5 μm, *Opt. Lett.* (2003) 28, 965–967.
3. M. W. Lee, C. Grillet, C. L. C. Smith, D. J. Moss, B. J. Eggleton, D. Freeman, B. Luther-Davies, S. Madden, A. Rode, Y. Ruan, Y.-H. Lee, Photosensitive post tuning of chalcogenide photonic crystal waveguides, *Opt. Exp.* (2007) 15, 1277–1285.

4. A. Klein, W. Jaegermann, Review-electronic properties of 2D layered chalcogenide surfaces and interfaces grown by (quasi) van der Waals Epitaxy, *ECS J. Solid-State Sci. Techno.* (2020) 9, 093012.

5. Q. V. Le, J.-Y. Choi, S. Y. Kim, Recent advances in the application of two-dimensional materials as charge transport layers in organic and perovskite solar cells, *FlatChem* (2017) 2, 54–66.

6. P. D. Matthews, P. D. McNaughter, D. J. Lewis, P. O'Brien, Shining a light on transition metal chalcogenides for sustainable photovoltaics, *Chem. Sci.* (2017) 8, 4177–4187.

7. B. Peng, P. K. Ang, K. P. Loh, Two-dimensional dichalcogenides for light-harvesting applications, *Nano Today* (2015) 10, 128–137.

8. A. L. Donne, V. Trifiletti, S. Binetti, New earth-abundant thin film solar cells based on chalcogenides, *Front. Chem.* (2019) 7, 297.

9. J. D. Cain, E. D. Hanson, F. Shi, V. P. Dravid, Emerging opportunities in the two-dimensional chalcogenide systems and architecture, *Curr. Opin. Solid State Mater. Sci.* (2016) 20, 374–387.

10. K. F. Mak, C. Lee, J. Hone, J. Shan, T. F. Heinz, Atomically thin MoS_2: A new direct-gap semiconductor, *Phys. Rev. Lett.* (2010) 105, 136805.

11. J. Zheng, H. Zhang, S. Dong, Y. Liu, C. T. Nai, H. S. Shin, H. Y. Jeong, B. Liu, K. P. Loh, High yield exfoliation of two-dimensional chalcogenides using sodium naphthalenide, *Nat. Commun.* (2014) 5, 1–7.

12. R. Woods-Robinson, Y. Han, H. Zhang, T. Ablekim, I. Khan, K. Persson, A. Zakutayev, Wide band gap chalcogenide semiconductors, *Chem. Rev.* (2020) 120, 4007–4055.

13. A. Kowsar, M. Rahaman, M. S. Islam, A. Y. Imam, S. C. Debnath, M. Sultana, M. A. Hoque, A. Sharmin, Z. H. Mahmood, S. F. U. Farhad, Progress in major thin-film solar cells: Growth technologies, layer materials and efficiencies, *Int. J. Renew. Energy Res.* (2019) 9, 579–597.

14. R. Manivannan, S. N. Victoria, Preparation of chalcogenide thin films using electrode-position method for solar cell applications: A review, *Sol. Energy* (2018) 173, 1144–1157.

15. B. M. Başol, V. K. Kapur, A. Halani, C. Leidholm, Copper indium diselenide thin film solar cells fabricated on flexible foil substrates, *Sol. Energy Mater. Sol. Cells,* (1993) 29, 163–173.

16. J. Tuttle, M. Contreras, T. Gillespie, K. Ramanathan, A. Tennant, J. Keane, A. M. Gabor, R. Noufi, Accelerated publication 17.1% efficient $Cu(In, Ga)Se_2$-based thin-film solar cell, *Prog. Photovolt. Res. Appl.* (1995) 3, 235–238.

17. T. Kato, J.-L. Wu, Y. Hirai, H. Sugimoto, V. Bermudez, Record efficiency for thin-film polycrystalline solar cells up to 22.9% achieved by Cs-treated $Cu(In, Ga)(Se, S)_2$, *IEEE J. Photovolt.* (2018) 9, 325–330.

18. G. Kakavelakis, A. E. D. R. Castillo, V. Pellegrini, A. Ansaldo, P. Tzourmpakis, R. Brescia, M. Prato, E. Stratakis, E. Kymakis, F. Bonaccorso, Size-tuning of WSe_2 flakes for high efficiency inverted organic solar cells, *ACS Nano* (2017) 11, 3517–3531.

19. T. Wang, F. Zheng, G. Tang, J. Cao, P. You, J. Zhao, F. Yan, 2D WSe_2 flakes for synergis-tic modulation of grain growth and charge transfer in tin-based perovskite solar cells, *Adv. Sci.* (2021) 8, 2004315.

20. D. Vikraman, A. A. Arbab, S. Hussain, N. K. Shrestha, S. H. Jeong, J. Jung, S. A. Patil, H.-S. Kim, Design of WSe_2/MoS_2 heterostructures as the counter electrode to replace Pt for dye-sensitized solar cell, *ACS Sustainable Chem. Eng.* (2019) 7, 13195–13205.

21. D. Vikraman, S. A. Patil, S. Hussain, N. Mengal, H.-S. Kim, S. H. Jeong, J. Jung, H.-S. Kim, H. J. Park, Facile and cost-effective methodology to fabricate MoS_2 counter electrode for efficient dye-sensitized solar cells, *Dyes Pigments* (2018) 151, 7–14.

22. K. Si, J. Ma, C. Lu, Y. Zhou, C. He, D. Yang, X. Wang, X. Xu, A two-dimensional MoS_2/WSe_2 van der Waals heterostructure for enhanced photoelectric performance, *Appl. Surf. Sci.* (2020) 507, 145082.

23. X. Yuan, X. Li, X. Zhang, Y. Li, L. Liu, MoS$_2$ vertically grown on graphene with efficient electrocatalytic activity in Pt-free dye-sensitized solar cells, *J. Alloys Compd.* (2018) 731, 685–692.

24. M. Xie, B. Cai, Z. Meng, Y. Gu, S. Zhang, X. Liu, L. Gong, X. Li, H. Zeng, Two-dimensional BAs/InTe: A promising tandem solar cell with high power conversion efficiency, *ACS Appl. Mater. Interfaces* (2020) 12, 6074–6081.

25. D. Wu, Y. Wang, F. Wang, H. Wang, Y. An, Z. Gao, F. Xu, K. Jiang, Oxygen-incorporated few-layer MoS$_2$ vertically aligned on three-dimensional graphene matrix for enhanced catalytic performances in quantum dot sensitized solar cells, *Carbon* (2017) 123, 756–766.

26. X. Zheng, H. Zhang, Q. Yang, C. Xiong, W. Li, Y. Yan, R. S. Gurney, T. Wang, Solution-processed graphene-MoS$_2$ heterostructure for efficient hole extraction in organic solar cells, *Carbon* (2019) 142, 156–163.

27. D. Krishnamoorthy, A. Prakasam, Reparation of MoS$_2$/graphene nanocomposite-based photoanode for dye-sensitized solar cells (DSSCs), *Inorg. Chem. Commun.* (2020) 118, 108016.

28. Z. Tian, Q. Chen, Q. Zhong, Honeycomb spherical 1T-MoS$_2$ as efficient counter electrodes for quantum dot sensitized solar cells, *Chem. Eng. J.* (2020) 396, 125374.

29. H. Lin, Study of molybdenum oxide as a back contact buffer for thin film n-CdS/p-CdTe solar cells, (2013) University of Rochester.

30. P. Tiwari, K. Patel, L. Krishnia, R. Kumari, P. K. Tyagi, Potential application of multilayer n-type tungsten diselenide (WSe$_2$) sheet as transparent conducting electrode in silicon heterojunction solar cell, *Compu. Mater. Sci.* (2017) 136, 102–108.

31. R. Huang, M. Yu, Q. Yang, L. Zhang, Y. Wu, Q. Cheng, Numerical simulation for optimization of an ultra-thin n-type WS$_2$/p-type c-Si heterojunction solar cells, *Comp. Mater. Sci.* (2020) 178, 109600.

32. L. Boudaoud, N. Boudaoud, Simulation of GaSe buffer layer on the CuInGaSe$_2$-based solar cells by SCAPs, *International Conference on Communications and Electrical Engineering (ICCEE)'s* (2018) 1, El Oued, Algeria.

33. B. Fleutot, D. Lincot, D. Jubault, Z. J. L. Kao, N. Naghavi, J.-F. Guillemoles, F. Donsanti, GaSe formation at the Cu(In, Ga)Se$_2$/Mo interface: A novel approach for flexible solar cells by easy mechanical lift-off, *Adv. Mater. Interfaces* (2014) 1, 1400044.

34. C. Routray, A. Panda, S. P. Singh, U. P. Singh, Progress in polycrystalline thin film CdTe solar cells, *Inv. J. Renew. Energy* (2011) 1, 75–82.

35. J. Barker, S. P. Binns, D. R. Johnson, R. J. Marshall, S. Oktik, M. E. Ozsan, M. H. Patterson, S. J. Ransome, S. Roberts, M. Sadeghl, J. Sherborne, A. K. Turner, J. M. Woodcock, Electrodeposited CdTe for thin film solar cells, *Int. J. Sol. Energy* (1992) 12, 79–94.

36. C. S. Ferekides, U. Balasubramanian, R. Mamazza, V. Viswanathan, H. Zhao, D.L. Morel, CdTe thin film solar cells: Device and technology issues, *Sol. Energy* (2004) 77, 823–830.

37. T. Song, A. Kanevce, J. R. Sites, Emitter/absorber interface of CdTe solar cells, *J. Appl. Phys.* (2016) 119, 233104.

38. U. Jahn, T. Okamoto, A. Yamada, M. Konagai, Doping and intermixing in CdS/CdTe solar cells fabricated under different conditions, *J. Appl. Phys.*, (2001) 90, 2552–2558.

39. D. Vikraman, S. A. Patil, S. Hussain, N. Mengal, S. H. Jeong, J. Jung, H. J. Park, H.-S. Kim, H.-S. Kim, Construction of dye-sensitized solar cells using wet chemical route synthesized MoSe$_2$ counter electrode, *J. Ind. Eng. Chem.* (2019) 69, 379–386.

40. X. Yuan, B. Zhou, X. Zhang, Y. Li, L. Liu, Hierarchical MoSe$_2$ nanoflowers used as highly efficient electrode for dye-sensitized solar cells, *Electrochim. Acta* (2018) 283, 1163–1169.

41. A.-J. Cho, K. Park, S. Park, M.-K. Song, K.-B. Chung, J.-Y. Kwon, Transparent solar cell based on mechanically exfoliated GaTe and InGaZnO p-n heterojuction, *J. Mater. Chem. C* (2017) 5, 4327–4334.

42. X. Liu, J.-C. Ren, T. Shen, S. Li, W. Liu, Lateral InSe p–n junction formed by partial doping for use in ultrathin flexible solar cells, *J. Phys. Chem. Lett.* (2019) 10, 7712–7718.
43. T. A. Gessert, A. R. Mason, P. Sheldon, A. B. Swartzlander, D. Niles, T. J. Coutts, Development of Cu-doped ZnTe as a back-contact interface layer for thin-film CdS/CdTe solar cells, *J. Vac. Sci. Techno. A* (1996) 14, 806.
44. J. Tang, D. Mao, L. Feng, W. Song, and J.U. Trefny, The properties and optimization of ZnTe:Cu back contacts on CdTe/CdS thin film solar cells, *Conference Record of the 25th IEEE Photovoltaic Specialists* (1996) 925–928. Washington, DC, USA.
45. H. Xu, L. Xin, L. Liu, D. Pang, Y. Jiao, R. Cong, W. Yu, Large area MoS$_2$/Si heterojunction-based solar cell through sol-gel method, *Mater. Lett.* (2019) 238, 13–16.
46. S. B. Kang, K. C. Kwon, K. S. Choi, R. Lee, K. Hong, J. M. Suh, M. J. Im, A. Sanger, I. Y. Choi, S. Y. Kim, J. C. Shin, H. W. Jang, K. J. Choi, Transfer of ultrathin molybdenum disulfide and transparent nanomesh electrode onto silicon for efficient heterojunction solar cells, *Nano Energy* (2018) 50, 649–658.
47. K. Kumar, A. Kumar, D, Kaur, Improved power conversion efficiency in n-MoS$_2$/AlN/p-Si (SIS) heterojunction based solar cells, *Mater. Lett.* (2020) 277, 128360.
48. Y. Zhang, Y. Zhuang, L. Liu, P. Qiu, L. Su, X. Teng, G. Fu, W. Yu, The microstructure evolution during MoS$_2$ films growth and its influence on the MoS$_2$ optical-electrical properties in MoS$_2$/p-Si heterojunction solar cells. *Superlattices Microstruct.* (2020) 137, 106352.
49. M. S. Hassan, S. Sapra, Cu$_2$S decorated MoSe$_2$ nanosheets as counter electrode for quantum dot sensitized solar cells, *Mater. Today: Proc.* (2021) 36, 605–608.
50. R. Ray, A. S. Sarkar, S. K. Pal, Improving performance and moisture stability of perovskite solar cells through interface engineering with polymer-2D MoS$_2$ nanohybrid, *Sol. Energy* (2019) 193, 95–101.
51. B. Sun, Z. Su, Y. Hao, J. Pei, Y. Li, Facile fabrication of MoS$_2$-P$_3$HT hybrid microheterostructure with enhanced photovoltaic performance in TiO$_2$ nanorod array based hybrid solar cell, *Solid State Sci.* (2019) 94, 92–98.
52. S. Hussain, S. A. Patil, A. A. Memon, D. Vikraman, B. A. Naqvi, S. H. Jeong, H.-S. Kim, H.-S. Kim, J. Jung, CuS/WS$_2$ and CuS/MoS$_2$ heterostructures for high performance counter electrodes in dye-sensitized solar cells, *Sol. Energy* (2018) 171, 122–129.
53. C. W. Jang, D. H. Shin, J. S. Ko, S.-H. Choi, Performance enhancement of graphene/porous Si solar cells by employing layer-controlled MoS$_2$, *Appl. Surf. Sci.* (2020) 532, 147460.
54. D. H. Shin, C. W. Jang, J. S. Ko, S.-H. Choi, Enhancement of efficiency and stability in organic solar cells by employing MoS$_2$ transport layer, graphene electrode, and graphene quantum dots-added active layer, *Appl. Surf. Sci.* (2021) 538, 148155.
55. S. Hussain, S. A. Patil, D. Vikraman, I. Rabani, A. A. Arbab, S. H. Jeong, H.-S. Kim, H. Choi, J. Jung, Enhanced electrocatalytic properties in MoS$_2$/MoTe$_2$ hybrid heterostructures for dye-sensitized solar cells, *Appl. Surf. Sci.* (2020) 504, 144401.
56. Y.-J. Huang, M.-S. Fan, C.-T. Li, C.-P. Lee, T.-Y. Chen, R. Vittal, K.-C. Ho, MoSe$_2$ nanosheet/poly(3,4-ethylenedioxythiophene): Poly(styrenesulfonate) composite film as a Pt-free counter electrode for dye-sensitized solar cells, *Electrochim. Acta* (2016) 211, 794–803.
57. D. Wang, N. K. Elumalai, M. A. Mahmud, H. Yi, M. B. Upama, R. A. L. Chin, G. Conibeer, C. Xu, F. Haque, L. Duan, A. Uddin, MoS$_2$ incorporated hybrid hole transport layer for high performance and stable perovskite solar cells, *Synth. Metals* (2018) 246, 195–203.
58. S. A. Pawar, D. Kim, A. Kim, J. H. Park, J. C. Shin, T. W. Kim, H. J. Kim, Heterojunction solar cell based on n-MoS$_2$/p-InP, *Opt. Mater.* (2018) 86, 576–581.

59. W. Li, X. Yan, A. G. Aberle, S. Venkataraj, Analysis of microstructure and surface mor-
    phology of sputter deposited molybdenum back contacts for CIGS solar cells, *Procedia
    Eng.* (2016) 139, 1–6.
60. K. H. Ong, R. Agileswari, B. Maniscalco, P. Arnou, C. C. Kumar, J. W. Bowers,
    M. B. Marsadek, Review on substrate and molybdenum back contact in CIGS thin
    film solar cell, *Int. J. Photoenergy* (2018) 2018, 9106269.
61. M. A. Green, E. D. Dunlop, J. Hohl-Ebinger, M. Yoshita, N. Kopidakis, A. W. Y. Ho-Baillie,
    Solar cell efficiency tables (version 57), *Prog. Photovolt.* (2020) 28, 3–15.

9 Metal Phosphide-Based 2D Nanomaterials for Solar Cells

Hamideh Mohammadian-Sarcheshmeh
and Mohammad Mazloum-Ardakani
Yazd University

CONTENTS

9.1 INTRODUCTION

Energy shortage and environmental contamination have been known as two essential problems for more advancement of human society. Many researchers have focused on introducing greatly effective and inexpensive photovoltaic systems to support all energy-critical subjects. Photovoltaics can efficiently convert light into electricity and display a great ability to create sustainable energy, which is useful to decrease fossil fuels consumption [1,2]. Recent progress in energy conversion researches, such as the suggestion of different low-cost compounds with remarkable electrochemical behavior, leads to creating energy requests for the future. Transition metal

DOI: 10.1201/9781003178422-9

phosphides (TMPs) have lately been introduced as great energy conversion compounds owing to some advantages such as electrical conductivity, thermal stability, and abundant active surface sites. Also, TMPs materials such as binary (Co_2P, Ni_2P, and Fe_2P), ternary (CoMnP and $Ni_{0.1}Co_{0.9}P$), and quaternary (NiCoFeP) compounds [3–8] have presented considerable properties in supercapacitors, CO_2 reduction, water splitting, and batteries [9–12]. TMPs are chemical compositions involving P and metals, with the formula of M_xP_y. In this structure, P can provide stable materials with d-group metals and rare earth metals. In addition, TMPs exhibit various stoichiometry producing different crystal compounds for binary TMPs. Compare to metal chalcogenide, TMPs dependent on the chemical compound, electronic state, and crystal structure are known as semiconductor, insulator, and conductor. These characterizations lead to TMPs with special properties, and also simplify various potential applications. TMPs show other properties, such as hardness and chemical stability as a result of rigid MP bond in the structure. The excellent electrochemical performance of TMPs is attributed to the synergistic properties of different combinations and surface compounds. Furthermore, the P and M-sites show good electrocatalytic behavior [13]. The electronegativity value of P is superior to most metals. Therefore, the combination of P elements with metals confines the electron delocalization in the metals, resulting in the electron transportation from metal toward P. The feature of M-P is ascribed to the M:P ratio and their electronegativity difference in the TMP structures. For example, in the TMP (M_xP_y formula), the strong M–P bonds provide good chemical and thermal stability. In metal-rich phosphides structure ($x{:}y \geq 1$), electrons are not limited and there are strong M–M interactions, resulting in a metallic property or good conductivity for TMP. For instance, nickel and cobalt phosphides exhibit metallic behavior, which provides good conductivity. However, in the phosphorus-rich structures ($x{:}y < 1$), there is no M–M bonding, leading to a low electrical conductivity [14].

The synthesis methods can be an important parameter in TMP morphologies and the contribution of P elements in the structures. The pore size of TMPs and their surface area mainly depend on various synthesis conditions, for example, the kind of utilized compounds, heating rate, time, and temperature. Different TMPs compounds are fabricated using various methods. The famous methods are the direct phosphorization of metals sources, solid-state interaction, decomposition of metal-organic compounds, and solvothermal and hydrothermal methods. Moreover, other procedures were reported such as the reduction of metal oxide NPs using PH_3/H_2, high-temperature annealing of nanoparticles, trioctylphosphine, and metal hypophosphites [15].

9.2 TMP PROPERTIES

The application of TMPs can mainly provide certain properties. They are listed in the following:

9.2.1 INCREASE LIGHT ABSORPTION

Light-harvesting ability is a significant property affecting photocatalytic performance because the increased light absorption capability promotes efficiency in the solar cells [16]. TMP compounds were introduced to remarkably increase light absorption owing to their dark color. Zhao et al. [17] showed an enhanced light absorption for iron phosphide/graphitic carbon nitride (Fe$_x$ P/g-C$_3$N$_4$), region of 450–800 nm, due to the well visible light absorption capability of Fe$_x$P. Also, the utilization of some nickel phosphides (Ni$_2$P, Ni$_{12}$P, and Ni$_3$P) can effectively promote the light absorption of g-C$_3$N$_4$ due to their dark appearance [18]. Some studies [19] reported that the cobalt phosphide (CoP$_3$) has a good light absorption, range of visible to near-infrared due to having a small bandgap. By using CoP$_3$, the CoP$_3$/Mn$_{0.2}$Cd$_{0.8}$S nanocomposite displayed enhanced light absorption. In addition, zinc phosphide (Zn$_3$P$_2$) has been shown considerable ability as a desirable absorber in photovoltaic devices, including a direct bandgap (1.5 eV), a big absorption coefficient, good light absorbance, and large carrier diffusion lengths [20]. These properties increase remarkably energy conversion efficiency. Solar cells by application of Zn$_3$P$_2$ were fabricated from p-n semiconductor heterojunctions, with Mg/Zn$_3$P$_2$ Schottky diodes revealed conversion efficiencies of more than 6% [21,22].

Other promising materials for light-harvesting are Cd$_3$P$_2$ colloidal quantum dots (CQDs) for light-harvesting in the region of visible and near-IR regions. Cd$_3$P$_2$ is a semiconductor with a small bandgap (0.55 eV) [23]. Cao et al. [24] introduced quantum heterojunction solar cell including ITO/PEDOT/PbS CQDs/Cd$_3$P$_2$ CQDs/Al. The Cd$_3$P$_2$ CQDs exhibited the adjustable optical and optoelectronic features in a large spectral range due to the narrow bandgap and big exciton Bohr radius.

ZnSnP$_2$ (ZTP) compounds have newly attracted a lot of attention due to their available compounds, great absorption coefficient in the visible light range (similar to CdTe and GaAs), and their controllable bandgap from 1.3 to 1.7 eV [25]. Application of a back buffer layer, which will be matched effectively with ZnSnP$_2$ as an absorber in the terms of atom configurations and band tunings, is efficient to provide an ohmic back contact with small resistance, resulting in improvement of device performance. Kuwano et al. [26] examined reactive diffusion at the interface between a Cu back electrode and ZnSnP$_2$ as an absorber. Photoelectron spectroscopy results displayed that the work function of Cu$_3$P is approximately similar to the ionization potential of ZnSnP$_2$, indicating that the potential barrier toward hole transport is low at the Cu$_3$P/ZnSnP$_2$ interface. The utilization of Cu$_3$P results in a larger current density. Therefore, a ZnSnP$_2$ solar cell with the improved efficiency (PCE = 3.87%) was obtained.

Indium phosphide colloidal nanocrystals (InP NCs) present low toxicity and adjustable emission regions from visible to near-infrared. Colloidal nanocrystals have attracted remarkable attention in different applications such as photovoltaics, optoelectronics, due to their special optical and electrical properties. Among various NCs, InP, Zn$_3$P$_2$, and Cd$_3$P$_3$ NCs are so important in various light emission and absorption-based utilizations. Especially, InP and Zn$_3$P$_3$ exhibit considerable attention because of popper bandgap and low toxicity [13].

9.2.2 Accelerate Charge Transfer

Loading of TMPs on semiconductors results in the fabrication of heterojunction. Therefore accelerates the separation and transportation of generated electron–hole pairs.

9.2.3 Strengthen Photostability

The introduction of TMPs on semiconductors effectively enhances the photostability of the semiconductors due to the timely transfer of charge carriers [27].

9.3 UTILIZATION OF TMPs IN DYE-SENSITIZED SOLAR CELLS (DSSCs)

Among the different devices for converting sunlight into electricity power, dye-sensitized solar cells (DSSCs) indicate special properties such as great power conversion efficiency (PCE), easy and low-cost construction, and environment-friendly. The standard DSSC involves a mesoporous titania photoanode layer as dye-absorber, an electrolyte consisting of a triiodide/iodide (I_3^- / I^-) redox couple, and a counter electrode (CE). The CE utilizes to gather electrons from the external circuit and catalyze the reduction of (I_3^- / I^-) [28]. TMPs compounds have been used in the fabrication of CE and photoanode materials in DSSCs. Herein, some of the studies are presented.

9.3.1 TMPs as a Counter Electrode

The CEs, as the most significant sections in DSSCs, are utilized to reduce redox electrolytes. Significantly, the CE will determine the PCE value in the DSSCs. As a CE for DSSCs, Pt has been extensively known owing to its great electrocatalytic behavior and good conductivity. However, the utilization of Pt is confined due to its expensive and low accessibility. Moreover, their scarcity and poor chemical stability offer Pt CEs inefficient for large-scale utilization [29]. Therefore, the introduction of low-cost and Pt-free electrodes with acceptable electrocatalytic performance to reduce I_3^- in DSSCs is so important. To overcome these problems, earth-plentiful, so active, and anticorrosive Pt-free CEs, metal combinations (sulfide, carbide, and nitride), conductive polymers, and carbon materials, were used. In comparison to these materials, the TMPs are very attractive to fabricate high-efficiency CE catalysts owing to their eco-friendly nature, low cost, availability, excellent electrocatalytic behavior, great conductivity, and high stability [30].

Di et al. [31] introduced a composite of TMPs/PEDOT polymers as CE. The optimized PEDOT-Ni_2P electrode exhibited the PCE value of 7.14% in comparison to Pt CE (PCE = 7.09%). This composite showed less charge carrier-transfer resistance and a larger decrease of current density, verifying its high catalytic capability. Also, the composite revealed acceptable chemical stability. The improved electrochemical behavior can be attributed to the increasing catalytic sites owing to the utilization of phosphide particles.

Recently, iodine-free redox couples-based electrolytes have been significantly exanimated as replacements for (I_3^- / I^-) redox in DSCs to overcome some disadvantages such as high energy loss in the dye regeneration, visible light absorption, and corrosiveness by current-collecting metal grids. Nevertheless, typical Pt cathodes exhibit low catalytic properties toward these iodine-free redox couples, leading to low fill factors and approximately moderate PCE. To solve this problem, economical Pt-free catalysts were developed. Recent investigations have reported that Pt can be dissolved in electrolytes consisting of (I/I⁻) redox couple, producing PtI_4 and therefore resulting in an enhancement in the activation energy and a decrement in the current density for (I_3^-) reduction. A composition of molybdenum phosphide (MoP), Ni_5P_4, and carbon incorporated Ni_5P_4 (Ni_5P_4/C) was suggested as cathode catalysts for both I^- / I_3^- redox couple and organic T_2/T^- redox couple (5,50-dithiobis(1-methyltetrazole)/1-methyltetrazole-5-thiolate). For I^- / I_3^- redox couple, DSSCs by using MoP and Ni_5P_4 as cathode showed a PCE value of 4.92% and 5.71%, and the DSSC by Ni_5P_4/C composite as a cathode exhibited a PCE value of 7.54%, in comparison with the PCE value of the DSSC with the Pt as a cathode (7.76%) [14].

9.3.1.1 Adding Carbon Materials

Among TMPs, the NiPs are known for their significant electrocatalytic performance, mechanical and magnetic resistance features, and great corrosion resistance. However, the most important problem of NiPs is their low electrical conductivities, which restrict electrochemical behavior. To solve this problem, TMPs can be incorporated on the carbon substrates including carbon fibers, carbon nanotubes (CNTs), carbon spheres, and reduced graphene oxide (rGO) [32]. Feng et al. [33] introduced a DSSC based on NiCoP-CNTs composite electrode with a PCE of 7.24% in comparison to that of Pt electrode (PCE = 7.12%) under the same condition. The increased catalytic performance is derived from the good catalytic property of NiCoP NPs and the great conductivity of the CNT. In addition, the fabricated DSSC by Ni_2P/carbon (10.4%) indicated a good PCE value (9.57%). Outstanding catalytic performance was ascribed to numerous $Ni^{\delta+}$ and $P^{\delta-}$ active centers, and the metal-like conductivity [29]. Cyclic voltammetry was utilized to examine the catalytic performance of Ni_2P/C_{10} and Ni_2P_5/C_{10} as CEs. As can be seen in Figure 9.1, these CEs show two pairs of redox peaks [34].

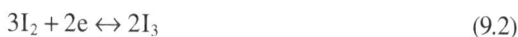

$$Nb_6O_{17}^{4-} \tag{9.1}$$

$$3I_2 + 2e \leftrightarrow 2I_3 \tag{9.2}$$

The higher reduction current (for reduction of I_3^- to I^-) was shown for Ni_2P/C_{10} in comparison to currents for $Ni_{12}P_5/C_{10}$ composite and Pt as CEs, indicating a higher catalytic reduction behavior. Furthermore, the lower peak-to-peak separation (ΔE_p) of Ni_2P/C_{10} than $Ni_{12}P_5/C_{10}$ and Pt as CEs verifies the faster reaction kinetics in catalyzing the reduction of I_3^- / I^-. Moreover, Figure 9.2 exhibits that the stability of DSSC based on Ni_2P/C_{10} was better than Pt-based cell, which indicates higher stability in I_3^- / I^- electrolyte [29].

Chen et al. [29] explained the catalytic mechanism of Ni_2P/C with the special hexagonal structure of Ni_2P (Figure 9.3). In this structure, the connection of P into the Ni provides the Ni–P bonds, resulting in a positive charge of nickel. Therefore, Ni_2P has metal-like conductivity such as conductors [35]. Furthermore, the negative ions (I_3^-) are adsorbed on the metal atoms with a positive charge instead of nonmetallic atoms. Generally, the Ni_2P involves two kinds of Ni active centers (Ni(1) and Ni(2)) in accordance with two kinds of P(1) and P(2) corresponding to their various occupancy sections. The active site of Ni(2) is known as the efficient active center in different

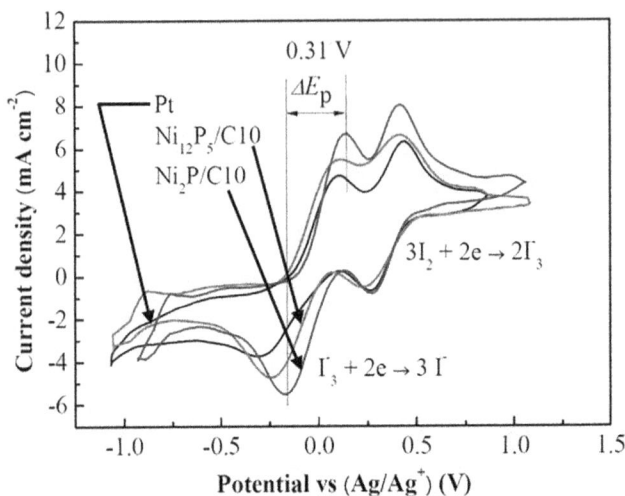

FIGURE 9.1 Cyclic voltammograms of Pt, $Ni_{12}P_5/C_{10}$ and Ni_2P/C_{10} as CEs. (Adapted with permission from Ref. [29]. Copyright (2017) American Chemical Society.)

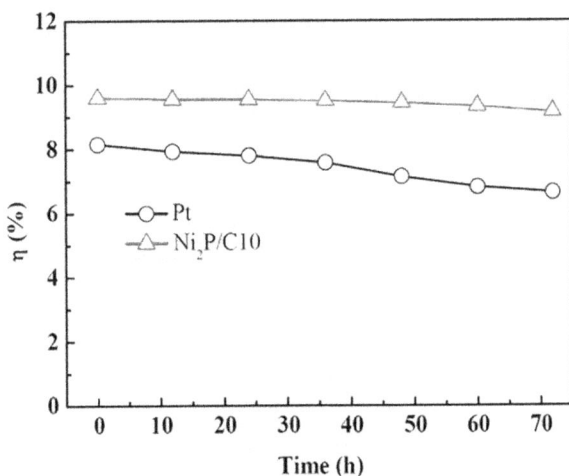

FIGURE 9.2 The investigation of photovoltaic stability for Pt and Ni_2P/C_{10} based devices after 72 hours. (Adapted with permission from Ref. [29]. Copyright (2017) American Chemical Society.)

other electrocatalysis [36]. Thus, the rapid adsorption of I_3^- onto the $Ni^{\delta+}$ (Ni(2)), the Easy I–I bond breakage, and also the high electron transportation in the Ni_2P and carbon, can result in remarkable catalytic performance in the Ni_2P/C composite.

In comparison with various carbon materials, rGO reveals some advantages such as good electrical conductivity, surface area, chemical stability, and the capability to have strong interaction with active compounds to increase charge transfer and enhance stability. A compound of nickel phosphide nanocrystals (NCs)/reduced

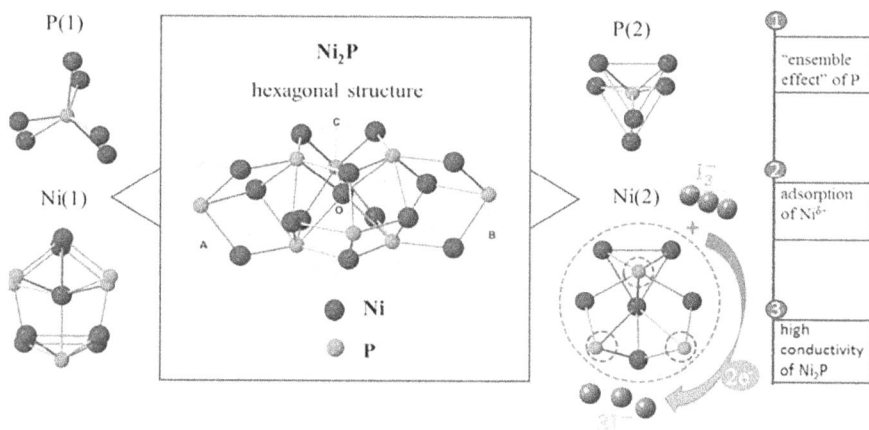

FIGURE 9.3 Illustration of catalytic mechanism for Ni_2P/C as a CE. (Adapted with permission from Ref. [29]. Copyright (2017) American Chemical Society.)

graphene oxide (rGO) was reported as the CE in DSSC. This composite included the catalytic activity of Ni_xP_y and the high conductivity of rGO, therefore showed greatly enhanced electrocatalytic activity with a PCE value of 8.19% [37].

Zhao et al. [38] introduced a DSSC based on the carbon nanofibers (CNs)-incorporated Pt and Ni_2P (Pt–Ni_2P/CNs) as a CE. It revealed a highly improved PCE value of 9.11% in comparison to the PCE value of Pt/CNs as a CE (8.35%). Thus, TMPs can be used as low-cost compounds to replace Pt for DSSCs with excellent efficiency. Some reported performances for DSSCs based on TMP as CE are summarized in Table 9.1.

9.3.1.2 Morphology

Generally, catalyst materials as a CE include some properties such as high surface area, excellent conductivity, and porous structures to provide rapid transportation of electrolytes and electrons. Therefore, TMPs with appropriate morphology have been a significant topic for the success of the CE. Nevertheless, a larger surface area can provide higher grain boundaries, resulting in high electron transport resistance.

TABLE 9.1

Some Reported Efficiencies for DSSCs Based on TMPs as CEs.

CE Material	PCE (%)	References
NiCoP-CNTs	7.24	[33]
Ni_2P-C	9.57	[29]
Ni_xP_y/rGO	8.19	[37]
Pt-Ni_2P/CNs	9.11	[38]
Ni_5P_4/C	7.54	[39]
Graphene/Ni_2P_5	5.70	[40]

Preparation of TMPs and their utilization on the surface of fluorine-doped tin oxide (FTO) are important to improve the efficiency of DSSCs. In addition, some CE materials indicate poor electrocatalytic behavior for electrolyte reaction. Therefore, films with suitable thicknesses (some micrometers) are necessary to compensate poor electrocatalytic properties. A pulse-reverse electrodeposition technique was used to deposit acceptable, ultrathin porous nanospheres Ni_2P on the surface of FTO to enhance the electrocatalytic behavior for the I^- / I_3^- reaction to provide DSSC with a PCE value of 7.32% [41].

A monolayer of highly spaced nickel phosphide (Ni_5P_4) clusters with nanoparticles was fabricated on the surface of FTO as a CE. These clusters consist of mesoporous NPs involved further channels for accelerating the electrolyte transportation and enlarging the surface area available to electrolyte ions. Consequently, the charge-transfer resistance and diffusion impedance of Ni_5P_4/FTO is low. The PCE value of DSSC with this CE (PCE = 7.6%) was better than the PCE value of DSSC based on Pt (PCE = 7.2%) [32].

9.3.1.3 Ternary Phosphides

In comparison to the binary TMPs, the ternary phosphides reveal higher catalytic behavior because the utilization of two transition metals leads to a redistribution of the valence electrons and provides two electron-donating active centers. The DSSC based on a ternary NiCoP as a CE indicated a good PCE value of 8.01%, which was remarkable to the binary TMPs due to its improved electrocatalytic behavior in comparison to that of the binary Ni_2P/Ni_2P_5 and CoP CEs [42].

9.3.2 TMPs as a Photoanode

TiO_2 semiconductor has been known as a conventional photoanode in DSSCs due to its special promising properties, large internal surface area to improved adsorption of sensitizers, and thus enhanced DSSC performance. Other utilized semiconductors are ZnO, SnO_2, Fe_2O_3, ZrO_2, Nb_2O_5, CeO_2, and Al_2O_3 [43]. However, their poor electrical conductivity considerably prevents their extensive utilization in DSSCs. Consequently, the suggestion and fabrication of photoanode materials with high electrical conductivity, excellent reaction capability, and stability for DSSC are important. TMPs are novel kinds of electrode materials with good performances. As n-type semiconductors, TMPs indicate good electrical conductivity and therefore are kinetically suitable for rapid charge transportation. The metalloid properties and great conductivity of TMPs provide superior rate ability and enhance the photoelectrochemical activity in DSSCs. Different TMPs have been investigated as useful co-catalysts owing to suitable stability and electrical conductivity [44–46]. Saravanan and Meeran [47] fabricated a heterostructure by coupling g-C_3N_4 with NiCoP as photoanode. The synthesized nanocomposite exhibited good performance owing to suitable visible-light adsorption intensity and pair electron–hole separation efficiency. A PCE of 11.24% was obtained by NiCoP/g-C_3N_4 hybrid composite, which was higher than the PCE of the NiCoP-based device (4.54%). This compound indicated a large surface area and porous structure.

9.4 SPACE SOLAR CELLS

9.4.1 INDIUM PHOSPHIDES MATERIALS

In comparison to Si as a known material in conventional photovoltaic applications, InP indicates a suitable bandgap (1.34 eV) in the optimized energy area for solar energy conversion. InP is known as a significant compound for superb efficiencies solar cells, such as InP-based homojunction and heterojunction solar cells. Furthermore, InP solar cells show better resistance to space radiation damage than GaAs and Si solar cells. Therefore, InP-based solar cells can be significant for space utilization [48]. Wang et al. [49] reported the PCE value of 3.3% by delicately designing a van der Waals heterostructure between graphene and InP A chemical doping resulted in a PCE = 5.6%. Yoshimura et al. [50] fabricated a heterojunction photovoltaic device with an InP nanowire array sputtered on the indium tin oxide (ITO) (Figure 9.4). The ITO served as a transparent electrode and provides a photovoltaic junction. This system indicated a PCE = 7.37%.

9.4.2 GALLIUM PHOSPHIDE MATERIALS

Gallium phosphides (GaPs) are good candidates for the top junction solar cells in multijunction solar cell systems. The GaPs include a wide bandgap (2.26 eV), great carrier mobility providing good collection ability, and excellent crystal quality. In addition, GaPs are a promising suggestion for solar cells in space utilization due to

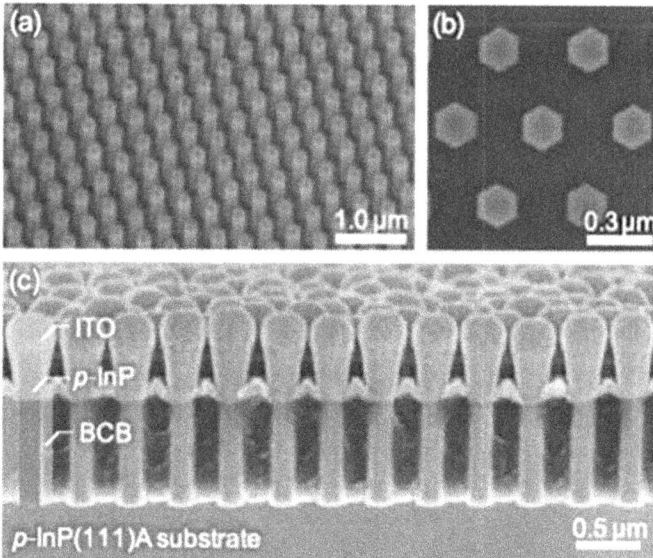

FIGURE 9.4 (a) A SEM image of p-InP NW array. (b) A top view SEM image of the NWs. (c) Cross-sectional SEM image of ITO/p-InP heterojunction NW solar cell. (Adapted with permission from Ref. [50]. Copyright (2013) AIP Publishing.)

their stability at high temperatures [51]. The efficiency of a five-junction solar cell system with a GaP solar cell as the top junction was 3.6% more than the four-junction solar cell device. All photovoltaic parameters of this solar cell are superior to the parameters presented in the previous work [52].

To enhance the thermal stability of solar cells in high-temperature utilizations, wide bandgap semiconductors such as GaP were being examined. The GaP was studied due to its wide bandgap and availability as substrate materials. The significant radiation tolerance of GaPs is owing to great bonding energies and lower temperature radiation damage annealing. The high-temperature function of the solar cells reduces the demand for active photovoltaic cooling, significantly decreasing both weight and device complexity. The large bandgap and temperature stability of GaP along with bandgap tuning by using InGaP multiple quantum wells (MQWs) is a new way to improve the performance of GaP solar cells. Application of MQW to a GaP solar cell leads to the enhancement of temperature stability of GaP and improving the current output of the device [53].

9.5   TMPs QUANTUM DOTS

Generally, some heavy metals such as Cd or Pb in quantum dots (QDs)-based electronics systems have shown environmental problems due to the very low concentrations. These topics can be decreased by popper encapsulation. Nevertheless, the utilization of heavy-metal materials in electrical devices confines some regulations in the EU [13]. Thus, some investigations have changed the attention from Cd and Pb QDs-based devices to InP QDs due to less toxicity. It was reported that charge carrier mobility and lifetimes for InP and InZnP QDs can be considered to those of PbS QDs, making it possible to fabricate solar cells based on InP and InZnP QDs with the PCE values of 0.65% and 1.2%, respectively [54].

9.6   CONCLUSION AND OUTLOOK

TMPs have been known as the most important materials to fabricate conversion energy devices, especially solar cells. Different suggestions to the progress of TMP materials are listed as follows:

1. Application of facile synthesis procedures, low toxic and low-cost compounds to prepare greatly effective TMPs are so important. These methods will result in widespread utilization of TMPs in energy conversion systems.
2. Most of the studies on TMP compounds were based on Ni compounds. New studies are necessary to introduce different types of TMPs.
3. To increase electrochemical property and stability, chemical doping and synthesis of various composites are suggested.
4. The electrochemical behavior of TMPs is widely dependent on their physicochemical characteristics. 3D structures compounds include more active surface sites and will exhibit improved electrochemical performance.
5. Fabrication of different compounds of TMPs by changing M:P will result in enhancement in the electrochemical performances.

By considering the all mentioned strategies, we believe that the progress of highly effective, low-cost, and high stable TMPs for a widespread field of energy conversion utilizations will be possible.

ACKNOWLEDGMENT

The authors wish to thank Yazd University Research Council for the financial support of this research.

REFERENCES

1. A. Hagfeldt, G. Boschloo, L. Sun, L. Kloo, H. Pettersson, Dye-sensitized solar cells, *Chemical Reviews* (2010) 110, 6595–6663.
2. Q. Xiang, J. Yu, M. Jaroniec, Preparation and enhanced visible-light photocatalytic H_2-production activity of graphene/C_3N_4 composites, *The Journal of Physical Chemistry C* (2011) 115, 7355–7363.
3. J. Park, B. Koo, K.Y. Yoon, Y. Hwang, M. Kang, J.-G. Park, T. Hyeon, Generalized synthesis of metal phosphide nanorods via thermal decomposition of continuously delivered metal−phosphine complexes using a syringe pump, *Journal of the American Chemical Society* (2005) 127, 8433–8440.
4. E.J. Popczun, J.R. McKone, C.G. Read, A.J. Biacchi, A.M. Wiltrout, N.S. Lewis, R.E. Schaak, Nanostructured nickel phosphide as an electrocatalyst for the hydrogen evolution reaction, *Journal of the American Chemical Society* (2013) 135, 9267–9270.
5. J. Park, B. Koo, Y. Hwang, C. Bae, K. An, J. Park, H.M. Park, T. Hyeon, Novel synthesis of magnetic Fe_2P nanorods from thermal decomposition of continuously delivered precursors using a syringe pump, *Angewandte Chemie* (2004) 116, 2332–2335.
6. D. Li, H. Baydoun, C.N. Verani, S.L. Brock, Efficient water oxidation using CoMnP nanoparticles, *Journal of the American Chemical Society* (2016) 138, 4006–4009.
7. L. Yu, I.K. Mishra, Y. Xie, H. Zhou, J. Sun, J. Zhou, Y. Ni, D. Luo, F. Yu, Y. Yu, Ternary $Ni_{2(1-x)}Mo_{2x}P$ nanowire arrays toward efficient and stable hydrogen evolution electrocatalysis under large-current-density, *Nano Energy* (2018) 53, 492–500.
8. X. Zheng, B. Zhang, P. De Luna, Y. Liang, R. Comin, O. Voznyy, L. Han, F.P.G. De Arquer, M. Liu, C.T. Dinh, Theory-driven design of high-valence metal sites for water oxidation confirmed using in situ soft X-ray absorption, *Nature Chemistry* (2018) 10, 149–154.
9. H. Gao, F. Yang, Y. Zheng, Q. Zhang, J. Hao, S. Zhang, H. Zheng, J. Chen, H. Liu, Z. Guo, Three-dimensional porous cobalt phosphide nanocubes encapsulated in a graphene aerogel as an advanced anode with high coulombic efficiency for high-energy lithium-ion batteries, *ACS Applied Materials & Interfaces* (2019) 11, 5373–5379.
10. N. Zhang, Y. Li, J. Xu, J. Li, B. Wei, Y. Ding, I. Amorim, R. Thomas, S.M. Thalluri, Y. Liu, High-performance flexible solid-state asymmetric supercapacitors based on bimetallic transition metal phosphide nanocrystals, *ACS Nano* (2019) 13, 10612–10621.
11. Z. Pu, J. Zhao, I.S. Amiinu, W. Li, M. Wang, D. He, S. Mu, A universal synthesis strategy for P-rich noble metal diphosphide-based electrocatalysts for the hydrogen evolution reaction, *Energy & Environmental Science* (2019) 12, 952–957.
12. K.U.D. Calvinho, A.B. Laursen, K.M.K. Yap, T.A. Goetjen, S. Hwang, N. Murali, B. Mejia-Sosa, A. Lubarski, K.M. Teeluck, E.S. Hall, Selective CO_2 reduction to C_3 and C_4 oxyhydrocarbons on nickel phosphides at overpotentials as low as 10 mV, *Energy & Environmental Science* (2018) 11, 2550–2559.
13. H. Li, C. Jia, X. Meng, H. Li, Chemical synthesis and applications of colloidal metal phosphide nanocrystals, *Frontiers in Chemistry* (2019) 6, 652.

14. Y. Shi, M. Li, Y. Yu, B. Zhang, Recent advances in nanostructured transition metal phosphides: Synthesis and energy-related applications, *Energy & Environmental Science* (2020) 13, 4564–4582.

15. Q. Guan, W. Li, A novel synthetic approach to synthesizing bulk and supported metal phosphides, *Journal of Catalysis* (2010) 271, 413–415.

16. Y. Yang, G. Zeng, D. Huang, C. Zhang, D. He, C. Zhou, W. Wang, W. Xiong, B. Song, H. Yi, In situ grown single-atom cobalt on polymeric carbon nitride with bidentate ligand for efficient photocatalytic degradation of refractory antibiotics, *Small* (2020) 16, 2001634.

17. H. Zhao, J. Wang, Y. Dong, P. Jiang, Noble-metal-free iron phosphide cocatalyst loaded graphitic carbon nitride as an efficient and robust photocatalyst for hydrogen evolution under visible light irradiation, *ACS Sustainable Chemistry & Engineering* (2017) 5, 8053–8060.

18. Z. Sun, M. Zhu, M. Fujitsuka, A. Wang, C. Shi, T. Majima, Phase effect of Ni_xP_y hybridized with g-C_3N_4 for photocatalytic hydrogen generation, *ACS Applied Materials & Interfaces* (2017) 9, 30583–30590.

19. Q.-Z. Huang, Z.-J. Tao, L.-Q. Ye, H.-C. Yao, Z.-J. Li, $Mn_{0.2}Cd_{0.8}S$ nanowires modified by CoP_3 nanoparticles for highly efficient photocatalytic H_2 evolution under visible light irradiation, *Applied Catalysis B: Environmental* (2018) 237, 689–698.

20. G.M. Kimball, N.S. Lewis, H.A. Atwater, Mg doping and alloying in Zn_3P_2 heterojunction solar cells, In *2010 35th IEEE Photovoltaic Specialists Conference*, IEEE, Honolulu, HI (2010) pp. 1039–1043.

21. F. Wang, A.L. Fahrenbruch, R.H. Bube, Transport mechanisms for Mg/Zn_3P_2 junctions, *Journal of Applied Physics* (1982) 53, 8874–8879.

22. M. Bhushan, A. Catalano, Polycrystalline Zn_3P_2 Schottky barrier solar cells, *Applied Physics Letters* (1981) 38, 39–41.

23. K. Wu, Z. Liu, H. Zhu, T. Lian, Exciton annihilation and dissociation dynamics in group II–V Cd_3P_2 quantum dots, *The Journal of Physical Chemistry A* (2013) 117, 6362–6372.

24. H. Cao, Z. Liu, X. Zhu, J. Peng, L. Hu, S. Xu, M. Luo, W. Ma, J. Tang, H. Liu, PbS/Cd_3P_2 quantum heterojunction colloidal quantum dot solar cells, *Nanotechnology* (2014) 26, 35401.

25. D.O. Scanlon, A. Walsh, Bandgap engineering of $ZnSnP_2$ for high-efficiency solar cells, *Applied Physics Letters* (2012) 100, 251911.

26. T. Kuwano, R. Katsube, K. Kazumi, Y. Nose, Performance enhancement of $ZnSnP_2$ solar cells by a Cu_3P back buffer layer, *Solar Energy Materials and Solar Cells* (2021) 221, 110891.

27. S. Chen, D. Huang, P. Xu, W. Xue, L. Lei, M. Cheng, R. Wang, X. Liu, R. Deng, Semiconductor-based photocatalysts for photocatalytic and photoelectrochemical water splitting: will we stop with photocorrosion, *Journal of Materials Chemistry A* (2020) 8, 2286–2322.

28. H. Mohammadian-Sarcheshmeh, R. Arazi, M. Mazloum-Ardakani, Application of bifunctional photoanode materials in DSSCs: A review, *Renewable and Sustainable Energy Reviews* (2020) 134, 110249.

29. M. Chen, L.-L. Shao, Z.-Y. Yuan, Q.-S. Jing, K.-J. Huang, Z.-Y. Huang, X.-H. Zhao, G.-D. Zou, General strategy for controlled synthesis of Ni_xP_y/carbon and its evaluation as a counter electrode material in dye-sensitized solar cells, *ACS Applied Materials & Interfaces* (2017) 9, 17949–17960.

30. T. Jayaraman, A.P. Murthy, S.J. Lee, K. Karuppasamy, S.R. Arumugam, Y. Yu, M.M. Hanafiah, H.-S. Kim, V. Mittal, M.Y. Choi, Recent progress on synthetic strategies and applications of transition metal phosphides in energy storage and conversion, *Ceramics International* (2020) 47, 4404–4425.

31. Y. Di, S. Jia, N. Li, C. Hao, H. Zhang, S. Hu, H. Liu, Electrocatalytic films of PEDOT incorporating transition metal phosphides as efficient counter electrodes for dye sensitized solar cells, *Solar Energy* (2019) 189, 8–14.

32. M.-S. Wu, C.-J. Chung, Z.-Z. Ceng, Cyclic voltammetric deposition of discrete nickel phosphide clusters with mesoporous nanoparticles on fluorine-doped tin oxide glass as a counter electrode for dye-sensitized solar cells, *RSC Advances* (2015) 5, 4561–4567.

33. Y. Di, Z. Xiao, Z. Zhao, G. Ru, B. Chen, J. Feng, Bimetallic NiCoP nanoparticles incorporating with carbon nanotubes as efficient and durable electrode materials for dye sensitized solar cells, *Journal of Alloys and Compounds* (2019) 788, 198–205.

34. X. Sun, L. Lu, Q. Zhu, C. Wu, D. Yang, C. Chen, B. Han, MoP nanoparticles supported on indium-doped porous carbon: Outstanding catalysts for highly efficient CO_2 electro-reduction, *Angewandte Chemie International Edition* (2018) 57, 2427–2431.

35. S.T. Oyama, Novel catalysts for advanced hydroprocessing: Transition metal phosphides, *Journal of Catalysis* (2003) 216, 343–352.

36. Y.-K. Lee, S.T. Oyama, Bifunctional nature of a SiO_2-supported Ni_2P catalyst for hydrotreating: EXAFS and FTIR studies, *Journal of Catalysis* (2006) 239, 376–389.

37. S. Wang, Y. Xie, K. Shi, W. Zhou, Z. Xing, K. Pan, A. Cabot, Monodispersed nickel phosphide nanocrystals in situ grown on reduced graphene oxide with controllable size and composition as a counter electrode for dye-sensitized solar cells, *ACS Sustainable Chemistry & Engineering* (2020) 8, 5920–5926.

38. K. Zhao, X. Zhang, M. Wang, W. Zhang, X. Li, H. Wang, L. Li, Electrospun carbon nanofibers decorated with Pt-Ni_2P nanoparticles as high efficiency counter electrode for dye-sensitized solar cells, *Journal of Alloys and Compounds* (2019) 786, 50–55.

39. M. Wu, J. Bai, Y. Wang, A. Wang, X. Lin, L. Wang, Y. Shen, Z. Wang, A. Hagfeldt, T. Ma, High-performance phosphide/carbon counter electrode for both iodide and organic redox couples in dye-sensitized solar cells, *Journal of Materials Chemistry* (2012) 22, 11121–11127.

40. Y.Y. Dou, G.R. Li, J. Song, X.P. Gao, Nickel phosphide-embedded graphene as counter electrode for dye-sensitized solar cells, *Physical Chemistry Chemical Physics* (2012) 14, 1339–1342.

41. M.-S. Wu, J.-F. Wu, Pulse-reverse electrodeposition of transparent nickel phosphide film with porous nanospheres as a cost-effective counter electrode for dye-sensitized solar cells, *Chemical Communications* (2013) 49, 10971–10973.

42. L. Su, H. Li, Y. Xiao, G. Han, M. Zhu, Synthesis of ternary nickel cobalt phosphide nanowires through phosphorization for use in platinum-free dye-sensitized solar cells, *Journal of Alloys and Compounds* (2019) 771, 117–123.

43. G. Shen, P.-C. Chen, K. Ryu, C. Zhou, Devices and chemical sensing applications of metal oxide nanowires, *Journal of Materials Chemistry* (2009) 19, 828–839.

44. D. Zeng, W. Xu, W.-J. Ong, J. Xu, H. Ren, Y. Chen, H. Zheng, D.-L. Peng, Toward noble-metal-free visible-light-driven photocatalytic hydrogen evolution: monodisperse sub–15 nm Ni_2P nanoparticles anchored on porous g-C_3N_4 nanosheets to engineer 0D-2D heterojunction interfaces, *Applied Catalysis B: Environmental* (2018) 221, 47–55.

45. J. Xu, Y. Qi, C. Wang, L. Wang, NH_2-MIL-101 (Fe)/Ni (OH)$_2$-derived C, N-codoped Fe_2P/Ni_2P cocatalyst modified g-C_3N_4 for enhanced photocatalytic hydrogen evolution from water splitting, *Applied Catalysis B: Environmental* (2019) 241, 178–186.

46. S. Hua, D. Qu, L. An, W. Jiang, Y. Wen, X. Wang, Z. Sun, Highly efficient p-type Cu_3P/n-type g-C_3N_4 photocatalyst through Z-scheme charge transfer route, *Applied Catalysis B: Environmental* (2019) 240, 253–261.

47. S.P. Saravanan, M.N. Meeran, Increasing the efficiency of dye-sensitized solar cells by NiCoP/g-C3 N4 hybrid composite photoanodes by facile hydrothermal approach, *Ionics* (2021) 27, 407–416.

48. P. Wang, X. Li, Z. Xu, Z. Wu, S. Zhang, W. Xu, H. Zhong, H. Chen, E. Li, J. Luo, Tunable graphene/indium phosphide heterostructure solar cells, *Nano Energy* (2015) 13, 509–517.

49. Y. Lu, T. Wang, X. Li, G. Zhang, H. Xue, H. Pang, Synthetic methods and electrochemical applications for transition metal phosphide nanomaterials, *RSC Advances* (2016) 6, 87188–87212.

50. M. Yoshimura, E. Nakai, K. Tomioka, T. Fukui, Indium tin oxide and indium phosphide heterojunction nanowire array solar cells, *Applied Physics Letters* (2013) 103, 243111.

51. X. Lu, S. Huang, M.B. Diaz, N. Kotulak, R. Hao, R. Opila, A. Barnett, Wide band gap gallium phosphide solar cells, *IEEE Journal of Photovoltaics* (2012) 2, 214–220.

52. C.R. Allen, J.M. Woodall, J.-H. Jeon, Results of a gallium phosphide photovoltaic junction with an AR coating under concentration of natural sunlight, *Solar Energy Materials and Solar Cells* (2011) 95, 2655–2658.

53. Z.S. Bittner, D. V Forbes, M. Nesnidal, S.M. Hubbard, Gallium phosphide solar cells with indium gallium phosphide quantum wells for high temperature applications, In *2011 37th IEEE Photovoltaic Specialists Conference*, IEEE (2011) pp. 1959–1964.

54. R.W. Crisp, N. Kirkwood, G. Grimaldi, S. Kinge, L.D.A. Siebbeles, A.J. Houtepen, Highly photoconductive InP quantum dots films and solar cells, *ACS Applied Energy Materials* (2018) 1, 6569–6576.

10 MXene-Based 2D Nanomaterials for Solar Cells

M.F. Aziz and Mohd Asyadi Azam
Universiti Teknikal Malaysia Melaka

CONTENTS

10.1 INTRODUCTION

The MXenes-based 2D nanomaterials are a novel material, which has been created in 2011 by Gogotsi research group. This titanium-carbide MXenes, $Ti_3C_2T_x$, (T signifies some surface-terminating functional groups ($-O$, $-OH$ and $-F$)), have triggered wide attention and have been explored in different research fields including sensors devices [1–7], catalysis [8–12], water purification [13–15], biomedical application [16–18], electromagnetic fields [19–23], batteries application [24–28], and supercapacitor [29–30]. This is due to the distinctive advantages shown by MXene-based 2D nanomaterials such as good electrical conductivity, high charge carrier transport, outstanding transparency in visible range, and tunable work function (WF) [31].

In the study of solar cell performance, the important parameters that are used to characterize the photovoltaic are including the open-circuit voltage V_{oc}, J_{sc}, fill factor (ff), and solar cell efficiency (η). At the maximum point for J-V curves shows the current density (J_{max}) and voltage (V_{max}). From this study, we can learn and investigate how good the MXene-based 2D nanomaterials influence or impact the solar cell performance.

This book chapter presents the utilization of MXene-based 2D nanomaterials for solar cells application such as perovskite solar cells (PSCs), silicon solar cells,

DOI: 10.1201/9781003178422-10

organic solar cells (OSCs), quantum dot-sensitized solar cells (QDSCs), and dye-sensitized solar cells (DSSCs). In this book chapter also provide the information about the role of MXene-based 2D nanomaterials in the solar cells application such as additive in the electrolyte or electrode, hole transport layer (HTL) or electron transport layer (ETL), and additive in HTL/ETL.

10.2 MXene-BASED 2D NANOMATERIALS FOR PEROVSKITE SOLAR CELLS

The exclusive and attractive properties of 2D MXenes-based nanomaterials that occur from terminating functional groups and oxidation of 2D MXenes-based nanomaterials, show good sign to use in solar cell fabrication such as PSCs. Saranin et al. [32] have proposed the usage of MXenes which has an advantageous role as perovskite absorber and ETL by doping these MXene nanomaterials in the nickel oxide-based inverted PSCs. In this research, the planned method opens uncountable ways for engineering inverted PSC mechanism by owing to the opportunity to excellently adjust the $Ti_3C_2T_x$ WF, which display a high PSC efficiency of ~20%. Another example is $Ti_3C_2T_x$ MXene hydrocolloid, which has got through the oxidation process has been used in the PSC, which is resulting the best PSCs efficiency of ~18% as reported by Yang et al. [33].

According to Wang et al. [34], the 2D MXene materials are expected to boost the charge transport in PSCs. Interface engineering is important to enhance the efficiency of PSCs. The authors offer a vowing method by introducing a 2D carbide (MXene) nanomaterials with great interface contact, which has improved the electron transportation and charge transfer ability of tin oxide (SnO_2) ETL. The 2D MXene nanomaterials-modified tin oxide ETL also proposes a better growth platform for the perovskite layer including decreased trap density. By using a spatially resolved imaging method, profoundly decreased non-radiative recombination and charge transport losses in PSCs based on 2D MXene nanomaterials-modified tin oxide are also recorded. As solar cells performance result, the PSC (Figure 10.1) with minor hysteresis displays an improved solar cell efficiency of ~20.7% (which is including ultralow saturated current density, J). In this research, the authors are offering in-depth systematic knowledge of the 2D MXene nanomaterials interface engineering, proposing an alternate method to achieve efficient PSCs.

Yang et al. [35] have proposed MXenes nanomaterials (a type of 2D Ti_3C_2) that can be employed in the PSCs application due to some advantages such as their exclusive electronic properties, optical character, and plasmonic properties. The authors investigate the usage of the MXene 2D nanomaterials in organic–inorganic lead halide (gaining from reacting halides of lead and organic ammonium with dopants) PSCs. In the methodology section, the composite of tin (IV) oxide (SnO_2) and Ti_3C_2 MXene nanomaterials have been prepared where the amounts of MXene nanomaterials was varied from 0, 0.5, 1.0, 2.0, and 2.5 (in weight percentage). In this study, the MXene has been applied as surface for electron mobility in the PSCs. In the solar cells result, by blending SnO_2 with 1.0 weight percentage of MXene nanomaterials showing improvement of the PSC efficiency, which started from ~17.2% to ~18.3%. However, PSCs utilizing with pure MXene as the ETL achieved a PSC efficiency

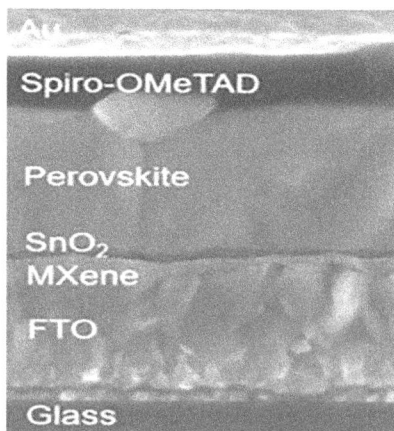

FIGURE 10.1 Configuration of 2D MXenes-based perovskite solar cell. (Adapted with permission from Ref. [34]. Copyright (2020) American Chemical Society.)

of ~5.3%. In the photoluminescence spectroscopy and electrochemical impedance spectroscopy (EIS) study, the results expose that MXene nanosheets offer higher charge transfer pathways. From this analysis, the presence of MXene nanomaterials shows improving the electron behavior at the ETL/PSC intercontact. This phenomenon is heading to greater photocurrents. Overall, this research highlights the usage of MXene nanomaterial and a favorable technique to improve the solar cells' conversion efficiency.

Some researchers have been utilizing the MXene nanosheets with the titanium dioxide doped for PSCs [36]. The inorganic lead-free $Cs_2AgBiBr_6$ double perovskite structure is the promising development path in PSCs device to work out the difficulty of the instability of the $APbX_3$ (A = MA, FA, and Cs; X = Cl, Br, and I) perovskite nanocrystals structure and lead toxicity. Though, the small short-circuit current (I) and power conversion efficiency affected by $Cs_2AgBiBr_6$ (with low crystallization) significantly reduce the optoelectronic research field. Here, the researcher approves an easy method to dope single-layered MXene nanosheets into titanium dioxide (titanium carbide MXene-TiO_2). This will perform as a multifunctional ETL for stability and effective $Cs_2AgBiBr_6$ double PSCs. The single-layered MXene nanosheets suggestively increased the electrical conductivity and electron extraction rate of titanium dioxide. At the same time, the single-layered MXene nanosheets shift the surface wettability of the ETL and boost the crystallization of the $Cs_2AgBiBr_6$ double PSCs assembly. Hence, the PSC (Figure 10.2) increased (hysteresis was eliminated) by approximately >40% to 2.81% (V_{oc} of 0.96 V, J_{sc} of 4.14 mA cm^{-2}, and ff of 0.70) compared to that of a titanium dioxide-based device (V_{oc} of 0.93 V, J_{sc} of 2.84 mA cm^{-2}, ff of 0.62, and η of 1.64%). Additionally, the PSCs based on titanium carbide MXene-titanium dioxide proved the long-term operational stability after keeping this PCS for 15 days with normal air conditions. This is resulting the PSCs efficiency still same showing a retention rate of ~90% of the primary one. This work showed the capability of titanium carbide MXene-titanium dioxide in ETL material for high-performance double PSCs.

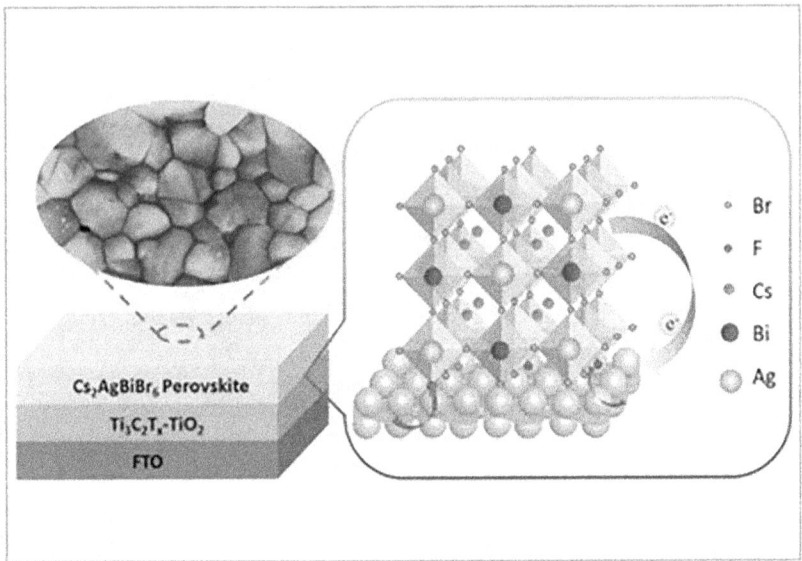

FIGURE 10.2 Schematic diagram for double perovskite solar cell fabrication. (Adapted with permission from Ref. [36]. Copyright (2021) American Chemical Society.)

10.3 MXene-BASED 2D NANOMATERIALS FOR SILICON SOLAR CELLS

In the silicon-type solar cells research, which is owing to their high-power conversion efficiency as attraction of global interest, a heterostructure solar cells (HSCs) utilizing donor and acceptor semiconductors can be constructed. According to Zhang et al. [37], a complete evaluation of 64 2D transition metal carbides (MXenes) was performed. The results emphasize that zirconium carbide (Zr_2CO_2) and transition metal carbides MXene are encouraging donor and acceptor materials, individually. Excitingly, the HSCs have a suitable bandgap (1,220 meV) and display a clear absorbance coefficient. Furthermore, the heterostructures (type-II nature) might generate valuable electron–hole separation. Moreover, the photocurrents of transition metal carbides/zirconium carbide and transition metal carbides/Hf_2CO_2 HSCs fabrication are competitive with the standard silicon devices. In detail, transition metal carbides/zirconium carbide and transition metal carbides/Hf_2CO_2 HSCs provide a high efficiency of ~23% and ~20%, individually. This study also exposes the method for enabling the potential application of 2D MXenes as solar cell materials.

10.4 MXene-BASED 2D NANOMATERIALS FOR ORGANIC SOLAR CELLS

As reported by Tang et al. [38], which is employing the MXene nanomaterials in the OSCs due to the excellent properties such as good metallic conductivity and hydrophilicity. In this research, the manufacture of visible, highly conductance, and suitable MXene and silver nanowire (AgNW) hybrid films has been reported. This results in the maximum figure of merit (a quantity used to characterize the performance of device) of ~162.50 in reported literature to date concerning the 2D MXene nanomaterial-based transparent electrode. The preparation of the hybrid films has been carried out through a straightforward and scalable solution-processed technique. By using this technique, the prepared hybrid films display high electrical conductivity, maximum transmittance, low toughness, suitable for WF, and strong mechanical properties. In the OSC assembly, the hybrid electrodes should exhibit to work as transparent electrodes in fullerene molecule (an allotrope of carbon) PTB7-Th:PC71BM and nonfullerene molecule PBDB-T-ITIC OSCs. In additional research to develop the performance of flexible OSCs, the PBDB-T:ITIC:PC71BM assembly has shown an efficiency of ~8%. The mechanical properties of this OCS have been studied. The flexible ternary OSCs can retain 84.6% of the original OSC efficiency after 1,000 bending and unbending cycles to a 5-mm bending radius. These OSCs and mechanical performance metrics describe a discovery in flexible OSC field.

According to Yu et al. [39], MXenes two-dimensional nanomaterials have fascinated good interest and potentially use in the OSCs. In this research, the utilization of MXene nanomaterials for the electron–hole collections of OSCs has been studied and shows the OSC of ~9.1%. This OSC efficiency is similar to that using standard charge-collection buffer layer materials.

10.5   MXene-BASED 2D NANOMATERIALS FOR QUANTUM DOT-SENSITIZED SOLAR CELLS

There are solar cells studies that are employing the MXene nanomaterials in the QDSCs as reported by Chen et al. [40]. This research shows that the copper selenide (CuSe) nanoparticles have been in situ growing on the MXene nanomaterials using one-step hydrothermal method to form the copper selenide – MXene composite. The copper selenide – MXene slurry was produced using a solvothermal method where the 20 mg of copper selenide – MXene and polyvinylidene fluoride (PVDF) (6 mg) has been mixed in N methyl-2-pyrrolidone (NMP) (300 µL). In the QDSCs, the counter electrode has been created using copper selenide – MXene slurry, which has been screen-printed (one layer) on a graphite sheet. The QDSCs have been configured with the copper selenide – MXene-based electrode, parafilm spacer (80 µm), and photoanodes (CdS/CdSe co-sensitized TiO_2). Then, the QDSC has been inserted with the polysulfide aqueous liquid electrolyte via the hole in the spacer. This electrolyte is containing sodium sulfide, sulfur, and potassium hydroxide. In the solar cell study, current density, J – voltage, V of QDSC utilizing copper selenide, copper selenide-MXene, and MXene counter electrode have been measured. The QDSC was done under illumination of 100×10^{-3} cm^2 (AM 1.5) with QDSC active area fixed at 1×10^{-1} cm^2. The authors reported that the QDSCs employing copper selenide-based counter electrode show solar cell parameter such as short-circuit current density (J_{sc}) of $\sim 17 \times 10^{-3}$ A cm^{-2}, V_{oc} of 547 mV, ff of 37%, and resulting QDSC efficiency of ~3.5%). Another test, when the QDSC using the copper selenide-0.3 mg of MXene counter electrode, it is resulting in an increment for the QDSC parameter such as J_{sc} of $\sim 19 \times 10^{-3}$ A cm^{-2}, V_{oc} of 565 mV, ff of 48%, and QDSC efficiency of 5.12%. This

QDSC improvement could be attributed to a 3D structure of copper selenide-MXene, which is leading to a large specific surface area (SSA) of the counter electrode. Another explanation from the author also described that the catalytic activity of counter electrode toward polysulfide electrolytes is increased, which is at the counter electrode-polysulfide the charge transport struggle is decreased. However, when the QDSC is utilizing the MXene counter electrode, the QDSC performance displays a decreasing trend where the short-circuit current density of ~10×10^{-3} A cm^{-2}, open-circuit voltage of 530 mV, fill factor of 39%, and QDSC efficiency achieved at ~2%. The incident photon to current efficiency patterns of the QDSC using various types of counter electrodes has been measured. The QDSC utilizing of copper selenide-MXene counter electrode shows an increment trend compared with QDSC using MXene counter electrode or copper selenide counter electrode, which is in good statement to support the increasing of QDSC's short-circuit current density, J_{sc}. The intercontact resistance and charge transfer at the interfaces in QDSC have been studied by carrying out the EIS analysis. From this EIS study, the Nyquist plot for all QDSC has been obtained, which has been fitted by an equivalent circuit resulting in charge transfer resistance at the counter electrode/electrolyte intercontact, R_{ct1} and the second charge transfer resistance at the photoanode/electrolyte intercontact, R_{ct2} [41]. The R_{ct1} and R_{ct2} shown by copper selenide-30 mg MXene counter electrode have the lowest value than both counter electrodes (copper selenide and MXene). This is because the combination of copper selenide and MXene resulting 3D structure, which provides more diffusion pathways and improves the maximum electrocatalytic area for redox reaction in the electrolyte.

According to Sharbirin et al. [42], the utilizing of metal carbides and nitrides or known as MXene nanomaterials is an evolving level of 2D nanolayer structures. $Ti_3C_2T_x$ nanomaterials were discovered by worldwide applications in sensor, batteries, supercapacitor, catalysis, and others. This is because the $Ti_3C_2T_x$ MXenes nanomaterials display good electronic conductivity and very broad range in optic properties. Though, the presence of $Ti_3C_2T_x$ nanomaterials has restricted their ability to correlate to light characterization. In this work, $Ti_3C_2T_x$ nanomaterials were potentially used in the light-emitting (LE)-MXene quantum dots (MQDs), which exhibited excellent advancement, and MQDs displayed multi-color photoluminescence emission along with maximum quantum yield. The synthesis techniques also act an important role in controlling the light emission properties of these MQDs. In this paper, a summary of LE-MQDs and their synthesis processes, optical characteristic, and applications in different optical fields, sensor devices, and imaging devices have been discussed. The next project of LE-MQDs is also reviewed to give an understanding that helps to further advance the development on MQDs.

According to Xue et al. [43], the Ti_3C_2 MXene nanomaterials can be utilized in the quantum dot. In this work, by using a facile hydrothermal method for the construction of photoluminescent Ti_3C_2 MQDs are described. This is considerably expanding the functions of MXene-based nanomaterials, which is potentially used in QDSCs fabrication. Excitingly, the as-prepared MQDs display excitation-dependent photoluminescence region with ~10% quantum yields due to powerful quantum confinement. This research is also demonstrating the applications of Ti_3C_2 MQD as biocompatible multicolor cellular imaging probes and zinc ion sensors.

10.6 MXene-BASED 2D NANOMATERIALS FOR DYE-SENSITIZED SOLAR CELLS

In the dye-sensitized solar cells (DSSCs) application, some research that is utilizing the MXene-based 2D nanomaterials as a shield layer of the counter electrode which is acting as conducting layer and catalyst as well [44]. Usually in the DSSC fabrication, the counter electrode is consisting of the transparent conducting oxide (TCO) layer and platinum (Pt)-based catalyst. The TCO layer of the counter electrode is responsible to assist the electron transport from the outer circuit to enter the DSSC. At the same time, the electrolyte will be oxidized by accepting the electron, which has been arrived through the counter electrode. The Pt-based catalyst will boost the efficient regeneration process in the electrolytes. From this research, both materials (TCO and Pt solution) have been replaced with the MXene-based 2D nanomaterials. The employing of MXene-based 2D nanomaterials in the counter electrode (substitute TCO-Pt) has been carried out by optimizing the layer thickness of MXene to offer a high DSSC efficiency. In this research, five samples counter electrode has been prepared with different thickness of MXene (2, 4, 6, 8, and 10 μm) and the standard counter electrode (TCO-Pt-based) as comparison reasons and to reach the highest possible implementation. Scanning electron microscope (SEM) and X-ray diffraction (XRD) have been performed to examine the MXene layer morphology and composition, respectively. From the SEM result, the MXene layer is in a microscale structure. XRD patterns of MXene show that the peaks appear in the right position approaching 2θ values (with related indices) of 10 (002), 20 (006), 40 (002), and 60 (110) represent of MXene.

The cyclic voltammetry has been done with the configuration of working electrode, counter electrode (Pt wire), reference electrode (Ag/AgCl), and electrolyte to test the catalytic process and stability purpose. By this CV analysis, no change in the shape of the curves for each variation thickness MXene-based electrode sample indicates the catalyst is more stable in the electrolyte. Fabrication of DSSC has been done by arranging the polymer sealing layer fill with the electrolyte in between of titanium dioxide (TiO_2)-dye photoanode and MXene-based counter electrode. EIS of the DSSC fabrication has been done to analyze the equivalent circuit. It shows that the R_s (serial resistance) and R_{ct} decrease as increasing is the thickness of MXene-based electrode, which has improved the surface area. The increasing of the MXene layer also advances the reduction rate of I_3^- ions and decreases the value of R_{ct}. The lowest R_{ct} value of ~1.8 Ω cm^{-2} was observed for counter electrode with MXene nanomaterials (8 μm thickness). This is because that the ideal contact surface area between electrolytes and catalytic sites of electrode affords for the good electronic path. The Z_w (Warburg diffusion resistance of I^- / I_3^- redox electrolytes) was obtained in the range between 1 and ~3 Ω cm^2. This shows that the contact surface for all MXene-based electrodes with electrolytes is in a good condition. The highest DSSC efficiency of ~8.7% was obtained by the electrode with the MXene nanomaterials (8 μm thickness). This is because the J_{sc} is increasing as the MXene thickness is increasing, which improves the ion transport between counter electrode and electrolytes. As comparison, the DSSC utilizing standard TCO-Pt counter electrode only shows the efficiency of ~8.3%. This proved that the replacement of TCO-Pt with the MXene-based 2D nanomaterials offers high conductivity, good catalytic process of MXene toward I^- / I_3^- electrolyte. The author concludes that the MXene nanomaterials have potential to replace the standard TCO and Pt in the counter electrode of DSSCs.

In Ref. [45], the researcher prepared the photoanode by using oxidizing process of 2D-MXene nanosheets (Ti_3C_2 MXene) with the nanoporous of TiO_2 for DSSCs fabrication. The electrodes preparation was started by using doctor-blade technique, the MXene slurry or TiO_2 paste was coated onto fluorine tin oxide (FTO) glass, which has been cleaned earlier. Then, FTO glass with coated MXene or TiO_2 was placed in the furnace with a temperature of 450°C for ~30 minutes. For the MXene-FTO glass electrode, this electrode has been cooled down to 100°C and then exposed to the outside air at various temperatures of 150°C–500°C for the oxidation process. Both MXene-FTO and TiO_2-fluorine tin oxide electrode have been soaked in the N719 (Ru-based complex dye sensitizer) for 18 hours at a temperature of 25°C under dark condition place. Then, MXene-FTO and TiO_2-FTO glass electrode were cleaned using ethanol solvent to remove the unbound dyes particles. All the electrodes prepared were tested in DSSCs. For the preparation of DSSCs fabrication, the MXene-FTO or TiO_2-FTO glass was located with the platinum electrode and sealed together using a thermal adhesive layer by supplying the heat at ~110°C on a hot surface. After the sealing process, the electrolyte was then inserted via a tiny gap in between both electrodes. DSSC characteristics were measured under illumination of 100×10^{-3} W cm^2 (AM1.5). In this work, the impacts of oxidation temperature and period time together with different thicknesses on the DSSC performance were explored. The best result has been obtained, which shows the DSSC efficiency of ~2.7%.

As reported by Adibah et al. [46], the new types of 2D transition metal carbides, carbonitrides, nitrides (known as MXene nanomaterials) have been found in the year of 2011. In this report, the synthesis MXene has been used for electrolyte filler for DSSCs fabrication. This is the MXene nanomaterials that have exclusive characteristic, which have made it fascinating for transparent conducting films, photocatalysts, charge–discharge batteries, electric double layer capacitor, catalysts, and flexible high-strength composites applications. The methodology section shows that the MXene nanomaterial has been synthesized via a discerning etching process by using either in situ or direct hydrofluoric acid (HF) techniques. This research details the influence of the in situ hydrofluoric acid and direct hydrofluoric acid etching processes on the morphology of the synthesis 2D MXene nanomaterials employing titanium aluminum carbide (Ti_2AlC_3) as a precursor. Finally, the results display that the Ti_2AlC_3 nanomaterial produced via the directly hydrofluoric acid process was effectively delaminated, which is in contrast with in situ hydrofluoric acid methods.

10.7 CONCLUSION

As conclusion, the employing of MXene-based 2D nanomaterials for solar cells application such as PSCs, silicon solar cells, OSCs, QDSCs, and DSSCs has been discussed in this book chapter. In the PSC, the researcher has presented the method of opens uncountable ways for engineering inverted PSC mechanism by owing to the opportunity to excellently adjust the $Ti_3C_2T_x$ WF, which display a high PSC efficiency of ~20%. Then the other example, $Ti_3C_2T_x$ MXene hydrocolloid, which has got through the oxidation process has been used in the PSC, which is resulting in the best PSCs efficiency of ~18%. Another research, the PSC based on 2D MXene nanomaterials-modified tin oxide and with minor hysteresis display an improved solar cell efficiency of ~20.7%. MXene has been applied as surface for electron mobility in the PSCs by blending SnO_2 with 1.0 weight percentage of MXene nanomaterials showing improvement of the PSC efficiency, which started from ~17.2% to ~18.3%. Another PSC analysis, the single-layered MXene nanosheets shift the surface wettability of the ETL and boost the crystallization of the $Cs_2AgBiBr_6$ double PSCs assembly. Hence, the PSC increased (hysteresis was eliminated) by approximately >40% to 2.81% (V_{oc} of 0.96V, J_{sc} of 4.14mA cm^{-2}, and ff of 0.70).

In the silicon-type solar cells research, which is owing to their high-power conversion efficiency as attraction of global interest, HSCs utilizing donor and acceptor semiconductors can be constructed. By completing the evaluation of 64 2D transition metal carbides (MXenes), it emphasizes the zirconium carbide and transition metal carbides MXene are encouraging donor and acceptor materials, individually. In detail, transition metal carbides/zirconium carbide and transition metal carbides/Hf_2CO_2 HSCs provide a high efficiency of ~23% and ~20%, individually. This study also exposes that the method for enabling the potential application of 2D MXenes as solar cell materials.

In the OSC assembly, the hybrid electrodes should exhibit to work as transparent electrodes in fullerene molecule (an allotrope of carbon) PTB7-Th:PC71BM and nonfullerene molecule PBDB-T-ITIC OSCs. In additional research to develop the performance of flexible OSCs, the PBDB-T:ITIC:PC71BM assembly has shown an efficiency of ~8%. The mechanical properties of this OCS have been studied. The flexible ternary OSCs can retain 84.6% of the original OSC efficiency after 1,000 bending and unbending cycles to a 5-mm bending radius. Another example of OSCs, the application of MXene nanomaterials for the electron–hole collections in OSCs displays the efficiency ~9.1%. This OSC efficiency is similar to that using standard charge-collection buffer layer materials.

In the QDSCs section, some researcher has prepared the copper selenide (CuSe) nanoparticles with in situ growing on the MXene nanomaterials using one-step hydrothermal method to form the copper selenide – MXene composite. This QDSC using the copper selenide-0.3 mg of MXene counter electrode is resulting in an increment for the QDSC parameter such as J_{sc} of ~19×10^{-3} A cm^{-2}, V_{oc} of 565mV, ff of 48%, and QDSC efficiency of 5.12%. This QDSC improvement could be attributed to a 3D structure of copper selenide-MXene, which is leading to large SSA of the counter electrode. In addition, the catalytic activity of counter electrode toward polysulfide electrolytes is increasing, which is at the counter electrode-polysulfide the charge transport struggle is decreased. Some researcher has proposed a review paper describing the employing of MXene in QDSC. $Ti_3C_2T_x$ nanomaterials has potentially used in the light emitting (LE)-MQDs which has exhibited excellent advancement, and MQDs displaying multi-color photoluminescence emission along with maximum quantum yield have been assembled. The synthesis techniques also act an important role in controlling the light emission properties of these MQDs. In this review paper, a summary of LE-MQDs and their synthesis processes, optical characteristic, and applications in different optical fields, sensor devices, and imaging devices has been discussed. The next project of LE-MQDs is also reviewed to give an understanding that helps to further advance the development on MQDs. Some researcher has utilized the facile hydrothermal method for the construction of photoluminescent Ti_3C_2 MQDs. This is considerably expanding the functions of MXene-based nanomaterials, which is potentially used in QDSCs fabrication. Excitingly, the as-prepared MQDs display excitation-dependent photoluminescence region with ~10% quantum yields due to powerful quantum confinement. This research is also showing the applications of Ti_3C_2 MQDs as biocompatible multicolor cellular imaging probes and zinc ion sensors.

In the DSSCs application, some research that is utilizing the MXene-based 2D nanomaterials as a shield layer of the counter electrode, which acts as conducting layer and catalyst as well. The employing of MXene-based 2D nanomaterials in the counter electrode (substitute TCO-Pt) has been carried out by optimizing the layer thickness of MXene to offer a high DSSC efficiency. The highest DSSC efficiency of ~8.7% was obtained by the electrode with the MXene nanomaterials (8 μm thickness). This is because the J_{sc} is increasing as there is an increase of MXene thickness, which is improving the ion transport between counter electrode and electrolytes. This proved that the replacement of TCO-Pt with the MXene-based 2D nanomaterials offers high conductivity, good catalytic process of MXene toward I^- / I_3^- electrolyte and completes that the MXene nanomaterials have the potential to replace the standard TCO and Pt in the counter electrode of DSSCs. In another example, the photoanode has been prepared by using oxidizing process of 2D-MXene nanosheets (Ti_3C_2 MXene) with the nanoporous of TiO_2 for DSSCs. In this study, the impacts of oxidation temperature and period time together with different thicknesses on the DSSC performance were explored. The best result has been obtained, which shows the DSSC efficiency of ~2.7%. Another DSSC report, the synthesis MXene has been used for electrolyte filler for DSSCs fabrication. This research details the influence of the in situ hydrofluoric acid and direct hydrofluoric acid etching processes on the morphology of the synthesis 2D MXene nanomaterials are employing titanium aluminum carbide (Ti_2AlC_3) as a precursor. Finally, the results display that the Ti_2AlC_3 nanomaterial produced via the directly hydrofluoric acid process was effectively delaminated, which is in contrast with in situ hydrofluoric acid methods.

REFERENCES

1. Zhao, Q., Yang, L., Ma, Y., Huang, H., He, H., Ji, H., Wang, Z., & Qiu, J. (2021). Highly sensitive, reliable and flexible pressure sensor based on piezoelectric PVDF hybrid film using MXene nanosheet reinforcement. *Journal of Alloys and Compounds*, 886, 161069. doi: 10.1016/j.jallcom.2021.161069.
2. Xu, S., Dall'Agnese, Y., Wei, G., Zhang, C., Gogotsi, Y., & Han, W. (2018). Screen-printable microscale hybrid device based on MXene and layered double hydroxide electrodes for powering force sensors. *Nano Energy*, 50, 479–488. doi: 10.1016/j.nanoen.2018.05.064.
3. Peng, Z., Zhang, X., Zhao, C., Gan, C., & Zhu, C. (2021). Hydrophobic and stable MXene/ reduced graphene oxide/polymer hybrid materials pressure sensors with an ultrahigh sensitive and rapid response speed pressure sensor for health monitoring. *Materials Chemistry and Physics*, 271, 124729. doi: 10.1016/j.matchemphys.2021.124729.

4. Zhang, D., Mi, Q., Wang, D., & Li, T. (2021). MXene/Co_3O_4 composite based form-aldehyde sensor driven by ZnO/MXene nanowire arrays piezoelectric nanogenerator. *Sensors and Actuators B: Chemical*, 339, 129923. doi: 10.1016/j.snb.2021.129923.

5. Su, T., Liu, N., Gao, Y., Lei, D., Wang, L., Ren, Z., Zhang, Q., Su, J., & Zhang, Z. (2021). MXene/cellulose nanofiber-foam based high performance degradable piezore-sistive sensor with greatly expanded interlayer distances. *Nano Energy*, 87, 106151. doi: 10.1016/j.nanoen.2021.106151.

6. Wang, X., Zhang, D., Zhang, H., Gong, L., Yang, Y., Zhao, W., Yu, S., Yin, Y., & Sun, D. (2021). In situ polymerized polyaniline/MXene (V2C) as building blocks of superca-pacitor and ammonia sensor self-powered by electromagnetic-triboelectric hybrid gen-erator. *Nano Energy*, 88, 106242. doi: 10.1016/j.nanoen.2021.106242.

7. Hao, S., Liu, C., Chen, X., Zong, B., Wei, X., Li, Q., Qin, H., & Mao, S. (2021). $Ti_3C_2T_x$ MXene sensor for rapid Hg2+ analysis in high salinity environment. *Journal of Hazardous Materials*, 418, 126301. doi: 10.1016/j.jhazmat.2021.126301.

8. Guo, Z., Li, Y., Sa, B., Fang, Y., Lin, J., Huang, Y., Tang, C., Zhou, J., Miao, N., & Sun, Z. (2020). M2C-type MXenes: Promising catalysts for CO_2 capture and reduction. *Applied Surface Science*, 521, 146436. doi: 10.1016/j.apsusc.2020.146436.

9. Wang, F., Lai, Y., Fang, Q., Li, Z., Ou, P., Wu, P., Duan, Y., Chen, Z., Li, S., & Zhang, Y. (2020). Facile fabricate of novel Co(OH)F@MXenes catalysts and their catalytic activity on bisphenol A by peroxymonosulfate activation: The reaction kinetics and mechanism. *Applied Catalysis B: Environmental*, 262, 118099. doi: 10.1016/j.apcatb.2019.118099.

10. Sun, J., Kong, W., Jin, Z., Han, Y., Ma, L., Ding, X., Niu, Y., & Xu, Y. (2020). Recent advances of MXene as promising catalysts for electrochemical nitrogen reduction reac-tion. *Chinese Chemical Letters*, 31(4), 953–960. doi: 10.1016/j.cclet.2020.01.035.

11. Li, K., Zhang, S., Li, Y., Fan, J., & Lv, K. (2021). MXenes as noble-metal-alternative co-catalysts in photocatalysis. *Chinese Journal of Catalysis*, 42(1), 3–14. doi: 10.1016/S1872-2067(20)63630-0.

12. Zhu, Q., Cui, Y., Zhang, Y., Cao, Z., Shi, Y., Gu, J., Du, Z., Li, B., & Yang, S. (2021). Strategies for engineering the MXenes toward highly active catalysts. *Materials Today Nano*, 13, 100104. doi: 10.1016/j.mtnano.2020.100104.

13. Fan, D., Lu, Y., Zhang, H., Xu, H., Lu, C., Tang, Y., & Yang, X. (2021). Synergy of photocatalysis and photothermal effect in integrated 0D perovskite oxide/2D MXene heterostructures for simultaneous water purification and solar steam generation. *Applied Catalysis B: Environmental*, 295, 120285. doi: 10.1016/j.apcatb.2021.120285.

14. Jin, Y., Fan, Y., Meng, X., Li, J., Li, C., Sunarso, J., Yang, N., Meng, B., & Zhang, W. (2021). Modeling of hydrated cations transport through 2D MXene ($Ti_3C_2T_x$) mem-branes for water purification. *Journal of Membrane Science*, 631, 119346. doi: 10.1016/j.memsci.2021.119346.

15. Sun, Y., Xu, D., Li, S., Cui, L., Zhuang, Y., Xing, W., & Jing, W. (2021). Assembly of multidimensional MXene-carbon nanotube ultrathin membranes with an enhanced anti-swelling property for water purification. *Journal of Membrane Science*, 623, 119075. doi: 10.1016/j.memsci.2021.119075.

16. Rafieerad, A., Sequiera, G. L., Yan, W., Kaur, P., Amiri, A., & Dhingra, S. (2020). Sweet-MXene hydrogel with mixed-dimensional components for biomedical applica-tions. *Journal of the Mechanical Behavior of Biomedical Materials*, 101, 103440. doi: 10.1016/j.jmbbm.2019.103440.

17. Lim, G. P., Soon, C. F., Ma, N. L., Morsin, M., Nayan, N., Ahmad, M. K., & Tee, K. S. (2021). Cytotoxicity of MXene-based nanomaterials for biomedical applications: A mini review. *Environmental Research*, 201, 111592. doi: 10.1016/j.envres.2021.111592.

18. George, S. M., & Kandasubramanian, B. (2020). Advancements in MXene-polymer com-posites for various biomedical applications. *Ceramics International*, 46(7), 8522–8535. doi: 10.1016/j.ceramint.2019.12.257.

19. Guo, S., Guan, H., Li, Y., Bao, Y., Lei, D., Zhao, T., Zhong, B., & Li, Z. (2021). Dual-loss $Ti_3C_2T_x$ MXene/$Ni_{0.6}Zn_{0.4}Fe_2O_4$ heterogeneous nanocomposites for highly efficient electromagnetic wave absorption. *Journal of Alloys and Compounds*, 887, 161298. doi: 10.1016/j.jallcom.2021.161298.

20. Lu, Y., Zhang, S., He, M., Wei, L., Chen, Y., & Liu, R. (2021). 3D cross-linked graphene or/and MXene based nanomaterials for electromagnetic wave absorbing and shielding. *Carbon*, 178, 413–435. doi: 10.1016/j.carbon.2021.01.161.

21. Iqbal, A., Sambyal, P., Kwon, J., Han, M., Hong, J., Kim, S. J., Kim, M.-K., Gogotsi, Y., & Koo, C. M. (2021). Enhanced absorption of electromagnetic waves in $Ti_3C_2T_x$ MXene films with segregated polymer inclusions. *Composites Science and Technology*, 213, 108878. doi: 10.1016/j.compscitech.2021.108878.

22. Yin, L., Kang, H., Ma, H., Wang, J., Liu, Y., Xie, Z., Wang, Y., & Fan, Z. (2021). Sunshine foaming of compact $Ti_3C_2T_x$ MXene film for highly efficient electromagnetic interference shielding and energy storage. *Carbon*, 182, 124–133. doi: 10.1016/j.carbon.2021.05.048.

23. Song, P., Liu, B., Qiu, H., Shi, X., Cao, D., & Gu, J. (2021). MXenes for polymer matrix electromagnetic interference shielding composites: A review. *Composites Communications*, 24, 100653. doi: 10.1016/j.coco.2021.100653.

24. Zhang, Y., Ma, C., He, W., Zhang, C., Zhou, L., Wang, G., & Wei, W. (2021). MXene and MXene-based materials for lithium-sulfur batteries. *Progress in Natural Science: Materials International*, 31 (4), 501–513. doi: 10.1016/j.pnsc.2021.07.003.

25. Giebeler, L., & Balach, J. (2021). MXenes in lithium–sulfur batteries: Scratching the surface of a complex 2D material: A minireview. *Materials Today Communications*, 27, 102323. doi: 10.1016/j.mtcomm.2021.102323.

26. Wang, X., Wang, X., Chen, J., Zhao, Y., Mao, Z., & Wang, D. (2021). Durable sodium battery composed of conductive $Ti_3C_2T_x$ MXene modified gel polymer electrolyte. *Solid State Ionics*, 365, 115655. doi: 10.1016/j.ssi.2021.115655.

27. Yin, T., Li, Y., Wang, R., Al-Hartomy, O. A., Al-Ghamdi, A., Wageh, S., Luo, X., Tang, X., & Zhang, H. (2021). Synthesis of $Ti_3C_2F_x$ MXene with controllable fluorination by electrochemical etching for lithium-ion batteries applications. *Ceramics International*, 47 (20), 28642–28649. doi: 10.1016/j.ceramint.2021.07.023.

28. Zhong, S., Ju, S., Shao, Y., Chen, W., Zhang, T., Huang, Y., Zhang, H., Xia, G., & Yu, X. (2021). Magnesium hydride nanoparticles anchored on MXene sheets as high capacity anode for lithium-ion batteries. *Journal of Energy Chemistry*, 62, 431–439. doi: 10.1016/j.jechem.2021.03.049.

29. Fan, Q., Zhao, R., Yi, M., Qi, P., Chai, C., Ying, H., & Hao, J. (2022). Ti_3C_2-MXene composite films functionalized with polypyrrole and ionic liquid-based microemulsion particles for supercapacitor applications. *Chemical Engineering Journal*, 428, 131107. doi: 10.1016/j.cej.2021.131107.

30. Ruan, C., Zhu, D., Qi, J., Meng, Q., Wei, F., Ren, Y., Sui, Y., & Zhang, H. (2021). MXene-modulated $CoNi_2S_4$ dendrite as enhanced electrode for hybrid supercapacitors. *Surfaces and Interfaces*, 25, 101274. doi: 10.1016/j.surfin.2021.101274.

31. Yin, L., Li, Y., Yao, X., Wang, Y., Jia, L., Liu, Q., Li, J., Li, Y., & He, D. (2021). MXenes for solar cells. *Nano-Micro Letters*, 13(1), 78. doi: 10.1007/s40820-021-00604-8.

32. Saranin, D., Pescetelli, S., Pazniak, A., Rossi, D., Liedl, A., Yakusheva, A., Luchnikov, L., Podgorny, D., Gostischev, P., Didenko, S., Tameev, A., Lizzit, D., Angelucci, M., Cimino, R., Larciprete, R., Agresti, A., & Carlo, A.D. (2021). Transition metal carbides (MXenes) for efficient NiO-based inverted perovskite solar cells. *Nano Energy*, 82, 105771. doi: 10.1016/j.nanoen.2021.105771.

33. Yang, L., Kan, D., Dall'Agnese, C., Dall'Agnese, Y., Wang, B., Jena, A. K., Wei, Y., Chen, G., Wang, X.-F., Gogotsi, Y., & Miyasaka, T. (2021). Performance improvement of MXene-based perovskite solar cells upon property transition from metallic

to semiconductive by oxidation of Ti$_3$C$_2$T$_x$ in air. *Journal of Materials Chemistry A*, 9(8), 5016–5025. doi: 10.1039/D0TA11397B.

34. Wang, Y., Xiang, P., Ren, A., Lai, H., Zhang, Z., Xuan, Z., Wan, Z., Zhang, J., Hao, X., Wu, L., Sugiyama, M., Schwingenschlögl, U., Liu, C., Tang, Z., Wu, J., Wang, Z., & Zhao, D. (2020). MXene-Modulated Electrode/SnO$_2$ interface boosting charge transport in perovskite solar cells. *ACS Applied Materials & Interfaces*, 12(48), 53973–53983. doi: 10.1021/acsami.0c17338.

35. Yang, L., Dall'Agnese, Y., Hantanasirisakul, K., Shuck, C. E., Maleski, K., Alhabeb, M., Chen, G., Gao, Y., Sanehira, Y., Jena, A. K., Shen, L., Dall'Agnese, C., Wang, X.-F., Gogotsi, Y., & Miyasaka, T. (2019). SnO$_2$–Ti$_3$C$_2$ MXene electron transport layers for perovskite solar cells. *Journal of Materials Chemistry A*, 7(10), 5635–5642. doi: 10.1039/C8TA12140K.

36. Li, Z., Wang, P., Ma, C., Igbari, F., Kang, Y., Wang, K.-L., Song, W., Dong, C., Li, Y., Yao, J., Meng, D., Wang, Z.-K., & Yang, Y. (2021). Single-layered MXene nanosheets doping TiO$_2$ for efficient and stable double perovskite solar cells. *Journal of the American Chemical Society*, 143(6), 2593–2600. doi: 10.1021/jacs.0c12739.

37. Zhang, Y., Xiong, R., Sa, B., Zhou, J., & Sun, Z. (2021). MXenes: Promising donor and acceptor materials for high-efficiency heterostructure solar cells. *Sustainable Energy & Fuels*, 5(1), 135–143. doi: 10.1039/D0SE01443E.

38. Tang, H., Feng, H., Wang, H., Wan, X., Liang, J., & Chen, Y. (2019). Highly conducting MXene-silver nanowire transparent electrodes for flexible organic solar cells. *ACS Applied Materials & Interfaces*, 11(28), 25330–25337. doi: 10.1021/acsami.9b04113.

39. Yu, Z., Feng, W., Lu, W., Li, B., Yao, H., Zeng, K., & Ouyang, J. (2019). MXenes with tunable work functions and their application as electron- and hole-transport materials of non-fullerene organic solar cells. *Journal of Materials Chemistry A*, 7, 11160–11169. doi: 10.1039/C9TA01195A.

40. Chen, Y., Wang, D., Lin, Y., Zou, X., & Xie, T. (2019). In suit growth of CuSe nanoparticles on MXene (Ti$_3$C$_2$) nanosheets as an efficient counter electrode for quantum dot-sensitized solar cells. *Electrochimica Acta*, 316, 248–256. doi: 10.1016/j.electacta.2019.05.132.

41. Kern, R., Sastrawan, R., Ferber, J., Stangl, R., & Luther, J. (2002). Modeling and interpretation of electrical impedance spectra of dye solar cells operated under open-circuit conditions. *Electrochimica Acta*, 47(26), 4213–4225. doi: 10.1016/S0013-4686(02)00444-9.

42. Sharbirin, A. S., Akhtar, S., & Kim, J. (2021). Light-emitting MXene quantum dots. *Opto-Electronic Advances*, 4(3), 200020–200077. doi: 10.29026/oea.2021.200077.

43. Xue, Q., Zhang, H., Zhu, M., Pei, Z., Li, H., Wang, Z., Huang, Y., Huang, Y., Deng, Q., Zhou, J., Du, S., Huang, Q., & Zhi, C. (2017). Photoluminescent Ti$_3$C$_2$ MXene quantum dots for multicolor cellular imaging. *Advanced Materials*, 29(15), 1604847. doi: 10.1002/adma.201604847.

44. Ahmad, M. S., Pandey, A. K., Abd Rahim, N., Aslfattahi, N., Mishra, Y. K., Rashid, B., & Saidur, R. (2021). 2-D Mxene flakes as potential replacement for both TCO and Pt layers for dye-sensitized solar cell. *Ceramics International*, 47 (19), 27942–27947. doi: 10.1016/j.ceramint.2021.06.225.

45. Dall'Agnese, C., Dall'Agnese, Y., Anasori, B., Sugimoto, W., & Mori, S. (2018). Oxidized Ti$_3$C$_2$ MXene nanosheets for dye-sensitized solar cells. *New Journal of Chemistry*, 42(20), 16446–16450. doi: 10.1039/C8NJ03246G.

46. Adibah, N. A., Azella, S. N., & Shukur, M. F. A. (2021). Synthesis of Ti$_3$C$_2$ Mxene through in situ HF and direct HF etching procedures as electrolyte fillers in dye-sensitized solar cell. *Materials Science Forum*, 1023, 15–20. doi: 10.4028/www.scientific.net/msf.1023.15.

11 Nouveau Advancements Achieved in the Performance Enchantment of Solar Cells Modified through MXene-Based 2D Nanomaterials Utilization

Khuram Shahzad Ahmad and Shaan Bibi Jaffri
Fatima Jinnah Women University

CONTENTS

11.1 INTRODUCTION

There has been an immense increment in the demand for the energy on a global scale in response to the uncontrolled global industrialization in response to the modern lifestyle. The way urbanization is increasing, there are chances of the global energy demands exceeding double demands by 2050. Therefore, energy requirements are moving toward a serious edge. In the current era, fossil fuels are responsible for the fulfillment of the major chunk of the global energy demands. However, fossil fuels are not infinite energy reserve. Rather, fossil fuels need hundreds of the years on a geological time scale to be formed. Another alarming situation in this regard is that the limited supply of the fossil fuels is reducing day by day. One of the main

DOI: 10.1201/9781003178422-11

concerns associated with the utilization of the fossil fuels for fulfillment of the energy demands is that they are indirectly playing an influential role in increasing the global warming due to carbon dioxide emissions [1,2]. The negative role of the fossil fuels in triggering global warming is comprehendible from the contribution of these fossil fuels reaching up to 75% of the total carbon dioxide emissions. This is just the percentage of the emissions specifically from the power production sector. There are other sources, which are also considerably associated with the increment in the global warming. Therefore, the overall role of the fossil fuels in global warming and air pollution is quite eminent.

In the current era of the increased air pollution and global warming augmented by the fossil fuels utilization, it is very essential to switch to the modes that save the earth in a sustainable manner. Only then the scientific community and the policy makers will be able to address the impinging issues of the energy requirements and climate change. In this regard, the first and essential step toward sustainable energy generation is the use of the inexhaustible, cleaner, and also sustainable forms of the energy [3]. Huge volumes of research have been done and published on the use of the renewable resources of the energy for replacing the conventional fossil fuels. Luckily, many successful attempts have been done and advancements in the energy sector have been achieved. In this regard, it is noteworthy that the contribution of the electrical energy or more specifically electricity in the energy mix is spanning around 19%. There are expectations of this figure reaching up to 29% by 2050 [4].

The major energy conversions devices that are predominating the current era are those based on the conversion of the solar energy to the electrical energy. That is the reason solar energy has excelled in terms of the sustainability and overall research in comparison to the other renewable resources, e.g., nuclear, biomass, tidal, wind, and geothermal. Therefore, the ultimate hopes of the global scientific community are associated with the solar energy. The main reason of focusing solar energy over other infinite forms of the energy are the wide-scale abundance, availability, and above all the sustainability achievement is much easier with it than other forms of the energy. Different kinds of devices have been used for the conversion of the solar energy and they are often referred to as the solar photovoltaic (PV) cells [5]. Solar PV cells are attracting a great deal of attention because of their economical cost that has an excellent affordability, simplicity, and easier installation. As a result of the rigorous working and research with the solar cells, the scientific community has defined three major generations for solar cells. They are the first generation, second generation, and the third generation of the solar cells. The first-generation solar cells are inclusive of the monocrystalline and polycrystalline solar cell. The second generation is predominantly based on thin films e.g., CdTe, and CIGS [6]. The last and the third generation of the solar cells is inclusive of the most advanced forms of the cells, e.g., perovskite, dye-sensitized, and organic solar cells. There has been a great deal of research done on the organic, dye-sensitized, and perovskite solar cells [7] for reaching higher efficiencies and overcoming the loopholes associated with them [8]. In short, the scientific community is doing efforts for making the cost-effective, efficient, cheap, lightweight, and sustainable solar cells.

When it comes to the fabrication of the solar and perovskite solar cells, then a myriad of the materials have been tried in efforts to improve the overall functionality and stability. There are thousands of the nanoscale materials that have been employed and advancements have been achieved in the sustainable conversion of solar energy. In terms of the materials, different oxides, sulfides, oxy-sulfides, nitrides, carbonitrides, and selenides have been employed in different manners [9]. Among these materials, the two-dimensional (2D) transition-metallic carbides, nitrides, and carbonitrides have expressed auspicious working in comparison to other chemical counterparts. These materials are referred to as the MXenes. MXenes are at the center of attention as the possible constitutional candidates for organic and perovskite solar cells. The main reason for opting MXenes for the different types of solar cells is the array of the unique characteristics that this group is offering. The unique intrinsic characteristics have made MXenes to be adopted at a large scale in the practical sense for the development of the eco-friendly solar cells and other PV devices. Previous studies are indicative of the auspicious role of the MXenes in conversion as well as storage of solar energy. Though there are some issues associated with these compounds like any other chemical group has, but the scientific community is actively engaged in finding a solution to these shortcomings. Therefore, the development in the field of MXenes-based solar cells has been fast and proposing different solutions [10].

Isolated researches have been done on the utilization of the MXenes in different solar cells, e.g., perovskite solar cells, organic cells, and dye-sensitized. The results of these studies have conspicuously signified the role of these materials in improving the performance of these cells. However, to the best of the information, no publication has reviewed the use of the nouveau advancements achieved in the performance enchantment of solar cells modified through MXene-based 2D nanomaterials utilization. That is the reason, we have undertaken the task of analyzing the most recent advancements done till August 2021 and compiled this chapter covering various things associated with the MXene-based 2D nanomaterials utilization in solar cells. The chapter will be covering the account of the auspiciousness of the MXene-based 2D nanomaterials for conversion of the solar energy to electrical energy. Excellent performing MXene-based 2D nanomaterials utilization from different research groups has been added. In addition to this, challenges and future outlook of these materials have also been added. On the basis of these analyses of the various researches in this regard, we see great potential in MXene-based 2D nanomaterials to be used at the practical scale for benefitting humanity after the resolution of the issues associated with them.

11.2 THE AUSPICIOUSNESS OF THE MXene-BASED 2D NANOMATERIALS

The scientific works have been highly impressed by the unique characteristics of the MXenes and MXene-based 2D nanomaterials. Especially after the synthesis of the $Ti_3C_2T_x$ nanosheets in 2011, there has been an augmentation in the interest in this unique class of the materials. The synthesis of these nanosheets was done by means of the selectively etching procedure. As the result of this major scientific achievement, the researchers in the associated fields did a number of researches and 150 various kinds of the MAX phases were successfully synthesized. In addition to this, 30 varied kinds of the MXenes were fabricated successfully and the systematic studies were conducted in the subsequent manner [11]. To be more specific, MXenes have the general formula $M_{n+1}X_nT_x$ ($n = 1$, 2, 3, or 4). In this general formula M is representative of the easier transition metals, e.g., Ti, Hf, V, Mo, Cr, and Nb. While X can be a carbon or nitrogen. T_x is representative of the terminal groups, which have been derived from the systematic synthetic methods, e.g., such as −F, and −OH [12]. In the wide range of the MXene compounds, MXene in the form of the layered 2D material has arrived greater attention because it comes with the excellent conductive core in addition to the remarkable dispersibility inside the aqueous solutions. This excellent behavior is attributable to the hydrophilic surficial characteristics of the MXene in the form of the layered 2D material.

The advancements achieved in the MXene have led to their use in the form of the 2D MXene nanosheets. These highly advanced nanosheets have been used in a myriad of applications, e.g., $Ti_3C_2T_x$ finds a useful employment in different batteries, e.g., lithium (Li)-ion or Li-sulfur (S), sensors, solar cells, and different photodetectors have also been formed with the Mo_2CT_x. In addition to these applications, MXenes and related compounds have also been used in the fabrication of the non-Li ion batteries having different constituents, e.g., Na^+, K^+, Mg^{2+}, Zn^{2+}, and Ca^{2+}. MXene plays the role of the excellent conductive substrate in these applications behind the chemical active and responsive material. The noteworthy application of the MXenes in terms of the energy and solar cells is their use because of their potential in enhancing the opto-electronic characteristics of the overall cells. That is the reason, they are foreseen as the suitable candidates for increasing efficiency of the different types of the solar cells including perovskite and organic cells. Effort has been done for the nanoengineering of 2D MXenes so that further improvement in the energy-related applications can be achieved [13].

11.3 MXene-BASED 2D NANOMATERIALS IN ENERGY APPLICATIONS

Cleaner energy is the utmost need of the hour. Achievement of the goals of the sustainability is only possible if cleaner and cheaper energy modes are devised. MXene has been widely studied for this purpose. An important thing to mention in terms of the utilization of the MXene-based 2D nanomaterials in energy applications is the understanding of the physical energy cycle. MXene can be an effective choice to use in the applications that mimic this physical energy cycle. This is associated with the excellent light-to-heat transformation efficiency of the MXene. Also, the thermal stability and excellent thermal conductivity are also seen as the important factors related with the MXene for use in the solar cells inclusive of the perovskite and organic cells [14]. In addition to this, other energy-related fields are catalysis, optics, energy conversion, and storage.

MXene is seen as one of the potential candidates on the energy horizon for the production of the sustainable energy in comparison to the conventional fossil fuels [15,16]. MXene-based materials are used in the energy conversion and storage devices. As a result of the novel device fabrication with the MXene, scientific groups have been able to achieve higher efficiencies, remarkable stability indices, and eco-friendly devices [17]. There are possible chances of these electrochemical energy devices to be converted to the practical use by means of the successful commercialization after sorting out some issues.

Number of MXene and MXene-based 2D nanomaterials have been used for modifying solar cells; however, the most famous one among them is $Ti_3C_2T_x$. This material has a surficial functional group in the form of the oxygen (−O), hydroxyl (−OH), and/or fluoride (−F). This functional group is obtained by the selective etching of the HF. It is due to the active terminal groups that lead to the efficient exfoliation of the MXene leading to the formation of the MXene-based 2D single layers possessing a thickness of ~1.6 nm. For the MXene-based 2D single layers possessing 2.8 nm thickness have excellently higher transparency, which is capable of 95% transmittance. Also, the chemical structure of the MXene-based 2D single layers has important aspects in terms of the work function (WF) [18]. The utilization of the MXene-based 2D nanomaterials is preferred for solar cells because of their efficient potential of energy conversion, hydrophilic surface, excellent metallic conductivity, and remarkable mechanical stability [19]. Nevertheless, there is an issue of the restacking in these layers just like the other individual 2D materials. Such restacking has serious implications on the electrochemical performance. In order to improve their electrochemical performance for use in solar cells, different strategies have been tried. For example, efforts have been done for the surficial modification of the 2D MXene in addition to the fabrication of the heterostructures with 2D MXene with the mixed dimensionality [20].

11.4 MXene-BASED 2D NANOMATERIALS AND SOLAR CELLS

11.4.1 PEROVSKITE SOLAR CELLS

MXene and MXene-based 2D nanomaterials have been used for the performance enhancement of the different solar cells [21] but their frequent use is done in the perovskite solar cells. That is why it is necessary to elucidate their role in the perovskite solar cells. Another reason for explaining the MXene-based 2D nanomaterials used in the perovskite solar cells is the potential of these cells as the best replacement for the costly silicon-based solar cells. Perovskite solar cells are third-generation cells and are a promising PV technology reaching up to the power conversion efficiency (PCE) up to 25% in 2021. The best thing about perovskite solar cells is their cost-effectiveness. Perovskite solar cells are new in the PV market with their inception in 2009 by Miyasaka and co-researchers. Yet there is a whooping augmentation in the PCE of these cells. Different research groups have explored perovskite solar cells by changing certain factors including champion device architecture, compositional materials, and deposition techniques.

An important consideration regarding perovskite solar cells is the selection of the charge transporting layers (CTLs). These CTLs are inclusive of the electron transport layer (ETL) and hole transport layer (HTL). The composition of these CTLs has serious impacts on the overall efficiency and stability of the perovskite solar cells. Therefore, a number of materials have been tried for making the perfect CTLs, however, with the present issues of the instability of the perovskite active absorber layer, much effort has to be done. MXene and MXene-based 2D nanomaterials have been actively employed for improving perovskite solar cells. There are different reports elucidating the suitability of using MXene-based 2D nanomaterials as solvents and as the constituent of the different layers [22,23].

Different 2D nanomaterials have been used for interfacial engineering of the perovskite solar cells for making them efficient and stable to be adopted at the practical scale. Scientific community is hoping to achieve efficient and stable perovskite solar cells on the larger area modules. In this regard, doping of the interlayer components has been done with various compounds, e.g., MXene-based 2D nanomaterials, graphene and associated materials, phosphorene and antimonene, and transition metal dichalcogenides . Also, these materials have been employed for the tuning of the morphological or optoelectronic features of the ETLs, HTLs, perovskite absorber layer, and their interfaces [24].

Among these 2D modifiers, MXene-based 2D nanomaterials have expressed the best candidacy for improving the charge collection potentials of the CTLs in addition to the enhancement of the other features. They are also associated with the speeding up of the charge addition at the interface of the active absorber layer and ETLs/HTLs [25]. Different studies are suggestive of the profound role of the MXene-based 2D nanomaterials in controlling the growth and morphological aspects of the perovskite crystalline growth [26]. Also, the results of some studies are indicating toward the role of the MXene-based 2D nanomaterials in the stabilization of the interfacial region. This stabilization is the utmost need of the hour and signifies one of the biggest objections being raised on the perovskite solar cells. So, the role of the MXene-based 2D nanomaterials in stability is also associated with the augmentation of the lifetime of the perovskite champion device [27].

MXene-based 2D nanomaterials are a great option to be used for improving perovskite solar cells. Their insertion can be done in the form of the interlayers or the buffer layer by means of transforming the physical or chemical characteristics of the participating layers at the interfacial region. As a result of the insertion of the MXene-based 2D nanomaterials, there is a consequent improvement in the charge injection in addition to the charge accumulation at the interfaces of the perovskite absorber layer and the ETL/HTLs. Another striking result obtained from different studies conducted with the MXene-based 2D nanomaterials is that they are also involved in the tailoring of the electronic structure, e.g., WF [28] or in simpler words, bandgap tailoring. These materials are involved in this tailoring by means of the accurate functionalization or quantum confinement.

MXene-based 2D nanomaterials are capable of the fine-tuning of the WF and thus obtaining the accurate alignment of the energy levels is easier with their use. They are known for the creation of the offset between the CTLs and the absorber layer in the perovskite solar cells. Consequently, they lead to the induction of the efficacious charge collection at the CTLs [29,30]. The rich chemistry and perfect surficial termination makes the MXene-based 2D nanomaterials the unique material for perovskite with endless possibilities for tuning of the electronic properties. Above all, the variation of the WF with these 2D nanomaterials is the best thing to use in the long term [31]. In addition to this, when MXene and MXene-based 2D nanomaterials are synthesized then their surfaces acquire functionalization in a natural way. This functionalization is associated with the potential alterations in the electrostatic potential in the close proximity of the surfaces. In addition to this, this change in the electronic structure ends up shifting the WF. Density functional theory was done in this regard and they were indicative of the surficial termination associated with the effect on the density of the states. In this regard, the WF of the MXene-based 2D nanomaterials to be spanning around 1.6 eV in case of the termination of the OH and 6.25 eV in case of the termination of the O [32]. These results are pointing toward the candidature of the MXene and MXene-based 2D nanomaterials for optoelectronics, especially in case of the PVs like perovskite solar cells [33].

The oxide-based ETLs in the perovskite solar cells have often been doped with the 2D materials and also their tailored counterparts and products, e.g., black phosphorene and zero-dimensional graphene quantum dot. For example, SnO_2 ETLs in the perovskite solar cells have been modified with the 2D materials and resultant products had higher electrical conductivity, energy level characteristics, and considerable alleviation in the surficial defects in addition to surficial electronic traps [34]. In a recent study, Ti_3C_2 MXene-doped SnO_2 electron extraction layer was used in the perovskite absorber layer. As the result of this doping, considerable improvement was seen in the perovskite champion device because it affected the charge transferal paths, electronic mobility, and also electronic extraction.

In another study, TiO_2 has been introduced in the MXene-doped SnO_2 ETL in the relevant crystalline phase for the formation of the efficacious hetero-junction in the perovskite solar cell. Through effective oxidation in the hydrothermal method, another MXene-based 2D nanomaterials modified material was produced. In a reaction with the reduction of the resulting hydrazine hydrate, Ti^{3+}-doped TiO_2 having rutile lattice was composited over the 2D Ti_3C_2 MXene sheets [35]. Also, another perovskite solar cell was developed by adopting hydrothermal route with the partial oxidation. 2D anatase TiO_2 was taken for in situ growth on the Ti_3C_2 for obtaining interface with the minimized defects [36]. The technique of the in situ hydrolysis and thermal treatment has also been adopted for the decoration of the nanoparticles of the 3D anatase TiO_2 on the surface of the Ti_3C_2 MXene for improving the performance of the perovskite solar cells [37].

In addition to these studies, there are other investigations as well that confirm the profound role of the MXene-based 2D nanomaterials in improving the perovskite solar cells. Some researchers have also used $Ti_3C_2T_x$ MXene nanosheets in a hydrothermal route where the assemblage of the 30 nm anatase TiO_2 has been done. It has been followed by the Schottky junction formation at the interfacial region of the TiO_2/MXene for the consequent promotion of the charge carrier separation through hole trapping mechanism [38].

In addition to the hydrothermal route, another effective method used for the employment of the MXene-based 2D nanomaterials in perovskite solar cells is the higher energy ball milling. In this method, the generation of the anatase TiO_2 has been done over the nanosheets of the carbon in addition to the raw 2D Ti_2CT_x MXene. Also, there are reports available on the Ti-based oxide. The growth of this Ti-based oxide is done on the MXene in the ambient environment considering is instability. Annealing at 250°C can be an effective strategy for getting excellent results for the perovskite solar cells [39]. The main reasons for preferring MXene-based 2D nanomaterials for perovskite solar cells modification can be mainly attributed to their higher transmittance, excellent conductivity, WF tunability, and also possibility for solution-based processing.

There are some of the challenges associated with the utilization of the MXene-based 2D nanomaterials in perovskite solar cells. A recent research has attempted to solve the critical problems associated with these nanomaterials at the interface. In a research by Wang et al. [40], $Ti_3C_2T_x$ MXene has been used as the ETL in the perovskite solar cells through ambient temperature solution processing. Also, this procedure was further aided by carrying out oxygen plasma treatment. The results of this study pointed toward the excellent tuning of the WFs. Also, there was a reduction in the trap states and consequent improvement in the electronic transport closer to the surface. The good thing explored through this research was the stability enhancement of the perovskite champion device [40]. Stability-based studies suggest the excellent potential of the MXene-based 2D nanomaterials for utilization in the perovskite solar cells. A recent study was indicative of the promising candidature of these materials for increasing the scalability and stability for larger area-based solar devices. The PCE of the MXene modified device was 19% showing an enhancement in the optoelectronic properties [41].

MXene-based 2D nanomaterials have also been used for the modification of the inorganic perovskite solar cells. A recent study successfully reached the PCE of 9.01% with 1,900 hours of the stability by the introduction of the Ti_3C_2-MXene nanosheets in the $CsPbBr_3$ solar cells in the form of the interlayers. The incorporation of the Ti_3C_2-MXene nanosheets exerted good impacts on the alignment of the energy level and as the result of this, there was a subsequent elimination of the mismatch between energy levels. In addition to these impacts, there was an acceleration in the hole extraction and alleviation in the agglomeration of the charges at the interfacial region of the perovskite absorber layer and CTLs [42].

11.4.2 Organic Solar Cells

MXene-based 2D nanomaterials are not only used for the performance enhancement of the perovskite solar cells. In fact, they have been used for the functionality enhancement of the organic solar cells as well. Utilization of the MXene-based 2D nanomaterials in these cells and other optoelectronic devices can be attributed to the higher conductivity and an excellent transparency [43]. In terms of the organic solar cells, these 2D nanomaterials have been specially employed in the form of charge collection and buffer materials. For example, the most frequently used compounds are the MXene-based 2D nanomaterials, graphene-oxide, and black phosphorus [44]. However, MXene-based 2D nanomaterials are advantageous over others because of their excellent dispersibility in water and also they exhibit the higher transmittance. The utilization of the MXene-based 2D nanomaterials is expressive of the excellent results. Especially, using them in the non-fullerene organic solar cells has attracted a great deal of interest. The organic solar cells are famous because of their excellent PV efficiency [45].

Surprisingly, the organic solar cell having composition of the poly[(2,6-(4,8-bis(5-(2- ethylhexyl)thiophen-2-yl)- benzo[1,2-b:4,5-b′]dithiophene))- alt- (5,5-(1′, 3′-di-2-thienyl-5′, 7′-bis(2-ethylhexyl)benzo[1′, 2′- c:4′, 5′-c′]dithiophene-4,8-dione))] and 3,9-bis(2-methylene-(3- (1,1-dicyanomethylene)-indanone))-5,5,11,11-tetrakis(4-hexylphenyl)-dithieno[2,3-d:2′, 3′-d′]-sindaceno[1,2-b:5,6- b′]dithiophene) (PBDB-T:ITIC) has acquired the PCE of the 12.7% [46]. In addition to such an increased efficiency, different researchers have reported the PCEs for MXene-based 2D nano-materials modified organic solar cells as high as 14% [47] and 17% [48] in case of the single junction and highly advanced tandem non-fullerene architecture, respectively. In a recent report, MXene/silver nanowire (AgNW) hybrid films with the excellent flexibility, transparence, and conductivity were fabricated. These hybrid films were created by the facile solution-processed technique. The MXene-based 2D nanomate-rials modified organic solar cells exhibited excellent electrical conductivity, remark-able transmittance, comparatively lower roughness, perfect matching of the WF, and efficient mechanical functionality. The results were indicative of the utilization of the MXene-based 2D nanomaterials modified organic solar cells in different types of the flexible optoelectronics.

Another study done with the MXene-based 2D nanomaterials used the $Ti_3C_2T_x$ for the collection of the electronic and holes species in an organic solar cell. The researches were capable of the tuning of the WF spanning over a range of the 4.08–4.95 eV. Films made with the MXene-based 2D nanomaterials possessed the different WFs. They have been used for the collection of the electrons and holes in the organic cells with nonfullerene composition. These nonfullerene-based organic solar cells were composed of the poly[(2,6-(4,8-bis(5-(2-ethylhexyl)thiophen-2-yl)- benzo[1,2-b:4,5-b′]dithiophene))- alt- (5,5-(1′, 3′-di-2-thienyl-5′, 7′-bis(2-ethylhexyl)benzo[1′, 2′-c:4′, 5′-c′]dithiophene-4,8-dione))] (PBDB-T) and 3,9-bis(2- methylene-(3-(1,1-dicyanomethylene)-indanone))-5,5,11,11-tetrakis(4-hexylphenyl)-dithieno[2,3-d:2′, 3′-d′]-sindaceno[1,2- b:5,6-b′]dithiophene) that served as the active materials. On the basis of the results obtained with different MXene-based 2D nanomaterials, it can be concluded that these materials are best for increasing the overall performance of the organic solar cells in comparison to other materials aimed at the collection of the charge and buffer materials.

11.5 CONCLUSION AND FUTURE PERSPECTIVE

Current review has analyzed the role and auspiciousness of the MXene-based 2D nanomaterials utilization for improving the functionality of the solar cells. All the recent advancements in this regard were summarized in a brief manner. From the detailed analysis, we can conclude that MXene-based 2D nanomaterials have mul-tiple roles in enhancing the efficiency and stability of the solar cells. However, there are three significant roles that they are playing in terms of the ongoing researches. First of all, MXene-based 2D nanomaterials are used as the additive for modifying different constituents. In addition to this, another role played by them is that they are used as an electrode material. Third, they are also used as a layer for the charge extraction. That means they are used as an ETL or HTL.

The results of the studies have also elucidated the profound role of the MXene-based 2D nanomaterials in interfacial engineering at the interface of the active absorber layers and the charge extraction layer. Scientific community has been intrigued by the excellent WF improvement of these compounds and thus they found MXene-based 2D nanomaterials to be increasing the lifespan of the solar cells. This is especially significant if we consider the perovskite solar cells. Perovskite solar cells have been cherished for their whooping efficiencies and potential of substituting the silicon-based expensive PV options. However, their commercialization will remain a dream until and unless they are not stable. Therefore, MXene-based 2D nanomaterials offering solution to this stability issue can be one of the biggest scientific breakthrough. However, we need some more research showing excellent results for the stability enhancement and only then the final conclusion can be reached for the use of the MXene-based 2D nanomaterials in solar cells including perovskite solar cells.

REFERENCES

1. Lipatov, A., Lu, H., Alhabeb, M., Anasori, B., Gruverman, A., Gogotsi, Y., & Sinitskii, A. (2018). Elastic properties of 2D $Ti_3C_2T_x$ MXene monolayers and bilayers. *Science Advances*, 4(6), eaat0491.
2. Jaffri, S. B., Ahmad, K. S., Thebo, K. H., & Rehman, F. (2021). Recent developments in carbon nanotubes-based perovskite solar cells with boosted efficiency and stability. *Zeitschrift für Physikalische Chemie*, 235, 1–10.
3. Elavarasan, R. M., Shafiullah, G. M., Padmanaban, S., Kumar, N. M., Annam, A., Vetrichelvan, A. M., & Holm-Nielsen, J. B. (2020). A comprehensive review on renewable energy development, challenges, and policies of leading Indian states with an international perspective. *IEEE Access*, 8, 74432–74457.
4. Kober, T., Schiffer, H. W., Densing, M., & Panos, E. (2020). Global energy perspectives to 2060–WEC's World Energy Scenarios 2019. *Energy Strategy Reviews*, 31, 100523.
5. Davis, V. L., Quaranta, S., Cavallo, C., Latini, A., & Gaspari, F. (2017). Effect of single-chirality single-walled carbon nanotubes in dye sensitized solar cells photoanodes. *Solar Energy Materials and Solar Cells*, 167, 162–172.
6. Ahmed, B., Anjum, D. H., Gogotsi, Y., & Alshareef, H. N. (2017). Atomic layer deposition of SnO_2 on MXene for Li-ion battery anodes. *Nano Energy*, 34, 249–256.
7. Ahmad, M. S., Pandey, A. K., Abd Rahim, N., Aslfattahi, N., Mishra, Y. K., Rashid, B., & Saidur, R. (2021). 2-D Mxene flakes as potential replacement for both TCO and Pt layers for dye-sensitized solar cell. *Ceramics International*, 47, 1–10.
8. Ahmad, M. S., Abd Rahim, N., Mehmood, S., & Khan, A. D. (2021a). Effect of WS_2 nano-sheets on the catalytic activity of polyaniline nano-rods based counter electrode for dye sensitized solar cell. *Physica E: Low-Dimensional Systems and Nanostructures*, 126, 114466.
9. Kwak, D., Wang, M., Koski, K. J., Zhang, L., Sokol, H., Maric, R., & Lei, Y. (2019). Molybdenum trioxide (α-MoO_3) nanoribbons for ultrasensitive ammonia (NH_3) gas detection: Integrated experimental and density functional theory simulation studies. *ACS Applied Materials & Interfaces*, 11(11), 10697–10706.
10. Zhang, A., Liu, R., Tian, J., Huang, W., & Liu, J. (2020). MXene-based nanocomposites for energy conversion and storage applications. *Chemistry–A European Journal*, 26(29), 6342–6359.
11. Wang, C., Chen, S., & Song, L. (2020). Tuning 2D MXenes by surface controlling and interlayer engineering: Methods, properties, and synchrotron radiation characterizations. *Advanced Functional Materials*, 30(47), 2000869.
12. Chen, X., Zhao, Y., Li, L., Wang, Y., Wang, J., Xiong, J., & Yu, J. (2021). MXene/polymer nanocomposites: Preparation, properties, and applications. *Polymer Reviews*, 61(1), 80–115.
13. Chen, Y., Wang, D., Lin, Y., Zou, X., & Xie, T. (2019). In suit growth of CuSe nanoparticles on MXene (Ti_3C_2) nanosheets as an efficient counter electrode for quantum dot-sensitized solar cells. *Electrochimica Acta*, 316, 248–256.
14. Liu, R., & Li, W. (2018). High-thermal-stability and high-thermal-conductivity $Ti_3C_2T_x$ MXene/poly (vinyl alcohol)(PVA) composites. *ACS Omega*, 3(3), 2609–2617.

15. Jun, B. M., Kim, S., Heo, J., Park, C. M., Her, N., Jang, M., & Yoon, Y. (2019). Review of MXenes as new nanomaterials for energy storage/delivery and selected environmental applications. *Nano Research*, 12(3), 471–487.

16. Xue, Q., Pei, Z., Huang, Y., Zhu, M., Tang, Z., Li, H., & Zhi, C. (2017). Mn_3O_4 nanoparticles on layer-structured Ti_3C_2 MXene towards the oxygen reduction reaction and zinc–air batteries. *Journal of Materials Chemistry A*, 5(39), 20818–20823.

17. Kim, H., Wang, Z., & Alshareef, H. N. (2019). MXetronics: Electronic and photonic applications of MXenes. *Nano Energy*, 60, 179–197.

18. Tang, H., Feng, H., Wang, H., Wan, X., Liang, J., & Chen, Y. (2019). Highly conducting MXene–silver nanowire transparent electrodes for flexible organic solar cells. *ACS Applied Materials & Interfaces*, 11(28), 25330–25337.

19. Yang, L., Dall'Agnese, Y., Hantanasirisakul, K., Shuck, C. E., Maleski, K., Alhabeb, M., & Miyasaka, T. (2019). SnO_2–Ti_3C_2 MXene electron transport layers for perovskite solar cells. *Journal of Materials Chemistry A*, 7(10), 5635–5642.

20. Malaki, M., Maleki, A., & Varma, R. S. (2019). MXenes and ultrasonication. *Journal of Materials Chemistry A*, 7(18), 10843–10857.

21. Jafarzadeh, M., Sipaut, C. S., Dayou, J., & Mansa, R. F. (2016). Recent progresses in solar cells: Insight into hollow micro/nano–structures. *Renewable and Sustainable Energy Reviews*, 64, 543–568.

22. Ahmad, M. S., Pandey, A. K., & Abd Rahim, N. (2018). Effect of nanodiamonds on the optoelectronic properties of TiO_2 photoanode in dye-sensitized solar cell. *Arabian Journal for Science and Engineering*, 43(7), 3515–3519.

23. Pandey, A. K., Ahmad, M. S., Alizadeh, M., & Abd Rahim, N. (2018). *Physica E: Low-Dimensional Systems and Nanostructures*, 101, 139–143.

24. Lipatov, A., Lu, H., Alhabeb, M., Anasori, B., Gruverman, A., Gogotsi, Y., & Sinitskii, A. (2018). Elastic properties of 2D $Ti_3C_2T_x$ MXene monolayers and bilayers. *Science Advances*, 4(6), eaat0491.

25. Najafi, L., Taheri, B., Martin-Garcia, B., Bellani, S., Di Girolamo, D., Agresti, A., & Bonaccorso, F. (2018). MoS_2 quantum dot/graphene hybrids for advanced interface engineering of a $CH_3NH_3PbI_3$ perovskite solar cell with an efficiency of over 20%. *ACS Nano*, 12(11), 10736–10754.

26. Biccari, F., Gabelloni, F., Burzi, E., Gurioli, M., Pescetelli, S., Agresti, A., & Vinattieri, A. (2017). Graphene-based electron transport layers in perovskite solar cells: A step-up for an efficient carrier collection. *Advanced Energy Materials*, 7(22), 1701349.

27. O'keeffe, P., Catone, D., Paladini, A., Toschi, F., Turchini, S., Avaldi, L., & Di Carlo, A. (2019). Graphene-induced improvements of perovskite solar cell stability: Effects on hot-carriers. *Nano Letters*, 19(2), 684–691.

28. Konios, D., Kakavelakis, G., Petridis, C., Savva, K., Stratakis, E., & Kymakis, E. (2016). Highly efficient organic photovoltaic devices utilizing work-function tuned graphene oxide derivatives as the anode and cathode charge extraction layers. *Journal of Materials Chemistry A*, 4(5), 1612–1623.

29. Hantanasirisakul, K., & Gogotsi, Y. (2018). Electronic and optical properties of 2D transition metal carbides and nitrides (MXenes). *Advanced Materials*, 30(52), 1804779.

30. Akuzum, B., Maleski, K., Anasori, B., Lelyukh, P., Alvarez, N. J., Kumbur, E. C., & Gogotsi, Y. (2018). Rheological characteristics of 2D titanium carbide (MXene) dispersions: A guide for processing MXenes. *ACS Nano*, 12(3), 2685–2694.

31. Khazaei, M., Ranjbar, A., Arai, M., Sasaki, T., & Yunoki, S. (2017). Electronic properties and applications of MXenes: A theoretical review. *Journal of Materials Chemistry C*, 5(10), 2488–2503.

32. Schultz, T., Frey, N. C., Hantanasirisakul, K., Park, S., May, S. J., Shenoy, V. B., & Koch, N. (2019). Surface termination dependent work function and electronic properties of $Ti_3C_2T_x$ MXene. *Chemistry of Materials*, 31(17), 6590–6597.

33. Huang, L., Zhou, X., Xue, R., Xu, P., Wang, S., Xu, C., & Liang, D. (2020). Low-temperature growing anatase TiO_2/SnO_2 multi-dimensional heterojunctions at MXene conductive network for high-efficient perovskite solar cells. *Nano-Micro Letters*, 12(1), 1–19.
34. Guo, Z., Gao, L., Xu, Z., Teo, S., Zhang, C., Kamata, Y., & Ma, T. (2018). High electrical conductivity 2D MXene serves as additive of perovskite for efficient solar cells. *Small*, 14(47), 1802738.
35. Peng, C., Wang, H., Yu, H., & Peng, F. (2017). (111) TiO_{2-x}/Ti_3C_2: Synergy of active facets, interfacial charge transfer and Ti^{3+} doping for enhance photocatalytic activity. *Materials Research Bulletin*, 89, 16–25.
36. Peng, C., Yang, X., Li, Y., Yu, H., Wang, H., & Peng, F. (2016). Hybrids of two-dimensional Ti_3C_2 and TiO_2 exposing {001} facets toward enhanced photocatalytic activity. *ACS Applied Materials & Interfaces*, 8(9), 6051–6060.
37. Zhu, J., Tang, Y., Yang, C., Wang, F., & Cao, M. (2016). Enhanced capacitive performance based on diverse layered structure of two-dimensional Ti_3C_2 MXene with long etching time. *Journal of the Electrochemical Society*, 163(5), A785.
38. Agresti, A., Pazniak, A., Pescetelli, S., Di Vito, A., Rossi, D., Pecchia, A., & Di Carlo, A. (2019). Titanium-carbide MXenes for work function and interface engineering in perovskite solar cells. *Nature Materials*, 18(11), 1228–1234.
39. Rakhi, R. B., Ahmed, B., Hedhili, M. N., Anjum, D. H., & Alshareef, H. N. (2015). Effect of postetch annealing gas composition on the structural and electrochemical properties of Ti_2CT_x MXene electrodes for supercapacitor applications. *Chemistry of Materials*, 27(15), 5314–5323.
40. Wang, J., Cai, Z., Lin, D., Chen, K., Zhao, L., Xie, F., & Zhu, R. (2021). Plasma oxidized $Ti_3C_2T_x$ MXene as electron transport layer for efficient perovskite solar cells. *ACS Applied Materials & Interfaces*, 13, 1–10.
41. Saranin, D., Pescetelli, S., Pazniak, A., Rossi, D., Liedl, A., Yakusheva, A., & Di Carlo, A. (2021). Transition metal carbides (MXenes) for efficient NiO-based inverted perovskite solar cells. *Nano Energy*, 82, 105771.
42. Chen, T., Tong, G., Xu, E., Li, H., Li, P., Zhu, Z., & Jiang, Y. (2019a). Accelerating hole extraction by inserting 2D Ti_3C_2-MXene interlayer to all inorganic perovskite solar cells with long-term stability. *Journal of Materials Chemistry A*, 7(36), 20597–20603.
43. Zhao, L., Dong, B., Li, S., Zhou, L., Lai, L., Wang, Z. & Huang, W. (2017). Interdiffusion reaction-assisted hybridization of two-dimensional metal–organic frameworks and $Ti_3C_2T_x$ nanosheets for electrocatalytic oxygen evolution. *ACS Nano*, 11(6), 5800–5807.
44. Liu, S., Lin, S., You, P., Surya, C., Lau, S. P., & Yan, F. (2017). Black phosphorus quantum dots used for boosting light harvesting in organic photovoltaics. *Angewandte Chemie*, 129(44), 13905–13909. 45. Li, W., Chen, M., Cai, J., Spooner, E. L., Zhang, H., Gurney, R. S., & Wang, T. (2019). Molecular order control of non-fullerene acceptors for high-efficiency polymer solar cells. *Joule*, 3(3), 819–833.
46. Bi, P., Xiao, T., Yang, X., Niu, M., Wen, Z., Zhang, K., & Liu, H. (2018). Regulating the vertical phase distribution by fullerene-derivative in high performance ternary organic solar cells. *Nano Energy*, 46, 81–90.
47. Li, H., Xiao, Z., Ding, L., & Wang, J. (2018). Thermostable single-junction organic solar cells with a power conversion efficiency of 14.62%. *Science Bulletin*, 63(6), 340–342.
48. Meng, L., Zhang, Y., Wan, X., Li, C., Zhang, X., Wang, Y., & Chen, Y. (2018). Organic and solution-processed tandem solar cells with 17.3% efficiency. *Science*, 361(6407), 1094–1098.

12 Nanocomposites of MXene for Photovoltaic Applications

Muhammad Bilal, Zia Ur Rehman,
Maryam Zaqa, and Jianhua Hou
Yangzhou University

Faheem K. Butt
University of Education Lahore

Asif Hussain
Yangzhou University

CONTENTS

12.1 INTRODUCTION

The combination of at least two constituent materials with various physical or synthetic properties is known as composites. When consolidated, they produce a material with qualities or properties unique in relation to their original properties. Nanocomposites

DOI: 10.1201/9781003178422-12

have gained great attraction because they show many unique properties on nanoscale. Due to attractive and promising structures of many materials, they have gained great potential for many applications such as optical, thermal, photovoltaic, and energy storage devices. Some other applications of nanocomposites are usage of thin film capacitors in computer CPU, solid polymer materials for battery electrolytes, and in packing of food for a long time [1]. Nanocomposite materials are categorized into three different types in respect of their matrix materials as given below

1. Ceramic Matrix Nanocomposites
2. Metal Matrix Nanocomposites
3. Polymer Matrix Nanocomposites

Ceramic-based nanocomposites consist of ceramic fibers embedded in a ceramic matrix and gained increase in potency, stiffness and abrasion through purifying the particle size. Modification in these materials can be produced by the conduction for electrical charges and magnetic features by escalating the disordered grain boundary line. Metal-based nanocomposites can also be described as metal matrix composites.

They have many applications due to the extraordinary mechanical, optical, and magnetic properties such as good mechanical strength, super plasticity, hardness due to metal composites, coercivity, and super magnetization. A polymer nanocomposite is made with the combination of polymer matrix and some kind of nanomaterial. What type of nanomaterial should be used? Three types of material are as follows:

- **Spherical nanomaterials**: which have all three dimensions in the nanoscale.
- **Nanotubes**: which are 2D at nanoscale.
- **Nanoplatelets**: which have just one dimension in the nanoscale.

Some examples are discussed here such as Nano-TiO_2, carbon nanotubes, halloysite clays or nanofibrillated cellulose, graphene, hexagonal boron nitride, or bentonite clays. These substances acquired tremendous consideration. This is in reality because of their radically upgraded or further developed thermal, mechanical just as the barricade characteristics when compared with the micro and ordinary composite materials.

The 2D nanomaterials are frequently referenced as the marvel materials and become promising materials for many applications to solve a lot of world's problems. MXenes were discovered in 2011. It consists of the ternary compound with a formulation of $M_{n+1}X_nT_x$, where M represents about the transition metals, X denotes the C or N, and T_x are functional groups terminating the M surface [2]. MXenes are a somewhat novel and energizing group of 2D nanomaterials, which gives a vast scope to use these materials with brilliant properties. They have some extraordinary applications like in sensors, catalysis, energy storage devices, biomedical, tribology, and photovoltaic devices. Specifically, photovoltaic applications have been strongly concentrated with a novel class of MXenes materials attributable to their good conductive, work function tunable property just as plasmonic and nonlinear optical properties [3]. Furthermore, the utilization of 2D MXene materials in photovoltaic has drawn increasing consideration from research publication of 2018 on MXene composite for solar cells. Among various MXene materials, $Ti_3C_2T_x$ leads the study of MXenes in the field of photovoltaic applications.

Nowadays, this novel group of Ti_3C_2 MXenes is the center of attention for research purpose such as Ti_3C_2/TiO_2 and TiO_2@C nanocomposites used in higher photo-catalytic degradation. Optical properties, thermal properties, bandgap alignment for various materials, light absorption coefficient, and energy level can be investigated from these nanocomposites [4]. For plasmonic MXene-based nanocomposites pho-tothermal treatment is considered as a noninvasive malignancy treatment wherein tumor cells killing is accomplished utilizing the warmth or heat created upon nano-composite materials. We can say that toxicity of MXene nanocomposites is less toxic as compared to the pure MXene, which are more appropriate materials in the field of biomedical applications [5]. Nanocomposites of MXenes possess various unique electronic and photonic properties. These are also more suitable for broadband opti-cal devices. Nanocomposites of MXene can also give such type of new functional groups for future production of nano-photonic device applications [6]. So due to the extraordinary and unique properties of MXene these nanocomposites can play a sig-nificant role in the photovoltaic applications.

12.2 SYNTHESIS TECHNIQUES

MXene is made of bulk crystal called MAX, where M stands for Metal transition from group 3rd, 4th, 5th, and 6th of the periodic table, the A element represents from 12th, 13th, 14th, 15th, and 16th group and the X element is carbon and nitrogen [7–9]. They have a 2D layered structure first discovered in 2011, and belong to a large family of 2D materials. MXene is synthesized by eliminating A element from parent MAX phase to form $M_{n+1}X_nT_x$, where T_x is surface termination groups (−O), (−F) and (−OH) [10]. The molecular 2D structure of MXene is made of carbides and nitrides, which provide it inherently good conductivity and volumetric capaci-tance. MXene also has many advantages in electronic, optical, and thermoelectric properties [11]. They are widely used for energy storage applications, photocataly-sis, catalysis, optical devices, and electromagnetic [12–15]. MXene nanocomposite has been manufactured by using different physical as well as chemical technique to improve the efficiency of material and for commercially synthesizing low-cost mate-rial. There are many methods for preparing MXene/polymer composites or com-plexes, such as dry mixing/thermal pressing, solution blending, emulsion mixing, in situ polymerization, and lamination stacking.

12.2.1 Ion Exchange Method

Besides, ion exchange path demand is substituting ions in which the residue flawless/intact as a "structure" framework of ionic particle, which is "Happens" formed due to the parent ionic particles to another form, which innovative ions [16]. When ion exchange method is compared with bulk samples, the nanoparticles of ion exchange can be finished constantly at specific temperature as well due to the lower activa-tion problem to the dispersing ions. In extension when the ion exchange path way is recognized, then it will be very easy to operate consequently the product by varying the reaction conditions. By this way ion exchange process was quickly discovered as a well-organized approach for post-synthetic chemical adaption. For nanoparticle, the great chemist Li et al. reported the synthesis of Nano hybrid of TiO_2/Ti_3C_2 by simultaneous alkalization and oxidation of MXene layers. After that ion exchange and calcinations process are forwarded as shown in Figure 12.1a.

For the synthesis of TiO_2/Ti_3C_2 hybrid nanocomposite 120 mg of Ti_3C_2 were dissolved into the solution having 1 M of sodium hydroxide (180 Mc) and 30% of H_2O_2 (3.6 Mc). After that, the resultant solution is inserted into 100 Mc Teflon-lined autoclaves and kept it at 140°C for 12 hours. The resulting product Ti_3C_2/Na_2Ti_3O photocatalyst was washed with deionized water and dried the sample in vacuum oven at 60°C for 12 hours. Finally to produce the nano hybrid of TiO_2/Ti_3C_2, it was attained by thermal decomposition of $Ti_3C_2/H_2Ti_3O_7$ in muffle furnace at different temperature (300°C–400°C and 500°C) [17].

12.2.2 HYDROTHERMAL METHOD

Hydrothermal pathway refers via chemical reactions to the synthesis of materials which are carried out by heating solution and keep it in ambient pressure and at a specific temperature. In 19th century the thinking of a hydrothermal initiate from earth science, where the system of high temperature and pressure implies. Moreover hydrothermal process is very cheap in terms of instrumentations; material precursors comparatively with other synthesis techniques. The hydrothermal method allows to synthesize the MXene-based semi-conductors photocatalyst with excellent performance. For example, $SrTiO_3/Ti_3C_2$ hybrid nanocomposite was prepared by hydrothermal procedure as shown in Figure 12.1b.

Typically the 1 mL of titanium isopropoxide $C_{12}H_{28}O_4Ti$ was dissolved to 50 mL of 2:3 acetonitrile and ethanol having $NH_3.H_2O$ (0.389 g) and H_2O (0.91 g), under/stir for 6 hours. The resultant mixture of TiO_2 was centrifuged, cleaned with ethanol and deionized H_2O, and heated at 60°C for 24 hours. Afterward 0.1 g of TiO_2, 1:1 M ratio of $Sr(NO_3)_2$, and 0.2 mL (5 mg) Ti_3C_2 were added into 25 mL of 2 mol L^{-1} NaOH solution. Finally, the obtained product was put into a 50 mL-Teflon lined autoclaves and heated for 4 hours at 40°C. The obtained product was centrifuged, systematically washed, and heated at 60°C [18]. Fang et al. reported Cds/Ti_3C_2-OH photocatalyst manufactured by hydrothermal process with excellent photocatalytic performance [20]. Li et al. reported the facile hydrothermal technique was adapted to synthesis of Bio Br/Ti_3C_2 Photo Catalyst [21].

12.2.3 SELF-ASSEMBLE

Soft materials have ability to assemble to make cell membrane and biopolymer at nanoscale. Spontaneous formation of ribosome of soft material is their natural property. Later researchers work on this process and finally they were able to synthesize nanoscale material by using bottom-up strategy, which is known as Self-assembly. It is a process by which materials at nanoscale are spontaneously arranged into order superstructure. This technique is used to make a controlled structure from an

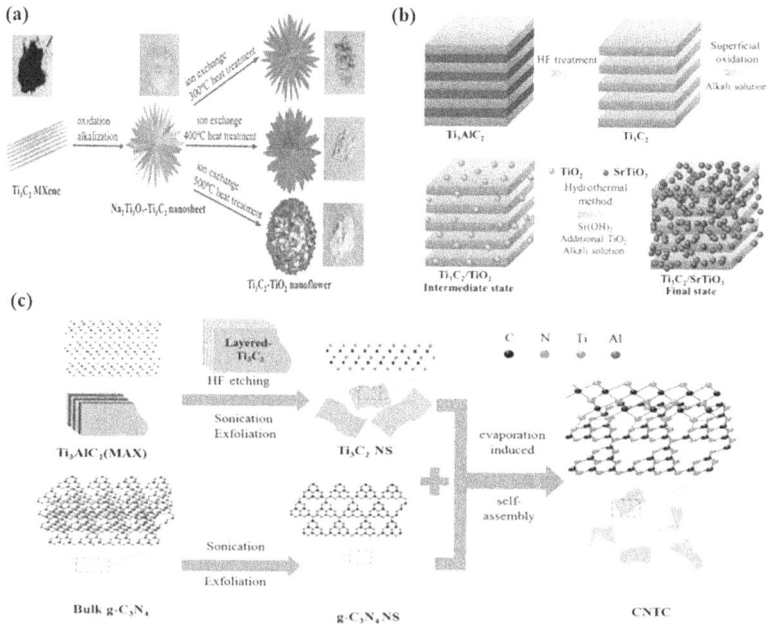

FIGURE 12.1 (a) Schematic diagram of the synthesis of TiO_2/Ti_3C_2 nanohybrid. (Adapted with permission from Ref. [17], Copyrights Intechopen limited (2020) IntechOpen.) (b) Preparation drawing of $SrTiO_3/Ti_3C_2$. (Adapted with permission from Ref. [18], Copyright *Applied Nano Materials* (2019) American Chemical Society.) (c) Schematic illustration for CNTC composite prepared by a combined HF etching and ultrasonic dispersion and evaporation induced self-assembly approach. (Adapted with permission from Ref. [19], Copyrights *Journal of Alloys and Compound* (2021) Elsevier.)

isolated sample. The electro-statics self-assemble is a process used to synthesize the heterocomposites to enhance the morphology management of materials. Huang et al. reported the synthesis of nanocomposite of BiOBr and Ti_3C_2 by electro-static self-assemble strategy [21].

The degradation ability of BiOBr enhances significantly making composite with Ti_3C_2. Typically, $Bi(NO_3).5H_2O$ (9.5 mg) is mixed into 200 mL of CH_3COOH with the continuous stirring to prepare solution A. Typically 20 mL solution of A was added into Ti_3C_2 dispersed solution and it was magnetically stirred for 2 hours. Afterward, NaBr (4 g) was mixed into 200 mL of distilled water to prepare solution B. And then 20 mL solution of B was taken and added into the above solution dropwise with continuous stirring for 2 hours to get BiOBr/Ti_3C_2 hybrid nanocomposite. Final product is washed with ethanol and with distilled water for several times, and the resultant obtained is dried at 60°C for 12 hours. Liu et al. studied the Ti_3C_2/g-C_3N_4 hybrid nanocomposite that was synthesized by evaporation-induced self-assembly process for ciprofloxacin photodegradation as shown in Figure 12.1c [19].

12.2.4 SOL-GEL METHOD

Sol–gel method is widely used for preparation of materials, makes a stronger compound, which consists of highly activated chemical component through a solution of sol or gel. In this method first raw materials disperse in solvent, which undergo a hydrolysis reaction to form monomer. This polymer then forms a polymer and begins to become a sol and then gel with a certain molecular structure. After drying and then heat treatment, the required nanoparticles are prepared. This technique has been successfully applied in synthesis of oxide coating, composite of materials, glass and to prepare superconducting materials that was prepared at high critical temperature. French chemist, M. Ebelmen was first who explained the sol gel preparation of Silica. He noticed that silica ester hydrolyzes in the presence of moistures to provide hydrated silica. In 1981, the first international workshop was organized for the preparation of glass and ceramics. Before that, thousands of papers on sol–gel synthesis technique have been reported, which opened a new door in the field of material science.

Sol–gel is a well-established synthesis procedure to make new composite of materials with metal oxides. This is a well-controlled procedure, which is used to control the structure and surface morphology of the nanocomposite [23]. Recently, MXene-based semiconductor photocatalyst has been prepared by using sol–gel technique. M. Abdullah Iqbal et al. reported nano hybrids of La- and Mn-codoped bismuth ferrite (BFO) nanoparticles implanted MXene-Ti_3C_2 sheets were synthesized by using sol–gel technique and investigated its photocatalytic activity. MXene-Ti_3C_2 has been prepared in distilled water having molarity of 0.5 mg mL^{-1} and then the solution was ultra-sonicated for 10 minutes. The nanoparticles of co-doped BFO dissolved into the solution of ethylene glycol and acetic acid having 0.01 M molarity with 1:1 ratio. The solution of $Bi_{1-x}La_xFe_{1-y}Mn_yO_3$ was ultra-sonicated at 60°C for 1 hour, and then this solution of $Bi_{1-x}La_xFe_{1-y}Mn_yO_3$ dissolved into Ti_3C_2 and the resultant solution was stirred for 2 hours at 80°C. The resultant product is washed several times with distilled water and dried at 60°C for 3 hours [24].

12.2.5 MECHANICAL MIXING

The thin layers of MXene are very small and can be easily stacked and make cluster, which makes it difficult to make large thin layers. Mechanical mixing of MXene and graphene is experienced directly by stirring to amalgamate the interlayers to make a hybrid structure. This will help to improve the performance of MXene and graphene by increasing the layer spacing and enhancing the fast transport of ions and ion sites. Feng et al. [25] reported the electrochemical spinning of MXene and graphene to prepare a nanocomposite thin film and used it as electrode in supercapacitor. As we can see in Figure 12.2a, when $Ti_3C_2T_x$ is mixed in rGO we can clearly see the Tyndall effect. Then a mixed solution of $Ti_3C_2T_x$/rGO is gained by mixing through ultra-sonication of two solutions, which is also clearly showing the Tyndall effect, which indicates that the resultant solution is uniformly dispersed as shown in Figure 12.2b. Finally, we can see in Figure 12.2c the membrane is obtained by vacuum filtration process, and the scanning electron microscopy (SEM) of sample is also shown in Figure 12.2e, which shows the clear hybrid structure of $Ti_3C_2T_x$/rGO.

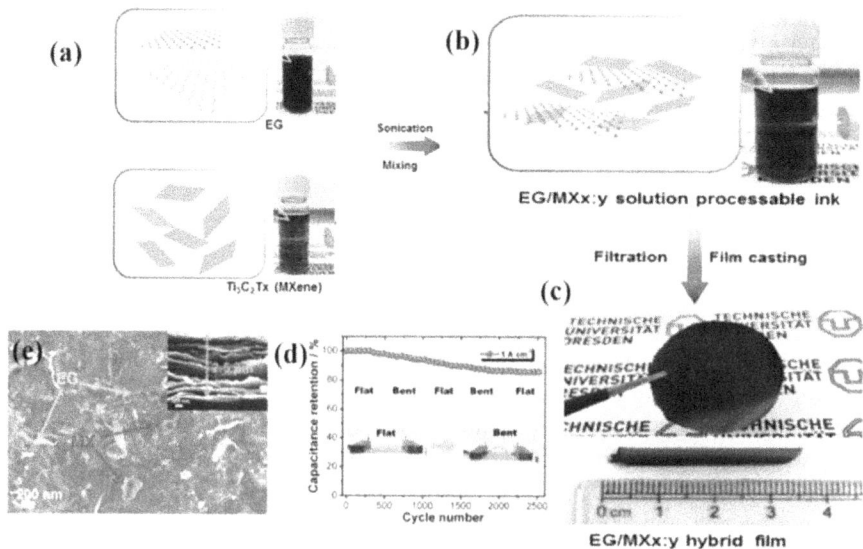

FIGURE 12.2   (a) Schematic diagram of layered rGO and MXene, (b) explanation of the preparation process of nanohybrid, (c) the photo of flexible and freestanding EGMX film, (d) capacitance retention rate for 2,500 cycles, (e) SEM images of the surface and cross-section of EGMX films. (Adapted with permission from Ref. [25], Copyrights *Chemosphere* (2021) Elsevier.)

In composite, MXene nanosheets are sandwiched between graphene sheets, which enhance layer spacing and are helpful in the insertion and delamination of ion during charging and discharging of supercapacitors. While graphene layers provide mechanical support to electrode, they enhance the electrical conductivity and stability of electrode. Therefore, the membrane-based electrode enhances the application of solid-state supercapacitors, which shows exceptional performance. These solid-state supercapacitors exhibit excellent life cycle as we can see in Figure 12.2d, the volumetric capacitance remains 82% of the initial capacitance after 2,500 cycles. Zhao et al. [26] synthesized the $Ti_3C_2T_x$/rGO composite electrode, they observed that in composite the graphene connected the different layers of the MXene and work as conductive bridge, which enhance the specific capacitance than a pure form of graphene and MXene.

12.2.6   HEAT TREATMENT METHOD

Besides the method of hydrothermal, there is another method to decrease graphene and this method is known as thermal reduction, which can discard groups of surface function in a better way than the method of chemical reduction. All oxygen groups can be discarded in the form of CO_2 and CO at high temperatures to heat the GO; without using the reductants. Gogotsi et al. [27] prepared $Ti_3C_2T_x$/rGO hybrid porous thin films by a mixture of MXene/GO with thermal reduction under a temperature at 300°C.

As well, Ma et al. proposed that both $Ti_3C_2T_x$ and rGO groups are hydrophilic, in which case $Ti_3C_2T_x$/rGO blended well, conforming to the theory of similarity. To synthesize the $Ti_3C_2T_x$/rGO composite 3D structure at 200°C for 2 hours, they use one-step freeze-drying method and annealing by Ar/H$_2$. With the property of pressure-sensitivity of Ti$_3$C$_2$T$_x$/rGO, we organized a peso-resistive sensor of ultra-elastic aerogel and ultra-light. So much rich porous architecture contributes to the fast electronic transmission channels and fast response with especially high sensitivity time. In this process, the temperature is kept low and it is carried out under a shielding gas to stop unwanted oxidation of layer of MXene.

12.3 PROPERTIES OF MXene NANOCOMPOSITE

In the age of industrialization and technology, we need clean and renewable energy resources instead of relying on fossil fuels. For this purpose, researchers focused on to synthesize and develop new novel material for solar cell, which are the best alternate of fossil fuels. As we discussed above, MXene belongs to 2D family of materials; its layered structure consists of carbides and nitrides of transition metal oxides. Gogosti group was the first to report in 2011 [28], after that MXene has been widely studied for different applications in different fields such as energy storage devices, photocatalysis, water purification, and biomedical field. MXene has a hexagonal crystal structure, in which M atoms belong to hexagonal close packing. While X atoms are filled in the octahedral gap, and its general formula (M$_{n+1}$X$_n$) is also consistent with MAX.

Recently it has been reported for solar cell, due to its good electrical conductivity, carrier mobility, and tunable work function [29,30]. They observe that the addition of Ti$_3$C$_2$T$_x$ increases the size of crystal and it is beneficial for accelerating charge through grain boundaries, which enhance the average power convergence efficiency from 15.18% to 16.6%. Agresti et al. reported the optimization in the energy level to improve the performance of solar cell by adding Ti$_3$C$_2$T$_x$ in MAPbI$_3$ film [31]. Generally, 2D MXene has been obtained from MAX phase different synthesis techniques such as chemical liquid-phase etching method to design the high-performance photocatalyst for solar fuel [32–35].

The 2D MXene has good corrosion resistance of alkali and acid due to trustworthy chemical liquid-phase etching process. As the MXene synthesis is a selective elemental process and has tunable bandgap as well as good layered structure make it good candidate for solar cell. Nanocomposite of MXene with other materials enhances the properties of MXene itself, for example, composite with graphene enhances the electrical conductivity and provides the large specific surface area and more absorption sites [36–38].

In addition with electrical conductivity, optical properties of MXene have been widely studied. MXene has good absorption ability, under low light Ti$_3$C$_2$T$_x$ has the advantage of saturation absorption and in strong light the saturation absorption ability makes MXene suitable for ultrafast laser application. The absorption properties and saturation flux of Ti$_3$C$_2$T$_x$ depend on the thickness of film. MXene has extraordinary optoelectronic properties, which are controlled by the electrochemical and chemical intercalation of cations. The excellent optoelectronic properties have been used to prepare transparent conductive film with low resistance and high electrical conductivity. The first transparent conductive film was prepared with conductivity of ~2,000–5,000 S cm^{-1} with transmittance of 14%–85% by etching of magnetron sputter Ti$_3$AlC$_2$ [39]. The photo thermal application of MXene has been studied as well; Li et al. reported that when the colloidal particle of MXene was irradiating on material, it starts absorbing light and converting it into heat energy [40].

12.4 PHOTOVOLTAIC APPLICATIONS OF MXene NANOCOMPOSITES

MXene is the group of two-dimensional materials, which have potential applications in the field of solar energy due to the unique electronic and optical properties. MXenes have gained great attraction due to good transmittance, high thermal stability, and better conductivity in the region of visible light [41]. Due to these properties, MXene becomes a promising material for Solar cells (SCs), Lithium-ion batteries (LIBs), sensors, and photovoltaic applications [42]. The work functions of the MXenes can play a significant role in the various applications for electrical devices. Through research, it is observed these work functions about MXenes depend on the making methodology and terminating groups [43]. MXenes have been explored with different work functions for photoelectron spectroscopy and electrodes of the transistor [44] but there is a need to explore more for better understanding about MXenes work functions. Furthermore good thermal conductivity and better transparency propose the potential application of MXenes in optoelectronic devices. There are no enormous publications in the field of photovoltaic applications as electrode of optoelectronics device with MXenes composites. Some 2D materials like GO and black phosphorous have been explored as charge collection material of organic solar cells but MXenes have some benefits because they have good ability of conductivity, high transmittance, and solvable in water [45]. The description about the role of MXenes composites in the field of photovoltaic application is given below (Figure 12.3).

12.4.1 ADDITIVE IN PEROVSKITE MATERIALS

For the first time in 2018, Guo et al. introduced the photovoltaic applications of MXenes described by the addition of $Ti_3C_2T_x$ onto the $MAPbI_3$-based perovskite absorber [46]. This addition increased the crystal size, which is shown in Figure 12.4a. This composite has good ability to escalate the electron transformation, reduced the resistance of charge carrier for the device through impedance spectra as shown in Figure 12.4b and power conversion efficiency (PCE) increased from 15.18% to 16.80%. Zhang et al. described about the heterostuctures of MXene with perovskite through in situ solution method. In these heterostuctures, $MAPbBr_3$ nanocrystals added with layers of $Ti_3C_2T_x$ MXene nanosheets as shown in Figure 12.4c [48]. Due to the matched energy levels, the electrons are easily transferred from the nanocrystals to MXene nanosheets, which is shown in Figure 12.4d and it is good for the better efficiency of solar cells.

12.4.2 ETLs/HTLs FOR SOLAR CELLS

Besides the study about the addition of MXene in the layer, some research articles have been published on implanting of $Ti_3C_2T_x$ MXene in electron transport layers (ETLs)/hole-transport layers (HTLs). Ti_3C_2 nanosheets is the MXene material, which are most studied in the previous years. The SnO_2-Ti_3C_2 MXene composite

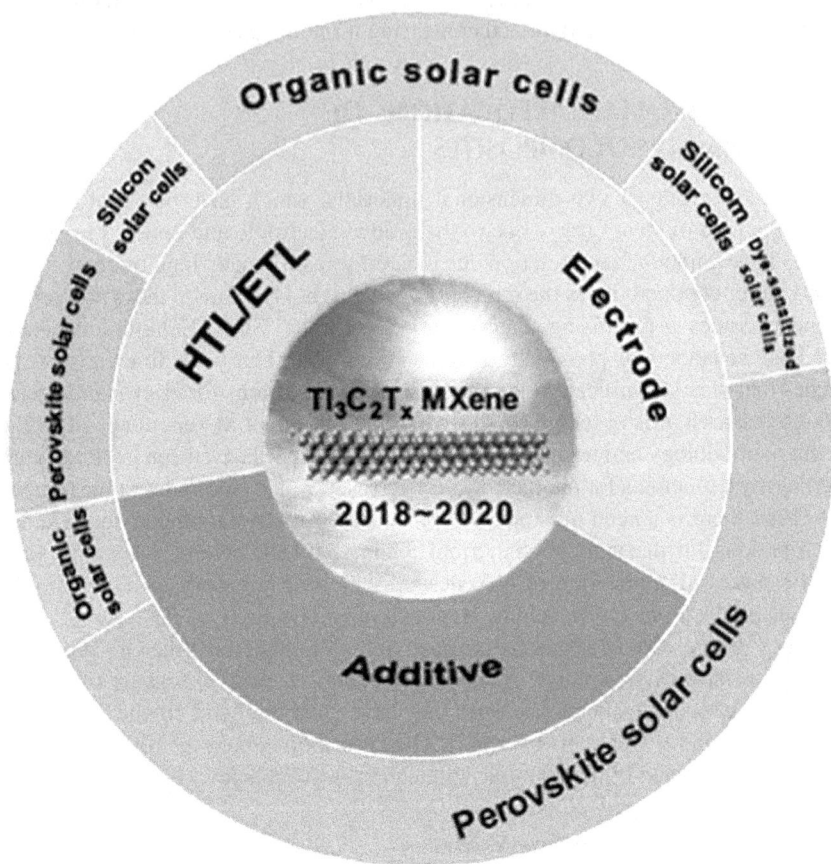

FIGURE 12.3 A schematic illustration of the roles of the $Ti_3C_2T_x$ MXene for application of different types of photovoltaic cells. (Adapted with permission from Ref. [46], Copyright (2021) *Polymer Reviews.*)

has potential application for the perovskite solar cells. The ETLs of SnO_2 with various Ti_3C_2 contents (0, 0.5, 1.0, 2.0, 2.5 wt.%) were synthesized by Yang et al. [48]. As compared to the SnO_2 the SnO_2-Ti_3C_2-based MXene composites showed better efficiency from 17.23% to 18.34% attained by adding the 1.0 wt.% Ti_3C_2 with SnO_2 layer.

The increment of PCE can be ascribed due to fast electron extraction, good electrical conductivity, and electron mobility in ETL [49]. The different characteristics of SnO_2-Ti_3C_2 MXene composite have been explored, some are given below. To increase the efficiency of perovskite solar cell (PSCs) and determine the potential of Ti_3C_2 the device with the architecture of ITO/ETL/$CH_3NH_3PbI_3$/ Spiro-OMeTAD/Ag based on SnO_2-Ti_3C_2 as ETL was fabricated, which is shown in Figure 12.4e. The UV-vis absorption spectrum of different films is shown in

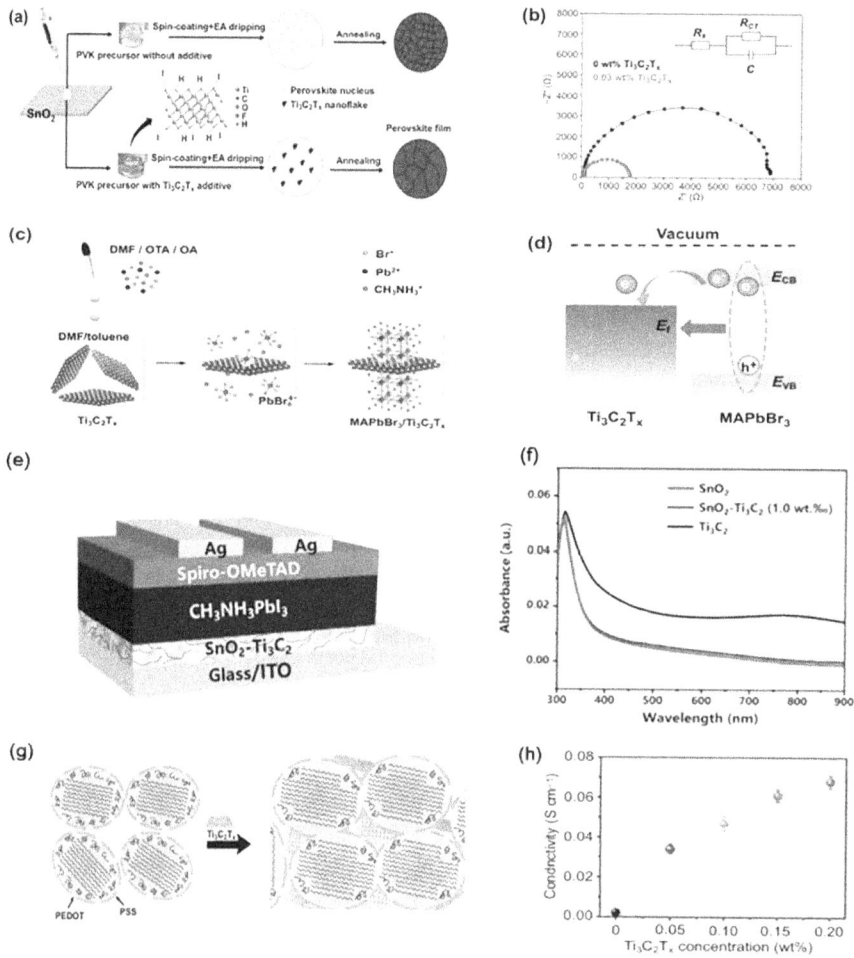

FIGURE 12.4 (a) Schematic diagram of growth of the MAPbI$_3$-based perovskite films by adding and without adding of Ti$_3$C$_2$T$_x$ MXene. (b) The Impedance spectra with and without 0.03 wt.% Ti$_3$C$_2$T$_x$. (Adapted with permission from Ref. [47], Copyrights *Coordination Chemistry Reviews* (2017) Elsevier.) (c) Schematic image of the preparation procedure MAPbBr$_3$ nanocrystals with Ti$_3$C$_2$T$_x$ MXene nanosheets, (d) energy-level configuration and electron transfer between nanocrystals and MXene nanosheets. (Adapted with permission from Ref. [48], Copyrights *Progress in Solid State Chemistry* (2000) Elsevier.) (e) Image of device architecture of ITO/ETL/CH$_3$NH$_3$PbI$_3$/Spiro-OMeTAD/Ag based on SnO$_2$-Ti$_3$C$_2$ as ETL, (f) UV-vis light absorption spectrum of SnO$_2$, SnO$_2$-Ti$_3$C$_2$ (1.0 wt.%), and Ti$_3$C$_2$ (Adapted with permission from Ref. [49], Copyrights *Chemistry of Materials*, (2017) ACS Publication.) (g) Diagram of morphological and structural modification in PEDOT:PSS, (h) electrical conductivity of PEDOT:PSS with various Ti$_3$C$_2$T$_x$ concentration. (Adapted with permission from Ref. [50], Copyrights *Thin Solid Film* (2010) Elsevier.)

(Continued)

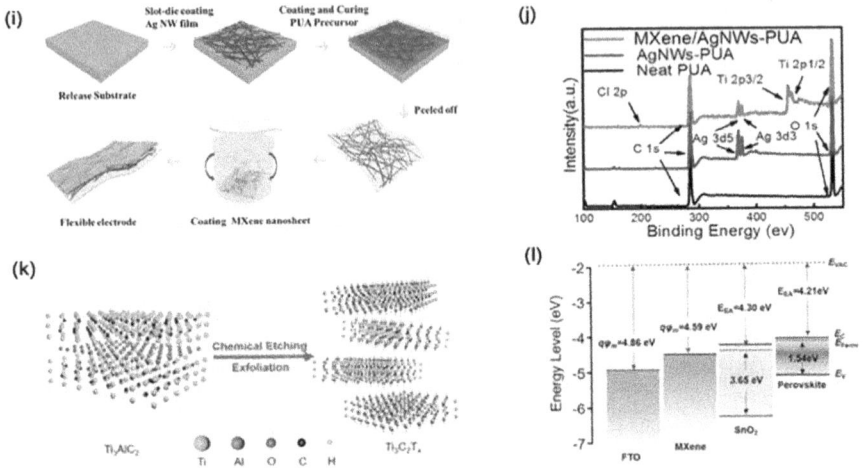

FIGURE 12.4 (Continued) (i) The illustration of synthesis process of flexible electrode, (j) XPS spectrum of neat PUA, AgNWs-PUA, and MXene/AgNWs-PUA films. (Adapted with permission from Ref. [51], Copyrights *Advance Materials* (2014) Wiley Online Library.) (k) Schematic diagram of $Ti_3C_2T_x$ MXene synthesis process. (l) Determined work function $(q\phi)$, E_{EA}, E_V, and E_{Fermi} of FTO, MXene, SnO$_2$, and perovskite layers. (Adapted with permission from Ref. [52], *Nature Reviews Materials*, (2017) Nature.)

Figure 12.4f. The curves of these films are near to each other, but the Ti_3C_2 curve shows some more absorption in UV light region, which is representing that Ti_3C_2 to SnO$_2$ layer does not disturb the transmission of light. The addition of $Ti_3C_2T_x$ into HTLs can also increase the device efficiency. The improvement in organic solar cells (OSCs) due to the combination of conductive polymer HTL with nanosheets of $Ti_3C_2T_x$ MXene was explored [50]. By the addition of nanosheets, the charge transfer channel can be established between the PEDOT nanocrystals as shown in Figure 12.4g. The PEDOT transformed from the coil structure into linear coil structure, which is better for the electrical conductivity of new modified PEDOT:PSS material. The increment in the electrical conductivity has been proved in the graph, which is shown in Figure 12.4h.

12.4.3 ELECTRODE OF DEVICE

MXene-silver nanowire was synthesized with MXene nanosheets with Ag nanowires to construct the transparent electrode for flexible solar cells. The Ag nanowire has high optoelectronic efficiency and zeta potential becomes negligible due to the absorbed electrostatic force between nanosheets of MXene and Ag nanowires. The Ag nanowire network is partially fixed with poly matrix, which lessens the roughness producing due to the loading of hybrid materials. In these types of solar cells, good PV efficiency was observed, which increase the photovoltaic applications [51]. The illustration about synthesis procedure of this flexible electrode is given in Figure 12.4i. The XPS spectra of various films are shown in Figure 12.4j.

The region 380–387 eV related to the Ag 3d XPS bands, which are associated to Ag nanowires and the Ti 2p XPS bands between 453 and 460 eV are ascribed to the MXene nanosheets. Additionally MXene layer has better conductivity as compared to the normally used conducting polymers and charges are easily transported in the region of Ag nanowires network.

An electrode is synthesized with the help of SnO_2 ELTs with $Ti_3C_2T_x$ MXene thin layer. This electrode is leading good ability of energy level, increased electron mobility, and fast charge transfer capability. Due to these reasons device represents an alleviated PCE of 20.65% with low current density [52]. To fabricate the $Ti_3C_2T_x$ MXene we etched aluminum from Ti_3AlC_2 powder and then exfoliated by centrifugation process. The schematic process of $Ti_3C_2T_x$ MXene nanoflakes synthesis is shown in Figure 12.4k. The CB minimum of perovskite (4.21 eV) and SnO_2 (4.30 eV) can be measured with this equation $E_C = q\phi + (E_{Fermi} - E_V) - E_g$. The description about work function of the materials is given in Figure 12.4l.

12.5   CONCLUSION

In recent times, MXene and its nanocomposite have been widely studied due to the unique properties of MXene. MXene belongs to a large family of 2D materials have a unique structure, inherited good electrical conductivity, and excellent photo absorption and mechanical properties. Its layered structure is made of carbide and nitrides, which make it very useful in the field of energy storage devices, photocatalyst, and biomedicine and electrocatalysis. In photocatalyst MXene provides fast photo generated charge carrier and separation and its terminal functional group enhances the light-harvesting and photo conversion efficiency. MXene also have large specific SA and surface functional group, which makes it very suitable for photocatalyst and for solar cell as it can act as photo generated electron acceptor. $Ti_3C_2T_x$ was the first MXene discovered, investigated, and exhibits excellent catalytic activity for the hydrogen evolution reaction. Moreover, $Ti_3C_2T_x$ has good electrical conductivity and optical bandgap, which is more importantly tunable in visible region. The addition of $Ti_3C_2T_x$ with BiOBr enhances the photocatalytic performance of BiOBr as compared to pure form of BiOBr. The nanocomposite of MXene with polymer acquires excellent functional properties such as it enhance the conductivity of polymer up to several order. The nanocomposites of MXene and polymer have attracted intense intention for tumor removal, solid-state capacitors, and electromagnetic shielding.

12.6   FUTURE RECOMMENDATIONS

Right now, nanotechnology is alluded to be one of the alluring research areas in a few countries as a result of its tremendous prospective and marketable impact. Nanotechnology incorporates the examination, investigation, improvement, manufacture, and preparing of designs the materials on a nanoscale in different fields of science, medical services, horticulture, innovation, technology, and industries. There are innovative applications of MXene in different zones such as for biomedicines, energy storage devices, solar cells, photocatalyst, and photochemical process. The improvements and development achieved in preparation, features, and highlights, together with the advanced uses of MXene, convey a vigorous stimulus to uttermost advancement in the depiction and utilization of these most recent 2D materials. We can suggest that these two-dimensional materials have surprising properties along massive impact on photovoltaics. The impact encourages and attracts young researchers and scholars to get some headings and progression in photovoltaic field in their future work. This research area can provide new and useful functional and structural building blocks for future production of optoelectronic and photovoltaic devices [53,54].

ACKNOWLEDGMENTS

Muhammad Bilal and J.H. Hou acknowledge the financial support from the *National Natural Science Foundation of China* (51602281), The "Qinglan Project" of Jiangsu Universities. Zia Ur Rehman and Faheem K. Butt acknowledge the funding support from HEC through grant number 7435/Punjab/NRPU/R&D/HEC/2017.

REFERENCES

1. M. Sen, Nanocomposite materials, chapter 6. In: B. Karn et al. (Eds), *Nanotechnology and the Environment*. IntechOpen: London, United Kindom, 2020: p. 12.
2. E.J.S. Jong Kwon Im, S. Kim, M. Jang, A. Son, K.-D. Zoh, & Y. Yoon, Review of MXene-based nanocomposites for photocatalysis. *Chemosphere*, 2021, 270, 129478.
3. N.J. Prakash, & B. Kandasubramanian, Nanocomposites of MXene for industrial applications. *Journal of Alloys and Compounds*, 2021, 862, 158547.
4. Y.Z. Xiaoyong Chen, L. Li, Y. Wang, J. Wang, J. Xiong, S. Du, P. Zhang, X. Shi & J. Yu, MXene/polymer nanocomposites: Preparation, properties, and applications. *Polymer Reviews*, 2021, 61(1), 35.
5. Yang, C, L.W., Yang Z, Gu L, & Yu Y., Nanoconfined antimony in sulfur and nitrogen co-doped three-dimensionally (3D) interconnected macroporous carbon for high performance sodium-ion batteries. *Nano Energy*, 2015, 18, 8.
6. J. Zhu, E. Ha, G. Zhao, Y. Zhou, D. Huang, G. Yue, L. Hu, N. Sun, Y. Wang, L.Y.S. Lee, Recent advance in MXenes: A promising 2D material for catalysis, *Sensor and Chemical Adsorption: Coordination Chemistry Reviews*, 2017, 352, 20.
7. M.W. Barsoum, The MN+ 1AXN phases: A new class of solids: Thermodynamically stable nanolaminates. *Progress in Solid State Chemistry*, 2000, 28(1–4), 201–281.
8. M. Alhabeb, K. Maleski, B. Anasori, P. Lelylukh, L. Clark, S. Sin, & Y. Gogotsi, Guidelines for synthesis and processing of two-dimensional titanium carbide ($Ti_3C_2T_x$ MXene). *Chemistry of Materials*, 2017, 29(18), 7633–7644.
9. P. Eklund, M. Beckers, U. Jansson, H. Hogberg, & L. Hultman, The $M_{n+1}AX_n$ phases: Materials science and thin-film processing. *Thin Solid Films*, 2010, 518(8), 1851–1878.
10. M. Naguib, V. Mochalin, W. Barsoum, & Y. Gogotsi, Two-dimensional materials: 25th anniversary article: MXenes: A new family of two-dimensional materials. *Advanced Materials*, 2014, 26(7), 982–982.
11. B. Anasori, M.R. Lukatskaya, & Y. Gogotsi, 2D metal carbides and nitrides (MXenes) for energy storage. *Nature Reviews Materials*, 2017, 2(2), 1–17.
12. B. Anasori, & Y. Gogotsi, Introduction to 2D transition metal carbides and nitrides (MXenes). In *2D Metal Carbides and Nitrides* (MXenes). Springer. 2019, pp. 3–12.

13. X. Zhan, C. Si, J. Zhou, & Z. Sun, MXene and MXene-based composites: Synthesis, properties and environment-related applications. *Nanoscale Horizons*, 2020, 5(2), 235–258.

14. X. Jiang, A.V. Kuklin, A. Baev, Y. Ge, H. Ågren, H. Zhang, & P.N. Prasad, Two-dimensional MXenes: From morphological to optical, electric, and magnetic properties and applications. *Physics Reports*, 2020, 848, 1–58.

15. L. Cheng, X. Li, & Q. Xiang, Two-dimensional transition metal MXene-based photocatalysts for solar fuel generation. *The Journal of Physical Chemistry Letters*, 2019, 10(12), 3488–3494.

16. J.B. Rivest, & P.K. Jain, Cation exchange on the nanoscale: An emerging technique for new material synthesis, device fabrication, and chemical sensing. *Chemical Society Reviews*, 2013, 42(1), 89–96.

17. Y. Li, X. Deng, J. Tian, Z. Liang, & H. Cui, Ti_3C_2 MXene-derived Ti_3C_2/TiO_2 nanoflowers for noble-metal-free photocatalytic overall water splitting. *Applied Materials Today*, 2018, 13, 217–227.

18. H. Deng, Z. Li, L. Wang, L.-Y. Yuan, J.-H. Lan, Z.-Y. Chang, Z.-F. Chai, & W.-Q. Shi, Nanolayered Ti_3C_2 and $SrTiO_3$ composites for photocatalytic reduction and removal of uranium (VI). *ACS Applied Nano Materials*, 2019, 2(4), 2283–2294.

19. H. Fang, Y. Pan, M. Yin, L. Xu, Y. Zhu, & C. Pan, Facile synthesis of ternary Ti_3C_2–$OH/In_2S_3/CdS$ composite with efficient adsorption and photocatalytic performance towards organic dyes. *Journal of Solid State Chemistry*, 2019, 280, 120981.

20. Z. Li, H. Zhang, L. Wang, M. Xiangchao, J. Shi, C. Qi, Z. Zhang, & L.F.C. Li, 2D/2D $BiOBr/Ti_3C_2$ heterojunction with dual applications in both water detoxification and water splitting. *Journal of Photochemistry and Photobiology A: Chemistry*, 2020, 386, 112099.

21. Q. Huang, Y. Liu, T. Cai, & X. Xia, Simultaneous removal of heavy metal ions and organic pollutant by $BiOBr/Ti_3C_2$ nanocomposite. *Journal of Photochemistry and Photobiology A: Chemistry*, 2019, 375, 201–208.

22. N. Liu, N. Lu, Y. Su, P. Wang, & X. Quan, Fabrication of g-C_3N_4/Ti_3C_2 composite and its visible-light photocatalytic capability for ciprofloxacin degradation. *Separation and Purification Technology*, 2019, 211, 782–789.

23. A. Rozmysłowska Wojciechowska, E. Karwowska, S. Poźniak, T. Wojciechowski, L. Chlubny, A. Olszyna, W. Ziemkowska, & A.M. Jastrzębska, Influence of modification of $Ti_3 C_2$ MXene with ceramic oxide and noble metal nanoparticles on its antimicrobial properties and ecotoxicity towards selected algae and higher plants. *RSC Advances*, 2019, 9(8), 4092–4105.

24. I.R. Shein, & A.L. Ivanovskii, Graphene-like nanocarbides and nanonitrides of d metals (MXenes): Synthesis, properties and simulation. *Micro & Nano Letters*, 2013, 8(2), 59–62.

25. H. Li, Y. Hou, F. Wang, M.R. Lohe, X. Zhuang, L. Niu, & X. Feng, Flexible all-solid-state supercapacitors with high volumetric capacitances boosted by solution processable MXene and electrochemically exfoliated grapheme, 2017.

26. C. Zhao, Q. Wang, H. Zhang, S. Passerini, & X. Qian, Two-dimensional titanium carbide/RGO composite for high-performance supercapacitors. *ACS Applied Materials & Interfaces*, 2016, 8(24), 15661–15667.

27. S. Xu, G. Wei, J. Li, W. Han, & Y. Gogotsi, Flexible MXene–graphene electrodes with high volumetric capacitance for integrated co-cathode energy conversion/storage devices. *Journal of Materials Chemistry A*, 2017, 5(33), 17442–17451.

28. M. Naguib, M. Kurtoglu, V. Presser, J. Lu, J. Niu, M. Heon, L. Hultman, Y. Gogotsi, & M.W. Barsoum, Two-dimensional nanocrystals produced by exfoliation of Ti_3AlC_2. *Advanced Materials*, 2011, 23(37), 4248–4253.

29. M. Xu, S. Lei, J. Qi, Q. Dou, L. Liu, Y. Lu, Q. Huang, S. Shi, & X. Yan, Opening magnesium storage capability of two-dimensional MXene by intercalation of cationic surfactant. *ACS Nano*, 2018, 12(4), 3733–3740.
30. M. Khazaei, A. Ranjbar, M. Arai, T. Sasaki, & S. Yunoki, Electronic properties and applications of MXenes: A theoretical review. *Journal of Materials Chemistry C*, 2017, 5(10), 2488–2503.
31. A. Agresti, A. Pazniak, S. Pescetelli, A. Di Vito, D. Rossi, A. Pecchia, M. Auf der Maur, A. Liedl, R. Larciprete, D.V. Kuznetsov, D. Saranin, & A. Di Carlo, Titanium-carbide MXenes for work function and interface engineering in perovskite solar cells. *Nature Materials*, 2019, 18(11), 1228–1234.
32. M. Naguib, O. Mashtalir, J. Carle, V. Presser, J. Lu, L. Hultman, Y. Gogotsi, & M.W. Barsoum, Two-dimensional transition metal carbides. *ACS Nano*, 2012, 6(2), 1322–1331.
33. M. Khazaei, M. Arai, T. Sasaki, C.-Y. Chung, N.S. Venkataramanan, M. Estili, Y. Sakka, & Y. Kawazoe, Novel electronic and magnetic properties of two-dimensional transition metal carbides and nitrides. *Advanced Functional Materials*, 2013, 23(17), 2185–2192.
34. Q. Liu, L. Ai, and J. Jiang, MXene-derived TiO_2@ $C/gC_3 N_4$ heterojunctions for highly efficient nitrogen photofixation. *Journal of Materials Chemistry A*, 2018, 6(9), 4102–4110.
35. X. Wang, C. Garnero, G. Rochard, D. Magne, S. Morisset, S. Hurand, P. Chartier, J. Rousseau, T. Cabioc'h, C. Coutanceau, V. Mauchamp, & S. Célérier, A new etching environment (FeF_3/HCl) for the synthesis of two-dimensional titanium carbide MXenes: A route towards selective reactivity vs. water. *Journal of Materials Chemistry A*, 2017, 5(41), 22012–22023.
36. X.-D. Zhu, Y. Xie, & Y.-T. Liu, Exploring the synergy of 2D MXene-supported black phosphorus quantum dots in hydrogen and oxygen evolution reactions. *Journal of Materials Chemistry A*, 2018, 6(43), 21255–21260.
37. H. Wang, Y. Wu, T. Xiao, X. Yuan, G. Zeng, W. Tu, S. Wu, H.Y. Lee, Y.Z. Tan, & J.W. Chew, Formation of quasi-core-shell In_2S_3/anatase TiO_2@ metallic $Ti_3C_2T_x$ hybrids with favorable charge transfer channels for excellent visible-light-photocatalytic performance. *Applied Catalysis B: Environmental*, 2018, 233, 213–225.
38. W. Yuan, et al., Laminated hybrid junction of sulfur-doped TiO_2 and a carbon substrate derived from Ti_3C_2 MXenes: Toward highly visible light-driven photocatalytic hydrogen evolution. *Advanced Science*, 2018, 5(6), 1700870.
39. W. Yuan, L. Cheng, Y. An, S. Lv, H. Wu, X. Fan, Y. Zhang, X. Guo, & J. Tang, Transparent conductive two-dimensional titanium carbide epitaxial thin films. *Chemistry of Materials*, 2014, 26(7), 2374–2381.
40. R. Li, L. Zhang, L. Shi, & P. Wang, MXene Ti_3C_2: An effective 2D light-to-heat conversion material. *ACS Nano*, 2017, 11(4), 3752–3759.
41. I. Persson, L.-Å. Näslund, J. Halim, M.W. Barsoum, V. Darakchieva, J. Palisaitis, J. Rosen, P.O.Å. Persson, On the organization and thermal behavior of functional groups on Ti_3C_2 MXene surfaces in vacuum. *2D Materials*, 2017, 5(1), 015002.
42. Z. Ma, X. Zhou, W. Deng, D. Lei, & Z. Liu, 3D porous MXene(Ti_3C_2)/reduced graphene oxide hybrid films for advanced lithium storage. *ACS Applied Materials & Interfaces*, 2018, 10(4), 3634–3643.
43. H.A. Tahini, X. Tan, & S.C.J.N. Smith, The origin of low workfunctions in OH terminated MXenes. *Nanoscale*, 2017, 9(21), 7016–7020.
44. Z. Wang, H. Kim, & H.N.J.A.M. Alshareef, Oxide thin-film electronics using all-MXene electrical contacts. *Advanced Materials*, 2018, 30(15), 1706656.
45. H.P. Kim, A.R.M. Yusoff, & J. Jang, Organic solar cells using a reduced graphene oxide anode buffer layer. *Solar Energy Materials and Solar Cells*, 2013, 110, 87–93.

46. Z. Guo, L. Gao, Z. Xu, S. Teo, C. Zhang, Y. Kamata, S. Hayase, & T. Ma, High electrical conductivity 2D MXene serves as additive of perovskite for efficient solar cells. *Small* 2018, 14(47), 1802738.

47. Z. Zhang, Y. Li, C. Liang, G. Yu, J. Zhao, S. Luo, Y. Huang, C. Su, & G. Xing, In situ growth of MAPbBr$_3$ nanocrystals on few-layer MXene nanosheets with efficient energy transfer. *Small*, 2020, 16(17), 1905896.

48. L. Yang, Y.D.K. Hantanasirisakul, C.E. Shuck, K. Maleski, M. Alhabeb, G.C.Y.G. Yoshitaka Sanehira, A.K. Jena, L. Shen, C. Dall'Agnese, X.-F. Wang, Y. Gogotsi, & T. Miyasaka, SnO$_2$–Ti$_3$ C$_2$ MXene electron transport layers for perovskite solar cells. *Journal of Materials Chemistry A,* 2019, 7(10), 5635–5642.

49. C. Houa, & H.J.J.M.C. Yu, Modifying the nanostructures of PEDOT: PSS/Ti$_3$C$_2$TX composite hole transport layers for highly efficient polymer solar cells. *Journal of Materials Chemistry C*, 2020, 100(8), 4169–4180.

50. H. Tang, H. Feng, H. Wang, X. Wan, J. Liang, & Y. Chen, Highly conducting MXene–silver nanowire transparent electrodes for flexible organic solar cells. *ACS Applied Materials & Interfaces*, 2019, 11(28), 25330–25337.

51. Y. Wang, P. Xiang, A. Ren, H. Lai, Z. Zhang, Z. Xuan, Z. Wan, J. Zhang, X. Hao, L. Wu, M. Sugiyama, U. Schwingenschlögl, C. Liu, Z. Tang, J. Wu, Z. Wang, & D. Zhao, MXene-modulated electrode/SnO$_2$ interface boosting charge transport in perovskite solar cells. *ACS Applied Materials & Interfaces*, 2020, 12(48), 53973–53983.

52. Y.Y. Jaeho Jeon, H. Choi, J.-H. Park, B.H. Lee, & S. Lee, MXenes for future nanophotonic device applications. *Nanophotonics*, 2020, 9(7), 23.

53. K.K.S. Karthik Kannan, A.M. Abdullah, & B. Kumar, Current trends in MXene-based nanomaterials for energy storage and conversion system: A mini review. *Catalysts*, 2020, 10, 28.

13 Graphene-Based Nanomaterials for Battery Applications

Arpana Agrawal
Shri Neelkantheshwar Government Post-Graduate College

CONTENTS

13.1 INTRODUCTION

For maintaining social development, there is a rapidly increasing demand for smart-phones, electric vehicles, and energy storage devices such as power banks which largely depends on renewable energy storage and hence the battery volume. Accordingly, it is highly imperative to develop novel materials that can be employed for energy storage applications, where the basic requirements will be the high energy density (ED), stability, cost effectiveness, and long cycle life. So far, several electrochemical (EC) energy storage devices have been invented including lithium-ion batteries (LIBs), lithium–selenium batteries, lithium–sulfur batteries (LSBs), sodium-ion batteries (SIBs), sodium–selenium batteries (SSeBs), sodium–sulfur batteries, magnesium-ion batteries, and magnesium–sulfur batteries. However, the efficiency/EC performances of such energy storage systems can be further enhanced by utilizing graphene-based nanomaterials as either anodes or cathodes for such devices. Since 2004, graphene has become one of the most extensively studied materials from fundamental physics

DOI: 10.1201/9781003178422-13

to device application view point, owing to its several fascinating properties and its utility in the world of EC energy-storage devices is not an exception [1].

It is worth stressing here that one of the main components of a battery system is its electrodes and the material utilized, which is a crucial decisive factor to obtain the high energy output [2]. The thumb rule that one has to keep in mind prior selecting any specific material to serve for such purposes is the choice of high energy/power systems and appropriate modification strategies on reaction kinetics [3]. The former can be examined on the basis of thermodynamics, which correlates the ED with that of the output voltage (V) and specific capacity (Q) for a specific EC reaction using the equation [4]

$$ED = V \times Q \tag{13.1}$$

Here, V ($V_{cathode} - V_{anode}$) is the voltage difference between the cathode and the anode with $V_{cathode}$ and V_{anode} being the chemical potential of cathode and anode materials, respectively. Hence, to obtain high current density, the chemical potential of cathode should have to be larger than the chemical potential of anode material. It should also be noted here that high power density also warrants fast diffusion time along with high electrical conductivity and is expressed as follows [4]:

$$\tau_{eq} = L^2 / 2D \tag{13.2}$$

where D is the diffusion coefficient of the respective ion in the host electrode material and L is the diffusion length. On the other hand, appropriate modification strategies on reaction kinetics can be achieved either by nanostructure engineering, conductive composite engineering, or binder-free array engineering. According to kinetic formulation, the power density (P) can be given by [4]

$$P = E^2 / 4R_s \tag{13.3}$$

where R_s is the average in an impedance of electrodes.

Several materials have been explored so far to act as anode or cathode electrode materials for EC energy storage devices, however, carbon isotopes particularly graphene-based nanomaterials are much more fascinating. Graphene-based nanomaterials are potential candidate materials to serve both as a cathode and an anode for several EC energy storage applications including LIBs, lithium–selenium batteries, LSBs, SIBs, SSeBs, sodium–sulfur batteries, magnesium-ion batteries, and magnesium–sulfur batteries.

Graphene-based nanomaterials can be used either in flexible or in wearable energy storage device applications where electrodes are in the form of thin films grown on suitable substrate or nonflexible applications where electrodes are in the form of powder material. Among the EC energy storage device, LIBs have gained immense research interest for EC energy storage devices because of the availability of lithium at lower cost. Apart from LIBs, LSBs and lithium–selenium batteries have also been fabricated. Among these, fabrication of LSBs are quite challenging due to low electrical conductivity of sulfur which leads to weaker EC responses. In contrast to this, lithium–selenium is much more fascinating and shows rapid and encouraging EC performance owing to the large intrinsic electrical conductivity of elemental selenium. However, there is a lack of Li resources which motivate researchers to look for other alternatives, and hence, SIBs, sodium–sulfur batteries, and SSeBs have progressively become potential energy storage systems. Also, the performances of all these energy storage devices are found to improve by employing graphene-based nanomaterials. Accordingly, in this chapter, a critical analysis of various EC energy storage applications employing graphene-based nanomaterials either as anode or as cathode has been presented along with the advantages, limitations, and few recent results and applications.

13.2 SYNTHESIS OF GRAPHENE-BASED NANOMATERIALS

Graphene-based nanomaterials can be synthesized either by adopting bottom-up or top-down approach depending upon the requirements and initial precursor material [1]. The former requires high processing temperature with slow cooling and heating processes to build the final product using atomic sized hydrocarbons as initial precursors, whereas the latter involves breaking of larger graphite (precursors) to produce nanosized graphene-based materials. Bottom-up approach involves sophisticated instrumentation along with expensive materials and flammable gases and hence is expensive as compared to top-down method. However, one of the most important advantages of this method is the growth of large area high quality graphene and includes chemical vapor deposition (CVD), thermal CVD, plasma-enhanced CVD, radio-frequency CVD, and laser CVD. On the other hand, electric wire explosion, arc-discharge method, unzipping of carbon nanotubes (CNTs), and exfoliation methods such as mechanical exfoliation, dry/wet ball milling, sonication-assisted liquid phase exfoliation, and EC exfoliation are top-down approaches. However, both the approaches have their own advantages and restrictions but can also be utilized for synthesizing graphene-based nanomaterials for EC energy storage applications.

13.3 GRAPHENE-BASED NANOMATERIALS FOR ELECTROCHEMICAL ENERGY STORAGE APPLICATIONS

Graphene-based nanomaterials are extensively utilized as an anode or a cathode material for several EC energy storage applications including LIBs, SIBs, lithium–selenium batteries, SSeBs, LSBs, and sodium–sulfur batteries. In this section, each of these systems will be discussed in detail.

13.3.1 LITHIUM ION BATTERY (LIB)

13.3.1.1 Graphene-Based Nanomaterials as Anode for LIBs

Several graphene-based nanomaterials have been reported to be utilized as anode materials for LIBs including reduced graphene oxide (RGO), hybrid structures such as CNT/graphene, nitrogen-doped graphene (N-G) nanosheets (NSs), and graphene/metal oxide composites. RGO synthesized using green method by treating as-synthesized graphene oxide with green solvent, i.e., supercritical ethanol, has been proposed as an anode material for LIBs by Liu et al. [5], which shows specific capacity of 89 mAh g^{-1} after 50 cycles at 50 mA g^{-1}. Single-walled CNTs/graphene grown by microwave autoclave method has been reported by Zhong et al. [6], which exhibits a high capacity of 303 mAh g^{-1} and no signature of performance deprivation even after 50 cycles. Liu et al. [7] have demonstrated the performance of N-GNS synthesized using mechanochemical method via ball milling of graphite with urea. The specific capacity was 895 mAh g^{-1} at 50 mA g^{-1} and retention capacity of 550 mAh g^{-1} after 100 cycles. Xing et al. [8] have also reported a high performance of hydrothermally synthesized N-G as anode material prepared using hexamethylenetetramine serving both as carbon and nitrogen source. EC performance was examined with CR2032 coin type cell where Li foil and N-doped graphene act as counter and anode electrodes, respectively. Working electrode was prepared from a mixture of active material (70 wt.%), super P carbon black (20 wt.%), and polyvinylidene fluoride (PVDF) (10 wt.%) which were ball milled in N-methylpyrrolidone (NMP) at 400 rpm for 5 hours. The obtained slurry was then coated on a copper foil and vacuum dried up for 12 hours at 80°C. Room temperature charging/discharging capacity as well as cyclic voltammetry (CV) was investigated in a voltage range of 0.01–3.0 V. Figure 13.1a depicts the EC performance of the synthesized N-G anode electrode by performing CV measurements for the first five cycles. The electrode shows two anodic peaks (designated as O$_1$ and O$_2$) and two cathodic peaks (labeled as R$_1$ (0.7 V) and R$_2$ (below 0.2 V)) during the first cycle. O$_2$ anodic and R$_2$ cathodic peaks may be assigned to intercalation of Li ions into graphene sheets, while O$_1$ anodic peak is associated with the bond rupturing of Li atoms via defects during the charging process at high voltages. Also, it can be seen that R$_1$ cathodic peak disappears after the first cycle which can be ascribed to the electrolyte decomposition taking place on the surface of graphene.

Charging/discharging profiles for a few cycles are shown in Figure 13.1b. In the first cycle, Li ion intercalation gets initiated around 2 V, and after this, the discharging curve becomes steeper. The initial discharge and charge specific capacities are found to be 1,420 and 950 mAh g^{-1}, respectively. During the second cycle, the discharge capacity reduces to 960 mAh g^{-1} and the irreversible capacity loss of 460 mAh g^{-1} was observed which may also be assigned to the electrolyte decomposition on graphene surface along with the adsorption of Li ions on vacancies or at the neighborhood of residual N groups. Figure 13.1c shows the discharge/charge cycling properties of N-doped graphene anode electrode at 100 mA g^{-1} where it was found that the specific capacity stabilizes after the fifth cycle and after 50 cycles; it remains above 600 mAh g^{-1} suggesting excellent capacity retention of the grown electrode. Upon increasing the current density to 200 mAh g^{-1}, stationary capacity maintains above 500 mAh g^{-1}. A large irreversible capacity in first discharge/charge process was evident from the columbic efficiency of 67% which was increased to 98% in the following cycles. The rate capacity examined at various current densities for the N-G anode has also been investigated as shown in Figure 13.1d where stable discharge capacities were observed upon increasing the current intensities. Also, the capacity retention is promising and can be seen from the recovering of initial capacity (~650 mAh g^{-1}) even when the current density reverses back to 100 mA g^{-1}.

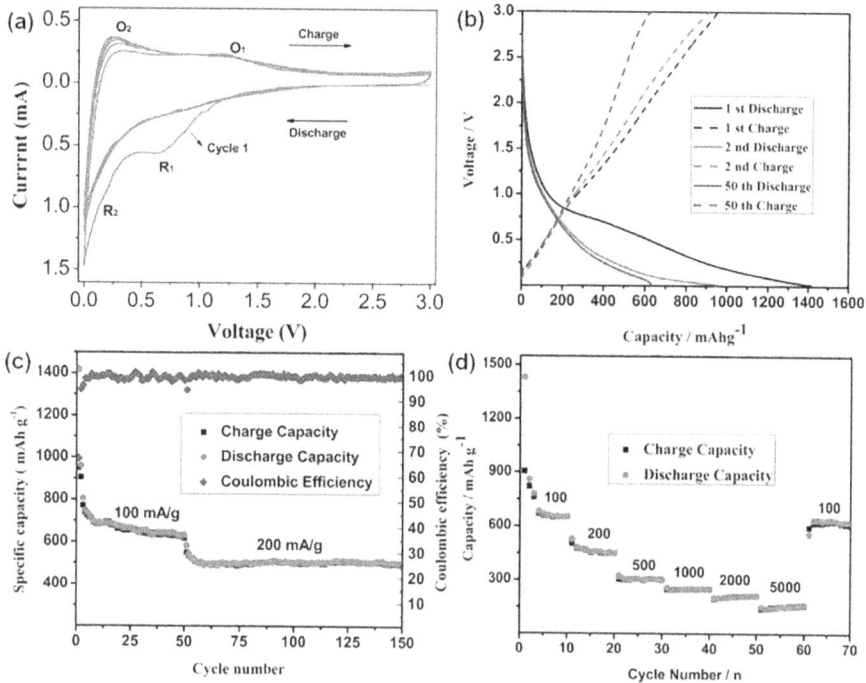

FIGURE 13.1 Electrochemical performance of N-doped graphene acting as an anode electrode for LIB: (a) cyclic voltammograms; (b) charge/discharge profile for few selected cycles; (c) discharge/charge capacity and Coulombic efficiency; and (d) rate performance. (Adapted with permission from Ref. [8]. Copyright (2016) Authors, some rights reserved; exclusive licensee [Springer Nature]. Distributed under a Creative Commons Attribution License 4.0 (CC BY) https://creativecommons.org/licenses/by/4.0/.)

Sandwich structure consisting of black phosphorus sandwich between two graphene layers has been illustrated by Liu et al. [9], which shows excellent cycle performance with high reversible capacities (1,401 mAh g^{-1}) during the 200th cycle at a current density of 100 mA g^{-1}. Hybrid structure of Si/GS grown using electrophoretic and radiofrequency magnetron deposition methods also shows efficient cycling stability of 87.7% after 150 cycles and enhanced reversible capacity (2,204 mAh g^{-1}) [10]. Several graphene/metal oxide composites serving as excellent anode materials for LIBs include NiO/graphene, SnO_2/graphene, graphene/ZnO, graphene-Fe_2O_3, graphene-Fe_3O_4, CuO/graphene, etc. Mai et al. [11] have also proposed a hybrid structure of NiO/graphene prepared using liquid phase deposition as anode for LIB which possesses high capacity of 646.1 mAh g^{-1} at a current density of 100 mA g^{-1} after 35 cycles and capacity retention of 86.3%. Zou et al. [12] have employed hydrothermal method to prepare NiO-GS composite exhibiting superior performance with a charge capacity of 1,056 mAh g^{-1} at 0.1 C which reduces to only 2.4% (1,031 mAh g^{-1}) after 40 cycles. Zhang et al. [13] have proposed a two-step method for uniform dispersion of SnO_2 (60 wt.%,) on both sides of graphene sheet and such structure is reported to exhibit a reversible capacity of 558 mAh g^{-1} after 50 cycles.

Zhao et al. [14] have demonstrated the growth and room temperature EC performance of graphene/ZnO composite as an anode (working electrode) for LIBs. They have prepared graphene oxide using modified Hummer's method which was then ultrasonically exfoliated in DI water for 90 minutes and then added with different composition of zinc acetate (1:0.5, 1:1, 1:2) followed by drying and heating. Figure 13.2a and b shows the transmission electron microscopy (TEM) images of graphene sheets and graphene/ZnO composite, respectively, which clearly shows that graphene possesses a flat lamellar structure and hence facilitates lithium storage because of large surface area (Figure 13.2a), while Figure 13.2b reveals that the ZnO nanoparticles (20–50 nm) uniformly wedge in flat graphene NSs. Figure 13.2c and d represents the two-dimensional (2D) and three-dimensional (3D) atomic force microscope (AFM) images of the grown graphene sheets which clearly shows that the thickness of the graphene is 3 nm which corresponds to eight graphene layers.

The battery is a coin type cell filled with a mixture of electrolyte- containing $LiPF_6$, ethylene carbonate and dimethylcarbonate, and metallic Li foil as counter electrode, which was investigated for its EC performance with charging and discharging voltages varying from 0.01 to 3.0 V. Figure 13.2e illustrates the cycle performance of graphene sheets, graphene/ZnO, and pure ZnO at 100 mA g^{-1} with the initial charge capacity and specific capacity of graphene/ZnO (1:1) of 650.79 and 401 mAh g^{-1}, respectively, while for graphene it was 537.9 and 357 mAh g^{-1}, respectively.

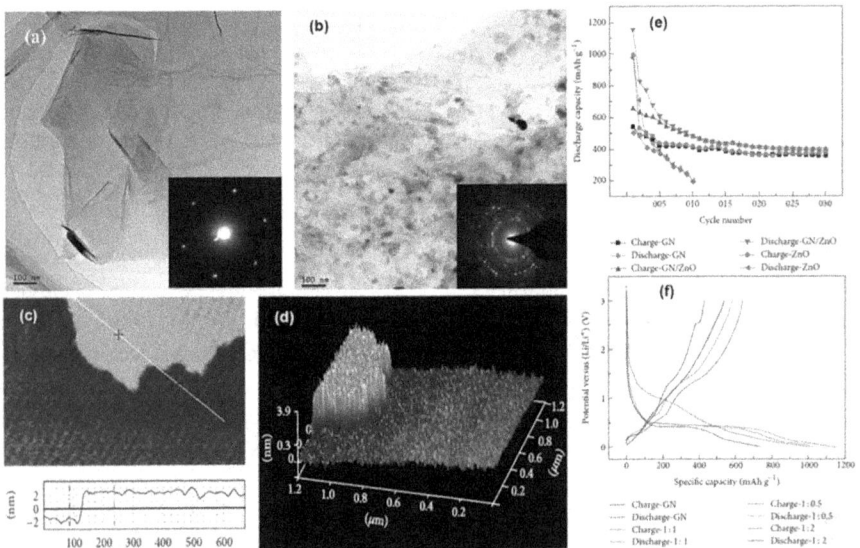

FIGURE 13.2   Transmission electron microscopy image of graphene sheets (a) and graphene/ZnO (b); 2D (c) and 3D (d) AFM image of graphene sheets with height profile; charging/discharging curves for graphene, ZnO, and graphene/ZnO composite (e); and cycle performance of graphene sheets, graphene/ZnO, and pure ZnO at 100 mA/g (f). (Adapted with permission from Ref. [14]. Copyright (2014) Authors, some rights reserved; exclusive licensee [Springer Nature]. Distributed under a Creative Commons Attribution License 4.0 (CC BY) https://creativecommons.org/licenses/by/4.0/.)

Additionally, it should be noted here that cycling performance of ZnO is very poor. Initial discharging/charging capacity of graphene/ZnO (1:0.5, 1:1, and 1:2) and graphene at the rate of $100 \, \text{mA g}^{-1}$ was examined as shown in Figure 13.2f where an enhanced reversible capacity of the grown graphene/ZnO composite (1028.8 and 582.3 mAh g^{-1}, 1155.27 and 650.79 mAh g^{-1}, 737.9 and 431.55 mAh g^{-1}) as compared to the graphene (992.4 and 537.9 mAh g^{-1}) has been observed and hence suggests its utility as an excellent alternative for anode material in LIBs. The encouraging capacity is ascribed to the novel structure of the prepared composite where the ZnO nanoparticles get uniformly distributed on the graphene surface and hence ensure easy electronic transition.

Graphene-Fe_2O_3 composite has also been reported to exhibit a high reversible specific capacity of 580 mAh g^{-1} after 100 cycles at $700 \, \text{mA g}^{-1}$ [15]. Lian et al. [16] further improve the reversible specific capacity (1198 mAh g^{-1} in the 115th cycle) by using graphene-Fe_3O_4 prepared by gas/liquid interface reaction process. EC properties of CuO/graphene composites were analyzed by Mai et al. [17], with a capacity retention of 75.5% and high reversible capability (583.5 mAh g^{-1}) after 50 cycles which was further improved (cyclic retention: 736.8 mAh g^{-1} after 50 cycles) by using CuO/RGO hybrid papers [18].

13.3.1.2 Graphene-Based Nanomaterials as Cathode for LIBs

Graphene-based nanomaterials can also be served as cathodes for LIBs and include graphene-modified $LiFePO_4$, $LiMn_{1-x}Fe_xPO_4$-graphene hybrid nanocomposite, and $LiMn_2O_4$ nanorods wrapped with graphene oxide. Xiong et al. [19] have reported N-G NSs as cathode material for LIBs where urea acts as a nitrogen source and reducing agent and displays high reversible capacity (146 mAh g^{-1}) at a current density of 1 A g^{-1} after 1,000 cycles. $LiMn_2O_4$ nanorods wrapped with graphene oxide when tested as cathode can exhibit specific charge capacity of 130 mAh g^{-1} at 0.05 C after 100 cycles and charge retention capacity of 87% [20]. Zhou et al. [21] have illustrated the battery performance of $LiFePO_4$ composite modified by graphene and prepared using spray drying and annealing methods with capacity retention of 85% under 10 C charging and 20 C discharging after 1,000 cycles. Wang et al. [22] have reported an encouraging specific capacity (155 mAh g^{-1}) of $LiMn_{1-x}Fe_xPO_4$ nanorods and graphene hybrid nanocomposites.

Ma et al. [23] have examined the EC performance of $LiFePO_4$/graphene nanocomposite prepared by rheological phase method, acting as cathode for LIB. They observed superior capacities and fine cyclabilities for $LiFePO_4$/graphene nanocomposite as compared to $LiFePO_4$ with high discharge capacity (156 mAh g^{-1} in the first cycle) and capacity retention of 96% after 15 cycles. Figure 13.3a depicts the X-ray diffraction (XRD) profiles of the nanocomposite grown at 600°C and 650°C where a much higher intensity was observed for the sample synthesized at 650°C, suggesting excellent crystallinity. Figure 13.3b and c shows the low and high magnification scanning electron microscope (SEM) images of $LiFePO_4$/graphene nanocomposite, respectively, which reveals the anchoring of several nanoscale particles on microplane. Low and high magnification TEM images of $LiFePO_4$/graphene were shown in Figure 13.3d and e, respectively, showing the non-uniformity in the particle size of phosphate.

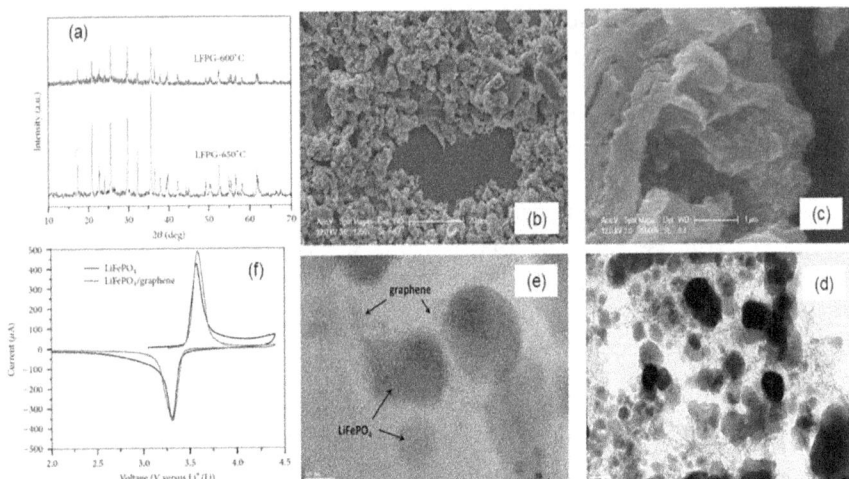

FIGURE 13.3 XRD profile of LiFePO$_4$/graphene nanocomposite (a). Low magnification (b) and high magnification (c) SEM image of LiFePO$_4$/graphene nanocomposite. Low magnification (d) and high magnification (e) TEM image of LiFePO$_4$/graphene nanocomposite. Cyclic voltammograms in the first charge/discharge cycle of LiFePO$_4$ and LiFePO$_4$/graphene (f). (Adapted with permission from Ref. [23]. Copyright (2015) Authors, some rights reserved; exclusive licensee [Springer Nature]. Distributed under a Creative Commons Attribution License 4.0 (CC BY) https://creativecommons.org/licenses/by/4.0/.)

EC performance was examined using CR2016 coin-type cell where the working electrode was made from a mixture of LiFePO$_4$/graphene composite, acetylene black, and PVDF binder coated on aluminum mesh, while Li metal dish serves as counter electrode. 1 M LiPF$_6$ in ethylene carbonate/dimethyl carbonate is used as an electrolytic medium and room temperature EC measurements were carried out in a voltage range of 2–4 V. Cyclic voltammograms in the first charge/discharge cycle of LiFePO$_4$ and LiFePO$_4$/graphene were shown in Figure 13.3f that clearly shows the presence of anodic (cathodic) peak at 3.5 V (3.3 V) which corresponds to oxidation (reduction) of Fe^{2+} (Fe^{3+}) to Fe^{3+} (Fe^{2+}) for LiFePO$_4$/graphene electrode.

13.3.2 SODIUM-ION BATTERIES (SIB)

It should be noted here that the availability of future LIBs is highly questionable because of the insufficient Li resources, and hence, SIBs have progressively become an important energy storage device owing to its easy availability and cost effectiveness. SIBs work on the same working mechanism as that of LIBs and rely mainly on the choice of materials that can easily intercalate and store the Na ions (Na$^+$ ions) having larger radius than that of Li ion (Li$^+$ ions). SIBs with encouraging EC performances including high specific capacity, excellent cyclic performances, and longer cycle life can be obtained by employing graphene-based nanomaterials as anode/cathode for SIBs.

13.3.2.1 Graphene-Based Nanomaterials as Anode for SIBs

Several graphene-based anode materials have been explored for SIBs, including pristine graphene, modified graphene electrodes, and graphene/oxides (such as TiO_2, Fe_2O_3, and Co_3O_4), flexible thin film based graphene electrodes, graphene/metal sulfides, etc. One way to adhesive graphene anodes for SIBs is to increase the sodium-ion storage capacity of graphene layers via intercalation of sodium ions. Luo et al. [24] have illustrated an increased interlayer spacing (~0.375 nm) of graphene NSs that could result in a capacity of 10^5 mAh g^{-1} even at a high current density of 5 mA g^{-1}. Contrary to this, modified graphene electrode shows much more encouraging EC performances. Quan et al. [25] have employed sulfur-doped graphene as an anode electrode for efficient SIBs. Graphene/TiO_2-based compounds are much more fascinating because TiO_2 facilitates easy intercalation of sodium ions at low voltages because of the reduction characteristics of Ti^{3+} ions and increased conductivity of graphene. Cha et al. [26] have demonstrated N-G/TiO_2 composite anode in SIBs which is having a high reversible capacity of 405 mAh g^{-1} at 50 mA g^{-1}. Chen et al. [27] have synthesized graphene/TiO_2 hybrid in a sandwich structure where TiO_2 sheet is sandwiched between two graphene sheets. Theoretical studies suggest that combining graphene with TiO_2 leads to reduced diffusion energy barrier which ensures easy intercalation of sodium ions into graphene sheets. It should be noted that the sodium storage capacity can be further enhanced by conversion reaction in metal-oxide. Fe_2O_3 nanocrystals (diameter = 2 nm) uniformly distributed onto graphene are reported by Jian et al. [28] which possess a very high capacity (400 mAh g^{-1}). Li et al. [29] have also illustrated and enhanced sodium storage capacity in graphene/amorphous Fe_2O_3 material where the enhanced capacity is attributed to the C–O–Fe oxygen bridge interaction between graphene and Fe_2O_3. Feasible EC sodium storage has also been reported using Co_3O_4 because of its high capacity; however, a voltage hysteresis has often been found in charging and discharging process. Appearance of such voltage hysteresis is attributed to the existence of an intermediate (CoO_{1-x}) phase [30]. Xie et al. [31] illustrated fabrication of SIBs using N-G/SnO_2 nanohybrids where N serves as an effective mediator for smooth transferring of electron between SnO_2 and graphene and shows an enhanced EC performance when used as anode material.

Wang et al. [32] have tested microwave plasma assisted Fe_2O_3/Fe_3O_4 nano-aggregates anchored on N-doped graphene (Fe_2O_3/Fe_3O_4/NG) as anode for SIBs. Initially, Fe_2O_3/NG was prepared by solvothermal method which was then treated with microwave plasma to produce Fe_2O_3/Fe_3O_4/NG. The overall growth process was schematically shown in Figure 13.4a. Figure 13.4b depicts the XRD profiles for both the Fe_2O_3/Fe_3O_4/NG and Fe_2O_3/NG electrodes confirming the existence (coexistence) of Fe_2O_3 (Fe_2O_3 and Fe_3O_4) phase in the respective composites. Raman spectra of microwave-assisted NG, NG, and RGO have also been analyzed as shown in Figure 13.4c which shows the appearance of Raman D and Raman G bands. Stable nitrogen doping was also confirmed after microwave plasma processing.

EC response was examined via CR2016 coin-type cell with Na metal as counter and reference electrodes. Mixture of 1 M $NaClO_4$, ethylene-carbonate/dimethyl-carbonate (1:1 volume), and fluoroethylene-carbonate serves as electrolyte. Galvanostatic charge/discharge performance was examined in the voltage range of 0.01–3 V (vs. Na/Na$^+$).

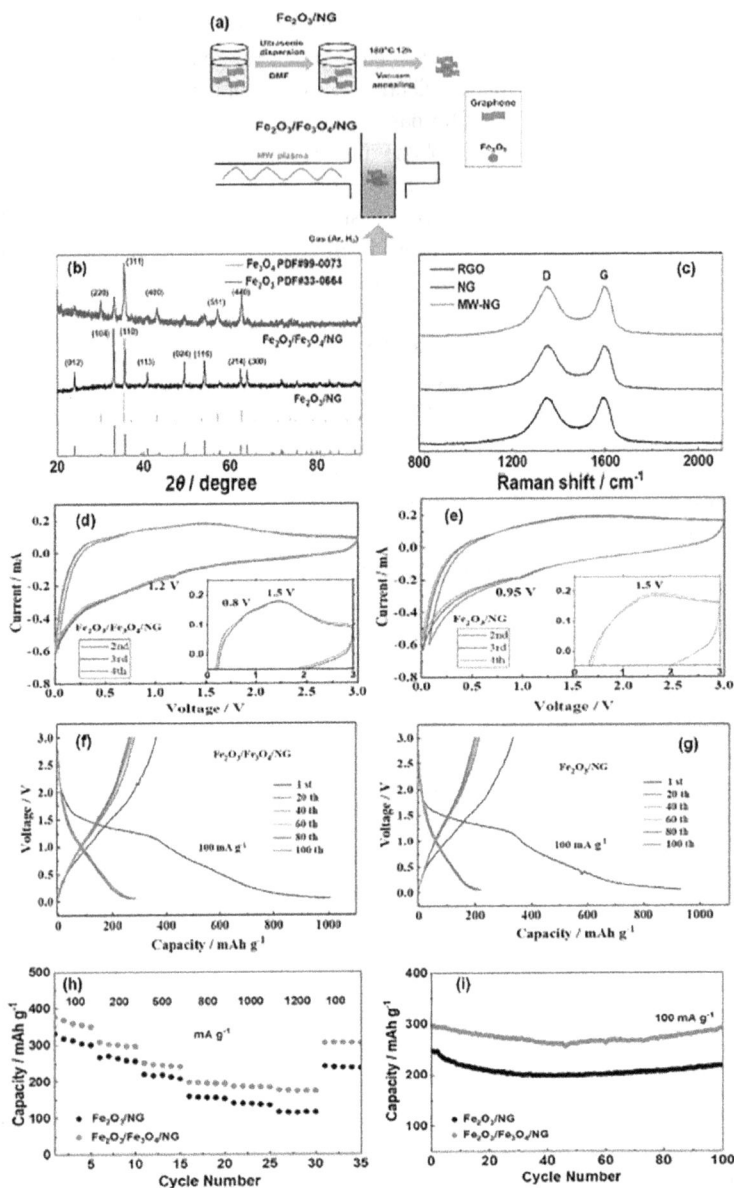

FIGURE 13.4 Schematic illustration for the overall growth process of $Fe_2O_3/Fe_3O_4/NG$ (a). XRD profiles of Fe_2O_3/NG and $Fe_2O_3/Fe_3O_4/NG$ (b). Raman spectra of microwave-assisted NG (MW-NG), NG, and RGO (c). Cyclic voltammogram from second to fourth cycle of $Fe_2O_3/Fe_3O_4/NG$ (d) and Fe_2O_3/NG (e). Charge/discharge curves for $Fe_2O_3/Fe_3O_4/NG$ (f) and Fe_2O_3/NG (g). Rate performances for both the $Fe_2O_3/Fe_3O_4/NG$ and Fe_2O_3/NG electrodes under 100–1,200mA g^{-1} (h). Cyclic performance of $Fe_2O_3/Fe_3O_4/NG$ and Fe_2O_3/NG electrodes at 100 mA/g (i). (Adapted with permission from Ref. [32]. Copyright (2020) Authors, some rights reserved; exclusive licensee [Springer Nature]. Distributed under a Creative Commons Attribution License 4.0 (CC BY) https://creativecommons.org/licenses/by/4.0/.)

Figure 13.4d and e shows cyclic voltammogram from second to fourth cycle of $Fe_2O_3/Fe_3O_4/NG$ and Fe_2O_3/NG, respectively, where a broad anodic peak at 1.5 V can be clearly visible for both the electrodes, corresponding to the oxidation of Fe^0 to Fe^{3+}. For $Fe_2O_3/Fe_3O_4/NG$ (Fe_2O_3/NG), cathodic peak appearing at 1.2 V (0.95 V) as shown in Figure 13.4d and e can be attributed to the reduction of Fe^{2+}/Fe^{3+} (expressed using Eqs. (13.4) and (13.5)) (Fe^{2+}) to Fe^0:

$$Fe_2O_3 + 6Na^+ + 6e^- \leftrightarrow 2Fe^0 + 3Na_2O \qquad (13.4)$$

$$Fe_3O_4 + 8Na^+ + 8e^- \leftrightarrow 3Fe^0 + 4Na_2O \qquad (13.5)$$

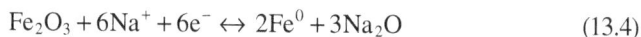

Charge/discharge curves for both the electrodes have also been examined as shown in Figure 13.4f and g, obtained at a current density of 100 mA g^{-1} which clearly shows that the initial specific discharge and charge capacities for Fe_2O_3/NG ($Fe_2O_3/Fe_3O_4/NG$) electrode are obtained as 931 and 334 mAh g^{-1} (1,004 and 363 mAh g^{-1}). However, the Columbic efficiency for both the electrodes remains almost the same (~ 36%). Figure 13.4h compares the rate performances for both the electrodes under the current density varying from 100 to 1,200 mA g^{-1}. It was observed that $Fe_2O_3/Fe_3O_4/NG$ electrode displays higher discharge capacity and a high reversible discharge capacity (305 mA g^{-1}) upon returning to a current density of 100 mA g^{-1}, as compared to Fe_2O_3/NG composite. Cyclic performance of both the electrodes in 100 cycles at 100 mA g^{-1} has also been shown in Figure 13.4i, where a capacity of 291 and 218 mAh g^{-1} was observed for $Fe_2O_3/Fe_3O_4/NG$ and Fe_2O_3/NG composites, respectively.

Wu et al. [33] have reported the fabrication of SIB using $NaTi_2(PO_4)_3$ (NTP) nanoparticles embedded in 3D graphene which shows high rate capacities up to 96 mAh g^{-1}. NTP/C/RGO composite has been employed by Song et al. [34], where an increased current density of 36.8 C has been observed. Graphene/metal sulfide such as graphene/$Co_{1-x}S$ nanocomposite has also been investigated as anode material for SIBs and shows excellent performance as compared to base $Co_{1-x}S$ [35]. Flexible graphene film based sodium battery has also been reported by Zhang et al. [36] with Sb/RGO and $Na_3V_2(PO_4)_3$/RGO (NVP/RGO) as anode and cathode, respectively, where Sb and NVP nanoparticles get uniformly distributed on RGO NSs and serve as stable host materials for the intercalation and deintercalation of Na ion. Wang et al. [37] have demonstrated high performance SIBs using reduced graphene-based nanomaterial wrapped with iron-disulfide (FeS_2/RGO) composite as anode material. The composite was prepared by hydrothermal method followed by sulfurization and the EC performance was investigated on CR2032 two electrode coin cells having Na metal and glass fiber as counter and reference electrode and separator. The working electrode of FeS_2/RGO was prepared by loading slurry onto Cu foil followed by vacuum drying at 80°C for 12 hours.

13.3.2.2 Graphene-Based Nanomaterials as Cathode for SIBs

Graphene-based cathode materials employed for SIBs generally include graphene/metal oxides, graphene/phosphates, graphene/pyrophosphate, fluorophosphate/graphene, graphene/organic compounds, and 2D graphene-based flexible cathodes. Zhang et al. [38] have proposed $Na_{2/3}Ni_{1/3}Mn_{5/9}Al_{1/9}O_2$ decorated with RGO as the cathode material that can exhibit much higher retention capacity (137.7 mAh g^{-1} at 0.1 C). RGO/$Na_{0.33}V_2O_5$

with sandwich-like structure has also been reported by Lu et al. [39], displaying rate capacity of 88.2 mAh g^{-1} at 6 A g^{-1}. Crystalline graphene/NVP as a cathode material for rechargeable SIBs has also been demonstrated by Jung et al. [40]. This battery exhibits higher reversible capacity of 86 mAh g^{-1} at 5 C. Xiang et al. [41] have prepared Na$_3$(VO$_{0.5}$)$_2$(PO$_4$)$_2$F$_2$ nanoparticles by solid state reaction method intercalated in graphene as cathode for SIB which displays rate capacity from 88 to 77 mAh g^{-1} at 10 C and 50 C, respectively, along with a high retention capacity of 73% after 1000 cycle. Flexible RGO/Na$_4$Mn$_9$O$_{18}$ composite has also been tested as cathode for SIB by Yuan et al. [42]. Chao et al. [43] have fabricated SIB using graphene form/VO$_2$@GQD composite which shows high capacity of 306 mAh g^{-1} and retention capacity of 88% after 1,500 cycles at 18 A g^{-1}.

13.3.3 GRAPHENE-BASED NANOMATERIALS FOR LITHIUM–SELENIUM AND SODIUM–SELENIUM BATTERIES

Lithium–sulfur and sodium–sulfur have also been proposed as the next-generation EC energy storage systems because of the availability of Li at lower cost. However, large scale commercialization of such batteries is quite challenging owing to the low electrical conductivity of sulfur, resulting in weaker EC performance [44]. Also, the downfall in the specific capacity and hence recycling performance are observed in such devices due to less solubility of higher order polysulfide intermediates during the charge–discharge process. Hence, research move forward to look for the alternative novel materials that can circumvent the aforementioned issues. Consequently, lithium–selenium batteries and SSeBs are therefore being proposed with much superior EC activities as compared to lithium–sulfur and sodium–sulfur batteries. As compared to sulfur, elemental Se possessed significantly larger intrinsic electrical conductivity leading to highly efficient and rapid EC reaction kinetics. However, there is a lack of reports focusing on graphene-based nanomaterials for SSeBs batteries. Also, the EC performances of such energy storage system can be enhanced manifold by utilizing graphene-based nanomaterials. Vapor-infiltration approach toward selenium/RGO composites enabling stable and high-capacity sodium storage has been illustrated by Yang et al. [45]. Han et al. [46] have demonstrated the performance of binder-free cathode made up of 3D structured graphene-Se@CNT composite film for rechargeable lithium–selenium which can display a specific discharge capacity of 315 mAh g^{-1} at 0.1 C after 100 cycles and an average Coulombic efficiency above 96%.

Gu et al. [47] have fabricated a rechargeable LSeB using nitrogen and sulfur co-doped graphene as a blocking layer. For the preparation of blocking layer, N- and S-co-doped graphene was prepared in NMP via sonication followed by vacuum filtration, vacuum heating and drying at room temperature, and for preparing free standing membrane, the N-S-G was peeled off from the filter membrane. For electrolytic medium, a solution of Li$_2$Se$_8$ was prepared serving as catholyte and the prepared free standing N-S-G membrane acts as current collector and blocking layer. Figure 13.5a shows the N, S-G membrane acting as an interlayer for trapping polyselenides. TEM images of N-S-G and the free standing membrane are shown in Figure 13.5b and d, respectively, while the top and side views of N, S-G membrane using SEM are depicted in Figure 13.5c and e, respectively. Figure 13.5f represents the image of free

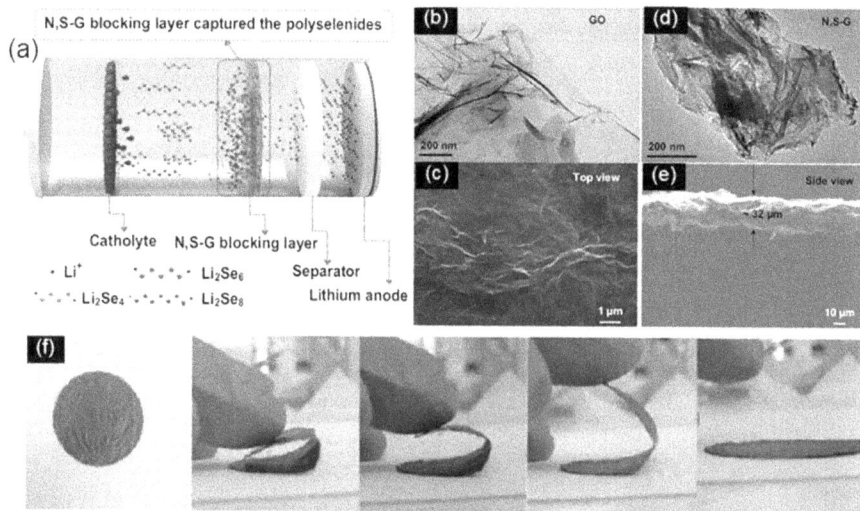

FIGURE 13.5 Schematic of N, S-G blocking layer for trapping polyselenides (a). TEM images of N-S-G interlayer (b) and free standing membrane (d). SEM image (top view (c) and side view (e) of N, S-G membrane. Images of free standing N, S-G membrane (f). (Adapted with permission from Ref. [47]. Copyright (2018) Authors.)

standing N, S-G membrane which clearly shows an excellent flexibility of the membrane. Room temperature EC performance was investigated on a coin-type cell in the potential ranging from 1.5 to 3.0 V where lithium–selenium with N, S-G interlayer shows superior battery performance as compared to the cell without the interlayer.

Room temperature galvanostatic charging/discharging voltage profile for lithium–selenium with and without N, S-G interlayer in the voltage range of 1.5–3.0 V for first to 500th cycles at a current density of 1 C was examined as shown in Figure 13.6a and b, respectively. For lithium–selenium with N, S-G interlayer, a short and long plateau at 2.25 and 1.97 V has been observed, respectively, in the discharge process for the catholyte and is attributed to the reduction reaction of high order polyselenide to Li_2Se. On the other hand, in the charge process, a short and long plateau at 2.4 and 2.25 V corresponds to reverse reaction ($Li_2Se \rightarrow$ high order polyselenides). For lithium–selenium without N, S-G interlayer (Figure 13.6b), no such voltage plateaus are visible and they show strong overcharge in different cycles. Figure 13.6c compares the rate performance of lithium–selenium with and without N, S-G interlayer at various charge rates. For the cell with interlayer, the initial discharge capacity was found to be 648.2 mAh g^{-1}, suggesting excellent reversibility as compared to the battery without the interlayer. The reversible capacities at 0.5 C, 1 C, 2 C, and 4 C were found to be 534.6, 473.5, 348.6, and 301.4 mAh g^{-1}, respectively. Additionally, upon switching the current density to 0.5 C, recovery of reversible discharge capacity of 512.2 mAh g^{-1} has been observed suggesting excellent stability of selenium cathode with the interlayer. Long-term cyclic performances of the battery with and without N, S-G interlayer have also been evaluated at 1 C for 500 cycles as shown in Figure 13.6d which reveals a high discharge capacity (638.5 mAh g^{-1}) and reversible

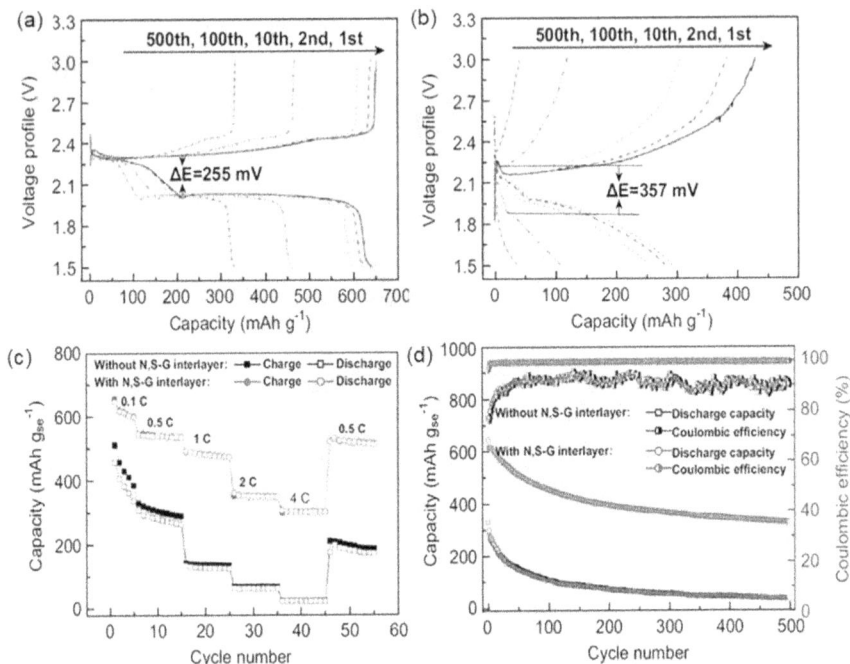

FIGURE 13.6 Galvanostatic charging/discharging voltage profile for lithium–selenium with (a) and without N, S-G interlayer (b) for first to 500th cycles at 1 C. Rate performance of lithium–selenium with and without N, S-G interlayer at various charge rates (c). Cyclic performance of lithium–selenium with and without N, S-G interlayer at 1 C (d). (Adapted with permission from Ref. [47]. Copyright (2018) Authors.)

capacity (330.7 mAh g^{-1}) after 500 cycles. Graphene-coated polymer separator of lithium–selenium has been reported by Fang et al. [48], where pure Se acts as a cathode electrode whose transport to the anode region can be suppressed via graphene coating on a commercial polymer separator. Haung et al. [49] discussed graphene oxide-protected 3D Se as a binder-free cathode for lithium–selenium.

As discussed, it has been found that the EC performances of all the graphene nanomaterial-based lithium-ion, lithium–sulfur, lithium–selenium, sodium-ion, sodium–sulfur, sodium–selenium batteries are superior as compared to the EC energy storage devices with graphene-based nanomaterials which itself suggest that graphene is an excellent material for EC energy storage applications.

13.4   CONCLUSION

There has been an immense progress in the field of EC energy storage devices and several materials have been employed for this purpose. However, the performances of such devices are not up to the mark before the invention of 2D counter part of carbon, i.e., graphene. Utilizing graphene-based nanomaterials in such EC energy storage devices has improved their performance manifold. Accordingly, the present chapter provides a comprehensive overview of several EC energy storage devices including lithium-ion, lithium–sulfur, lithium–selenium, sodium-ion, sodium–sulfur, and sodium–selenium employing graphene-based nanomaterials along with recent results. It should be also noted that inspite of having an explosive evolution in the field of graphene nanomaterial based EC energy storage device applications, there still exist immatureness in the development of such devices for commercialization. Consequently, efforts should have to be devoted for the realization of highly efficient commercial energy storage device applications. Hence, more novel graphene-based nanomaterials are required to be developed to be utilized for the fabrication of such devices which will be one of the most promising future research directions.

REFERENCES

1. A. Agrawal, G.-C. Yi, Sample pretreatment with graphene materials, In: Volume Editor: C.M. Hussain, *Analytical Applications of Graphene for Comprehensive Analytical Chemistry*, Volume 91, 2020, Elsevier: Cambridge.
2. M. Winter, R.J. Brodd, What are batteries, fuel cells, and supercapacitors? *Chem. Rev.* 104 (2004) 4245–4270.
3. K. Zhang, X. Han, Z. Hu, X. Zhang, Z. Tao, J. Chen, Nanostructured Mn-based oxides for electrochemical energy storage and conversion, *Chem. Soc. Rev.* 44 (2015) 699.
4. J. Chen, F.Y. Cheng, Combination of lightweight elements and nanostructured materials for batteries, *Acc. Chem. Res.* 42 (2009) 713.
5. S.Y. Liu, K. Chen, Y. Fu, S.Y. Yu, Z.H. Bao, Reduced graphene oxide paper by supercritical ethanol treatment and its electrochemical properties, *Appl. Surf. Sci.* 258 (2012) 5299.
6. C. Zhong, J.Z. Wang, D. Wexler, H.K. Liu, Microwave autoclave synthesized multilayer graphene/single-walled carbon nanotube composites for free-standing lithium-ion battery anodes, *Carbon* 66 (2014) 637.
7. C. Liu, X.G. Liu, J. Tan, Q.F. Wang, H. Wen, C.H. Zhang, Nitrogen-doped graphene by all-solid-state ball-milling graphite with urea as a high-power lithium ion battery anode, *J. Power Sources* 342 (2017) 157.
8. Z. Xing, Z. Ju, Y. Zhao, J. Wan, Y. Zhu, Y. Qiang, Y. Qian, One-pot hydrothermal synthesis of nitrogen-doped graphene as high performance anode materials for lithium ion batteries, *Sci. Rep.* 6 (2016) 26146.
9. H.W. Liu, Y.Q. Zou, L. Tao, Z.L. Ma, D.D. Liu, P. Zhou, H.B. Liu, S.Y. Wang, Sandwiched thin-film anode of chemically bonded black phosphorus/graphene hybrid for lithium-ion battery, *Small* 13 (2017) 1700758.
10. Y.Q. Zhang, X.H. Xia, X.L. Wang, Y.J. Mai, S.J. Shi, Y.Y. Tang, L. Li, J.P. Tu, Silicon/graphene-sheet hybrid film as anode for lithium ion batteries, *Electrochem. Commun.* 23 (2012) 17.
11. Y.J. Mai, S.J. Shi, D. Zhang, Y. Lu, C.D. Gu, J.P. Tu, NiO–graphene hybrid as an anode material for lithium ion batteries, *J. Power Sources* 204 (2012) 155.
12. Y.Q. Zou, Y. Wang, NiO nanosheets grown on graphene nanosheets as superior anode materials for Li-ion batteries, *Nanoscale* 3 (2011) 2615.
13. L.S. Zhang, L.Y Jiang, H.J. Yan, W.D. Wang, W. Wang, W.G. Song, Y.G. Guo, L.J. Wan, Mono dispersed SnO_2 nanoparticles on both sides of single layer graphene sheets as anode materials in Li-ion batteries, *J. Mater. Chem.* 20 (2010) 5462.
14. Y. Zhao, G. Chen, Y. Wang, Facile synthesis of graphene/ZnO composite as an anode with enhanced performance for lithium ion batteries, *J. Nanomater.* 2014 (2014) 1–6.
15. G. Zhou, D.W. Wang, F. Li, L. Zhang, N. Li, Z.S. Wu, L. Wen, G.Q. Lu, H.M. Cheng, Graphene-wrapped Fe_3O_4 anode material with improved reversible capacity and cyclic stability for lithium ion batteries, *Chem. Mater.* 22 (2010) 5306.

16. P.C. Lian, S.Z. Liang, X.F. Zhu, W.S. Yang, H.H. Wang, A novel Fe_3O_4-SnO_2-graphene ternary nanocomposite as an anode material for lithium-ion batteries, *Electrochim. Acta.* 58 (2011) 81.

17. Y.J. Mai, X.L. Wang, J.Y. Xiang, Y.Q. Qiao, D. Zhang, C.D. Gu, J.P. Tu, CuO/graphene composite as anode materials for lithium-ion batteries, *Electrochim. Acta* 56 (2011) 2306.

18. Y. Liu, W. Wang, L. Gu, Y.W. Wang, Y.L. Ying, Y.Y. Mao, L.W. Sun, X.S. Peng, Flexible CuO nanosheets/reduced-graphene oxide composite paper: Binder-free anode for high-performance lithium-ion batteries, *ACS Appl. Mater. Interfaces* 5 (2013) 9850.

19. D.B. Xiong, X.F. Li, Z.M. Bai, H. Shan, L.L. Fan, C.X. Wu, D.J. Li, S.G. Lu, Superior cathode performance of nitrogen-doped graphene frameworks for lithium ion batteries, *ACS Appl. Mater. Interfaces* 9 (2017) 10643.

20. N. Kumar, J.R. Rodriguez, V.G. Pol, A. Sen, Facile synthesis of 2D graphene oxide sheet enveloping ultrafine 1D $LiMn_2O_4$ as interconnected framework to enhance cathodic property for Li-ion battery, *Appl. Surf. Sci.* 463 (2018) 132.

21. X.F. Zhou, F. Wang, Y.M. Zhu, Z.P. Liu, Graphene modified $LiFePO_4$ cathode materials for high power lithium ion batteries, *J. Mater. Chem.* 22 (2011) 3353.

22. H.L. Wang, Y. Yang, Y.Y. Liang, L.F. Cui, H. Sanchez Casalongue, Y.G. Li, G.S. Hong, Y. Cui, H.J. Dai, $LiMn_{1-x}Fe_xPO_4$ nanorods grown on graphene sheets for ultrahigh-rate-performance lithium ion batteries, *Angew. Chem.* 50 (2011) 7364.

23. X. Ma, G. Chen, Q. Liu, G. Zeng, T. Wu, Synthesis of $LiFePO_4$/graphene nanocomposite and its electrochemical properties as cathode material for Li-ion batteries, *J. Nanomater.* 2015 (2015) 1–6.

24. X.-F. Luo, C.-H. Yang, Y.-Y. Peng, N.-W. Pu, M.-D. Ger, C.-T. Hsieh, J.-K. Chang, Graphene nanosheets, carbon nanotubes, graphite, and activated carbon as anode materials for sodium-ion batteries, *J. Mater. Chem. A* 3 (2015) 10320.

25. B. Quan, A. Jin, S.H. Yu, S.M. Kang, J. Jeong, H.D. Abruna, L. Jin, Y. Piao, Y.E. Sung, Solvothermal-derived S-doped graphene as an anode material for sodium-ion batteries, *Adv. Sci.* 5 (2018) 1700880.

26. H.A. Cha, H.M. Jeong, J.K. Kang, Nitrogen-doped open pore channeled graphene facilitating electrochemical performance of TiO_2 nanoparticles as an anode material for sodium ion batteries, *J. Mater. Chem. A* 2 (2014) 5182.

27. C. Chen, Y. Wen, X. Hu, X. Ji, M. Yan, L. Mai, P. Hu, B. Shan, Y. Huang, Na^+ intercalation pseudocapacitance in graphene-coupled titanium oxide enabling ultra-fast sodium storage and long-term cycling, *Nat. Commun.* 6 (2015) 6929.

28. Z. Jian, B. Zhao, P. Liu, F. Li, M. Zheng, M. Chen, Y. Shi, H. Zhou, Fe_2O_3 nanocrystals anchored onto graphene nanosheets as the anode material for low-cost sodium-ion batteries, *Chem. Commun.* 50 (2014) 1215.

29. D. Li, J. Zhou, X. Chen, H. Song, Amorphous Fe_2O_3/graphene composite nanosheets with enhanced electrochemical performance for sodium-ion battery, *ACS Appl. Mater. Interfaces* 8 (2016) 30899.

30. H. Kim, H. Kim, H. Kim, J. Kim, G. Yoon, K. Lim, W.-S. Yoon, K. Kang, Understanding origin of voltage hysteresis in conversion reaction for Na rechargeable batteries: The case of cobalt oxides, *Adv. Funct. Mater.* 26 (2016) 5042.

31. X. Xie, D. Su, J. Zhang, S. Chen, A.K. Mondal, G. Wang, A comparative investigation on the effects of nitrogen-doping into graphene on enhancing the electrochemical performance of SnO_2/graphene for sodium-ion batteries, *Nanoscale* 7 (2015) 3164.

32. Q. Wang, Y. Ma, L. Liu, S. Yao, W. Wu, Z. Wang, P. Lv, J. Zheng, K. Yu, W. Wei, K. (Ken) Ostrikov, Plasma enabled Fe_2O_3/Fe_3O_4 nano-aggregates anchored on nitrogen-doped graphene as anode for sodium-ion batteries, *Nanomaterials* 10 (2020) 782.

33. C. Wu, P. Kopold, Y. Ding, P.A. van Aken, J. Maier, Y. Yu, Synthesizing porous $NaTi_2(PO_4)_3$ nanoparticles embedded in 3D graphene networks for high-rate and long cycle-life sodium electrodes, *ACS Nano* 9 (2015) 6610.

34. J. Song, S. Park, J. Gim, V. Mathew, S. Kim, J. Jo, S. Kim, J. Kim, High rate performance of a $NaTi_2(PO_4)_3$/rGO composite electrode via pyro synthesis for sodium ion batteries, *J. Mater. Chem. A* 4 (2016) 7815.

35. T. Chen, Y. Ma, Q. Guo, M. Yang, H. Xia, A facile sol–gel route to prepare functional graphene nanosheets anchored with homogeneous cobalt sulfide nanoparticles as superb sodium-ion anodes, *J. Mater. Chem. A* 5 (2017) 3179.

36. W. Zhang, Y. Liu, C. Chen, Z. Li, Y. Huang, X. Hu, Flexible and binder-free electrodes of Sb/rGO and $Na_3V_2(PO_4)_3$/rGO nanocomposites for sodium-ion batteries, *Small* 11 (2015) 3822.

37. Q. Wang, C. Guo, Y. Zhu, J. He, H. Wang, Reduced graphene oxide-wrapped FeS_2 composite as anode for high-performance sodium-ion batteries, *Nano-Micro Lett.* 10 (2018) 30.

38. X.H. Zhang, W.L. Pang, F. Wan, J.Z. Guo, H.Y. Lu, J.Y. Li, Y.M. Xing, J.P. Zhang, X.L. Wu, P2–$Na_{2/3}Ni_{1/3}Mn_{5/9}Al_{1/9}O_2$ microparticles as superior cathode material for sodium-ion batteries: Enhanced properties and mechanism via graphene connection, *ACS Appl. Mater. Interfaces* 8 (2016) 20650.

39. Y. Lu, N. Su, L. Cheng, J. Liu, L. Yang, H. Yang, Q. Yang, S. Li, J. Min, M. Lei, $Na_{0.33}V_2O_5$ nanosheet@graphene composites: Towards high performance cathode materials for sodium ion batteries, *Mater. Lett.* 183 (2016) 346.

40. Y.H. Jung, C.H. Lim, D.K. Kim, Graphene-supported $Na_3V_2(PO_4)_3$ as a high rate cathode material for sodium-ion batteries, *J. Mater. Chem. A* 1 (2013) 11350.

41. X. Xiang, Q. Lu, M. Han, J. Chen, Superior high-rate capability of $Na_3(VO_{0.5})_2(PO_4)_2F_2$ nanoparticles embedded in porous graphene through the pseudocapacitive effect, *Chem. Commun.* 52 (2016) 3653.

42. G. Yuan, J. Xiang, H. Jin, Y. Jin, L. Wu, Y. Zhang, A. Mentbayeva, Z. Bakenov, Flexible free-standing $Na_4Mn_9O_{18}$/reduced graphene oxide composite film as a cathode for sodium rechargeable hybrid aqueous battery, *Electrochim. Acta* 259 (2018) 647.

43. D. Chao, C. Zhu, X. Xia, J. Liu, X. Zhang, J. Wang, P. Liang, J. Lin, H. Zhang, Z.X. Shen, H.J. Fan, Graphene quantum dots coated VO_2 arrays for highly durable electrodes for Li and Na ion batteries, *Nano Lett.* 15 (2015) 565.

44. T. Liu, Y. Zhang, J.K. Hou, S.Y. Lu, J. Jiang, M.W. Xu, High performance mesoporous C@Se composite cathodes derived from Ni-based MOFs for lithium-selenium batteries, *RSC Adv.* 5 (2015) 84038–84043.

45. X. Yang, J. Wang, S. Wang, H. Wang, O. Tomanec, C. Zhi, R. Zboril, D.Y.W. Yu, A. Rogach, Vapor-infiltration approach toward selenium/reduced graphene oxide composites enabling stable and high-capacity sodium storage, *ACS Nano* 12 (2018) 7397–7405.

46. K. Han, Z. Liu, H.Q. Ye, F. Dai, Flexible self-standing graphene-Se@CNT composite film as a binder-free cathode for rechargeable lithium-selenium batteries, *J. Power Sources* 263 (2014) 85–89.

47. X. Gu, L. Xin, Y. Li, F. Dong, M. Fu, Y. Hou, Highly reversible Li–Se batteries with ultralightweight N,S codoped graphene blocking layer, *Nano-Micro Lett.* 10 (2018) 1–10.

48. R.P. Fang, G.M. Zhou, S.F. Pei, F. Li, H.M. Cheng, Localized polyselenides in a graphene-coated polymer separator for high rate and ultra-long life lithium-selenium batteries, *Chem. Commun.* 51 (2015) 3667–3670.

49. D.K. Huang, S.H. Li, Y.P. Luo, X. Xiao, L. Gao, M.K. Wang, Y. Shen, Graphene oxide-protected three dimensional Se as a binder-free cathode for Lithium-selenium battery, *Electrochim. Acta* 190 (2016) 258–263.

14 Chalcogenide-Based 2D Nanomaterials for Batteries

*Sachin Kumar Yadav, Shiv Kumar Pal,
and Neeraj Mehta*
Banaras Hindu University

CONTENTS

14.1 INTRODUCTION

One may note that chalcogen elements (S, Se, and Te) form compounds easily with other elements. And thus, they have a variety of structures, compositions exhibiting a wide range of physical–chemical properties.

The experimentation methods of nanomaterial employ such tactics to nanotechnology that is based on material science. For bulk samples of nanostructured materials, the characteristics of the substances are sturdily influenced by the nanoparticles. The substances having structural properties lying under the nanoscale frequently demonstrate exceptional electronic, mechanical, and optical properties. In this context, various researchers used nanostructured metal chalcogenides (MCs) extensively in a diversity of energy conversion and storage (ECS) devices. Examples are solar or fuel cells, LEDs, supercapacitors, and Li/Na-ion batteries [1–5]. Many MCs and transition metal dichalcogenides (TMDs) have become prevalent adoptions for solar absorber substances and device constructions [1–5]. Some best examples of TMDs are MoS_2, $MoSe_2$, $MoTe_2$, WS_2, WSe_2, and WTe_2. In recent years, various research groups have synthesized and developed the layered metal dichalcogenides (e.g., MoS_2) and reported that these substances are potential candidates for the fabrication of electrode substances in electrochemical supercapacitors and Li-ion batteries (LIBs) [1–3].

DOI: 10.1201/9781003178422-14

With this background, the attention now is on the design of devices using synthetic structures. One deposits atoms in the layer by layer (ALD) form and then one creatively employs redesigned architectures and manipulates molecules individually. This has the potential of creating systems with novel and unusual properties. It is satisfying to note that a variety of substances can be produced to form nanostructures. This may be attempted by initiating change in their intrinsic properties at the nano-scale level. Alternatively, the strategy is to get extrinsic properties such as sensitivity to the ambient environment. The outcome critically depends on size and shape and may show new behavior due to variation in nanostructure morphology. There is renewed theoretical interest in understanding the role of particle size in determining important properties such as electronic structure, conductivity, melting temperature, and mechanical characteristic. Indeed, it is a major challenge to monitor and control size, shape, and morphology during synthesis. Mention must be made on optical properties of nanostructured substances.

Since the finding of the first two-dimensional (2D) substance, i.e., graphene (an allotrope of carbon), researchers have identified TMDs as an alternative group of layered 2D material. The pragmatic chemical formula to represent TMDs is MX_2 (M: transition metal, X: chalcogen) differ from the semi-metallic nature of graphene, have drawn great attention of technologists because of their potential applications in electrochemical energy storage (EES) systems and their novel physicochemical properties [1,2]. We can characterize TMDs by feeble, non-covalent bonding between atomic layers and strong in-plane covalent bonds. An expanse in interlayer spacing in 2D material is available for the accommodation of intercalating ions, which regulates their physical and chemical properties and contributes to additional reachable active reaction locations for the interaction of ions, and making them highly efficient energy storage technologies [3,4]. The electrochemical functionality of TMDs-based anodes is limited by natural accumulation throughout the conversion reaction, which can be tailored by the development of hierarchical nanostructures. The development of composite substances as graphene/TMD nanostructure enhances electrical conductivity and structural steadiness due to the inherent properties of graphene [5,6]. In recent years, the energy density of LIBs having a theoretical specific capacity of 670 mAh g^{-1} [7] is high in comparison to other alkali metal ion batteries (Na, K), which have a large ionic radius and their very short operational life. Therefore, significant efforts are needed by the scientific community for improving the electrochemical capability of rechargeable batteries concerning high specific capacity, low cost, and safety issues. The incorporation of inorganic glass-ceramics solid electrolytes boosts the cycling capability of Li/S batteries. A solid-state electrochemical cell In/$LiCoO_2$ with a glass-ceramic solid electrolyte ($80Li_2S.20P_2S_5$) maintains a high capacity of 1,000 mAh g^{-1} with coulombic efficiency of 100% for 200 cycles. The ionic conductivity of glass electrolytes (Li_2O-SiO_2, Li_2O-P_2O_5, Li_2S-SiS_2, and Li_2S-P_2S_5) increases with the increment of Li-ion concentration and also by the change of the glass matrix [8–10]. The historical overview of ECS applications of TMDs is shown in Figure 14.1a.

14.2 PROPERTIES OF METAL CHALCOGENIDES (MCs) AND TRANSITION METAL DICHALCOGENIDES (TMDs)

The storage mechanism of cyclic voltammetry of MCs is shown in Figure 14.1b. It plays an important role when we want to choose chalcogenide-based substances for

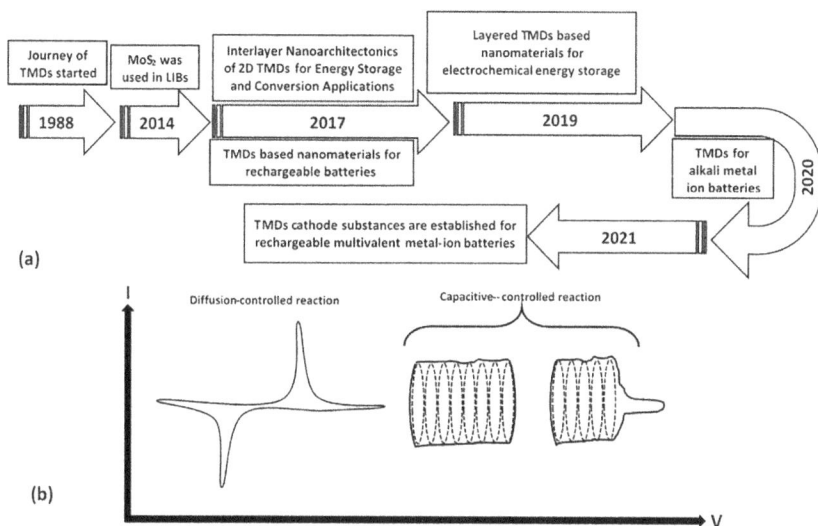

FIGURE 14.1 (a) Historic outline of energy conversion and storage (ECS) applications of TMDs and (b) illustration of different storage mechanisms behind the cyclic voltammetry of MCs.

battery or capacitor applications. Chalcogenide-based substances having the capability of alloys are used for the frequent transportation of a huge amount of charge capacity by making alloys. Some of them are conversion-kind that experience multielectron shifting when the metal cations are reduced into neutral metal. The progressions of bond breaking and formation facilitate them more vulnerability to reduction, at the price of long-standing stability. When the alkali is adjusted to form a binary alloy, the volume change is significantly more grievous to make alloys of the chalcogenides. Consequently, the structure becomes very deformed, which leads to short cycle life. Conversely, the basic kinetic limits linked with traditional solid-state diffusion procedures may be improved by pseudo-capacitance, irrespective of intercalative or redox-type reaction. The first panel of Figure 14.1b represents the diffusion-controlled reaction storage mechanism that is possible in all-MCs and it is governed by a phase change. The second/third panel of Figure 14.1b shows capacitive-controlled reaction storage mechanisms having pseudo-capacitance of redox-type and intercalation-type, respectively. These two mechanisms experience no phase change during the reaction. The second storage mechanism is applicable only for nanostructured MCs while the third one is possible for MCs with a layered structure.

Among all TMDs semiconductor family, MoS_2 (molybdenum disulfide) as a model electrode compound having different structural phases and coordination geometry is widely accepted for energy-related applications owing to its high surface area and low diffusion distance and able to accommodate intercalating ions into its geometry. The electrical characteristics of TMDs are governed by the filling state of d orbital, partially filled configuration contributes to metallic nature, whereas filled configuration leads to semiconducting nature. MoS_2 monolayers are bonded by the interaction of feeble Van der Waals forces as well as their electrical conducting nature. TMD can be altered by phase control mechanism, e.g. MoS_2, which has two common polymorph phases with different characteristics – (i) $1T\text{-}MoS_2$ metallic phase because coordination of Mo atoms with six adjacent atoms in an octahedron, (ii) 2H-MoS$_2$ semiconducting phase because of prismatic coordination of each Mo atom with six neighboring S atoms [11,12]. The two phases of MoS$_2$ have different electronic properties because of differences in the crystal structure. The interplanar spacing for Mo-S length is 0.24 nm and the lattice constant is 0.32 nm [13]. Further 0.31 nm is the gap between the sulfur atoms located at the lower position and upper position [13]. When the thickness of bulk MoS$_2$ reduces to monolayer, the nature of the electronic bandgap reveals a wide range of bandgap, changes from indirect (1.2 eV) bandgap to direct bandgap (1.8 eV) [14].

14.3 SYNTHESIS APPROACHES OF MCs

The overview of the synthetic routes for MCs is shown in Figure 14.2. The hydro/solvothermal method is one of the renowned wet-chemical approaches. In this method, the boiling temperature of the solvents is typically lesser than that of the chemical reaction. Organic liquids or water is frequently used as solvents in this method. With the temperature rise, the creation of enough pressure takes place in the locked system. This helps in expanding the crystallinity by facilitating the reaction rate and so MCs having diverse features, phases, and shapes may be prepared. The mutual interfacing of surfactants and host species is critically controlled in hydro/solvothermal for the development of nanostructured MCs. We can prepare MCs of different geometries by varying the surfactants/solvents and other additives (atoms/ions). The second convenient approach is the hot-injection method for the preparation of MCs in the form of monodisperse colloidal nanocrystals. This method helps in controlling the shape/size and configuration. Usually, the first step of the hot-injection method is the quick dose of the reactants into surfactants that contain the hot reaction solution. In the self-assembly approach, the pre-synthesized nanocrystals are automatically arranged into an ordered structure. The nanocrystals in the MCs prepared by this technique consist of bonding of hydrogen, electrostatic, and Van der Waals kinds. It is an effective technique to produce nanoscale designs with collective or unified catalytic performance. The next method is the ion exchange method that is employed to match the composition of the MCs by substituting the prevailing ions with other ions. This is a general approach that is utilized for the production of a sequence of mixed structured nano-sized substances like mixed organic–inorganic polymers, metal oxides, metal phosphides, and MCs.

The ion exchange method generally creates defects, nanointerface, or strain in the nanocrystals due to which some modifications may take place in their electrocatalytic features. To synthesize TMDs, semiconductor-based electrodes substances for fundamental research in novel portable devices to achieve excellent performance several synthesis approaches (exfoliation, hydrothermal route, Chemical vapor deposition (CVD), and atomic layer deposition) were explored. CVD is a vacuum deposition method of depositing solid substance at a high temperature to form a thin film (a layer or coating having a maximum thickness of 1 μm or less whose properties are different from bulk substance) from the volatile gaseous precursors as a consequence of chemical reaction and decomposition on the substrate surface. Zhu et al. used MoO$_3$ and S powder as precursors to synthesize MoS$_2$ nanosheets on SiO$_2$/Si substrate [15]. CVD is the most reliable method for uniform film growth with good reproducibility and having the strength to regulate crystal arrangement. CVD deposited high-quality, uniform

Hydro/solvothermal approach

Hot-injection approach

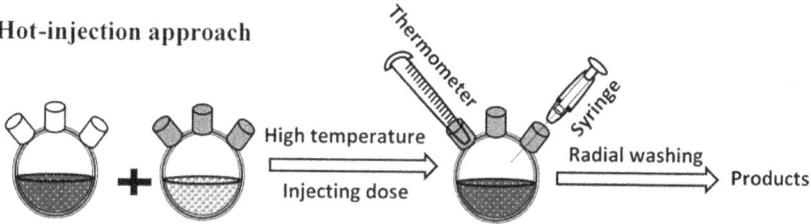

Self-assembly approach

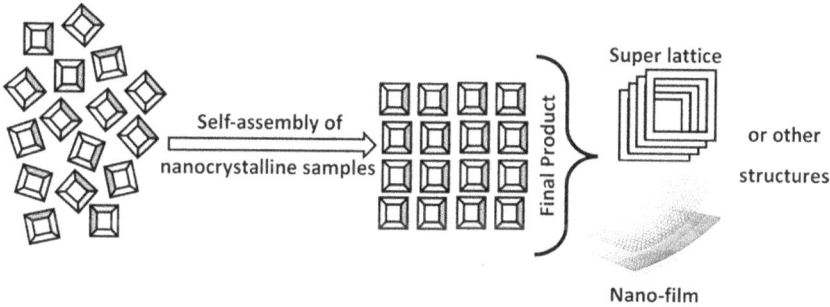

Ion exchange method

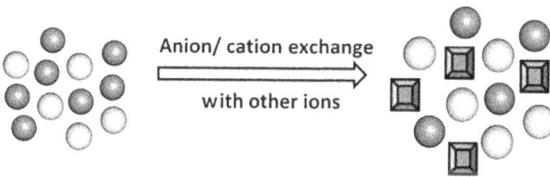

FIGURE 14.2 Different approaches used for the synthesis of 2D MCs.

and continuous atomic thin MoS_2 film increased relative surface area, which is a possible application in the arena of EES and translation. Ren et al. fabricated graphene foam that was coated by carbon nanotube and bedecked with MoS_2 nanoparticles. The electrodes of this assembly (i.e., GF@CNT/MoS_2) were unified by hydrothermal reaction and tracked by nickel foam pattern through the CVD method. The working protocol of the proposed battery was unable to transport a specific capacity of 935 mAh g^{-1} at a current density of 0.1 A g^{-1} with a high reversible capacity of 606 mAh g^{-1} after 200 cycles at 0.2 A g^{-1}. This group reported that GF@CNT/MoS_2 electrode might be an ideal platform as a flexible anode for LIBs [16]. Yu and co-workers reported the synthesis of graphene/MoS$_2$ heterostructure by a controlled CVD process. They studied the lithium storage characteristics of the manufactured substance and found that a volumetric capacity of 1,260 mAh cm^{-3} and a working voltage of 1.31 V as compared to monolayer graphene electrode (1.46 V) and promotes this fundamental research as an ideal platform for energy storage applications [17]. Wang's group systematically demonstrated MoS$_2$-coated vertical graphene nanosheet (MoS$_2$/VGNS) hybrid electrodes synthesized by solvothermal method pursued by plasma-assisted chemical vapor deposition for LIBs, delivering a specific capacity of 1,277 mAh g^{-1}. The corresponding current density and coulombic efficiencies were 100 mA g^{-1} and 76.65%, respectively. After 100 cycles of operation, the heterostructure retains a capacity of 1,109 mAh g^{-1} at 200 mA g^{-1} [18]. As a class of 2D-TMDs energy storage substances, graphene-decorated TMDs demonstrated a very high potential platform as the anode substance for LIBs. The hydrothermal approach is widely accepted for 2D-TMDs/graphene heterostructure owing to low cost, easy synthesis, and large-scale production. The expanded 1T-rich MoSe$_2$ (e-MoSe$_2$) with NH^{4+} intercalation was synthesized by Zhou's group via a hydrothermal route having an interlayer distance of 9.8 Å as the electrode substance in LIBs. They demonstrated that rapid diffusion of ions and electron transportation leads to enhancement in the electrochemical functionality of lithium-ion storage. The corresponding primary coulombic efficiency and specific capacities were 70% and 1,217 mAh g^{-1}, respectively, in the initial charge/discharge cycle at 0.2 C [19].

14.4 PAST AND PRESENT STATUS

To fulfill the daily needs of human beings the storage of electrical energy is a crucial task. For that in 1859, the first lead-acid-based rechargeable battery was invented by French physicist Gaston Plante [20]. Figure 14.3 shows the flow chart of the applications of TMDs in batteries.

A patent on MoS$_2$ as a cathode substance for rechargeable LIBs was credited to Haering's research group [21]. The first commercial LIB was developed by Sony with layered LiCoO$_2$ as a cathode substance in 1991 [22]. Recently, in the scientific community, researchers are interested in TMDs for utilization in energy storage devices because of their interesting features (e.g., high electrical conductivity, large relative area, and multivalent redox activity).

Nguyen et al. examined the electrochemical functionality of Ag-decorated 1T MoS$_2$ nanosheet synthesized by liquid exfoliation method and reported delivery of specific capacity of 510 mAh g^{-1} after 100 cycles. This value was almost 73% times the initial capacity and so the sample proved strongly his candidature for high lithium storage by good performance [23]. Wang et al. prepared MoS$_2$ nanosheet in a mixed solution of ethanol and octylamine via solvothermal reaction as a portable electrode substance for LIBs. The achieved values of current density and discharge capacities were 0.1 A g^{-1} and 960 mAh g^{-1}, respectively. The sample also retains a capacity of 606 mAh g^{-1} after 50 cycles [24]. To achieve the high energy density electrode substance, Kato et al. synthesized vanadium disulfide (VS$_2$) via hydrothermal reaction route for Li-ion storage mechanism and point out that the lithium intercalation mechanism involves the two-electron transfer and carries a charge capacity of 400 mAh g^{-1} at a current density

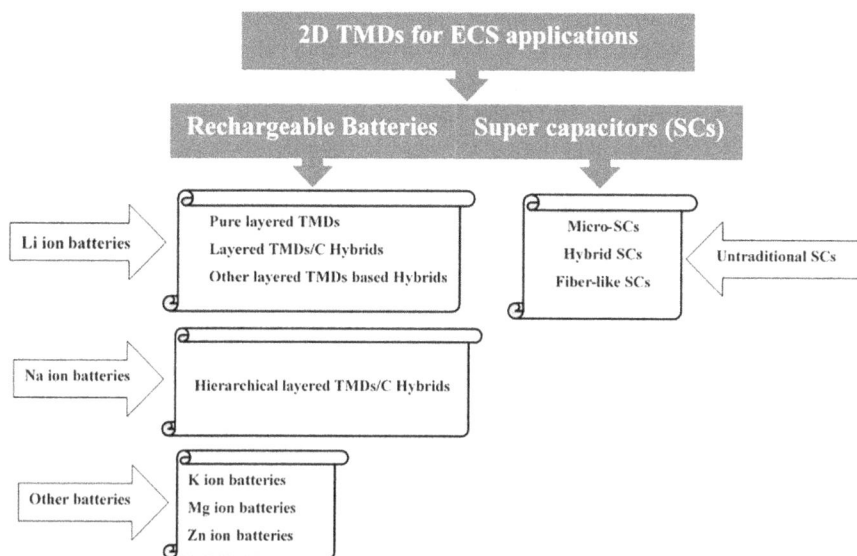

FIGURE 14.3   Flow charts showing the applications of TMDs in batteries.

of 0.1 A g^{-1} [25]. The improved electrochemical characteristics of VS$_2$ are due to the 2D metallic nature, which results in improved ionic conductivity. Ju et al. [26] studied the hybridization of TMDs substance with carbon allotropes to achieve a stable power output, synthesized Fe$_2$O$_3$/MoS$_2$/rGO via hydrothermal route followed by subsequent heating. They demonstrated that the values of the current density and discharge capacity were 0.2 A g^{-1} and 906 mAh g^{-1}, respectively, corresponding to 100 cycles. The increased value of current density (1 A g^{-1}) retains a capacity value of 711 mAh g^{-1} after 500 cycles [26]. The experimental results support the substance to be an efficient anode substance for LIBs. The MoS$_2$/TiO$_2$ nanohybrid composite material manufactured by the hydrothermal route and electrochemical characteristics are studied by Zhu et al. after 100 cycles [27]. The reported values of the current density and discharge capacity were 0.1 A g^{-1} and 604 mAh g^{-1}, respectively. These features promote the substance as a suitable candidate for lithium storage and retain the MoS$_2$ structure upon cycling mechanism [27]. Bai's research group developed a new composite material (MoS$_2$/Co) as an anode substance for LIBs storage. They used the hydrothermal method and a subsequent step of ball milling for the synthesis of their samples. They reported that the initial value of discharge capacity was 1,130 mAh g^{-1} corresponding to a current density of 0.1 A g^{-1}. The electrochemical capacity was raised to 1,340 mAh g^{-1} after 50 cycles [28]. The composite structure of the substance reduces the path length of Li$^+$ ions and accelerates the kinetics of electron transport to improve electrochemical performance.

Rui et al. [29] also studied the MoS$_2$/Co composite as a possible applicant for LIB anodes. They fabricated their samples by hydrothermal route followed by carbonization. They achieved a high value of 568.5 mAh g^{-1} for the specific capacity corresponding to the 0.2 A g^{-1} value of the current density. The rate capability was 291.9 mAh g^{-1} at 5 A g^{-1} while the cycling stability was found for 500 cycles. Li et al. [30] fabricated E-MoS$_2$ (i.e., expanded MoS$_2$) by glucose-assisted hydrothermal approach as an aqueous Zn-ion battery (ZIBs). They were able to deliver a 202.6 mAh g^{-1} value of the specific capacity for 0.1 A g^{-1} value of current density. The corresponding values of the capacity retention ratio and the energy density were 98.6% and 148.2 Wh kg^{-1}, respectively, while the cycle stability was found for 600 cycles [30]. The highly conductive carbon allotropes enhance the electrochemical performance due to large intrinsic conductivity, promoting anode substance having extraordinary characteristics for LIBs. Lui et al. [31] investigated the electrochemical functionality of hydrated MoS$_2$ nanosheet prepared by carbon fiber-assisted hydrothermal method delivers a sufficient reversible capacity and current density having values 182 mAh g^{-1} and 0.1 A g^{-1}, respectively. The value of corresponding capacity retention was 94% over 1,500 cycles at 2 A g^{-1} for aqueous ZIBs [31]. The enlarged capacity is accredited because of the embolism of water molecules to expand the interplanar spacing and accelerate the charge transfer phenomena for the charge storage mechanism. With the limited reserves of lithium, the scientist is working for a replacement of LIBs, to fulfill their demands alkali metal ion (Na, K) batteries tend to store huge amounts of electrochemical energy. A detailed comparative study of electrode substance for SIBs was investigated based on different TMDs for high storage capacity and expected that a translation kind of mechanism is accountable. Zhou et al. fabricated tungsten disulfide (WS$_2$) nanosheet via NaCl template-assisted in situ process for Na$^+$ ion anode substance and recorded a high value (453.2 mAh g^{-1}) of reversible capacity corresponding to 0.1 A g^{-1} value of current density [32]. Layered MoSe$_2$ nanoplates as anode substances for SIBs were investigated by a research group by synthesizing the samples via the thermal decomposition route. Their samples demonstrated the values 440/513 mAh g^{-1} for first charge/discharge capacity. The first principles result obtained by Density functional theory (DFT) calculations give a good agreement to the Na ion diffusion mechanism and demonstrate that the intrinsic electrical conductivity of the substance is the fundamental aspect of electrochemical performance [33]. Mukherjee et al. studied the different TMD electrode substances for SIBs, out of which MoSe$_2$ flakes demonstrate a stable initial dissociation capacity of 399.10 mAh g^{-1} with ~100% coulombic efficiency [34]. Metallic 1T-MoS$_2$ of interlayer spacing (1.025 nm), integrated by Cai et al. via hydrothermal reaction delivers a capacity of 125 mAh g^{-1} at 2 A g^{-1} with a capacity retention rate of 100% after 500 cycles as a potential candidate for ZIBs [35]. Zhang et al. fabricated MoS$_2$ nanotubes via ionic-liquid-assisted hydrothermal method and point out the first discharge/charge capacities of 2,065/1,334 mAh g^{-1} with a coulombic efficiency of 64% and second and third discharge capacities are 1,467/1,473, respectively, with coulombic efficiency of 93% and 95% [36]. Wang et al. developed SnS/SnSe$_2$ composite material by employing the CVD process followed by the template-assisted calcination method. They were able to demonstrate the values 624.9/1,131.3 mAh g^{-1} of preliminary charge/discharge capacities corresponding to value 0.1 A g^{-1} of current density. Subsequently, they achieved the value 580 mAh g^{-1} for the stable capacity after an insignificant falloff that is continued for 100 cycles through 90.2% capacity retention [37].

Wu et al. confirmed that the FeS$_2$ and FeSe$_2$ incorporated with N-doped carbon (FeS$_2$@NC) are utilized as anode for potassium-ion batteries over 5,000 cycles. With FeS$_2$@NC anode, they demonstrated the huge values 525.5 and 154.7 mAh g^{-1} of reversible capacity corresponding to respective values 0.1 and 10 A g^{-1} of current density. They also tested this anode for full-cell configuration and reported 88% value for its capacity

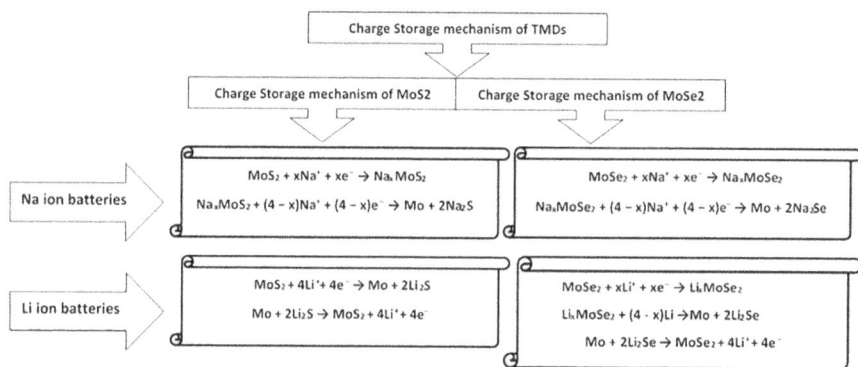

FIGURE 14.4 Flow charts showing the charge storage mechanism of TMDs-based batteries.

retention corresponding to 120 cycles having 99.9% value of Coulombic efficiency [38]. Tatsumisago et al. conveyed that when sulfur is utilized as an energetic substance having a high reversible capacity of more than 10^3 mAh g^{-1} for better cyclability in solid-state batteries then sulfide electrolyte coated on $LiCoO_2$ enhanced their energy density [8]. Sun et al. successfully developed first time a novel anode for SIBs by synthesizing the open-framework Cu-Ge-based chalcogenide anode material for sodium-ion batteries (SIBs) having a reversible capacity of 463.3 mAh g^{-1} [39]. The theoretical capacities of several TMDs substances (TiS_2, ZrS_2, NbS_2, and MoS_2) utilized as anode substances for SIBs are ranging between 260 and 339 mAh g^{-1} demonstrated by Yang et al. [40].

Li et al. developed an anode for LIBs using CoSe and doping of nitrogen in carbon. They finally prepared CoSe/NC composites and reported reversible capacities 1,244 and 310.11 mAh g^{-1} after 190 and 500 cycles. The corresponding values of current density were 0.1 and 1.0 A g^{-1}, respectively [41]. Panda et al. [42] reported an initial specific capacity of 432 mAh g^{-1} at 1.0 A g^{-1} current density and reversible specific capacity 291 mAh g^{-1} after 250 cycles for $MoTe_2$ electrode utilized in LIBs. They also found, the 1T phase-$MoTe_2$/LCO shows a reversible capacity of 388.4 mAh g^{-1} at 0.1 Ag^{-1} for 100 cycles, which can retain 74% of its earlier capacity with Coulombic efficiency of $\approx$96% [42]. Zhang et al. developed a heterostructure using facile single-walled carbon nanotube gallium chalcogenide, i.e., GaX NS/SWCNT (X = S or Se), which is utilized as anode substances for LIBs. They reported a significant specific capacity of 838 and 713 mAh g^{-1} in GaS NS and GaSe NS, respectively [43]. Liu et al. conveyed that 2D Ti_2PTe_2 monolayer having metallic character is a substance that can work as anodic substances utilized in SIBs with the large power density and fast charge/discharge rates. This substance as an anode showed a high theoretical capacity of 280.72 mAh g^{-1} [44]. The flow chart for charge storage mechanisms of two main TMDs-based batteries (LIBs and SIMs) is shown in Figure 14.4.

14.5 CONCLUSIONS

However, limited countries are consecrated with natural sources of fuels and gas reserves but these natural assets will not exist persistently. Consequently, the hunt for a novel and competent photovoltaic substance is an ongoing research activity over the globe. In the family of TMDs, MoS_2 and $MoSe_2$ are the highly demanding members because of their numerous applications in electrochemistry. The layered structuring of these members facilitates a decent opening for them to wide utilization in EES devices. In the last decade, numerous research groups examined different samples of TMDs to check their candidature as an electrode substance. Various TMDs are found suitable for use in electrochemical batteries such as LIBs and NIBs. Though, there is a necessity to resolve some intrinsic restrictions that are still existed in several other TMDs substances. The small magnitude of electrical conductivity restricts the electrochemical functionality of TMDs. Further, when we use TMDs-based electrochemical devices then the amendments are observed in intrinsic volume and high mechanical stress of TMDs during the process of insertion/extraction of ions in LIBs and NIBs. Consequently, the accumulation of electrode substances occurs followed by a reduction in the rate performance and the cyclic stability. Many scientific groups have dedicated their attempts to fabricate electrodes by using $MoSe_2$ so that the cycling stability, rate capability, and electrochemical capacitance can be improved. Thus, TMDs are attracting significant attention as electrode substances for energy storage batteries because of their atomically layered structure, high surface area, and outstanding electrochemical features.

ACKNOWLEDGMENTS

Shiv Kumar Pal is grateful to UGC, New Delhi, India for providing fellowship under the JRF scheme for National Eligibility Test (NET) qualified scholars. Neeraj Mehta is thankful to his university for providing an incentive IOE scheme (Dev. Scheme No. 6031).

REFERENCES

1. G. Du, Z. Guo, S. Wang, R. Zeng, Z. Chen, H. Liu, Superior stability and high capacity of restacked molybdenum disulfide as anode material for lithium-ion batteries, *Chem. Commun.*, 2010, 46, 1106–1108.
2. K. Chang, W. Chen, L. Ma, H. Li, H. Li, F. Huang, Z. Xu, Q. Zhang, J. Y. Lee, Graphene-like MoS_2/amorphous carbon composites with high capacity and excellent stability as anode materials for lithium-ion batteries, *J. Mater. Chem.* 2011, 21, 6251–6257.
3. L. Lin, S. Zhang, D. A. Allwood, Transition metal dichalcogenides for energy storage applications. In: N. S. Arul, V. D. Nithya (Eds), *Two-Dimensional Transition Metal Dichalcogenides Synthesis, Properties, and Applications* (pp. 173–201), 2019, Springer: Singapore.
4. W. Choi, N. Choudhary, G. H. Han, J. Park, D. Akinwande, Y. H. Lee, Recent development of two-dimensional transition metal dichalcogenides and their applications, *Mater. Today*, 2017, 20, 116–130.
5. S. Wu, Y. Du, S. Sun, Transition metal dichalcogenide based nanomaterials for rechargeable batteries, *Chem. Eng. J.*, 2017, 307, 189–207.
6. Q. Yun, Q. Lu, X. Zhang, C. Tan, H. Zhang, Three-dimensional architectures constructed from transition-metal dichalcogenide nanomaterials for electrochemical energy storage and conversion, *Angew. Chem. Int. Ed.*, 2018, 57, 626–646.

7. Y. Shi, Y. Wang, J. I. Wong, A. Y. S. Tan, C. L. Hsu, L. J. Li, Y. C. Lu, H. Y. Yang, Towards ultrahigh volumetric capacitance: Graphene derived highly dense but porous carbons for supercapacitors, *Sci. Rep.*, 2013, 3, 1–8.

8. M. Tatsumisago, A. Hayashi, Chalcogenide glasses as electrolytes for batteries. In: J. L. Adam, X. Zhang (Eds), *Chalcogenide Glasses, Preparation, Properties and Applications* (pp. 632–654), 2014, Woodhead Publishing, Cambridge.

9. M. Nagao, A. Hayashi, M. Tatsumisago, Sulfur–carbon composite electrode for all-solid-state Li/S battery with $Li_2S–P_2S_5$ solid electrolyte, *Electrochim. Acta*, 2011, 56, 6055–6059.

10. A. Hayashi, R. Ohtsubo, T. Ohtomo, F. Mizuno, M. Tatsumisago, All-solid-state rechargeable lithium batteries with Li_2S as a positive electrode material, *J. Power Sources*, 2008, 183, 422–426.

11. A. Gigot, M. Fontana, M. Serrapede, M. Castellino, S. Bianco, M. Armandi, B. Bonelli, C. F. Pirri, E. Tresso, P. Rivolo, Mixed 1T–2H phase MoS_2/reduced graphene oxide as active electrode for enhanced supercapacitive performance, *ACS Appl. Mater. Interfaces*, 2016, 8, 32842–32852.

12. D. Wang, X. Zhang, S. Bao, Z. Zhang, H. Fei, Z. Wu, Phase engineering of a multiphasic 1T/2H MoS_2 catalyst for highly efficient hydrogen evolution, *J. Mater. Chem. A*, 2017, 5, 2681–2688.

13. B. Han, Y. H. Hu, MoS_2 as a co-catalyst for photocatalytic hydrogen production from water, *Energy Sci. Eng.*, 2016, 4, 285–304.

14. B. Radisavljevic, A. Radenovic, J. Brivio, V. Giacometti, A. Kis, Single-layer MoS_2 transistors, *Nat. Nanotechnol.*, 2011, 6, 147–150.

15. Z. Zhu, J. You, D. Zhu, G. Jiang, S. Zhan, J. Wen, Q. Xia, Effect of precursor ratio on the morphological and optical properties of CVD-grown monolayer MoS_2 nanosheets, *Mater. Res. Express*, 2021, 8, 045008.

16. J. Ren, R. P. Ren, Y. K. Lv, A flexible 3D graphene@ CNT@ MoS_2 hybrid foam anode for high-performance lithium-ion battery, *Chem. Eng. J.*, 2018, 353, 419–424.

17. X. Yu, J. Tang, K. Terabe, T. Sasaki, R. Gao, Y. Ito, K. Nakura, K. Asano, M. Suzuki, Fabrication of graphene/MoS_2 alternately stacked structure for enhanced lithium storage, *Mater. Chem. Phys.*, 2020, 239, 121987.

18. Y. Wang, B. Chen, D. H. Seo, Z. J. Han, J. I. Wong, K. Ostrikov, H. Zhang, H. Y. Yang, MoS_2-coated vertical graphene nanosheet for high-performance rechargeable lithium-ion batteries and hydrogen production, *NPG Asia Mater.*, 2016, 8, 268.

19. R. Zhou, H. Wang, J. Chang, C. Yu, H. Dai, Q. Chen, J. Zhou, H. Yu, G. Sun, W. Huang, Ammonium intercalation induced expanded 1T-rich molybdenum diselenides for improved lithium-ion storage, *ACS Appl. Mater. Interfaces*, 2021, 13, 17459–17466.

20. P. Kurzweil, Gaston Planté and his invention of the lead–acid battery-the genesis of the first practical rechargeable battery, *J. Power Sources*, 2010, 195, 4424–4434.

21. R. R. Haering, J. A. R. Stiles, K. Brandt, Lithium molybdenum disulphide battery cathode, US Patent US4224390A, 1980.

22. T. Nagaura, K. Tozawa, Lithium-ion rechargeable battery, *Prog. Batteries Sol. Cells*, 1990, 9, 209.

23. T. P. Nguyen, I. T. Kim, Ag nanoparticle-decorated MoS2 nanosheets for enhancing electrochemical performance in lithium storage, *Nanomaterials*, 2021, 11, 626.

24. P. P. Wang, H. Sun, Y. Ji, W. Li, X. Wang, Three-dimensional assembly of single-layered MoS_2, *Adv. Mater.*, 2014, 26, 964–969.

25. K. Kato, F. N. Sayed, G. Babu, P. M. Ajayan, All 2D materials as electrodes for high power hybrid energy storage applications, *2D Mater.*, 2018, 5, 025016.

26. W. Ju, C. Dong, B. Jin, Y. Zhu, Z. Wen, Q. Jiang, Composites of reduced graphene oxide and Fe₂O3 nanoparticles anchored on MoS_2 nanosheets for lithium storage, *ACS Appl. Nano Mater.*, 2020, 3, 9009–9015.

27. X. Zhu, C. Yang, F. Xiao, J. Wang, X. Su, Synthesis of nano-TiO_2-decorated MoS_2 nanosheets for lithium-ion batteries, *New J. Chem.*, 2015, 39, 683–688.

28. J. Bai, B. Zhao, J. Zhou, Z. Fang, K. Li, H. Ma, J. Dai, X. Zhu, Y. Sun, Improved electrochemical performance of ultrathin MoS_2 nanosheet/Co composites for lithium-ion battery anodes, *Chem. Electro. Chem.*, 2019, 6, 1930–1938.

29. B. Rui, J. Li, L. Chang, H. Wang, L. Lin, Y. Guo, P. Nie, Engineering MoS_2 nanosheets anchored on metal organic frameworks derived carbon polyhedra for superior lithium and potassium storage, *Front. Energy Res.*, 2019, 7, 142.

30. H. Li, Q. Yang, F. Mo, G. Liang, Z. Liu, Z. Tang, L. Ma, J. Liu, Z. Shi, Chunyi Zhi, MoS_2 nanosheets with expanded interlayer spacing for rechargeable aqueous Zn-ion batteries, *Energy Storage Mater.*, 2021, 19, 94–101.

31. H. Liu, J. G. Wang, W. Hua, Z. You, Z. Hou, J. Yang, C. Wei, F. Kang, Boosting zinc-ion intercalation in hydrated MoS_2 nanosheets toward substantially improved performance, *Energy Storage Mater.*, 2021, 35, 731–738.

32. J. Zhou, J. Qin, L. Guo, N. Zhao, C. Shi, E. Liu, F. He, L. Ma, J. Lia, C. He, Scalable synthesis of high-quality transition metal dichalcogenide nanosheets and their application as sodium-ion battery anodes, *J. Mater. Chem. A*, 2016, 4, 17370–17380.

33. H. W. X. Lan, D. Jiang, Y. Zhang, H. Zhong, Z. Zhang, Y. Jiang, Sodium storage and transport properties in pyrolysis synthesized $MoSe_2$ nanoplates for high performance sodium-ion batteries, *J. Power Sources*, 2015, 283, 187–194.

34. S. Mukherjee, J. Turnley, E. Mansfield, J. Holm, D. Soares, L. David, G. Singh, Exfoliated transition metal dichalcogenide nanosheets for supercapacitor and sodium ion battery applications, *R. Soc. Open Sci.*, 2019, 6, 190437.

35. C. Cai, Z. Tao, Y. Zhu, Y. Tan, A. Wang, H. Zhou, Y. Yang, A nano interlayer spacing and rich defect 1T-MoS_2 as cathode for superior performance aqueous zinc-ion batteries, *Nanoscale Adv.*, 2021, 3, 3780–3787.

36. G. Zhang, H. Feng, C. Ma, J. Chen, Z. Wang, W. Zheng, MoS_2 nanotubes via ionic-liquid-assisted assembly of MoS_2 nanosheets for lithium storage, *ACS Appl. Nano Mater.*, 2021, 4, 3397–3405.

37. T. Wang, D. Legut, Y. Fan, J. Qin, X. Li, Q. Zhang, Building fast diffusion channel by constructing metal sulfide/metal selenide heterostructures for high-performance sodium ion batteries anode, *Nano Lett.*, 2020, 20, 6199–6205.

38. H. Wu, S. Lu, S. Xu, J. Zhao, Y. Wang, C. Huang, A. Abdelkader, W. A. Wang, K. Xi, Y. Guo, S. Ding, G. Gao, R. V. Kumar, Blowing iron chalcogenides into two-dimensional flaky hybrids with superior cyclability and rate capability for potassium-ion batteries, *ACS Nano*, 2021, 15, 2506–2519.

39. Q. Sun, L. Fu, C. Shang, A novel open-framework Cu-Ge-based chalcogenide anode material for sodium-ion battery, *Scanning*, 2017, 27, 3876525.

40. E. Yang, H. Ji, Y. Jung, Two-dimensional transition metal dichalcogenide monolayers as promising sodium ion battery anodes, *J. Phys. Chem. C*, 2015, 119, 26374–26380.

41. Z. Li, L. Y. Zhang, L. Zhang, J. Huang, H. Liu, ZIF-67-derived CoSe/NC composites as anode materials for lithium-ion batteries, *Nanoscale Res. Lett.*, 2019, 14, 1–11.

42. M. R. Panda, R. Gangwar, D. Muthuraj, S. Sau, D. Pandey, A. Banerjee, A. Chakrabarti, A. Sagdeo, M. Weyland, M. Majumder, Q. Bao, S. Mitra, High performance lithium-ion batteries using layered 2H-$MoTe_2$ as anode, *Small*, 2020, 16, 2002669.

43. C. Zhang, S. H. Park, O. Ronan, A. Harvey, A. S. Ascaso, Z. Lin, N. McEvoy, C. S. Boland, N. C. Berner, G. S. Duesberg, P. Rozier, J. N. Coleman, V. Nicolosi, Enabling flexible heterostructures for Li-ion battery anodes based on nanotube and liquid-phase exfoliated 2D gallium chalcogenide nanosheet colloidal solutions, *Small*, 2017, 13, 1701677.

44. J. Liu, M. Qiao, X. Zhu, Y. Jing, Y. Li, Ti_2PTe_2 monolayer: A promising two-dimensional anode material for sodium-ion batteries, *RSC Adv.*, 2019, 9, 15536–15541.

15 Metal Phosphide-Based 2D Nanomaterials for Batteries

Jay Singh
University of Delhi (North Campus)

Rajesh Kumar Singh
Central University of Himachal Pradesh

Alok Kumar Rai
University of Delhi (North Campus)

CONTENTS

15.1 INTRODUCTION

Continuous advancements in the field of energy conversion and storage including the development and evaluation of abundant and inexpensive materials with good electrochemical performances have only aim to meet the future energy demands. Since the energy demands of modern society are significantly increasing from the last decades, rechargeable batteries are emerging as one of the finest alternatives toward the energy storing devices due to the uncertainty and irregular distribution of renewable energy sources [1]. On the other hand, the overuse of the fossil fuels is also causing serious concern toward the harmful environment due to the release of greenhouse gases like CO_2, resulting in global climate change worldwide [2,3]. Hence, the technological revolution is increased with development of flexible and portable advanced electronic devices [4]. Thus, numerous rechargeable batteries such as Ni-Cd, Ni-MH, lead-acid, and lithium-ion batteries (LIBs) are designed to fulfil the modern energy requirements [3]. However, among all the invented rechargeable batteries till now, LIBs are found to be the most promising energy storing devices that have been widely used in the various applications including hybrid electrical vehicles, mobile communication, portable electronics, several power backup systems, and medical appliances (Figure 15.1a) [4]. More precisely, LIBs have attracted so much attention due to their prominent features of environmental benignity, suitable working voltages, high energy/power density, lower self-discharge rate, and long rate capability. It is well-known that the higher gravimetric and volumetric density of LIBs can be improved by constructing the novel high-capacity electrode materials [3].

Since LIBs are currently a leading technology for energy storage devices; however, the dearth of lithium (~20 ppm in the earth's crust) notably increases the cost of battery, which is one of the major issues to limit LIBs applications in large-scale energy storage systems [5]. Therefore, to develop the LIBs market, the unsteady supply and increasing price of lithium are the potential risks that need to be settled down first. As a result, in order to find an alternative of LIBs, various investigations have emerged. Recently, since the substantial abundance of sodium on the earth's crust is high (~2.5%) as well as sodium has similar electrochemical properties as lithium, Na-ion batteries (SIBs) have been regarded as a promising alternative to LIBs particularly at low cost for large-scale energy storage systems [5]. In addition, the higher standard reduction potential of Na^+/Na than Li^+/Li and the heavier atomic mass with large ionic radius exhibit large energy density for SIBs over LIBs, which finally makes SIBs as next-generation energy storage devices for large-scale applications (Figure 15.1b). On the other hand, in order to replace the graphite anode from LIBs due to its low theoretical capacity (372 mAh g^{-1}) and limited fast-charging capability as well as to find a suitable anode material for SIBs, various electrode materials have been explored such as metal sulfides [6], phosphides [7], nitrides [8], chalcogenides [9], phosphates [10], transition metal oxides [11], non-graphitic carbon [12], lithium-alloy materials [13], and their composites [14]. Among all these anodes, transition metal phosphides (MPs) have attracted much attention as they are naturally abundant, innocuous, and electrochemically active with high theoretical

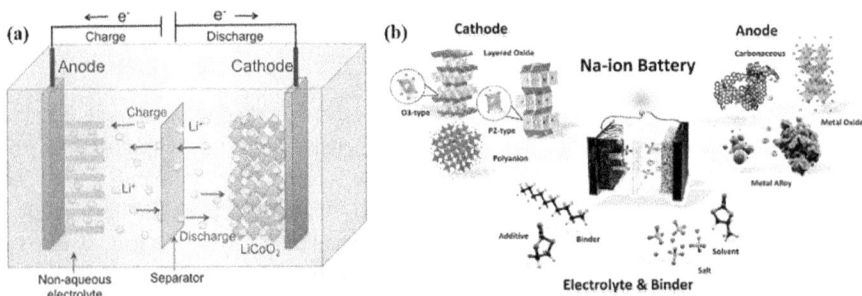

FIGURE 15.1   (a) Rechargeable lithium ion battery. (Adapted with permission from Ref. [2]. Copyright (2015) Royal Society of Chemistry.) (b) Illustration of a sodium ion battery system. (Adapted with permission from Ref. [15]. Copyright (2017) Royal Society of Chemistry.)

capacities [11]. However, they are still under investigations due to their structural instability because of drastic volume changes during charging/discharging cycles, resulting in poor cycling stability.

Nanostructures are mainly categorized as: zero-dimensional (0D), one-dimensional (1D), two-dimensional (2D), and three-dimensional (3D) nanostructures. 0D nanostructures mainly nanoparticles have three dimensions restrained on the nanoscale of ~1–100 nm, possessing short diffusion lengths with minimum surface area, which finally has tendency of agglomeration during cycling. Since 1D nanostructures such as nanotubes, nanowires, and nanorods have one dimension outside the nanoscale, they exhibit fast electron transport at 1D direction with short ion diffusion lengths along the radial direction. However, their fixed size and steady structure limit the nonadjustable specific surface area and porosity properties.

In contrast, the 2D nanostructures such as nanosheets and nanofilms displayed two dimensions outside the nanoscale with thickness of few to tens of nanometers, indicating unique advantages for LIB and SIB applications. However, the atomic or molecular thickness with infinite planar lengths of 2D nanomaterials has different atomic structures from their bulk counterparts such as differences in atomic arrangement, chemical valences, coordination number, and bond lengths [16]. Thus, the large surface area and open structures of 2D nanomaterials not only increase the contact area between the active materials and electrolytes but also reduce the path length toward the diffusion of Li and Na ions; therefore, an electrode with 2D nanomaterials provides a higher specific capacity than its bulk form (Figure 15.2a).

The 2D nanomaterials exhibit an ultrathin atomic/molecular layered structure with infinite planar lengths as well as highly exposed interior atoms, resulting in tunable chemical and physical properties through the control of defects [17]. In addition, since 2D nanomaterials are stacked by weak van der Waals forces, it can exhibit fascinating energy storage properties such as high mechanical flexibility, short ion diffusion lengths, and a large exposed surface area for electrochemical processes [16]. There are numerous 2D nanomaterials such as graphene nanosheets and its derivatives as

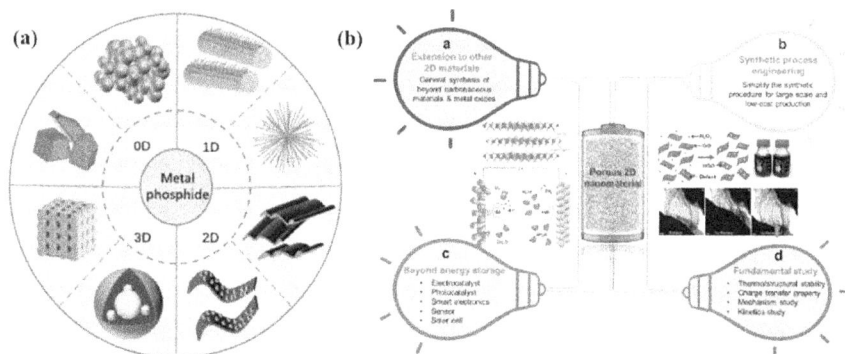

FIGURE 15.2 (a) A schematic diagram of the typical morphologies. (Adapted with permission from Ref. [18]. Copyright (2020) Royal Society of Chemistry.) (b) Prospects for future research on porous/holey 2D nanomaterials. (Adapted with permission from Ref. [19]. Copyright (2017) John Wiley & Sons.)

well as elemental nanosheets of phosphorene, antimonene, transition metal oxides, dichalcogenides, MPs, and MXene have been reported for high-performance LIBs and SIBs anodes. As can be seen that 2D nanomaterials have several extraordinary advantages, however, its certain impediments such as complex synthetic processes, indistinct insertion/de-insertion mechanisms, and severe layer-stacking and aggregation problems have to be solved before they can be commercially utilized as electrodes in both LIBs and SIBs (Figure 15.2b) [18].

Among all the investigated anode materials, MPs have been recently emerged as excellent energy storage candidates due to their highly active surface sites, excellent electrical conductivity, high thermal and structural stability [20]. MPs also exhibit numerous other desirable properties like hardness and chemical stability, which can be ascribed to the presence of strong M–P bonds. Hence, MPs have been deeply investigated as prominent anodes for LIBs, which delivered the high capacities in the range of ~500–1,800 mAh g^{-1} and high electrical conductivity (10^{-4} S cm^{-1}) at 298 K [20]. In contrast, MPs have also showed promising characteristics toward SIBs anodes due to their noticeable high gravimetric and volumetric specific capacities at appropriate redox potentials (~0.3 V vs. Na/Na$^+$) and higher electrical conductivities as compared with pure phosphorus [21]. The higher performance of MPs as SIB anodes can mainly be attributed to the presence of metal atoms acting as electronic pathways that significantly enhance the electronic conductivity [22]. More precisely, the formation of conductive metal nanoparticles during discharge of battery further improves the conductivity of electrodes [21]. It has been also reported that the metal nanocrystals are generally embedded in phosphorus matrices during cycling and thus inhibit the agglomeration of metals, resulting in enhancement in the cycling performance of MPs anodes. However, there are also some drawbacks of MPs anodes like phosphorous such as large volume variations and poor diffusion kinetics during cycling. In general, the reaction mechanisms of MPs are more complex than those of pure phosphorus. MPs can be divided into two groups on the basis of their reaction mechanism toward LIBs. The first one belongs to the insertion/de-insertion mechanism of Li$^+$ without breaking the metal–phosphorous bond (Eq. 15.1):

$$M_x P_y + zLi^+ + ze^- \rightarrow Li_z M_{x-z} P_y \qquad (15.1)$$

While the second one exhibits a conversion reaction mechanism, where metal–phosphorous bonds are broken during electrochemical reactions (Eq. 15.2) [20]:

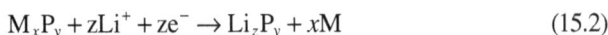

$$M_x P_y + zLi^+ + ze^- \rightarrow Li_z P_y + xM \qquad (15.2)$$

On the other hand, MPs can be again divided into metal-active phosphides and metal-inactive phosphides based on the electrochemical reactivity of metal ions with Na. For metal-active phosphides, metal element and P phase can directly react with Na. Initially, Na$_y$M and Na$_3$P are formed during the charging process (Eq. 15.3).

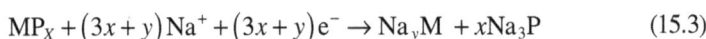

$$MP_X + (3x + y)Na^+ + (3x + y)e^- \rightarrow Na_y M + xNa_3 P \qquad (15.3)$$

Thereafter, the crystalline metal and phosphorous phases are formed through a conversion reaction as expressed in Eqs. (15.4) and (15.5) [22,23]:

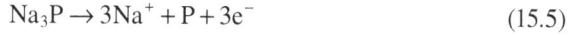

$$Na_yM \rightarrow M + yNa^+ + ye^- \tag{15.4}$$

$$Na_3P \rightarrow 3Na^+ + P + 3e^- \tag{15.5}$$

However, the discharging (Na extraction) of metal-active phosphides is again two types:

i) The metal and phosphorous phases, which are formed during the charging process (Eqs. 15.4 and 15.5), further react with sodium during the subsequent cycles through a reversible alloying process, respectively.

ii) Second, the crystalline metal and phosphorous, which are formed during the charging process (Eqs. 15.4 and 15.5), can reform the original phase of MPs after de-sodiation and the reaction can be expressed as follows (Eq. 15.6) [24]:

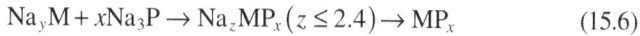

$$Na_yM + xNa_3P \rightarrow Na_zMP_x (z \leq 2.4) \rightarrow MP_x \tag{15.6}$$

Similarly, the metal (M) and Na_xP phases are first formed for metal-inactive phosphides also during sodiation and the reaction can be expressed as given in Eq. (15.7):

$$MP_X + 3xNa^+ + 3xe^- \rightarrow xNa_3P + M \tag{15.7}$$

Then metal-inactive phosphides undergo two main reactions for discharging (de-sodiation): first one allows the re-conversion of the original MPs phase after de-sodiation, whereas the second one does not act in the same way.

i) Formation of original MPs during de-sodiation reaction by the re-bonding of metal and P as shown in Eq. (15.8) [21,25]:

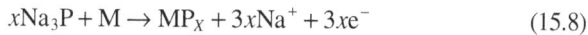

$$xNa_3P + M \rightarrow MP_X + 3xNa^+ + 3xe^- \tag{15.8}$$

ii) Whereas in the second reaction, the obtained Na_xP phase and metallic nano-crystals as shown in Eq. (15.7) further led to the formation of pure P phase during de-sodiation and the reaction can be expressed as follows (Eq. 15.9) [26]:

$$Na_3P \rightarrow 3Na^+ + P + 3e^- \tag{15.9}$$

Overall, the details of Li- and Na-storage mechanisms in MPs electrodes are yet to be fully understood, resulting in numerous advanced characterization techniques that are needed.

In this book chapter, a comprehensive study of MPs as an anode based on their 2D nanostructures for both LIBs and SIBs applications are discussed. The fundamental studies based on the various MPs lithiation/de-lithiation, sodiation/de-sodiation mechanism are also described. At the end of the chapter, the conclusions and future perspectives of MPs including major problems and opportunities are also summarized.

This chapter provides detailed information enabling in-depth understanding and rational design of 2D MPs as an anode material toward battery performances.

15.2 2D NANOSTRUCTURED METAL PHOSPHIDES: EVOLUTION AND CHALLENGES

When phosphorus reacts with lithium, it forms Li_3P by three-electron transfer and exhibits the theoretical capacity of 2,596 mAh g^{-1}, which is almost seven times higher than that of graphite [27]. However, phosphorous shows low electronic conductivity, small coulombic efficiency, and the large volume changes (~300%) during alloying with Li and Na, which actually limits its practical applications [28]. Similarly, phosphorous-based MPs also suffer from remarkable capacity decline during cycling due to the large volume variations, resulting in the pulverization of active electrode materials, and possess poor ion and charge diffusion kinetics, which is identical to phosphorus. In order to overcome these issues of MPs anodes and make them as a potential candidate for both LIBs and SIBs applications, the several extensive investigations have been conducted to synthesize numerous nanostructures of MPs such as yolk–shell Sn_4P_3@C nanospheres [22], CoP nanowires [29], monodisperse Ni_2P particles [30], and hollow FeP@carbon nanocomposites [31], which have unique morphology as well as conductive matrices that could accommodate the volume expansion/contractions of electrode materials and offer more active sites for Li and Na ion insertion/extraction, resulting in the high electrochemical performances.

15.3 ELECTROCHEMICAL PERFORMANCES OF 2D NANOSTRUCTURED METAL PHOSPHIDES

LIBs are currently leading the market with various portable electronic devices, smart grids, and hybrid electric vehicles due to their long shelf life, slow discharge rate, high power, and energy density. More importantly, LIBs can deliver a maximum cell potential of 4 V with high specific energies of 120 Wh kg^{-1} [32]. However, the small theoretical capacity of graphite anode (372 mAh g^{-1}) inhibits its commercial applications in advanced high power electronic devices [3]. Therefore, to find a new alternative anode material with improved life-time, high specific capacity, and excellent rate capability at high current density is now the priority for researcher. There are promising anode materials such as oxides, nitrides, and intermetallic compounds, which reversibly uptake lithium by alloying reactions (Li_xM) followed by reduction to metallic state [33]. Thereafter, Nazar et al. was the first group, who have demonstrated MPs as an anode for LIBs and reported its low-potential intercalation behavior [33]. On the other hand, SIBs are also under the substantial evaluation due to the sodium, which is the second-lightest and smallest alkali metal next to lithium. Additionally, it is also one of the most cheap and abundant elements on Earth's crust with an average abundance of 28,400 ppm [34]. As a battery point of view, sodium also provides a suitable electrochemical redox potential of ~2.71 V vs. SHE, which is only −0.3 V less than that of lithium. Unfortunately, the ionic radius of Na is larger than that of Li (~1.02 Å for Na^+ vs. ~0.76 Å for Li^+), resulting in the sluggish reaction kinetics and poor electrochemical performances. However, MPs and their alloys are still under deep investigation as anode materials for SIBs due to their

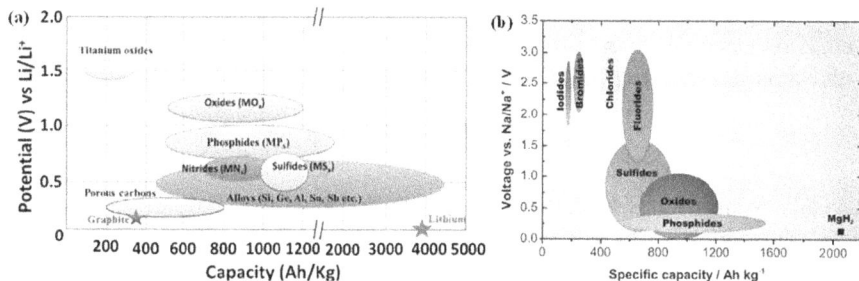

FIGURE 15.3   Potential vs. capacity for (a) Li/Li$^+$ and (b) Na/Na$^+$. (Adapted with permission from Ref. [20]. Copyright (2014) Authors, some rights reserved; exclusive licensee [Elsevier]. (a) and adapted with permission from [35]. Copyright (2013) Royal Society of Chemistry (b).)

high theoretical gravimetric/volumetric energy densities and relatively low intercalation potentials vs. Na/Na$^+$ (Figure 15.3a and b). MPs as anode materials exhibit high degree of electrons delocalization, low formal oxidation state, strong covalent bond between metal and phosphorous as well as lower insertion potential than their oxides [20]. Since MPs usually have low electrical conductivity and high-volume changes upon charge/discharge cycling, the significant efforts have been undertaken in recent years to use them as an electrode material and check their suitability toward battery applications [23–26]. Various MPs electrodes are discussed below and used for both LIBs and SIBs.

15.3.1   NICKEL PHOSPHIDES

Nickel phosphides have attracted considerable attention for LIBs, owing to their changeable components and formation of phosphorus-rich ($x > 1$, e.g., NiP_2) and metal-rich ($x \leq 1$, e.g., Ni_2P and $Ni_{12}P_5$) phases [36]. Though the phosphorus-rich phases exhibit higher theoretical Li storage capacities, their stability is an issue. Thus, the high nucleation energy will be finally required to convert them into metal-rich phases. The various morphologies of nickel phosphides such as nanocrystals, particles, nanoarrays, nanosheets, nanorods, and films were reported as anodes for Na-ion and Li-ion batteries. The theoretical capacity of Ni_2P is about 542 mAh g^{-1}.

A 2D porous nanosheets of nickel phosphide as a LIBs anode was first synthesized by Lu et al. using facile organic phase strategy [37]. Furthermore, NiP films were also fabricated using electrodeposition methods at room temperature [38]. It was found that the surface morphologies of these films were highly affected by the initial concentration of Ni^{2+}. The different phases of NiP such as $Ni_{12}P_5$ and Ni_2P can be also obtained after the phosphorization process. The conversion reaction mechanism of Ni_2P with Li$^+$ can be expressed by following Eq. (15.10):

$$Ni_2P + 3Li^+ + 3e^- \rightarrow Li_3P + 2Ni \qquad (15.10)$$

Since pulverization of electrode materials during long-range cycling is still a key issue that restricts the commercial development of anode materials, Zhao et al. [39] synthesized hierarchical 2D nanosheets of Ni_2P anchored on graphene by chemical vapor deposition method, followed by hydrothermal treatment and gas phase phosphating for SIBs applications. Herein, the effects of pH on the morphology as well as on electrochemical properties are studied. The conversion reaction of Ni_2P with Na^+ is shown in Eq. (15.11):

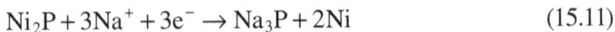

$$Ni_2P + 3Na^+ + 3e^- \rightarrow Na_3P + 2Ni \tag{15.11}$$

Recently, Wang et al. fabricated Ni_2P nanosheets on carbon cloth to overcome the huge volume expansion/contraction issue followed by loss of electrical contacts during the sodiation/de-sodiation processes [40]. It was found that Ni_2P nanosheets on carbon cloth act as a remarkable anode for SIBs applications.

To sum up, nickel phosphides (Ni_2P and $Ni_{12}P_5$) showed great potential as an anode material for both LIBs and SIBs applications. It is believed that the high content of Ni may play an important role to achieve long cycle stability. Unfortunately, all the NiP anodes illustrated small capacities with low coulombic efficiency, which is far from the practical values (1,333 mAh g^{-1} for Ni_2P). Hence, more efforts should be made to design and synthesis of novel 2D nanostructures of nickel phosphides with exploration of facile and large-scale synthesis methods to attain excellent electrochemical performances.

15.3.2 TIN PHOSPHIDE

Tin phosphides (Sn_xP_y) have also been tested for battery applications due to their high theoretical volumetric specific capacity (6,650 mAh cm^{-3} for Sn_4P_3 vs. 5,710 mAh cm^{-3} for P) and large electronic conductivity (~30.7 S cm^{-1} for Sn_4P_3 vs. ~10^{-14} S cm^{-1} for Red P) (Figure 15.4a) [41]. There are numerous reports about successful synthesis of Tin phosphides prepared by various approaches such as solution method [42], solvothermal method [22], high-energy ball-milling [41], and pulsed laser deposition method [43].

Since Sn_4P_3 has a high theoretical capacity of 1,255 mAh g^{-1}, 2D nanosheets of Sn_4P_3 was synthesized as an anode for LIBs [44]. It is well-known that Sn_4P_3 has a layered structure stacked via weak van der Waals forces, which is highly beneficial for Li^+ ion diffusions. More importantly, the mechanism involves the insertion of Li^+ ions into Sn_4P_3 to form the main amorphous intermediate product of $Li_xSn_4P_3$ as well as the crystalline-$Li_{4.4}Sn$ (c-$Li_{4.4}Sn$) and amorphous-Li_3P (a-Li_3P). The lithiation/de-lithiation mechanism of Sn_4P_3 during the discharge followed by subsequent charge processes can be described as follows:

During discharge:

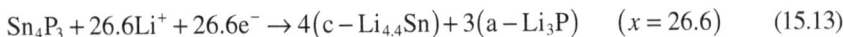

$$Sn_4P_3 + xLi^+ + xe^- \rightarrow a - Li_xSn_4P_3 \quad (0 < x < 26.6) \tag{15.12}$$

$$Sn_4P_3 + 26.6Li^+ + 26.6e^- \rightarrow 4(c - Li_{4.4}Sn) + 3(a - Li_3P) \quad (x = 26.6) \tag{15.13}$$

Subsequent charge:

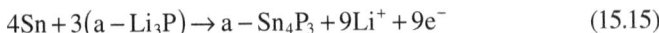

$$c - Li_{4.4}Sn \rightarrow Sn + 4.4Li^+ + 4.4e^- \tag{15.14}$$

$$4Sn + 3(a - Li_3P) \rightarrow a - Sn_4P_3 + 9Li^+ + 9e^- \tag{15.15}$$

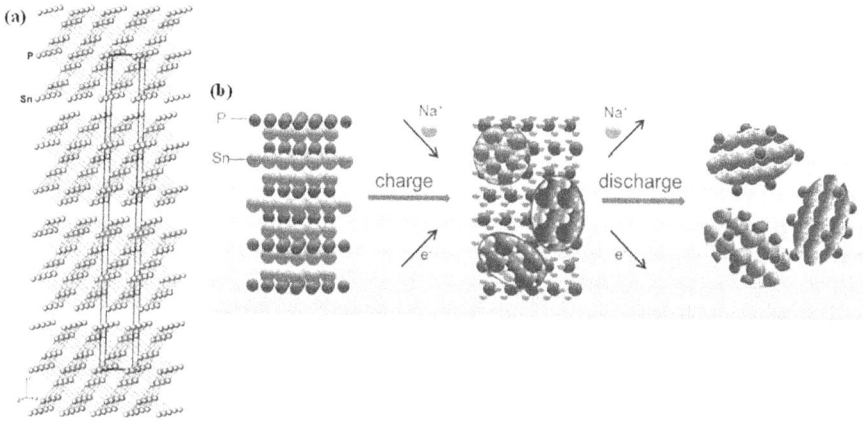

FIGURE 15.4 (a) A general view of the crystal structure of layered tin phosphide Sn_4P_3 (hexagonal, $R\,\bar{3}\,m$, $a = 3.9677$ Å, $c = 35.331$ Å) [46]. (Adapted with permission from Ref. [47] Copyright (2006) Elsevier.) (b) Na-storage mechanism in Sn_4P_3 electrode. (Adapted with permission from Ref. [23]. Copyright (2014) American Chemical Society.)

Moreover, Wu et al. have reported the fabrication of Sn_4P_3 thin-film electrode by pulsed laser deposition method for LIBs applications [43]. In order to further understand the reaction mechanism, the electrochemical reactions are given below:

During discharge:

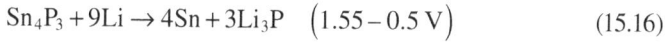

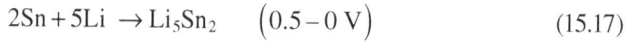

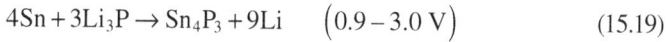

$$Sn_4P_3 + 9Li \rightarrow 4Sn + 3Li_3P \quad \left(1.55 - 0.5\ V\right) \tag{15.16}$$

$$2Sn + 5Li \rightarrow Li_5Sn_2 \quad \left(0.5 - 0\ V\right) \tag{15.17}$$

During charge:

$$Li_5Sn_2 \rightarrow 2Sn + 5Li \quad \left(0 - 0.9\ V\right) \tag{15.18}$$

$$4Sn + 3Li_3P \rightarrow Sn_4P_3 + 9Li \quad \left(0.9 - 3.0\ V\right) \tag{15.19}$$

In contrast, Tin phosphide and its composites have also demonstrated excellent electrochemical performances for SIBs (Figure 15.4b). Thus, Xu et al. have reported a nanocomposite of Sn_4P_3-P@graphene synthesized by a novel and facile mechanochemical transformation of SnP_3@graphene composite [45].

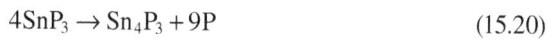

$$4SnP_3 \rightarrow Sn_4P_3 + 9P \tag{15.20}$$

where SnP_3 is also initially mechanochemically synthesized from Sn and red P:

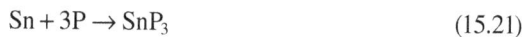

$$Sn + 3P \rightarrow SnP_3 \tag{15.21}$$

The complete sodiation reaction of Sn_4P_3 is depicted below:

$$24Na + Sn_4P_3 \rightarrow 3Na_3P + Na_{15}Sn_4 \tag{15.22}$$

During the de-sodiation process:

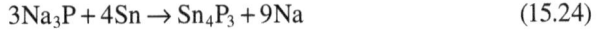

$$Na_{15}Sn_4 \rightarrow 4Sn + 15Na \tag{15.23}$$

$$3Na_3P + 4Sn \rightarrow Sn_4P_3 + 9Na \tag{15.24}$$

Additionally, Qian et al. have also demonstrated a green approach for the synthesis of Sn_4P_3/C nanocomposite using mechanochemical reaction of metallic Sn, elemental phosphorus and carbon, and tested it as an anode for Na-storage performances [23]. This work shows the accommodation of 24 Na^+ by Sn_4P_3 and delivered the total capacity of 1,132 mAh g^{-1}, which are mainly contributed by 4 Sn atoms (708 mAh g^{-1}) and 3 P atoms (424 mAh g^{-1}). The Na insertion in Sn_4P_3 can be described as follows:

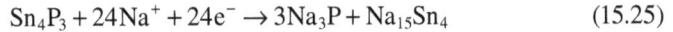

$$Sn_4P_3 + 24Na^+ + 24e^- \rightarrow 3Na_3P + Na_{15}Sn_4 \tag{15.25}$$

During the subsequent Na extraction reaction, it can be represented as follows:

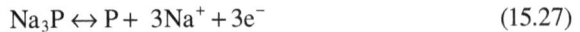

$$Na_{15}Sn_4 \leftrightarrow 4Sn + 15Na^+ + 15e^- \tag{15.26}$$

$$Na_3P \leftrightarrow P + 3Na^+ + 3e^- \tag{15.27}$$

15.3.3 COBALT PHOSPHIDE

Among all the investigated MPs, cobalt phosphides are also considered as a promising anode material for high-performance LIBs due to its high theoretical capacity (894 mAh g^{-1}), low charge/discharge potential, and good thermal stability (Figure 15.5a). Yang et al. have synthesized a nanocomposite of cobalt phosphide nanowires with reduced graphene oxide (CoP/rGO) via a facile hydrothermal method [48]. Since the conductivity of the nanocomposite electrode was greatly improved because of the presence of reduced graphene oxide in comparison to pure CoP, it exhibits better electrochemical performances. In this continuation, Yang et al. have also prepared a robust high-capacity CoP/reduced graphene oxide anode for enhanced Li-ion storage properties using a versatile strategy, which includes an oil bath, freeze drying, and phosphidation processes [49]. Generally, the reactions involved during the lithiation/de-lithiation processes can be expressed as follows:

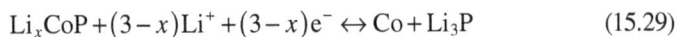

$$CoP + xe^- + xLi^+ \leftrightarrow Li_xCoP \tag{15.28}$$

$$Li_xCoP + (3-x)Li^+ + (3-x)e^- \leftrightarrow Co + Li_3P \tag{15.29}$$

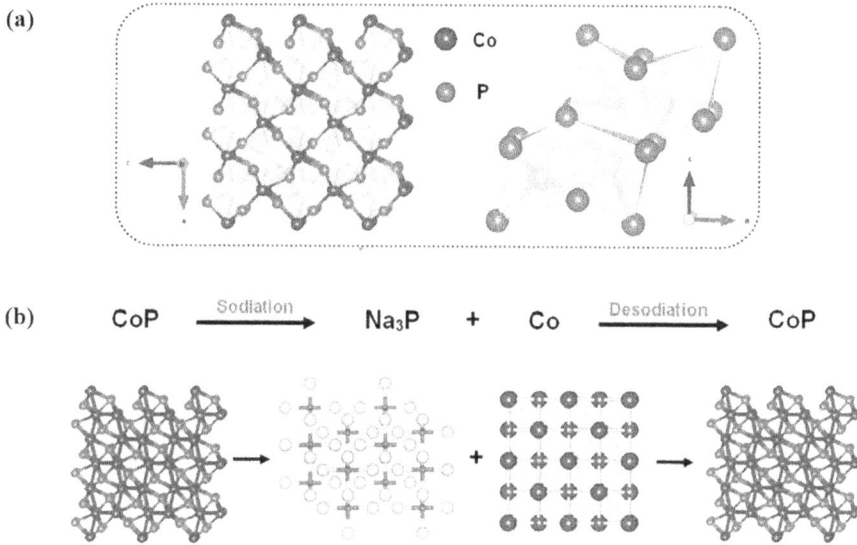

FIGURE 15.5 (a) Crystal structure of orthorhombic CoP. (Adapted with permission from Ref. [51]. Copyright (2019) Royal Society of Chemistry.) (b) Schematic of the Na insertion/extraction mechanism in CoP. (Adapted with permission from Ref. [51]. Copyright (2019) Royal Society of Chemistry.)

Unfortunately, cobalt phosphide has not shown till now very satisfactory Na ion storage performances due to its huge volume expansion/contraction and low Na ion diffusivity (Figure 15.5b). Recently, Qun Li et al. have fabricated a nanohybrid of Cobalt phosphide/reduced graphene oxide through a facile chemical precipitation method for better electrochemical performances and the corresponding sodiation/de-sodiation reactions are given below [50]:

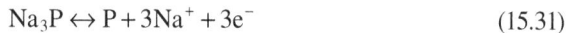

$$CoP + 3Na^+ + 3e^- \rightarrow Co + Na_3P \qquad (15.30)$$

$$Na_3P \leftrightarrow P + 3Na^+ + 3e^- \qquad (15.31)$$

15.3.4 GERMANIUM PHOSPHIDE

Since germanium-based anodes show excellent rate performances with good cycling stability due to their large Li^+ ion diffusivity as well as high electronic conductivity, it is believed that they could meet the ever-growing demands of high energy and power density energy storage devices [52]. Thus, in the extension of MPs research, germanium phosphides were also investigated for battery applications because of their high theoretical capacity (1,033 mAh g^{-1}) and relatively large initial coulombic efficiency. Since germanium phosphides also have a layered structure like black phosphorus, which are stacked by weak van der Waals forces, each Ge atom is attached to three P atoms through covalent bonds in every internal layer. The interlayer spacing of the two adjacent layers is ~2.3 Å, which significantly favors easy and fast Li-ion diffusion. More importantly, the layered structure of germanium phosphide is almost similar to black P and graphite. The layered GeP has a high electrical conductivity of 5×10^3 S m^{-1}, which is nearly one order of magnitude high to black phosphorus (3×10^2 S m^{-1}) [53].

Recently, Shen et al. have published an article on the development of 2D germanium phosphide anode with study of narrowing the working voltage window and found promising performances toward LIBs [54]. Furthermore, germanium phosphides nanoflakes were also exfoliated from bulk material by using a high-power ultrasonicator, though it shows poor cycling performances. However, when the voltage window narrows between 0.001 and 0.85 V, the cycling behavior is significantly improved. The obtained enhancement in the cyclability can be credited to the efficient dealloying reaction of Li$_3$P [54]. The schematic illustration (Figure 15.6a) followed by the detailed reactions of GeP for LIB can be summarized as given below:

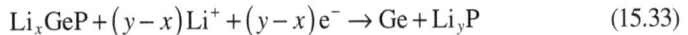

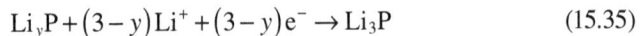

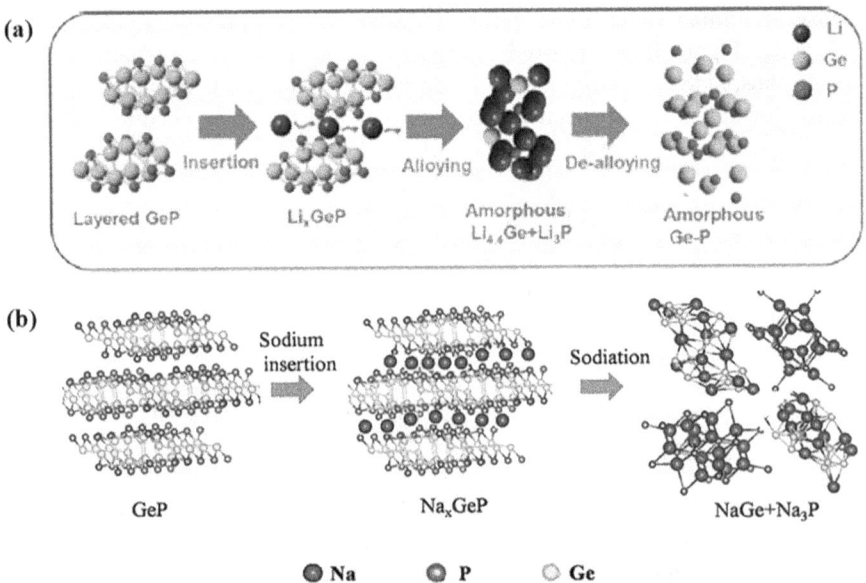

$$GeP + xLi^+ + e^- \rightarrow Li_xGeP \tag{15.32}$$

$$Li_xGeP + (y-x)Li^+ + (y-x)e^- \rightarrow Ge + Li_yP \tag{15.33}$$

$$Ge + 4.4Li^+ + 4.4e^- \rightarrow Li_{4.4}Ge \tag{15.34}$$

$$Li_yP + (3-y)Li^+ + (3-y)e^- \rightarrow Li_3P \tag{15.35}$$

FIGURE 15.6 (a) Lithium storage mechanism in GeP. (Adapted with permission from Ref. [54]. Copyright (2020) American Chemical Society.) (b) Sodium storage mechanism in GeP. (Adapted with permission from Ref. [55]. Copyright (2019) Elsevier.)

Additionally, Shen et al. have also synthesized layered germanium phosphides nano-flakes by high-temperature and pressure-oriented growth technique for SIBs applications [55]. By narrowing the voltage window between 0.15 and 1.5 V, the fabricated electrode material exhibits better electrochemical performances. It has been also reported that germanium phosphide follows multi-step reactions, i.e., intercalation and alloying, accompanied with sequentially formation of Na_xGeP ($0 < x < 1/3$), layered $NaGe_3P_3$, and amorphous $NaGe$ and Na_yP ($0 < y \leq 3$). The illustration of sodiation mechanism (Figure 15.6b) of GeP along with reactions is as follows:

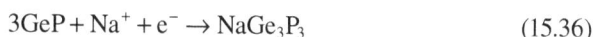

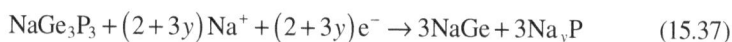

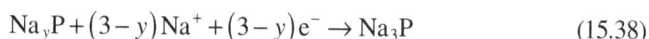

$$3GeP + Na^+ + e^- \rightarrow NaGe_3P_3 \tag{15.36}$$

$$NaGe_3P_3 + (2 + 3y)Na^+ + (2 + 3y)e^- \rightarrow 3NaGe + 3Na_yP \tag{15.37}$$

$$Na_yP + (3 - y)Na^+ + (3 - y)e^- \rightarrow Na_3P \tag{15.38}$$

15.3.5 OTHER METAL PHOSPHIDES

Many other 2D nanostructured MPs electrodes have also been studied as anode materials for LIBs and SIBs applications. For example, recently Li et al. have prepared a ternary $Cu_2P_7/CuP_2/C$ nanocomposite using facile ball milling method as a multiphase high-performance anode material for both LIBs and SIBs applications [56]. In addition, Jiang et al. have synthesized FeP@reduced graphene oxide nanocomposite anode by using low-temperature phosphorization method for LIB application [57]. While Qun Li et al. have synthesized mesoporous FeP/rGO nanocomposite anode for SIBs application through precipitation reaction process and low-temperature phosphorization technology and found high specific capacity at a high current density [58]. A general representation of lithiation and sodiation reaction of FeP is displayed in Eq. (15.39).

$$FeP \xrightarrow{Li/Na- \text{ insertion}} LiFeP / NaFeP \xrightarrow{Conversion} Li_3P / Na_3P + Fe \tag{15.39}$$

15.4 CONCLUSIONS AND FUTURE PERSPECTIVES

In this chapter, we have discussed the recent research progress on 2D nanostructured MPs including their fabrication processes followed by their electrochemical performances for both LIBs and SIBs. Since the electrochemical performances of an electrode material are strongly dependent on the morphology, it can be improved or tuned by specifically designed numerous nanoarchitectures. As can be seen from the above discussion that MPs anode has great potential toward battery performances due to their unique physical properties such as highly active surface sites, high thermal, and structural stability. However, there are also numerous issues, which should be fixed before their commercial use like poor electronic conductivity, poor diffusion kinetics, low initial coulombic efficiency, and the large volume changes (~300%) during insertion/de-insertion, resulting in the fast degradation of battery performances.

Hence, in order to overcome the issues of MPs to make them superior electrode materials, there are various future perspectives, which can be focused such as:

1. The enrichment of phosphorous content in MPs to improve the gravimetric and volumetric specific capacities could be an effective strategy to achieve the aim.
2. Surface engineering, doping, and fabrication of novel designed nanoarchitectures will be another best strategy to enhance the poor diffusion kinetics and rate performances of MPs.
3. Exploration of new MPs fabricated by facile unique synthesis methods to attain novel nanostructures to accommodate the large volume changes during long-range cycling will be also one of the milestone approaches toward final goal.
4. A combination of theoretical studies and experimental characterizations may also help to deeply understand the fundamentals of lithium and sodium storage mechanism in 2D nanomaterials along with the corresponding changes in their crystal structure.
5. Since the majority of investigated MPs belong to Ni and Co, it would be excellent to find a novel method, which should be highly feasible, less expensive, and environment friendly for clean innovation.
6. Further tuning the synergy between phosphorous-based anode with suitable cathode and electrolyte could be another possible option to fabricate a battery of choice for large-scale energy storage devices.
7. The pulverization of electrode materials during long-range cycling could be investigated more for the development of mitigation strategies to achieve high durability and large stability of the batteries.

It is believed that 2D nanostructures of MPs anodes have the capability to open a new window toward energy storage devices for advanced applications.

REFERENCES

1. G. Li, Y. Li, J. Chen, P. Zhao, D. Li, Y. Dong, L. Zhang, Synthesis and research of egg shell-yolk NiO/C porous composites as lithium-ion battery anode material, *Electrochim. Acta* 245 (2017) 941–948.
2. P. Roy, S. K. Srivastava, Nanostructured anode materials for lithium ion batteries, *J. Mater. Chem. A* 3 (2015) 2454–2484.
3. J. M. Tarascon, M. Armand, Building better batteries, *Nature* 451 (2008) 652–657.
4. Y. Ai, X. Geng, Z. Lou, Z. M. Wang, G. Shen, Rational synthesis of branched CoMoO$_4$@CoNiO$_2$ core/shell nanowire arrays for all-solid-state supercapacitors with improved performance, *ACS Appl. Mater. Interfaces* 7 (2015) 24204–24211.
5. N. Yabuuchi, K. Kubota, M. Dahbi, S. Komaba, Research development on sodium-ion batteries, *Chem. Rev.* 114 (2014) 11636–11682.
6. C. H. Lai, M. Y. Lu, L. J. Chen, Metal sulfide nanostructures: Synthesis, properties and applications in energy conversion and storage, *J. Mater. Chem.* 22 (2012) 19–30.

7. J. Theerthagiri, A. P. Murthy, S. J. Lee, K. Karuppasamy, S. R. Arumugam, Y. Yu, M. M. Hanafiah, H. Kim, V. Mittal, M. Y. Choi, Recent progress on synthetic strategies and applications of transition metal phosphides in energy storage and conversion, *Ceram. Int.* 47 (2021) 4404–4425.

8. J. L. C. Rowsell, V. Pralong, L. F. Nazar, Layered lithium iron nitride: a promising anode material for Li-ion batteries, *J. Am. Chem. Soc.* 123 (2001) 8598.

9. J. W. Seo, J. T. Jang, S. W. Park, C. Kim, B. Park, J. Cheon, Two-Dimensional SnS_2 nanoplates with extraordinary high discharge capacity for lithium ion batteries, *Adv. Mater.* 20 (2008) 4269–4273.

10. Y. Lu, J. P. Tu, J. Y. Xiang, X. L. Wang, J. Zhang, Y. J. Mai, S. X. Mao, Improved electrochemical performance of self-assembled hierarchical nanostructured nickel phosphide as a negative electrode for lithium ion batteries, *J. Phys. Chem. C* 115 (2011) 23760–23767.

11. X. Ma, N. Wang, Y. Qian, Z. Bai, Large-scale synthesis of NiO polyhedron nanocrystals as high-performance anode materials for lithium ion batteries, *Mater. Lett.* 168 (2016) 5–8.

12. X. Wang, J. Wang, H. Chang, Y. Zhang, Preparation of short carbon nanotubes and application as an electrode material in Li-ion batteries, *Adv. Funct. Mater.* 17 (2007) 3613–3618.

13. W. J. Zhang, A review of the electrochemical performance of alloy anodes for lithium-ion batteries, *J. Power Sources* 196 (2011) 13–24.

14. M. V. Reddy, G. V. Subba Rao, B. V. R. Chowdari, Metal oxides and oxysalts as anode materials for Li ion batteries, *Chem. Rev.* 113 (2013) 5364–5457.

15. J. Y. Hwang, S. T. Myung, Y. K. Sun, Sodium-ion batteries: Present and future, *Chem. Soc. Rev.* 46 (2017) 3529.

16. C. L. Tan, X. H. Cao, X. J. Wu, Q. Y. He, J. Yang, X. Zhang, J. Z. Chen, W. Zhao, S. K. Han, G.-H. Nam, M. Sindoro, H. Zhang, Recent advances in ultrathin two-dimensional nanomaterials, *Chem. Rev.* 117 (2017) 6225–6331.

17. L. Shi, T. Zhao, Recent advances in inorganic 2D materials and their applications in lithium and sodium batteries, *J. Mater. Chem.* 5 (2017) 3735–3758.

18. Z. Li, Y. Zheng, Q. Liu, Y. Wang, D. Wang, Z. Li, P. Zheng, Z. Liu, Recent advances in nanostructured metal phosphides as promising anode materials for rechargeable batteries, *J. Mater. Chem. A* 8 (2020) 19113–19132.

19. L. Peng, Z. Fang, Y. Zhu, C. Yan, G. Yu, Holey 2D nanomaterials for electrochemical energy storage, *Adv. Energy Mater.* (2017) 1702179.

20. S. Goriparti, E. Miele, F. D. Angelis, E. D. Fabrizio, R. P. Zaccaria, C. Capiglia, Review on recent progress of nanostructured anode materials for Li-ion batteries, *J. Power Sources* 257 (2014) 421–443.

21. F. Zhao, N. Han, W. Huang, J. Li, H. Ye, F. Chen, Y. Li, Nanostructured CuP_2/C composites as high-performance anode materials for sodium ion batteries, *J. Mater. Chem. A* 3 (2015) 21754–21759.

22. J. Liu, P. Kopold, C. Wu, P. A. van Aken, J. Maier, Y. Yu, Uniform yolk–shell Sn_4P_3@C nanospheres as high-capacity and cycle-stable anode materials for sodium-ion batteries, *Energy Environ. Sci.* 8 (2015) 3531–3538.

23. J. Qian, Y. Xiong, Y. Cao, X. Ai and H. Yang, Synergistic Na-storage reactions in Sn_4P_3 as a high-capacity, cycle-stable anode of Na-ion batteries, *Nano Lett.* 14 (2014) 1865–1869.

24. K. H. Nam, K. J. Jeon, C. M. Park, Layered germanium phosphide-based anodes for high-performance lithium and sodium-ion batteries, *Energy Storage Mater.* 17 (2019) 78–87.

25. M. Kong, H. Song, J. Zhou, Metal–organophosphine framework-derived N, P-Co doped carbon-confined Cu_3P nanoparticles for superb Na-ion storage, *Adv. Energy Mater.* 8 (2018) 1801489.

26. W. J. Li, S. L. Chou, J. Z. Wang, H. K. Liu, S. X. Dou, A new, cheap, and productive FeP anode material for sodium-ion batteries, *Chem. Commun.* 51 (2015) 3682–3685.

27. L. Wang, X. He, J. Li, W. Sun, J. Gao, J. Guo, C. Jiang, Nano-structured phosphorus composite as high-capacity anode materials for lithium batteries, *Angew. Chem.* 124 (2012) 9168–9171.

28. J. W. Hall, N. Membreno, J. Wu, H. Celio, R.A. Jones, K.J. Stevenson, Low-temperature synthesis of amorphous FeP_2 and its use as anodes for Li ion batteries, *J. Am. Chem. Soc.* 134 (2012) 5532–5535.

29. J. Zhang, K. Zhang, J. Yang, G. H. Lee, J. Shin, V. W. Lau, Y. M. Kang, Bifunctional conducting polymer coated CoP core–shell nanowires on carbon paper as a free-standing anode for Sodium ion batteries, *Adv. Energy Mater.* 8 (2018) 1800283.

30. S. Shi, Z. Li, Y. Sun, B. Wang, Q. Liu, Y. Hou, S. Huang, J. Huang, Y. Zhao, A covalent heterostructure of monodisperse Ni_2P immobilized on N, P-Co-doped carbon nanosheets for high performance sodium/lithium storage, *Nano Energy* 48 (2018) 510–517.

31. X. Wang, K. Chen, G. Wang, X. Liu, H. Wang, Rational design of three-dimensional graphene encapsulated with hollow FeP@Carbon nanocomposite as outstanding anode material for lithium ion and sodium ion batteries, *ACS Nano* 11 (2017) 11602–11616.

32. C. M. Park, J. H. Kim, H. Kim, H. J. Sohn, Li-alloy based anode materials for Li secondary batteries, *Chem. Soc. Rev.* 39 (2010) 3115–3141.

33. D.C.S. Souza, V. Pralong, A.J. Jacobson, L.F. Nazar, A reversible solid-state crystalline transformation in a metal phosphide induced by redox chemistry, *Science* 296 (2002) 2012–2015.

34. M. D. Slater, D. Kim, E. Lee and C. S. Johnson, Sodium-ion batteries, *Adv. Funct. Mater.* 23 (2013) 947–958.

35. F. Klein, B. Jache, A. Bhide, P. Adelhelm, Conversion reactions for sodium-ion batteries, *Phys. Chem. Chem. Phys.* 15 (2013) 15876.

36. Y. Feng, H. Zhang, Y. Mu, W. Li, J. Sun, K. Wu, Y. Wang, Monodisperse sandwich-like coupled quasi-graphene sheets encapsulating Ni_2P nanoparticles for enhanced lithium-ion batteries, *Chem. Eur. J.* 21 (25) (2015) 9229–9235.

37. Y. Lu, J. Tu, Q. Xiong, H. Zhang, C. Gu, X. Wang, S. X. Mao, Large-scale synthesis of porous Ni_2P nanosheets for lithium secondary batteries, *Cryst. Eng. Comm.* 14 (2012) 8633–8641.

38. Y. Lu, C. D. Gu, X. Ge, H. Zhang, S. Huang, X. Y. Zhao, X. L. Wang, J. P. Tu, S. X. Mao, Growth of nickel phosphide films as anodes for lithium-ion batteries: Based on a novel method for synthesis of nickel films using ionic liquids, *Electrochim. Acta* 112 (2013) 212–220.

39. Z. Zhao, H. Li, Z. Yang, S. Hao, X. Wang, Y. Wu, Hierarchical Ni_2P nanosheets anchored on three-dimensional graphene as self-supported anode materials towards long-life sodium-ion batteries, *J. Alloys Compds.* 817 (2020) 152751.

40. Y. Wang, Q. Pan, K. Jia, H. Wang, J. Gao, C. Xu, Y. Zhong, A. A. Alshehri, K. A. Alzahrani, X. Guo, X. Sun, Ni_2P nanosheets on carbon cloth: An efficient flexible electrode for sodium-ion batteries, *Inorg. Chem.* 58 (2019) 6579–6583.

41. Y. Kim, Y. Kim, A. Choi, S. Woo, D. Mok, N. S. Choi, Y. S. Jung, J. H. Ryu, S. M. Oh, K. T. Lee, Tin phosphide as a promising anode material for Na-ion batteries, *Adv. Mater.* 26 (2014) 4139–4144.

42. Y. Kim, H. Hwang, Chong S. Yoon, Min G. Kim, J. Cho, Reversible lithium intercalation in teardrop-shaped ultrafine SnP0.94 particles: an anode material for lithium-ion batteries, *Adv. Mater.* 19 (2007) 92–96.

43. J. Wu, Z. Fu, Pulsed-laser-deposited Sn_4P_3 electrodes for lithium-ion batteries, *J. Electrochem. Soc.* 156 (1) (2009) A22–A26.
44. J. Liu, W. Sun, Y. Ran, S. Zhou, L. Zhang, A. Wu, H. Huang, M. Yao, Progressive lithiation mechanism of Sn_4P_3 nanosheets as anodes for Li-ion batteries, *Appl. Surf. Sci.* 550 (2021) 149247.
45. Y. Xu, B. Peng, F. M. Mulder, A high-rate and ultrastable sodium ion anode based on a novel Sn_4P_3-P@Graphene nanocomposite, *Adv. Energy Mater.* 8 (2018) 1701847.
46. O. Olofsson, X-ray investigation of the tin-phosphorus system, *Acta Chem. Scand.* 21 (1967) 1659–1660.
47. K. A. Kovnira, Y. V. Kolen'ko, S. Ray, J. Li, T. Watanabe, M. Itoh, M. Yoshimura, A. V. Shevelkov, A facile high-yield solvothermal route to tin phosphide Sn_4P_3, *J. Solid State Chem.* 179 (2006) 3756–3762.
48. J. Yang, Y. Zhang, C. Sun, H. Liu, L. Li, W. Si, W. Huang, Q. Yan, X. Dong, Graphene and cobalt phosphide nanowire composite as an anode material for high performance lithium-ion batteries, *Nano Res.* 9 (2016) 612–621.
49. Y. Yang, Y. Jiang, W. Fu, X. Liao, Y. He, W. Tang, F. M. Alamgir, Z. Ma, Cobalt phosphide embedded in a graphene nanosheet network as a high-performance anode for Li-ion batteries, *Dalton Trans.* 48 (2019) 7778–7785.
50. Q. Li, S. Dong, Y. Zhang, S. Feng, Q. Wang, J. Yuan, Ultrafine CoP nanoparticles anchored on reduced graphene oxide nanosheets as anodes for sodium ion batteries with enhanced electrochemical performance, *Eur. J. Inorg. Chem.* 2018 (2018) 3433–3438.
51. B. Wang, K. Chen, G. Wang, X. Liu, H. Wang, J. Bai, A multidimensional and hierarchical carbon-confined cobalt phosphide nanocomposite as an advanced anode for lithium and sodium storage, *Nanoscale* 11 (2019) 968.
52. J. Graetz, C. C. Ahn, R. Yazami, B. Fultz, Nanocrystalline and thin film germanium electrodes with high lithium capacity and high rate capabilities, *J. Electrochem. Soc.* 151 (2004) A698–A702.
53. J. Sun, H. W. Lee, M. Pasta, H. Yuan, G. Zheng, Y. Sun, Y. Li, Y. Cui, A phosphorene–graphene hybrid material as a high-capacity anode for sodium-ion batteries, *Nat. Nanotechnol.* 10 (2015) 980–985.
54. H. Shen, Y. Huang, Y. Chang, R. Hao, Z. Ma, K. Wu, P. Du, B. Guo, Y. Lyu, P. Wang, H. Yang, Q. Li, H. Wang, Z. Liu, A. Nie, Narrowing working voltage window to improve layered GeP anode cycling performance for lithium-ion batteries, *ACS Appl. Mater. Interfaces* 12 (2020) 17466–17473.
55. H. Shen, Z. Ma, B. Yang, B. Guo, Y. Lyu, P. Wang, H. Yang, Q Li, H. Wang, Z. Liu, A. Nie, Sodium storage mechanism and electrochemical performance of layered GeP as anode for sodium ion batteries, *J. Power Sources* 433 (2019) 126682.
56. X. Li, J. Liao, P. Shen, X. Lia, Y. Li, N. Li, Z. Li, Z. Shi, H. Zhang, W. Li, Ternary Cu_2P_7/CuP_2/C composite: A high-performance multi-phase anode material for Li/Na-ion batteries endowed by heterointerfaces, *J. Alloys Compds.* 803 (2019) 804–811.
57. H. Jiang, B. Chen, J. Pan, C. Li, C. Liu, L. Liu, T. Yang, W. Li, H. Li, Y. Wang, L. Chen, M. Chen, Strongly coupled FeP@ reduced graphene oxide nanocomposites with superior performance for lithium-ion batteries, *J. Alloys Compds.* 728 (2017) 328–336.
58. Q. Li, J. Yuan, Q. Tan, G. Wang, S. Feng, Q. Liu, Q. Wang, Mesoporous FeP/RGO nanocomposites as anodes for sodium ion batteries with enhanced specific capacity and long cycling life, *New J. Chem.* 44 (2020) 5396.

16 MXene-Based 2D Nanomaterials for Batteries

Zhaolin Tan, Jingxuan Wei, Yang Liu,
Linrui Hou, and Changzhou Yuan
University of Jinan

CONTENTS

16.1 INTRODUCTION

The contradiction between the growth of human demand for energy and the declining existing energy makes the energy problem a significant issue that affects human development. To cope with the ever-decreasing fossil energy and the following greenhouse effect and other issues, the use of renewable resources of wind, solar energy, and so forth has become urgent. The effective use of energy requires advanced electrochemical energy storage devices. Nowadays, lithium-ion batteries (LIBs) are widely used in various industries [1], as well as sodium-ion batteries (SIBs) and potassium-ion batteries (PIBs) [2,3] that are similar in mechanism and have great potential. Low-cost, green, and environmentally friendly multivalent metal (Zn, Mg, Ca, and Al)-ion batteries (MIBs) [4–6] and metal (Li and Na)-sulfur batteries (MSBs) [7] with high specific capacity and high performance are also the researching hotpot in the energy field. However, the electrochemical performance mainly depends on the inherent properties of the internal electrode materials. In general, the ideal electrode materials should have good conductivity, high specific surface area

(SSA), lightweight, tunable surface wettability, and excellent electrochemical properties. Two-dimensional (2D) materials can meet the above requirements.

In 2012, a material with particular properties and morphology gradually attracted peoples' attention; this is MXene, a kind of 2D layered transition metal carbides, nitrides, or carbonatites, different from graphene that is composed of material from one element. These materials have a unified formula of $M_{n+1}X_nT_x$ ($n = 1$–3) (Figure 16.1) in which M is the early transition metal (such as Sc, Ti, V, Mn, Zr, Nb, and Mo in the periodic table of groups 3–7), X represents carbon (C) or nitrogen (N), and T_x is the surface termination (−O, −OH, −F) that is generally derived from the synthetic process. A few articles report some special surface termination such as −Cl, −Br, and −S related to some specific properties. It is important to note that introducing elements from other positions in the periodic table into the MAX phase is already being explored. For instance, Ti_3AuC_2, $Ti_3Au_2C_2$, and Ti_3IrC_2 can be obtained by high-temperature noble metal substitution reaction [8], which also marks the first introduction of the subgroup Au and Ir elements into the MAX phase interlayer. Therefore, it is very likely that the definition of MXene will be further expanded in the future.

Generally, MXene is obtained by chemically etching the A element from its precursor. The first discovery of MXene could date back to 2011 when Gogotsi and his coworkers used 50% hydrofluoric acid (HF), realizing the selective etch to Ti_3AlC_2 [6]. Given the high risk of HF, the need for safer and more convenient methods is urgent; until now, researchers have made many efforts to prepare MXene. We will introduce them in detail in the following section. Except that, MXene material itself also has many intrinsic defects such as aggregation, ease to oxidize, and the overshadowing effect of surface functional groups [9]. Researchers have taken tremendous actions to revise its microstructure and surface properties by derivation and composition. This chapter will elaborate on the MXene-based nanomaterials for batteries such as alkali metal-ion batteries (AMIBs), MIBs, and MSBs. At first, we

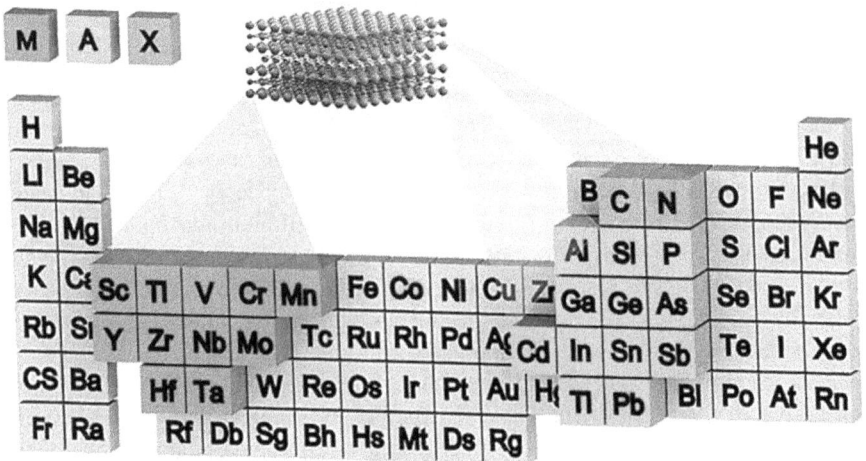

FIGURE 16.1   The position of the MAX phase in the periodic table of elements.

briefly summarize fundamental properties and synthetic strategies, a sort of modification, derivation, and composition methods of MXene, and corresponding applications in batteries are then followed. Finally, we will develop our opinions on the challenges and prospects that remained in MXene materials.

16.2 FUNDAMENTALS FOR MXene

16.2.1 STRUCTURE AND STABILITY

Unlike traditional 2D materials, MXenes are mainly synthesized by selectively etching MAX phases, which inherited the close-packed hexagonal structure from MAX [10]. After removing the A atoms, these $M_{n+1}X_n$ ($n = 1-3$) layers are attached by $-O$, $-OH$, $-F$, and terminations. In general, MXene is arranged in an $(MX)_nM$ structure. Namely, M layers interleaved with X layers [11]. According to the parameter "n" values, the structure and stability of MXene will also change. The M_2X is AB stacking, while M_3X_2 and M_4X_3 have ABC arrangement, which could account for some of the MXenes that are not stable with high "n" indexes. For example, Mo_2C is more durable than both Mo_3C_2 and Mo_4C_3. The structure of MXene is relatively simple, but due to the wide variety of transition metals and the in-depth development of MXene, more and more MXenes containing two transition metals are fabricated (e.g., Mo_2TiC_2, TiVC, and $Mo_2Ti_2C_3$) [12–14]. At the same time, the M element also exists in two forms: ordered phases and solid solution, making MXene materials more diverse [15].

It has been confirmed that functional groups are closely related to the properties of MXene [16] and the most studied Ti_3C_2 is taken for example, theoretically, compared to other structures with $-OH$, $-F$, and $-O$ as the group, as $Ti_3C_2O_2$ is the most stable structure. The stable order of the functional group is followed: $Ti_3C_2O_2$ > $Ti_3C_2F_2$ > $Ti_3C_2(OH)_2$ > Ti_3C_2 [17]; this is because oxygen molecules are adsorbed on the surface of MXene and then dissociate to form $Ti_3C_2O_x$. Once the $Ti_3C_2O_2$ is generated, it will make repulsion for more oxygen according to the First-principles calculation [18]. However, in actual conditions, MXene materials are not stable but are very easy to oxidize. It is reported that the $Ti_3C_2T_x$ would quickly degrade; after exposure to air for 70 hours, the initial conductivity will decrease by about 20% [19].

16.2.2 ELECTRONIC STRUCTURE

The pristine MXene has good metallic conductivity. However, up to now, non-terminated MXenes are yet to be synthesized; the existence of terminations somewhat impacts the electronic property of MXene. For instance, the $-O$ terminations exhibit the lowest electrical conductivity compared with $-F$ and $-OH$, this is because the terminations can significantly reduce the density of states (DOS) as well as create a bandgap that could be as large as 0.194 eV (W_2CO_2) [20]. According to Berdiyorov's simulation on Ti_3C_2, the electron transmission almost decreases in most electron energies. Only in the deep of the valence band, a better transmission can be observed. At the same time, $-F$ and $-OH$ show similar effects on electron transmission; they have better transmission than bare MXene for almost all the electron energies and only a small range near the Fermi level [20]. Thus, the MXene electronic property mainly depends on the "M," "X" elements and type of terminations; once the "M" and "X" have been confirmed, it is reasonable to tune the MXene's surface chemistry by varying the type and quantity of terminations. Presently, adjusting the terminations mainly employs post-processing; since MXene is generally obtained by HF etching, its surface is covered by a large number of −F functional groups; however, the existence of −F will impede the transmission of lithium ions. Thus, researchers usually replace −F with −O and −OH, which have better lithium storage characteristics and stability [21–23]. During the process of neutralizing the pH of the MXene, part of the −F will be replaced by −O and −OH initially. Alkalization treatment can significantly increase the proportion of oxygen-containing groups (−OOH, −OH, and −O), and the SSA of MXene before and after treatment will be dramatically increased, resulting in higher specific capacitance, which is beneficial to the application of materials in the field of energy storage [4]. Naguib et al. [24] used tetrabutyl-ammonium hydroxide to intercalate multilayer MXene. After treatment, the content of the F element is significantly reduced, while the O element increased via energy dispersive spectroscopy analysis.

16.2.3 Synthesis Methods

As we mentioned before, the first synthesis of MXene was selectively etched by HF (Figure 16.2a) because of the differentials of bond strength between M-X and M-A. Compared to M-X, the M-A is weaker and can be easily broken [5]; thus, the basic idea of obtaining MXene is based on removing the weakly bonded layers of A atoms. Due to the high corrosion of HF, researchers attempt to use less aggressive reactants to replace HF. Wu et al. [25] explored a new way to generate HF in situ: as depicted in Figure 16.2b, by agitating a solution with hydrochloric acid (HCl) and fluoride

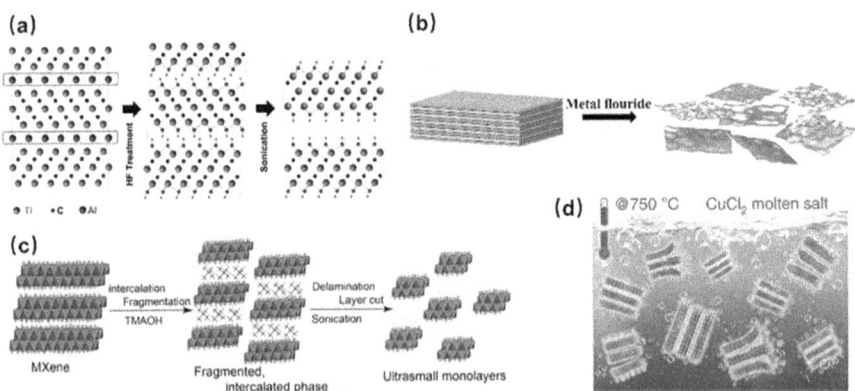

FIGURE 16.2 (a) Schematic of the exfoliation process for Ti_3AlC_2. (Adapted with permission from Ref. [5]. Copyright (2011) Wiley-VCH.) (b) Schematic of a one-step method to synthesize few-layered MXene. (c) Schematic illustration of the synthetic process for preparing ultrasmall Ti_3C_2 sheets. (Adapted with permission from Ref. [26]. Copyright (2017) American Chemical Society.) (d) Schematic of $Ti_3C_2T_x$ MXene preparation by immersing in $CuCl_2$ Lewis molten salt at 750°C. (Adapted with permission from Ref. [27]. Copyright (2020) Springer.)

salt (LiF, NaF, and KF), which reduce the risk of direct contact with HF, and at the same time, the alkali metal ions can act as an intercalant during the etching process, achieving a one-step method to obtain few-layered MXenes. Apart from that, this route can introduce fewer defects but get a high exfoliation yield, and such advantages have also made it a mainstream method that is still in use today. Nevertheless, this method still has HF, and there are many −F functional groups covering the MXene surface after etching, which is highly disadvantageous to lithium storage. Therefore, for the sake of safety and efficiency, more and more researchers have invested in exploring fluorine-reduced and fluorine-free synthetic methods.

Wang et al. [26] discovered that except for the corrosive HF, TMAOH could also extract Al layers in the base environment, and TMA ions can act as an intercalant to delaminate MXene (Figure 16.2c). It will further break the strong M–X bonds resulting in many ultrasmall sheets. Through this method, Wang and his coworkers have successfully synthesized Ti_3C_2, Nb_2C, and Ti_2C ultrasmall sheets; these products are 4 nm in lateral dimensions and 1 nm in thickness on average and have well-maintained host layer structure demonstrating a facile and general strategy to fabricate nanoscale materials. As MAX phases continue to increase, a specific MXene etching method may not suit another MXene. At the same time, the etching of the A atomic layer is no longer limited to Al atoms anymore. Some conventional etching methods may not achieve the expectation, inspiring people to explore etching methods constantly. Quite recently, Huang et al. [27] reported a general Lewis acidic etching route expected to realize large-scale production (Figure 16.2d). More importantly, the etching of the A-site element from MAX further expanded into Si, Ga, Zn, etc. Other than Al, which are difficult or even impossible to etch with conventional etching methods. Take Ti_3SiC_2, for example; according to the Gibbs free mapping, the Si^{4+}/Si couple has a redox potential as low as −1.38 V versus Cl_2/Cl. In comparison, the Cu^{2+}/Cu couple is −0.43 V indicating $CuCl_2$ molten salt can easily oxidize Si into Si^{4+}. At a specific temperature, the mechanism is analogous to that of chemical etching. Cu^{2+} and Cl serve as H^+ and F^-, respectively. But far from the latter, the Lewis acidic etching route has a better universality and safety, which offers a more diverse space to optimize such etching methodology. At the same time, choosing different molten salt can help tune the surface chemistry of MXene. Using Lewis acidic etchants can lead to the MXene materials with −Br and −I surface groups which have confirmed that these halogen terminations will improve energy density further [26]. There is no doubt that the general Lewis acid etching route drastically expands the range of MAX-phase precursors and provides more opportunities to customize the surface chemistry, which laid a solid foundation for the commercialization of MXene materials.

16.3   MXenes, THEIR DERIVATIVES, AND COMPOSITES FOR BATTERIES

As mentioned earlier, MXene material has a high SSA, good electrical conductivity, and appropriate interlayer spacing for cations to intercalate [27]. Furthermore, since the physicochemical properties of MXene are greatly affected by surface functional groups, it offers the possibility to modify its surface properties and arouse the researcher's significant interest in investigating its storage mechanism among different metal ions. Theoretical calculations show that the types of MXene surface functional groups and transition metal ions will directly affect the electrochemical performance of the material [14]. MXene containing oxygen functional groups has the highest specific capacity, while −F and −OH will cause the specific capacity decay [28]. At the same time, the exfoliated MXene often has a higher SSA, more active sites, and larger interlayer spacing, which is more conducive to ion transmission, intercalation/deintercalation, and electrolyte penetration. If MXenes are composed of other materials or self-derived methods to form 2D electrode materials, it will improve the electrochemical performance, thus having great potential in energy storage.

16.3.1 ALKALI METAL-ION BATTERIES (AMIBs)

In 2012, Naguib et al. [1] first used Ti_2CT_x MXene as a negative electrode material for LIBs; this is also the first time MXene material has been used as an energy storage material. Since then, MXene materials have demonstrated broad prospects in the energy storage field. Under the C/10 rate (~20 mA·g^{-1}), Ti_2CT_x could deliver a specific capacity of 150–200 mAh·g^{-1} with good retention, which is five times higher than pristine Ti_2AlC. Due to the wide variety of MXene, Naguib and his coworkers subsequently extended their work into Nb_2CT_x and V_2CT_x MXenes [29]; their study suggested that both Nb_2CT_x and V_2CT_x are also promising electrode materials for Li-ion batteries. At 1 C (1 C = 372 mA·g^{-1}), they could deliver reversible capacities of 170 and 260 mAh·g^{-1}, respectively; even at a high cycling rate (10 C), up to 110 and 125 mAh·g^{-1} can be obtained, implying MXene materials have great potential in high power applications. It was also found that M_2C structure MXenes generally have high theoretical gravimetric capacities [27] but experimentally confirmed that the actual specific capacity deviates from theoretical value. For example, Nb atoms are heavier than Ti atoms, whereas their particular capacity is higher than Ti_2CT_x. Except for the complicated feature of ion storage, the surface functional groups can partially shed light on it. According to the density functional theory (DFT) calculation, Xie et al. [18] discovered that compared to −F and −OH, the −O terminated MXenes have lower adsorption energies indicating a better interaction with Li. Therefore, maximizing the −O terminations could improve the Li capacity. Apart from tuning the surface functional groups, the simple and reasonable design of composite materials is also an important strategy to improve the performance of the batteries.

Sun et al. [30] prepared a large-size 2D composite (Co_3O_4 NCs@s-$Ti_3C_2T_x$) with Co_3O_4 nanocrystals (NCs) anchoring a monolayer of $Ti_3C_2T_x$ by simple freeze-drying combined with high-temperature annealing method (Figure 16.3a). By changing the synthesis conditions, the small size (<5 nm) of Co_3O_4 NCs can be uniformly dispersed on the $Ti_3C_2T_x$ MXene substrate (Figure 16.3b and c). For one thing, the uniformly distributed Co can minimize the stacking tendency of Ti, and for the other things, the functional groups on the surface of MXene can anchor and inhibit the growth and aggregation of Co. Benefiting from the flexible substrate of $Ti_3C_2T_x$ and small size structure of Co_3O_4 NCs, the internal stress caused by lithium-ion insertion/desertion is significantly reduced, which results in a good rate (~223 mAh·g^{-1} at a current density of 10 A·g^{-1}) and long-cycle performance (a reversible capacity of 550 mAh·g^{-1} at 1 A·g^{-1} after 700 cycles).

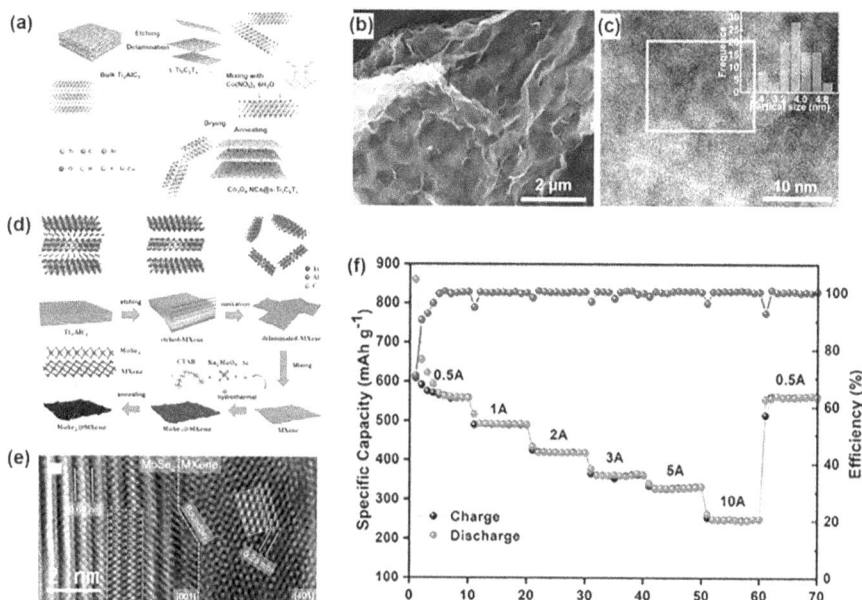

FIGURE 16.3 (a) Schematic illustration of the fabrication process of the Co_3O_4@s-$Ti_3C_2T_x$. (b) FESEM image of the Co_3O_4@s-$Ti_3C_2T_x$. (c) HRTEM image of the Co_3O_4@s-$Ti_3C_2T_x$ and lateral size distribution of the Co_3O_4 NCs inset in the panel. (Adapted with permission from Ref. [30]. Copyright (2019) The Royal Society of Chemistry.) (d) Schematic single-layer MXene and $MoSe_2$/MXene heterojunction. (e) Single-layer MXene and $MoSe_2$/MXene heterojunction. (f) Discharge–charge capacities of $MoSe_2$/MXene at different cycling current rates. (Adapted with permission from Ref. [32]. Copyright (2020) Elsevier.)

Besides, a large number of modifications have been proposed for MXene materials and applied to LIBs, resulting in a new level of performance. Whereas limited metal lithium resources and high costs urge people to explore new energy storage devices. Under these circumstances, SIBs and PIBs, which are similar to the operating principles of the "rocking-chair" mechanism of LIBs, gradually arouse people's interest. Na^+ and K^+ have a larger ion radius than Li^+, dramatically affecting the mass transport and electrochemical reaction. Such problems paralyze part of the anode materials used in LIBs. For example, graphite cannot perform very well as usual. On the contrary, benefiting from the laminar structure and inherent large interlayer spacing, MXene materials could accommodate these large cations. Both experimentally and theoretically the feasibility of functionalized MXene electrodes for sodium/potassium-ion storage, especially the M_2C structure MXenes with oxygen terminations were reported. In addition, a majority of experimental studies have confirmed that the MXene materials have an excellent sodium and potassium storage capacity. Xie et al. [31] tested the multilayer $Ti_3C_2T_x$ nanosheet electrochemical performance in SIBs and PIBs. The result showed that under 0.1 $A \cdot g^{-1}$, the initial discharge capacity of Na-ion intercalation and K-ion intercalation can reach 370 and 260 $mAh \cdot g^{-1}$, respectively, which is higher than the prediction. Noteworthy, the initial charge of Na-ion and K-ion intercalation will fall to 164 and 146 $mAh \cdot g^{-1}$, respectively, and the reason behind it may be attributed to some irreversible process, such as the formation of SEI, or reaction between Na^+ and K^+ with interlayer water or impurities. At the same time, the size of multilayer MXenes is generally more extensive, and the active sites exposed are more limited than that of delaminated MXenes.

Apart from the above reasons, the shortcomings of MXene being easy to aggregate and restack, the overshadowing effect of surface functional groups is likewise detrimental to MXene's application. Therefore, reducing the size of MXenes into nano size, compounding them with other materials, and derivation to form heterogeneous structures are practical solutions to improve electrochemical performance. Quite recently, Xu et al. [32] constructed a $MoSe_2$/MXene heterojunction by a simple hydrothermal method for solving the sluggish redox kinetics problem from SIBs (Figure 16.3d). $MoSe_2$, as a classic transition metal dichalcogenide, has an adjustable interlayer spacing favorable for sodium storage. Before the discovery of MXene, the common modification methods are compounding with graphene/carbon nanotubes (CNTs), Fe_3O_4, or TiO_2. In this work, $MoSe_2$ was successfully modified on the surface of MXene forming the heterojunction (Figure 16.3e). MXene as the matrix could improve the electron transport and enhance the structural stability of $MoSe_2$, while $MoSe_2$ with a large interlayer distance (0.64 nm) would resist huge stress/strain. Consequently, the volumetric change during the sodium ions insertion/extraction was restrained, and the reaction kinetics was greatly enhanced further. Even at a high current of 10 $A \cdot g^{-1}$, about 250 mAh g^{-1} can be obtained with a Coulombic efficiency (CE) of up to 99.8% (Figure 16.3f).

From the abovementioned literature and research progress, it is obvious that the electrochemical performance is controlled by several factors such as electrode structure, terminations, type of MXene, and interlayer spacing. A larger interlayer spacing or pillared structure generated by composite strategy is significantly beneficial to improving the cycle stability and energy density of SIBs. Recently, aiming for opening the interlayer spacing for long-term capacity, Zhao et al. [33] demonstrated a rapid, simpler, and effective flocculation strategy to fabricate porous 3D $Ti_3C_2T_x$ networks. Due to the fact that the interlayer spacing can be tuned by the intercalation of alkali ions, it is reported that by simply adding MOH solution (M refers to Na, Li, K, and tetrabutylammonium), a kind of foam-like, 2D crumpled flakes (denoted as M-c-$Ti_3C_2T_x$) can be easily fabricated. Since these actions will end up between the 2D sheets, the interlayer spacing is significantly expanded, resulting in excellent capacity retention, especially for the Li-c-$Ti_3C_2T_x$. When used as the SIB electrode, the capacity is ≈160 $mAh \cdot g^{-1}$ under 0.1 $A \cdot g^{-1}$ after 300 cycles. Also employing the large layer spacing to store sodium, Maughan et al. [34] applied amine-assisted silica pillaring to create the first example of a porous Mo_2TiC_2 (Figure 16.4a). The pillared Mo_2TiC_2 has a surface area of 202 $m^2 \cdot g^{-1}$, among the highest reported for any MXene, and has a variable gallery height between 0.7 and 3 nm. The expanded interlayer spacing leads to significantly enhanced cycling performance for Na-ion storage, with superior capacity, rate capability, and cycling stability compared to the non-pillared counterpart. As depicted in Figure 16.4b, the pillared Mo_2TiC_2 can deliver reversible capacities up to 109 $mAh \cdot g^{-1}$ at 20 $mA \cdot g^{-1}$, which is over twice that of the non-pillared.

For storing potassium ions with a larger ionic radius, Tian et al. [35] innovated facile one-step electrodeposition to grow hierarchical Sb, Sn, and Bi on MXene

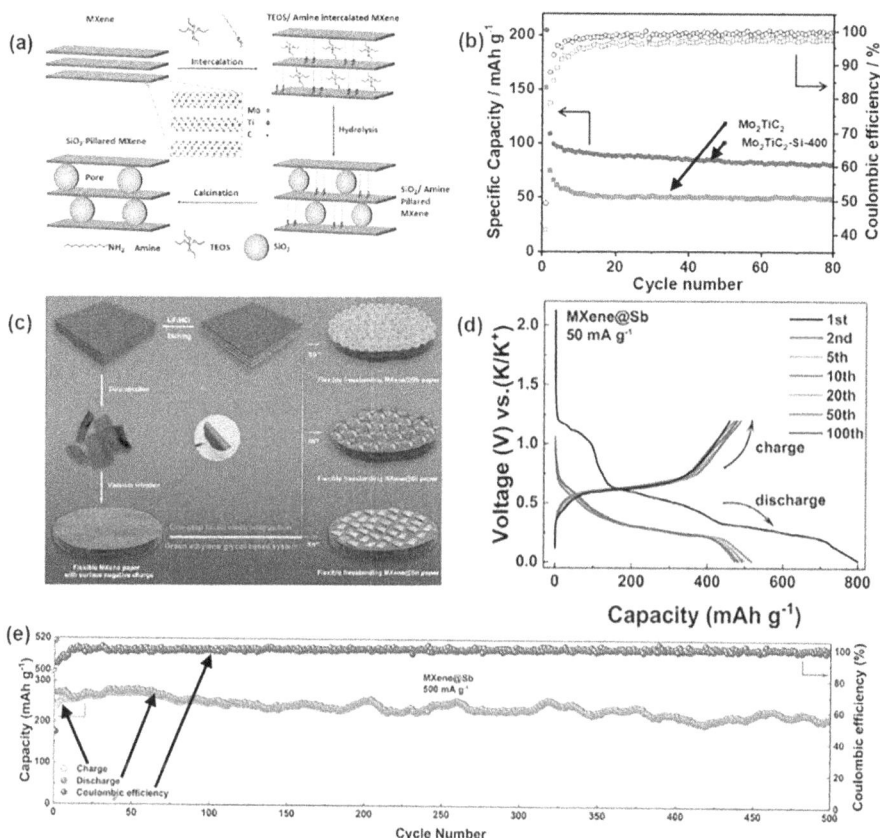

FIGURE 16.4 (a) Schematic representation of the amine-assisted silica pillaring method. (b) The comparison of cycling stability between Mo_2TiC_2 and Mo_2TiC_2-Si-400. (Adapted with permission from Ref. [34]. Copyright (2021) Copyright The Authors, some rights reserved; exclusive license [The Royal Society of Chemistry].) (c) A schematic illustration of the overall fabrication process for flexible free-standing MXene@Sb, MXene@Bi, and MXene@Sn paper. (d) Galvanostatic charge/discharge curves of MXene@Sb. (e) Long-term cycling capabilities of MXene@Sb. (Adapted with permission from Ref. [35]. Copyright (2019) The Royal Society of Chemistry.)

paper to solve volume expansion (Figure 16.4c). Take antimony (Sb), for example; it is abundant in the earth's crust and has the advantages of low price and environmental friendliness. Furthermore, Sb anodes can possess up to a high specific theoretical capacity of 660 mAh·g^{-1} in PIBs, which enable them to be a promising candidate anode material for potassium storage. Nevertheless, a considerable volume expansion (~300%) severely hinders its application. Before this work, Sb nanoparticles, 3D Sb NPs@C, Sb-C [36–38], etc., have been investigated to tackle that issue. Inspired by previous research and considering that most of the PIBs' electrodes are made by slurry-casting, Tian et al. attempted to fabricate a free-standing configuration with Sb involved to improve energy density and cut down cost. Benefiting from the outstanding conductivity, high SSA, and particular excellent flexibility, it has been reported that $Ti_3C_2T_x$ sheets are feasibly assembled into additive-free and flexible films. Naturally, combining the advantages of both, a hierarchical porous Sb on MXene paper (MXene@Sb) is generated. Based on the practical design, such MXene@Sb anodes successfully integrate the unique properties of MXene and hierarchical porous Sb. The porous Sb can provide a short diffusion distance for ion's fast transport. At the same time, the porous structure will offset partial volume expansion to a certain degree. Moreover, MXene films can serve as elastic collectors providing an electrical highway for electrons transport due to their high conductivity and flexibility, which contribute to a reversible capacity of 516.8 mAh·g^{-1} at 50 mA·g^{-1}, and the capacity fading rate of only 0.042% per cycle at 500 mA·g^{-1} (Figure 16.4d and e).

16.3.2 METAL-ION BATTERIES (MIBS)

For exploring one potential alternative with low cost, safe, and prosperous materials to replace lithium-based energy storage systems, some other metal anode materials such as zinc, magnesium, calcium, and aluminum have been intensively investigated. These materials can effectively reduce the cost of the electrochemical energy storage system and enable a higher theoretical capacity. Therefore, numerous academic studies were undergoing in these batteries. Some early studies generally believed that the kinetics of insertion of multivalent ions into host solid-state electrode is slower than that of univalent ions [39]. Meanwhile, the lower standard redox potential compared with alkali metal results in a lower energy density; hence, finding a suitable cathode material with high capacity and high voltage and a large tunnel structure for fast ion transport is also urgent. MXene-based nanomaterials with practical design inherit advantages of MXene 2D materials and solve some of the intrinsic defects of MXene itself, which have been confirmed as suitable cathode materials for hosting multivalent metal ions [40].

Take zinc-ion batteries (ZIBs), for example. At present, layered vanadium-based compounds play a predominant role in being the cathode in aqueous ZIBs. Therefore, vanadium-based MXene has a natural advantage in applying zinc-ion batteries with its layered structure, variable oxidation state, and tunable surface properties. Narayanasamy et al. [41] synthesized a morphology retained growth of layered V_2O_5 on V_2CT_x (denoted as $V_2O_5@V_2C$) (Figure 16.5a). By using hydrothermal method at an optimum temperature, V_2O_5 nanorods firmly embed internally/externally on the surface of V_2CT_x (Figure 16.5b). It turns out that such heterostructure would prevent the volume expansion and provide an adequate electroactive surface for Zn-ion intercalation. Besides, the phase stability and integrity of $V_2O_5@V_2C$ nanohybrid will bring synergism-enhanced kinetics and higher rate capacity. Even at a high current density of 4 A·g^{-1} for 2,000 cycles, 87% capacity can be retained, suggesting that the $V_2O_5@V_2C$ nanohybrid has excellent potential for Zn-ion storage.

Similarly, MXene-based nanomaterials also show great potential in magnesium-ion batteries (MIBs) and calcium-ion batteries (CIBs), which have the exact bivalent nature compared with ZIBs. It is reported that constructing MXene@C nanospheres, for example, $Ti_3C_2T_x@C$ cathode for magnesium-ion batteries, can achieve a remarkable capacity of 198.7 mAh·g^{-1} at 10 mA·g^{-1}; after 400 cycles, up to 85% capacity can be retained [5]. Furthermore, this strategy can be extended into $V_2CT_x@C$, delivering a capacity of 75.4 mAh·g^{-1} under 10 mA·g^{-1}. In contrast, the bare V_2C electrode can only reach 5.9 mAh·g^{-1}.

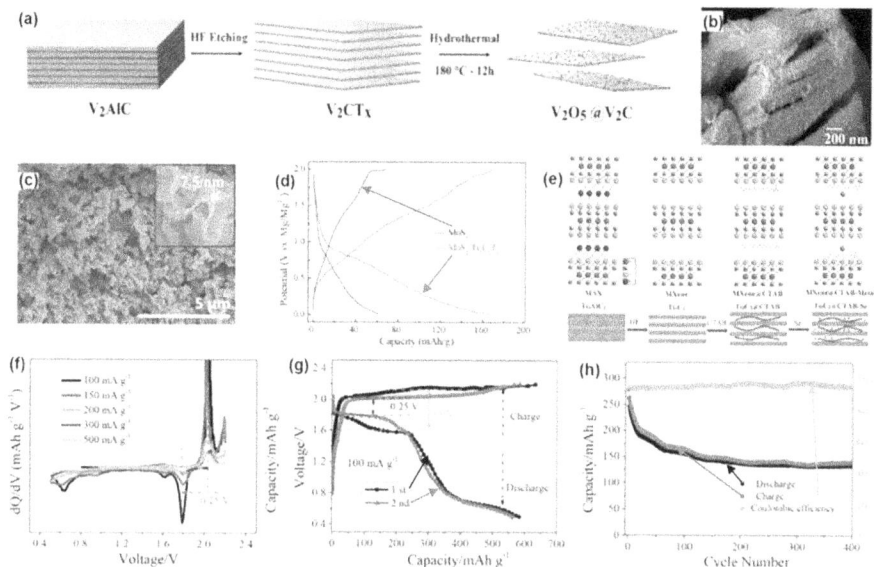

FIGURE 16.5 (a) Schematic representation of $V_2O_5@V_2C$ nanohybrid. (b) FESEM image of $V_2O_5@V_2C$ nanohybrid. (Adapted with permission from Ref. [41]. Copyright (2020) The Royal Society of Chemistry.) (c) FESEM image of $MoS_2/Ti_3C_2T_x$ composite. (d) Charge–discharge curves of magnesium batteries utilizing MoS_2 and $MoS_2/Ti_3C_2T_x$ electrodes. (Adapted with permission from Ref. [42]. Copyright (2020) The Royal Society of Chemistry.) (e) Schematic diagram of the preparation process of $Ti_3C_2@CTAB-Se$. (f) dQ/dV differential curves of $Ti_3C_2@CTAB-Se$ at different current densities. (g) The first two charge–discharge curves and (h) cycling performance of $Ti_3C_2@CTAB-Se$. (Adapted with permission from Ref. [44]. Copyright (2020) Elsevier.)

Xu et al. [42] demonstrated a kind of composites in which MoS_2 nanosheets are vertically loaded on $Ti_3C_2T_x$ nanosheets (Figure 16.5c). MoS_2, as one of the cathode materials in MIBs, has been proved to be a suitable host for the storage and diffusion of Mg^{2+}.

In contrast, the reported MoS_2-based cathodes generally suffer from either moderate capacity or poor cycling stability. Using MXene as the conductive substrate can effectively improve the electrical conductivity of MoS_2 for high capacities utilization. Consequently, a total of 165 mAh·g^{-1} can be obtained at 50 mA·g^{-1}, which is better than the pure MoS_2 electrode (Figure 16.5d).

Same as MIBs, some reported chalcogenides-based or metal oxides cathodes could not well withstand the fast insertion and desertion of Ca^{2+} without structural collapse. 2D MXene nanosheets possess large interlayer spacing and high conductivity, expecting to break the dilemma of CIBs. It was predicted that V_3C_2 monolayers have excellent calcium storage performance owing to their low diffusion barrier (0.04 eV for Ca), according to DFT calculations [43]. The discharge–charge rate significantly depends on the mobility of the intercalating ions. By simulating three possible diffusion paths (denoted path 1, path 2, and path 3, respectively), path 2 is proven to have the lowest diffusion barrier. Notably, in all three possible paths, the diffusion barriers are far smaller than the typical 2D materials such as MoN_2, graphene, and phosphorene.

In a pioneer study, V_2CT_x MXene, as the cathode material for aluminum-ion batteries (AIBs), exhibited an excellent reversible discharge specific capacity (over 300 mA g^{-1}); regretfully, its capacity would rapidly decrease below 100 mAh·g^{-1} [4]. Inspired by the result mentioned above, Li et al. [44] chose to optimize MXene materials based on $Ti_3C_2T_x$. As the most extensively studied MXene, it has been used in the fields of various batteries and capacitors, benefiting from its large interlayer spacing (0.977 nm). To further improve its rate and cycle performance, CTAB was first used to expand the interlayer structure, subsequently selenization to form a composite (Ti_3C_2@CTAB-Se) (Figure 16.5e), which will help reduce the surface functional groups of Ti_3C_2, further improving the CE. More importantly, according to the dQ/dV differential curves (Figure 16.5f), there is a prominent peak in 1.8 V, which is contributed to the fact that Se will participate in electrochemical reactions. Consequently, Ti_3C_2@CTAB-Se exhibited excellent electrochemical performance. At 100 mA·g^{-1}, its CE can reach up to 92.15%, which is two times that of Ti_3C_2, and it showed a reversible discharge specific capacity of 583.7 mAh·g^{-1}, and after 400 cycles, 132.6 mAh·g^{-1} is still retained under a current of 200 mA·g^{-1} (Figure 16.5g and h).

16.3.3 METAL-SULFUR BATTERIES (MSBS)

In 2015, Nazar et al. [45] first used MXene materials (Ti_2C) as the cathode for lithium–sulfur batteries (LSBs); the result turned out that the high underlying metallic conductivity and the strong interaction between the polysulfide species and Ti atoms will bring a good specific capacity. It is well known that the shuttle effect of LSBs results in poor long-term and rate performance. The application of MXene provides a new chance to solve that issue. As reported by Lv et al., a kind of nanotube bridged hierarchical Mo_2C-based MXene nanosheets are fabricated for LSBs (Figure 16.6a) [46]. In this work, hydroxyl-functionalized Mo_2C-based MXene nanosheets were first synthesized by etching the Mo_2SnC (Figure 16.6b), and these hydroxyl functional groups can play an important role in anchoring the polysulfides (LiPSs). Meanwhile, the Mo atoms on the surface of MXene also have strong interaction with LiPSs, which all benefit to suppress the shuttle effect. Due to the fact that Mo_2C MXene has a relatively low electronic conductivity, CNTs were then introduced, and this simple and reasonable composite can realize high sulfur loading. Even at a high areal loading up to 5.6 mg·cm^{-2}, an initial reversible capacity of 959 mAh·g^{-1} can be obtained at 0.1 C (Figure 16.6c).

Not only that, MXene's great potential in energy storage can even be extended to the more difficult room temperature sodium–sulfur batteries (NSBs). Compared with LSBs, sodium metal electrode has the advantages of natural richness and low cost. Still, the much more severe shuttle effect results in low reversible capacity, self-discharging, and poor cyclability. An ideal NSBs cathode should possess strong chemisorption properties and sufficient SSA to encapsulate sulfur. Because of MXene-based nanomaterials that have been extensively used in LSBs, it is expected that the solid chemical affinity of MXene can also work in NSBs. Based on a previous report, Bao et al. [7] innovatively utilize A-site doping to modify MXene properties; using a mixture of Al and S during the synthesis of MAX, they achieved in situ sulfur doping MAX structure ($Ti_3AlC_2S_x$) with a minor change in lattice parameters

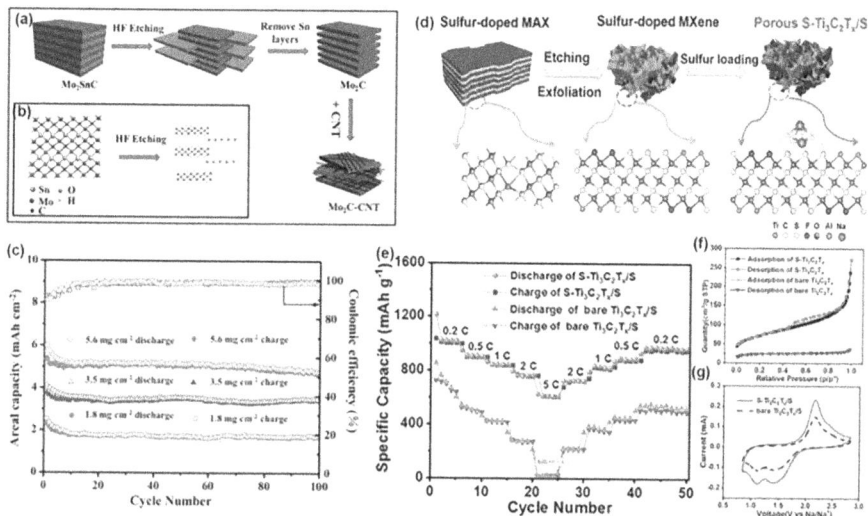

FIGURE 16.6 Schematic display of (a) the growth routine of Mo$_2$C-CNT composites and (b) the formation process of Mo$_2$C. (c) Cycling performance of Mo$_2$C-CNT/S cells with various S loading amounts of 1.8, 3.5, and 5.6 mg cm^{-2} at 0.1 C. (Adapted with permission from Ref. [46]. Copyright (2018) Wiley-VCH.) (d) Schematic illustration of synthesis of sulfur-doped Ti$_3$C$_2$T$_x$ MXene nanosheets. (e) Nitrogen adsorption−desorption curve of the S−Ti$_3$C$_2$T$_x$ and bare Ti$_3$C$_2$T$_x$ sample. (f) Comparison of rate capability between S−Ti$_3$C$_2$T$_x$/S and bare Ti$_3$C$_2$T$_x$/S. (g) Cyclic voltammograms of Na−S cells with S−Ti$_3$C$_2$T$_x$/S and Ti$_3$C$_2$T$_x$/S cathodes. (Adapted with permission from Ref. [7]. Copyright (2019) Copyright The Authors, some rights reserved; exclusive license [American Chemical Society].)

(Figure 16.6d). After the etching and freeze-drying process, the S−Ti$_3$C$_2$T$_x$ has a bigger interlayer spacing due to the solidification and sublimation of water, which leads to a wrinkled structure. The nitrogen adsorption−desorption isotherms indicate that the SSA and the pore volume of the S−Ti$_3$C$_2$T$_x$ are up to 258.1 m^2·g^{-1} and 0.523 cm^3; in contrast, the bare Ti$_3$C$_2$T$_x$ only has 38 m^2·g^{-1} surface area and 0.008 cm^3·g^{-1} pore volume (Figure 16.6e). Furthermore, the electrochemical performance turns out that the S−Ti$_3$C$_2$T$_x$/S electrode exhibited a discharge capacity of 610.3 mAh·g^{-1} at 5C. In contrast, the bare Ti$_3$C$_2$T$_x$/S electrode could not withstand such a high current rate dropping to almost zero (Figure 16.6f), which can be boiled down to the existence of S−Ti bonds in S−Ti$_3$C$_2$T$_x$ sheets. The high polarity caused by the S−Ti bonds will accelerate the transfer of sodium polysulfides and restrict the diffusion of sodium polysulfides. It also agrees well with the cyclic voltammetry (CV) (Figure 16.6g), confirming that the S-Ti$_3$C$_2$T$_x$/S sulfur cathode has good dynamic characteristics.

16.4 SUMMARY AND OUTLOOK

Since the first synthesis to the extensive application in various energy storage devices, MXene materials have experienced nearly explosive growth in less than a decade due to their outstanding properties, including high thermal and electrical conductivity, hydrophilic nature, high thermal conductivity SSA, and compositional variability.

More importantly, as a large and emerging family of 2D materials, apart from dozens of typical mono-M elements MXenes, latent MXenes with two and more transition metal elements are waiting to be explored further. Let alone, there are numerous derivative and composite methods to customize it with some specific properties. Based on that, we elaborate on the progress of MXene-based nanomaterials in the field of energy storage in recent years, like AMIBs, MIBs, and MSBs. Extensive studies have shown considerable potential in MXenes; furthermore, we are still far from realizing the full performance of the MXene materials.

It is essential to note that many issues need to be addressed for their diverse application:

(1) There is an urgent need to find an environmentally friendly, efficient, and easily scalable synthesis method; at present, the dominant etching method is still based on fluorine-containing top-down preparation which is highly risky; meanwhile, the strong corrosion will introduce a large number of defects and the MXene surface will be covered by a large number of −F functional groups, inducing the performance of MXene far from the theoretical value. Although efforts are also being made to explore other green and efficient synthesis methods, for example, CVD, which has been used to fabricate graphene and transition metal chalcogenides, can synthesize large-area and high-quality MXenes; what's more, this method can even get MXenes with MC structure such as WC, MoC, and NbC. Whereas, the severe requirements for sophisticated devices, high temperature and pressure reaction conditions, as well as slow deposition speed hinder its large-scale commercial preparation in the future. Since the preparation method significantly affects the properties of MXene, finding an optimum methodology from the source would be highly beneficial to the development of MXene.

(2) Ti_3C_2 MXene is still the extensively studied hot spot. It is undeniable that the mild synthesis conditions and excellent performance of Ti_3C_2 are still in the lead among many MXene materials. With the continuous exploration of MXene materials, such as V_2C, Nb_2C, and Mo_2C, it has also attracted many researchers' attention. However, compared to the enormous MXene family, our exploration is still relatively rudimentary.

(3) The cost of MXene is relatively high; it is crucial to find solutions that combine reasonable cost and exceptional performance. In other words, capacity, lifetime, cost and safety, etc., should be taken well into consideration.

(4) Rational surface chemical modulation is one of the critical factors affecting the performance of MXene materials. As mentioned before, various functional groups usually cover the etched MXene, which significantly impairs its stability, electronic structure, and other physicochemical properties. For example, researchers typically choose to remove the −F and −OH functional groups that hinder lithium-ion transport for making electrode materials with high lithium storage and high-rate performance. The halogen groups introduced by Lewis acidic etching route can expand the interlayer spacing and even increase energy density and power density. Nevertheless, it is still hard to realize the precise control of surface functional groups at the present stage.

(5) The chemical stability of MXene material is not good, and it is straightforward to oxidize when exposed to the air environment, especially the aqueous solution. Even if researchers have taken the strategy of storing it in low temperature and oxygen-free conditions, its degradation rate is just slowed down. Whether for large-scale preparation in the laboratory or future commercial production, how to prolong the storage time of MXene is also an urgent problem to be solved.

In summary, although MXene materials have made massive progress in energy storage, MXene-based nanomaterials opened a new pathway to the advancement in various energy systems, our current research is still in its infancy. With the introduction of subgroup elements into the MAX phase, the definition of MXene will be further expanded in the future, which brings unlimited potential and more challenges. To balance the co-existing challenges and opportunities, novel methods and techniques are highly needed.

REFERENCES

1. Naguib M, Come J, Dyatkin B, Presser V, Taberna P-L, Simon P, Barsoum MW, Gogosti Y (2012) MXene: A promising transition metal carbide anode for lithium-ion batteries. *Electrochem. Commun.* 16:61–64.
2. Xie Y, Dall'Agnese Y, Naguib M, Gogosti Y, Barsoum MW, Zhuang HL, Kent PRC (2014) Prediction and characterization of MXene nanosheet anodes for non-lithium-ion batteries. *ACS Nano.* 8:9606–9615.
3. Zhao R, Di H, Hui X, Zhao D, Wang R, Wang C, Yin L (2020) Self-assembled Ti3C2 MXene and N-rich porous carbon hybrids as superior anodes for high-performance potassium-ion batteries. *Energy. Environ. Sci.* 13:246–257.
4. Yoo HD, Liang Y, Dong H, Lin J, Wang H, Liu Y, Ma L, Wu T, Li Y, Ru Q, Jing Y, An Q, Zhou W, Guo J, Lu J, Pantelides ST, Qian X, Yao Y (2017) Fast kinetics of magnesium monochloride cations in interlayer-expanded titanium disulfide for magnesium rechargeable batteries. *Nat. Commun.* 8:339.
5. Xu M, Lei S, Qi J, Dou Q, Liu L, Lu Y, Huang Q, Shi S, Yan, X (2018) Opening magnesium storage capability of two-dimensional MXene by intercalation of cationic surfactant. *ACS Nano.* 12:3733–3740.
6. VahidMohammadi A, Hadjikhani A, Shahbazmohamadi S, Beidaghi M (2017) Two-dimensional vanadium carbide (MXene) as a high-capacity cathode material for rechargeable aluminum batteries. *ACS Nano.* 11:11135–11144.
7. Bao W, Shuck CE, Zhang W, Guo X, Gogosti Y, Wang G (2019) Boosting performance of Na-S batteries using sulfur-doped $Ti_3C_2T_x$ MXene nanosheets with a strong affinity to sodium polysulfides. *ACS Nano.* 13:11500–11509.
8. Fashandi H, Dahlqvist M, Lu J, Palisaitis J, Simak SI, Abrikosov IA, Rosen J, Hultman L, Andersson M, Spetz AL, Eklund P (2017) Synthesis of Ti_3AuC_2, $Ti_3Au_2C_2$ and Ti_3IrC_2 by noble metal substitution reaction in Ti_3SiC_2 for high-temperature-stable Ohmic contacts to SiC. *Nat. Mater.* 16:814–818.
9. Hart JL, Hantanasirisakul K, Lang AC, Anasori B, Pinto D, Pivak Y, Omme T, May SJ, Gogosti Y, Taheri ML (2019) Control of MXenes' electronic properties through termination and intercalation. *Nat. Commun.* 10:522.
10. Tang X, Guo X, Wu W, Wang G (2018) 2D metal carbides and nitrides (MXenes) as high-performance electrode materials for lithium-based batteries. *Adv. Energy. Mater.* 8:1801897.

11. Zheng L, Wang J, Lu X, Li F, Wang J, Zhou Y (2010) ($Ti_{0.5}Nb_{0.5})_5AlC_4$: A new-layered compound belonging to MAX phases. *J Am Ceram Soc.* 93:3068–3071.

12. Tan TL, Jin HM, Sullivan MB, Anasori B, Gogotsi Y (2017) High-throughput survey of ordering configurations in MXene alloys across compositions and temperatures. *ACS Nano.* 11:4407–4418.

13. Anasori B, Xie Y, Beidaghi M, Lu J, Hosler BC, Hultman L, Kent PRC, Gogotsi Y, Barsoum MW. (2015) Two-dimensional, ordered, double transition metals carbides (MXenes). *ACS Nano.* 9:9507–9516.

14. Cheng Y-W, Dai J-H, Zhang Y-M, Song Y (2018) Two-dimensional, ordered, double transition metal carbides (MXenes): A new family of promising catalysts for the hydrogen evolution reaction. *J. Phys. Chem. C.* 122:28113–28122.

15. Persson I, El Ghazaly A, Tao Q, Halim J, Kota S, Darakchieva V, Palisatis J, Barsoum MW, Rosen J, Persson POÅ (2018) Tailoring structure, composition, and energy storage properties of MXenes from selective etching of in-plane, chemically ordered MAX phases. *Small* 14:1703676.

16. Srivastava P, Mishra A, Mizuseki H, Lee KR, Singh AK (2016) Mechanistic insight into the chemical exfoliation and functionalization of Ti_3C_2 MXene. *ACS Appl. Mater. Interfaces.* 8:24256–24264.

17. Wang X, Shen X, Gao Y, Wang Z, Yu R, Chen L (2015) Atomic-scale recognition of surface structure and intercalation mechanism of Ti_3C_2X. *J. Am. Chem. Soc.* 137:2715–2721.

18. Xie Y, Naguib M, Mochalin VN, Barsoum MW, Gogotsi Y, Yu X, Nam K-W, Yang X-Q, Kolesnikov A, Kent PRC (2014) Role of surface structure on Li-ion energy storage capacity of two-dimensional transition-metal carbides. *J. Am. Chem. Soc.* 136:6385–6394.

19. Lipatov A, Alhabeb M, Lukatskaya MR, Boson A, Gogotsi Y, Sinitskii A (2016) Effect of synthesis on quality, electronic properties and environmental stability of individual monolayer Ti_3C_2 MXene flakes. *Adv. Electron. Mater.* 2:1600255.

20. Berdiyorov GR (2015) Effect of surface functionalization on the electronic transport properties of Ti_3C_2 MXene. *EPL (Europhys. Lett.)* 111:67002.

21. Tang Q, Zhou Z, Shen P (2012) Are MXenes promising anode materials for Li ion batteries? Computational studies on electronic properties and Li storage capability of Ti_3C_2 and $Ti_3C_2X_2$ (X = F, OH) monolayer. *J. Am. Chem. Soc.* 134:16909–16916.

22. Zang X, Wang J, Qin Y, Wang T, He C, Shao Q, Zhu H, Cao N (2020) Enhancing capacitance performance of Ti3C2Tx MXene as electrode materials of supercapacitor: From controlled preparation to composite structure construction. *Nanomicro. Lett.* 12:77.

23. Li G, Tan L, Zhang Y, Wu B, Li L (2017) Highly efficiently delaminated single-layered MXene nanosheets with large lateral size. *Langmuir* 33:9000–9006.

24. Naguib M, Unocic RR, Armstrong BL, Nanda J (2015) Large-scale delamination of multi-layers transition metal carbides and carbonitrides "MXenes". *Dalton. Trans.* 44:9353–9358.

25. Wu M, Wang B, Hu Q, Wang L, Zhou A (2018) The synthesis process and thermal stability of V_2C MXene. *Mater (Basel).* 11:2112.

26. Wang Z, Xuan J, Zhao Z, Li Q, Geng F (2017) Versatile cutting method for producing fluorescent ultrasmall MXene sheets. *ACS Nano.* 11:11559–11565.

27. Li Y, Shao H, Lin Z, Lu J, Liu L, Duployer B, Persson OÅ, Eklund P, Hultman L, Li M, Chen K, Zha X-H, Du S, Rozier P, Chai Z, Raymundo-Piñero E, Taberna P-L, Simon P, Huang Q (2020) A general Lewis acidic etching route for preparing MXenes with enhanced electrochemical performance in non-aqueous electrolyte. *Nat. Mater.* 19: 894–899.

28. Wang Z, Xu Z, Huang H, Chu X, Xie Y, Xiong D, Yan C, Zhao H, Zang H, Yang W (2020) Unraveling and regulating self-discharge behavior of $Ti_3C_2T_x$ MXene-based supercapacitors. *ACS Nano.* 14: 4916–4924.

29. Naguib M, Halim J, Lu J, Cook KM, Hultman L, Gogotsi Y, Barsoum MW (2013) New two-dimensional niobium and vanadium carbides as promising materials for Li-ion batteries. *J. Am. Chem. Soc.* 135:15966–15969.

30. Sun X, Tan K, Liu Y, Zhang J, Denis DK, Zaman FU, H L, Yuan C (2019) A two-dimensional assembly of ultrafine cobalt oxide nanocrystallites anchored on single-layer $Ti_3C_2T_x$ nanosheets with enhanced lithium storage for Li-ion batteries. *Nanoscale* 11:16755–16766.
31. Xie Y, Xiong X, Han K (2021) Organic molecule intercalated multilayer $Ti_3C_2T_x$ MXene with enhanced electrochemical lithium and sodium storage. *Ionics* 27:3373–3382.
32. Xu E, Zhang Y, Wang H, Zhu Z, Quan J, Chang Y, Li P, Yu D, Jiang Y (2020) Ultrafast kinetics net electrode assembled via $MoSe_2$/MXene heterojunction for high-performance sodium-ion batteries. *Chem. Eng. J.* 385:123839.
33. Zhao D, Clites M, Ying G, Kota S, Wang J, Natu V, Wang V, Pomerantseva E, Cao M, Barsoum MW (2018) Alkali-induced crumpling of $Ti_3C_2T_x$ (MXene) to form 3D porous networks for sodium ion storage. *Chem. Commun (Camb).* 54:4533–4536.
34. Maughan PA, Bouscarrat L, Seymour VR, Shao S, Haigh SJ, Dawson R, Tapia-Ruiz N, Bimbo N (2021) Pillared Mo_2TiC_2 MXene for high-power and long-life lithium and sodium-ion batteries. *Nanoscale. Adv.* 3:3145–3158.
35. Tian Y, An Y, Xiong S, Feng J, Qian Y (2019) A general method for constructing robust, flexible and freestanding MXene@metal anodes for high-performance potassium-ion batteries. *J. Mater. Chem. A. Mater.* 7:9716–9725.
36. McCulloch WD, Ren X, Yu M, Huang Z, Wu Y (2015) Potassium-ion oxygen battery based on a high capacity antimony anode. *ACS Appl. Mater. Interfaces* 7:26158–26166.
37. Liu Q, Fan L, Ma R, Chen S, Yu X, Yang H, Xie Y, Han X, Lu B (2018) Super long-life potassium-ion batteries based on an antimony@carbon composite anode. *Chem. Commun (Camb).* 54:11773–11776.
38. Han C, Han K, Wang X, Wang C, Li Q, Meng J, Xu X, He Q, Luo W, Wu L, Mai L (2018) Three-dimensional carbon network confined antimony nanoparticle anodes for high-capacity K-ion batteries. *Nanoscale* 10:6820–6826.
39. Xu C, Chen Y, Shi S, Li J, Kang F, Su D (2015) Secondary batteries with multivalent ions for energy storage. *Sci Rep.* 5:14120.
40. Dong Y, Shi H, Wu ZS (2020) Recent advances and promise of MXene-based nano-structures for high-performance metal ion batteries. *Adv. Funct. Mater.* 30: 2000706.
41. Narayanasamy M, Kirubasankar B, Shi M, Velayutham S, Wang B, Angaiah S, Yan C (2020) Morphology restrained growth of V_2O_5 by the oxidation of V-MXenes as a fast diffusion controlled cathode material for aqueous zinc ion batteries. *Chem. Commun (Camb).* 56:6412–6415.
42. Xu M, Bai N, Li H-X, Hu C, Qi J, Yan X-B (2018) Synthesis of MXene-supported layered MoS_2 with enhanced electrochemical performance for Mg batteries. *Chin. Chem. Lett.* 29:1313–1316.
43. Fan K, Ying Y, Li X, Luo X, Huang H (2019) Theoretical investigation of V_3C_2 MXene as prospective high-capacity anode material for metal-ion (Li, Na, K, and Ca) batteries. *J. Phys. Chem C.* 123:18207–18214.
44. Li Z, Wang X, Zhang W, Yang S (2020) Two-dimensional Ti_3C_2@CTAB-Se (MXene) composite cathode material for high-performance rechargeable aluminum batteries. *Chem. Eng. J.* 398:125679.
45. Liang X, Garsuch A, Nazar LF (2015) Sulfur cathodes based on conductive MXene nanosheets for high-performance lithium-sulfur batteries. *Angewandte. Chemie.* 127:3979–3983.
46. Lv LP, Guo CF, Sun W, Wang Y (2019) Strong surface-bound sulfur in carbon nano-tube bridged hierarchical Mo_2C-based MXene nanosheets for lithium-sulfur batteries. *Small* 15:e1804338.

17 Vanadium Dichalcogenides-Based 2D Nanomaterials for Batteries

Zeeshan Tariq
Minzu University of China
Chinese Academy of Sciences

Sajid Ur Rehman
Minzu University of China

Faheem K. Butt
University of Education Lahore

Chuanbo Li
Minzu University of China

CONTENTS

17.1 INTRODUCTION

Energy in any form is an indispensable commodity worldwide. It is one of the crucial and widely used ingredients for global development [1]. One of the most significant tasks in recent decades is to overwhelm the escalating energy demand. Technological revolutions and inventions of new compact electronic devices (laptops, camcorders, and cell phones) are driving the demand for energy [2]. Also, electric cars and hybrid electric vehicles use energy storage and conversion devices, and therefore, enhancing the requirement of energy [3]. Mostly, fossil fuels (oil, coal, and natural gas) are being used to generate energy for the needs of mankind [4]. These fossil fuels have

DOI: 10.1201/9781003178422-17

limited deposits, and with constant use, they will diminish one day. Also, these fossil fuel resources (oil and coal) are doing considerable harm to our environment by emitting toxic gases into the atmosphere during their burning [5–7]. Renewable energy resources that are environmentally benign, in recent decades, earned huge attention [3]. In this situation, batteries [8–10], supercapacitors [7,11], and fuel cells [12,13] can function as efficient and capable renewable energy storage systems [14]. There is a pressing demand for potential materials which can be used in energy storage systems for improved and efficient storage and conversion of energy.

Like graphite, two-dimensional (2D) transition metal chalcogenides have distinctive monolayered and multilayered structures possessing greater interlayer spacing unlike graphite [15]. Also, 2D-transition metal dichalcogenides (TMDs) are scientifically attractive and have industrial importance. These 2D TMDs possess distinctive structural characteristics and noteworthy properties like electronics, optoelectronics, optical, mechanical, thermal, energy storage, and superconducting properties [16–22]. In layered metal chalcogenides, MX_2 (M = Mo, W, Sn and X= S, Se), layers are held together with van der Waals forces [23,24], and by using exfoliation, monolayers and nanolayers can be obtained [25]. Among many metal chalcogenides, vanadium-based dichalcogenides are receiving the interest of researchers since vanadium shows many valence states during electrochemical reactions. Among these layered metal chalcogenides, VSe_2 has the advantage of being lighter over other metal chalcogenides like $MoSe_2$, WSe_2, and $SnSe_2$. Also, the literature shows that VSe_2 has ultrathin nanosheets that are metallic. These two advantages make VSe_2 a suitable candidate for electrode material in rechargeable batteries [23]. Also, Vs_2 and VTe_2 and their composites such as carbon, graphene, blue phosphorene, and titanium are studied for different batteries like Li-ion, Na-ion, Mg-ion, K-ion, and Zn-ion batteries. In this chapter, we have given an overview (computational and experimental) of 2D vanadium dichalcogenides in the abovementioned batteries' applications.

17.2 APPLICATIONS IN BATTERIES

Vanadium dichalcogenide-based 2D nanomaterials are being used in various metal ion batteries as electrode material. The following figure (Figure 17.1) is showing the applications of vanadium dichalcogenide-based 2D nanomaterials in rechargeable batteries. Here, we will discuss the computational [density functional theory (DFT)-based] and experimental applications of vanadium-based 2D nanomaterials as an electrode in different rechargeable batteries.

17.2.1 COMPUTATIONAL STUDY

Jing et al. [26] used DFT computations to study the performance of VS_2 monolayer as a promising anode material for lithium-ion batteries (LIBs). Jing and coauthors implemented generalized gradient approximation (GGA) by using the DMol code. Accurate measurements for electrostatic interactions (long-range) between lithium atoms in high concentrations are carried out by using Perdew–Burke–Ernzerhof (PBE) + D_2 method and by taking into account grimme vdW correction. A k-point set of $4 \times 4 \times 1$ is used for the integration of Brillouin zone. They systematically explored the adsorption and

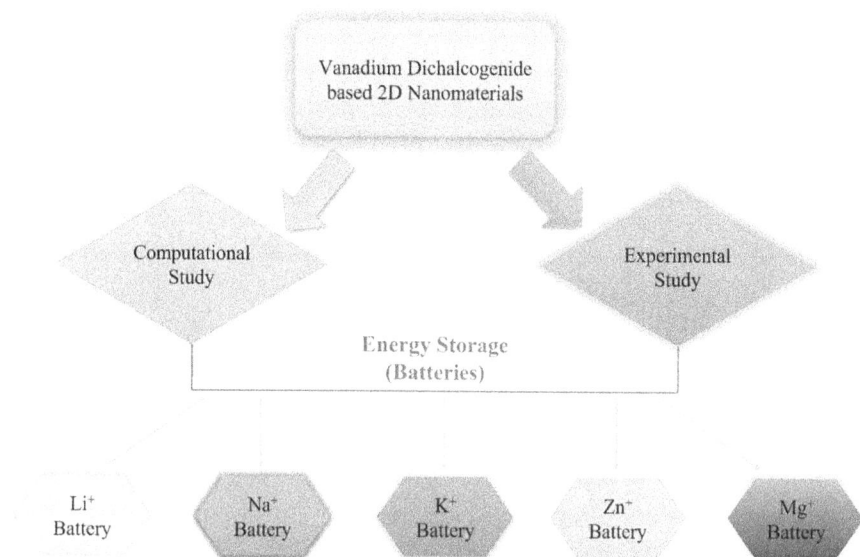

FIGURE 17.1 Displaying application of vanadium dichalcogenide-based 2D nanomaterials for energy storage (batteries).

diffusion of lithium ions for VS_2 monolayer and compared it with MoS_2 and graphite. Figure 17.2a (inset) is displaying the optimized VS_2 monolayer and the two stable sites for lithium adsorption, i.e., hollow site H and top site T on the monolayer.

The binding energy of lithium is calculated by using the following expression:

$$E_b = E_{VS_2\text{-}Li} - E_{VS_2} - \mu_{Li} \qquad (17.1)$$

Binding energy of adsorbed lithium at sites T and H is -2.13 and $-2.01\,eV$, respectively, where $E_{VS_2\text{-}Li}$ and E_{VS_2} are the total energies of adsorbed lithium on the surface of VS_2 and VS_2 monolayers, respectively, and μ_{Li} is the chemical potential of Li. In comparison to the VS_2 monolayer, binding energy of adsorbed lithium on the MoS_2 monolayer is $-0.6\,eV$ indicating that the lithium atom is more stable on VS_2 monolayers as compared to MoS_2. In LIBs, the rate performance is calculated by determining the mobility of the Li^+, so it is important to measure the diffusion of lithium atoms on VS_2 monolayer surface. Figure 17.2a is illustrating the pathways of the diffusion of lithium on the surface of VS_2 monolayer during its motion from T_1 to T_2 site via H site.

It can be seen (Figure 17.2a) that Li is only required to overcome a little energy barrier of $0.22\,eV$ while moving between two T sites. In comparison, the value of this barrier is $0.25\,eV$ for MoS_2 when calculated at the identical parameters. Diffusion constant (D) is directly proportional to $\exp(-E_a / k_BT)$, where E_a and k_B represent the activation energy and Boltzmann constant, respectively. At room temperature, it is estimated that on the surface of VS_2 monolayer, the mobility of Li is three times faster than that on MoS_2. Jing et al. also calculated the theoretical capacity of VS_2

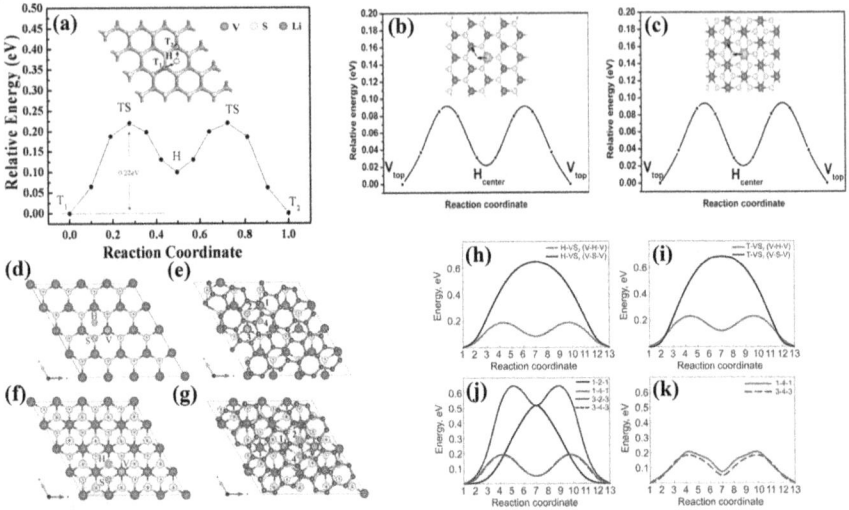

FIGURE 17.2 Depicting (a) pathways for the diffusion of lithium (T_1-H-T_2) along with the barrier of lithium motion from T to H on the VS_2 monolayer surface (the inset is lithium adsorbed VS_2 monolayer); (b) 1H and (c) 1T VS_2 monolayers. Lithium atom sites: (d) H-VS_2 and (e) T-VS_2 on VS_2 surface; (f) H-VS_2 and (g) T-VS_2 in between VS_2 and graphene. Transition barriers: (h) H-VS_2 and (i) T-VS_2 monolayer; (j) H-VS_2 and (k) T-VS_2 in the space of interlayers of graphene and VS_2. (Adapted with permission from Refs. [26], [27], and [28]. Copyright (2013) (2017) and (2016), American Chemical Society.)

monolayer which is 466 mAh g^{-1}, whereas for MoS_2 it is 335 mAh g^{-1}. They deduced that VS_2 monolayer has the higher theoretical capacity, high lithium mobility, and higher lithiation stability as compared to MoS_2 and graphite. Thus, VS_2 monolayer could be a potential candidate as an anode material for LIBs. Mikhaleva et al. [27] explored the properties of VS_2/graphene heterostructure within the framework of DFT by implementing the VASP code. They studied the sorption and diffusion of lithium atoms in the interface layer and on the VS_2/graphene heterostructure surface by using both trigonal-prismatic (H) and octahedral (T) configurations. The highly desirable sites for sorption of lithium on the surface of VS_2 (H, S, and V) and in the interface area sandwiched between the graphene and VS_2 (1–4) layers were established (Figure 17.2d–g).

On impartial monolayer of VS_2 or heterostructure, sorption energy of lithium is calculated by using the following equation:

$$E_{sorp} = \left(E_{total} - E_{substrate} - nE_{Li}\right)/n \qquad (17.2)$$

Additionally, for the interface area of heterostructure of VS_2/graphene, diffusion of single Li was studied (Figure 17.2j and k) and is compared with the surface of VS_2 monolayer (Figure 17.2h and i). The easiest and simplest way for diffusion of Li atom on both H and T phases surface is V–H–V. As compared to the phase T-VS_2 (0.23 eV), the height of the barrier V–H–V transition for H-VS_2 phase is little lower (0.18 eV). T-VS_2/graphene heterostructure interface shows that lithium atom diffusion is limited by the 1–4 transition (Figure 17.2k). The preferable path for diffusion is 1–4–3 for H-VS$_2$/graphene heterostructure (Figure 17.2j). Diffusion may carry on with equal rates on the surface and between the layers because 1–4, 4–3, and V–H transition barriers are nearly same.

In between VS$_2$ and graphene layers, the maximum amount of Li "x" sorbed is calculated by applying the bigger cell simulation for $x < 0.25$, and for high concentration, small cell is used. It is believed that atoms of Li reside in the most desirable locations in the interface region, such as, on top of vanadium atoms. It is observed that the high concentration of lithium is energetically unfavorable, and per VS$_2$ unit, 1.75 Li atoms are the maximum amount of lithium which is sorbed between the interface of heterostructure of VS$_2$/graphene. This amount of lithium is higher as compared to its bulk counterpart (1 Li atom per formula unit). Mikhaleva et al. also measured its theoretical capacity which is 569 mAh g^{-1}, showing good consistency with the experimental results. Consequently, the combination of VS$_2$ with graphene improves the lithium storage and makes it a potential choice as anode material in LIBs.

Polytypes of metallic VS$_2$ monolayer are examined computationally (DFT based) by Putungan et al. [28] for sodium-ion (Na$^+$) battery. They implemented PBE(GGA) spin polarization-based DFT calculations by employing VASP code. For the integration of Brillouin zone, $5 \times 5 \times 1$ k-points are used with cutoff energy of 550 eV. Vacuum of 20 Å^{-1} is applied in the z-direction to avoid the artificial interactions. Ab initio random structure searching (AIRSS) method is employed to assess the stable adsorption phases at high concentrations of Na. Two distinct adsorption sites are predicted by applying AIRSS on both polytypes of VS$_2$ such as 1H and 1T (the insets of Figure 17.2b and c). Binding energies can be calculated by replacing the values of Li with Na in Eq. (17.1). The calculated binding energies of 1H and 1T for V$_{top}$ are −2.17 and −1.86 eV, respectively. Whereas, in the case of H$_{center}$, 1H and 1T have binding energies −2.15 and −1.84 eV, respectively. Further, it is observed that 1H phase shows more stability than 1T phase for V$_{top}$ adsorption site. Also, on VS$_2$ monolayer, single Na adsorption shows more stability when compared to other 2D materials like monolayer of MoS$_2$ (−1.27 eV), graphene doped with boron (−0.79 eV) and phosphorene (−1.59 eV). Therefore, ions have very little tendency to form clusters, thus making VS$_2$ monolayer a potential candidate for electrode material.

Diffusion energy barriers and diffusion pathways of Na on different polytypes of VS$_2$ are also examined as charge/discharge rates depend on the metal–ions transportation. Diffusion pathways are explored for 1H and 1T phase on the most suitable binding sites (V$_{top}$). The minimum energy paths (MEPs) estimated are depicted in Figure 17.2b and c. Along the V$_{top}$–V$_{top}$ migration pathway, Na passes through H$_{center}$ site for MEP of different polytypes of VS$_2$. The calculated values of barrier during the diffusion are 85 and 88 meV for 1H and 1T phases, respectively. This results in the increased efficiency of electrode due to the fast diffusion kinetics of Na.

Putungan et al. also evaluated the theoretical capacity of VS$_2$ monolayer for Na storage. Theoretical capacity C in units mAh g^{-1} is calculated by using the following equation:

$$C = \frac{1}{MV_{VS_2}}\left(x_{max}\upsilon F \cdot 10^3\right) \tag{17.3}$$

In the above equation, MV_{VS_2}, x_{max}, υ, and F are denoting VS_2 molecular weight, maximum concentration of sodium which can be stored in VS_2, Na valence electron, and Faraday's constant (26.801 Ah mol^{-1}), respectively. Maximum concentration of sodium achieved by 1H phase is $x = 1.0$ and by 1T is $x = 0.5$. Putting all these values in the above equation gives theoretical capacities of 232.91 and 116.45 mAh g^{-1} for 1H and 1T phases, respectively. This study revealed that different polytypes of VS_2 monolayer make it a promising candidate for electrode material in sodium-ion battery (NIB) because of better structural stability, Na adsorption stability, fast kinetics of Na diffusion, and reasonable high specific capacity as compared to some other 2D materials like MoS_2, graphene doped with boron and phosphorene.

Yang et al. [29] carried out detailed study on VS_2 used as an electrode for magnesium (Mg)-ion battery (MIB). They computationally studied monolayer, double layer, and bulk VS_2 symbolized as M-VS_2, D-VS_2 and B-VS_2, respectively, for MIB. VASP code is used for the DFT calculations by implementing GGA-PPE approximation along with van der waals DFT-3 correction. k-Point grid of $4 \times 4 \times 1$ is used for M and D VS_2 and $4 \times 4 \times 2$ k-point grid is used for B-VS_2. 400 eV Cut-off energy is used for basis of plane wave and vacuum of >15 Å is applied to avoid the artificial interactions. Side and top views of monolayer, double layer, and bulk VS_2 are displayed in Figure 17.3 along with the different suitable adsorption sites for Mg cation. It is observed that three possible sites are considered for M-VS_2: Mg$^+$ on top of hollow site is referred as C cite, whereas V atom and S atom with Mg$^+$ on top of them are referred as V-site and S-site, respectively. There are four probable sites of adsorption for D-VS_2, sites sandwiched between layers are denoted as C_i and V_i

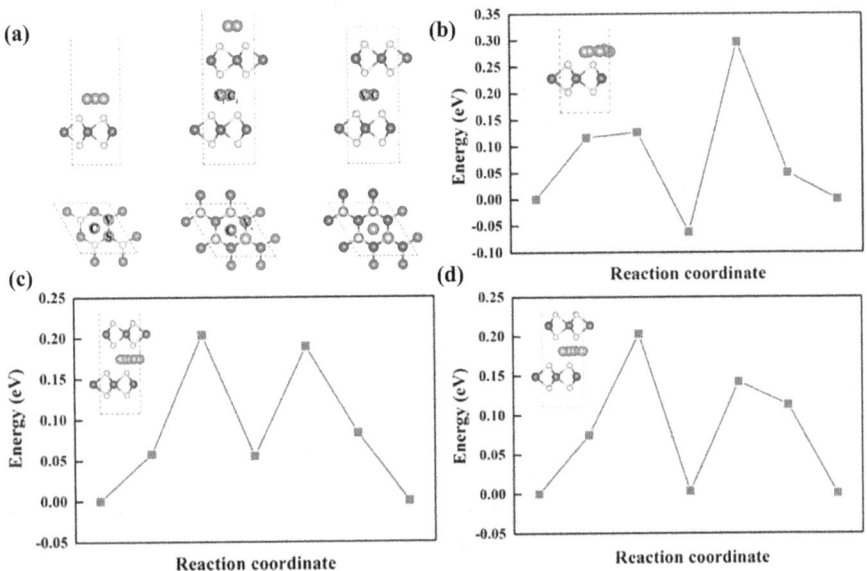

FIGURE 17.3 Illustrating (a) side and top view illustration of M-VS_2, D-VS_2, and B-VS_2. The diffusion barriers along most feasible path for (b) M-VS_2 (c) D-VS_2, and (d) B-VS_2. (Adapted with permission from Ref. [29]. Copyright (2021), Elsevier.)

sites, whereas sites present on the surface are represented as C_s and V_s sites. C and V sites are considered for B-VS$_2$.

Charge/discharge rate properties of MIB can be explored by significant factor known as Mg cation's diffusion kinetic performance. So, it is vital to calculate the diffusion barrier of magnesium ion for diffusion pathways between adjoining adsorption sites for M-VS$_2$, D-VS$_2$, and B-VS$_2$. Out of three possible pathways, Figure 17.3b displays the optimized one for M-VS$_2$ diffusion barrier. The path between the neighboring C sites for Mg$^+$ migration is named as path 1, pathway between the neighboring V sites for Mg$^+$ movement is named as path 2, and the diffusion path for Mg$^+$ migration between C sites of the upper and lower surface is named as path 3. It is observed that path 1 (0.36 eV) and path 2 (0.37 eV) have lower values of diffusion barrier as compared to path 3 (4.78 eV). These results indicate that most stable adsorption site is V.

C_i and V_i interlayer sites cause five different paths for D-VS$_2$ and the best path is displayed in Figure 17.3c. Path 1 displays the smallest barrier (0.21 eV) for diffusion among five different diffusion pathways. This indicates that the most feasible path for migration of Mg$^+$ is path 1 (C_i to nearest C_j). The most advantageous path for diffusion of Mg$^+$ is from the nearest C–C site (path 1 with 0.20 eV) among three possible diffusion paths for B-VS$_2$. These suitable paths can be seen in Figure 17.3d.

M-VS$_2$, D-VS$_2$, and B-VS$_2$ theoretical capacities are also calculated by using Eq. (17.3) and found to be 233 mAh g^{-1}. Results of Yang et al. predict that Mg$^+$ adsorbed on all VS$_2$ is stable, and has low diffusion barrier and high theoretical capacity, which makes VS$_2$ a promising alternative electrode material for Mg-ion batteries.

17.2.2 Experimental Study

Chen et al. [15] synthesized the vanadium disulfide and explored its electrochemical performance for sodium- and potassium-ion batteries. Template free growth approach is used for the synthesis of VS$_2$. Scanning electron microscope (SEM) and transmission electron microscope (TEM) characterizations are used to study the morphology of the resultant product and revealed that the synthesized VS$_2$ is in the shape of spherical nanoflowers (SNF-VS$_2$). Nanoplates are assembled in the shape of a spherical flower-like structure. Stacked layers of vanadium disulfide (SL-VS$_2$) are also synthesized to compare their performance with SNF-VS$_2$. Morphology (SEM and TEM; Figure 17.4a–d) of the SNF-VS$_2$ along with EDX elemental mapping is studied. Spherical nanoflowers are mainly discussed in this study because they show better performance than stacked layer structure.

Electrochemical performance for Na- and K-ion batteries is evaluated by assembling the half coin cells. Synthesized spherical nanoflowers and stacked layers of vanadium disulfide are used to make electrodes to be used as anode with counter electrodes of Na and K. Coins cells are assembled in glove box in the absence of air. Cycling performance, rate capabilities, and long cycling performance of SNF-VS$_2$ for Na-ion battery are evaluated. Cycling performance of spherical nanoflowers and stacked layer vanadium disulfide at 0.2 A g^{-1} for NIB is displayed in Figure 17.4e. It is observed that SNF-VS$_2$ shows better specific capacity (292 mAh g^{-1}) than SL-VS$_2$ (149 mAh g^{-1}) up to 300 cycles, which is almost double. Rate capabilities of SNF-VS$_2$

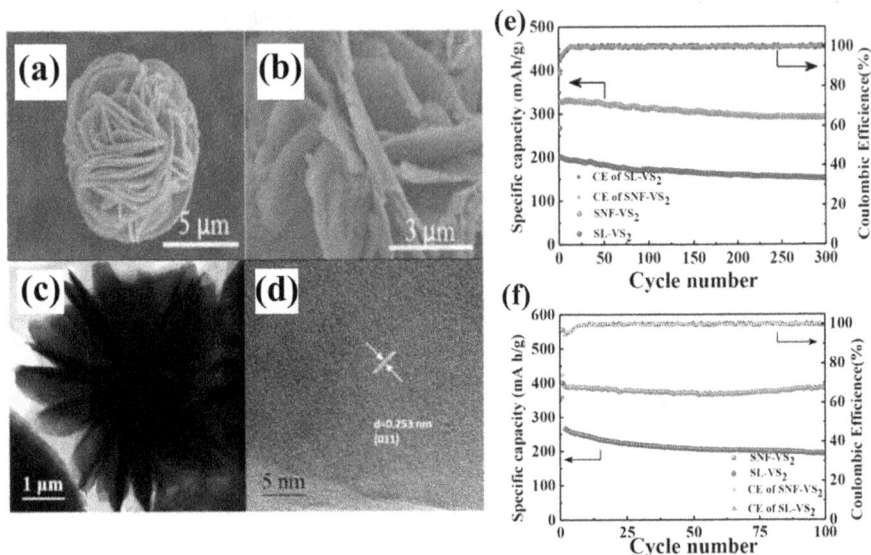

FIGURE 17.4 Displaying (a and b) SEM images and (c and d) TEM images of SNF-VS$_2$. Cycling performance of SNF and SL VS$_2$ at (e) 0.2 A g^{-1} and (f) 25 mA g^{-1}. (Adapted with the permission from Ref. [15]. Copyright (2020), Elsevier.)

at various different current densities such as 0.1, 0.2, 0.5, 1, and 2 Ah g^{-1} are superior to those of SL-VS$_2$. Long cycling performance of SNF-VS$_2$ is also studied for 800 cycles at high current density of 2 Ah g^{-1} and its value is around 185 mAh g^{-1} with almost 100% Coulombic efficiency. This indicates greater cycling stability for storage of sodium ions.

Additionally, SNF-VS$_2$ and SL-SV$_2$ electrochemical performance is also examined for K-ion battery. Cycling performance, rate capabilities, and long cycling of SNF-VS$_2$ and SL-VS$_2$ are explored. SNF-VS$_2$ shows better specific capacity as compared to SL-VS$_2$ for 100 cycles at 25 mAh g^{-1} (Figure 17.4f). The value of specific capacity for flower-like vanadium disulfide is ~383 mAh g^{-1}, and for stacked layer, VS$_2$ is ~ 194 mAh g^{-1}. Rate performance is also evaluated at different current densities such as 25, 50, 100, 200, 500, 800, 1,000, and 2,000 mAh g^{-1}. SNF-VS$_2$ long cycling performance for 800 cycles at 500 mA g^{-1} is measured and the value of reversible capacity is maintained at 98 mAh g^{-1}.

Porous spherical flower-like structure along with high surface area might be the reason for this enhanced cycling performance of SNF-VS$_2$ for Na-ion and K-ion batteries. Also, the space between neighboring nanosheets limits the damage to the structure during the exchange of ions, and tolerates changes in the volume, thus improving the cycling performance. Li et al. [30] explored the potential of titanium disulfide coated vanadium disulfide flakes as a cathode for LIBs. Li and coauthors synthesized the flakes of VS$_2$ by using atmospheric pressure chemical vapor deposition and then these flakes are coated with titanium disulfide in atomic layered deposition chamber at ~400°C. SEM and TEM characterizations are carried out to investigate the structure of synthesized VS$_2$-TiS$_2$ flakes. Smooth flakes are appeared

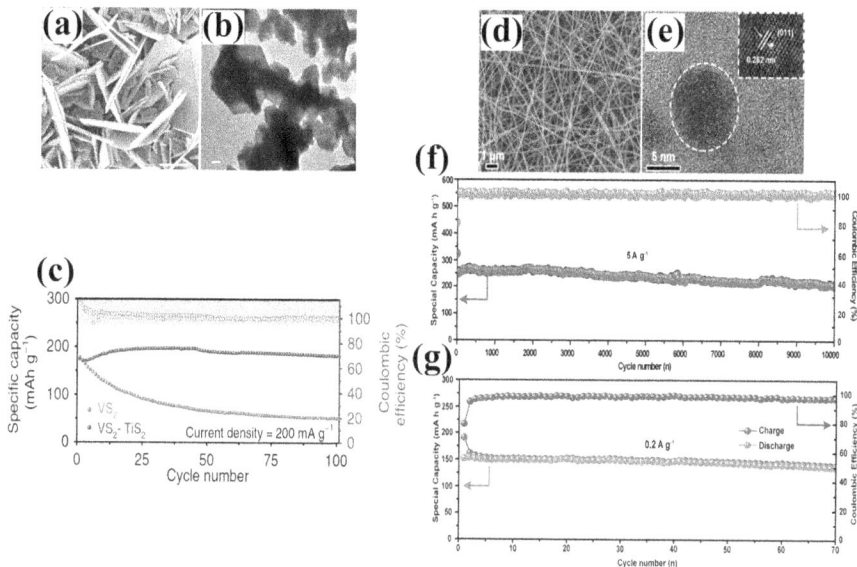

FIGURE 17.5 Illustrating (a) SEM image and (b) TEM image (200 nm) of VS_2-TiS_2 flakes. (c) Cyclic performance at 200 mA g^{-1} of pure VS_2 and VS_2-TiS_2 electrodes. (d) FESEM image and (e) HRTEM images of synthesized VSe_2/NCNFs. Cycling performance of (f) VSe_2/ NCNFs. and (g) full battery (VSe_2/NCNFs ‖ $Na_3V_2(PO_4)_3$/C). (Adapted with the permission from Refs. [30] and [31]. Copyright (2019), Springer Nature and Copyright (2020), Elsevier.)

to be hexagonal in shape having high crystallinity. Figure 17.5a and b is displaying SEM and TEM images of VS_2-TiS_2 flakes.

Electrochemical performance is explored by assembling the half coin cells in argon filled gloves box. VS_2 and VS_2-TiS_2 along with current collector are cut in circular shape to be used as anode, and lithium foil is used as counter electrode. During the assembly of half coin cells, no binder or conductive additives were used. Arbin BT2000 battery testing system was used for the measurement of charging–discharging in LIBs. Rate capabilities at different current densities, VS_2 and VS_2-TiS_2 cycling performance at 200 mA g^{-1}, and long cycling performance at high current density of 1,000 mA g^{-1} are measured. VS_2 and VS_2-TiS_2 electrodes rate performance is measured at different current densities such as 200, 500, 1,000, and 2,000 mA g^{-1}. At 200 mA g^{-1}, both pure VS_2 and VS_2-TiS_2 have specific capacities of 150 and 180 mAh g^{-1}. VS_2-TiS_2 displays a specific capacity of 40 mAh g^{-1} even at 2,000 mA g^{-1} (~10C). This indicates that VS_2-TiS_2 is capable of operating at high C-rate values. VS_2-TiS_2 electrode shows a very good specific capacity of 180 mAh g^{-1} up to 100 cycles at 200 mA g^{-1}, whereas VS_2 electrode shows a rapid loss in the specific capacity under the same conditions (Figure 17.5c). A high current density of 1000 mA g^{-1} is also used to investigate the long cyclic performance of VS_2-TiS_2 electrode for 400 cycles. The retention in capacity is almost 100% for titanium-coated vanadium disulfide electrode and is almost 40% for pure VS_2 electrode under the same conditions. Results of Li et al. indicate that coating of titanium disulfide on vanadium disulfide improves the overall specific capacity of the electrode and shows stability for long cycling performance, thus making VS_2-TiS_2 electrode a potential candidate to be used in Li-ion batteries.

Wu et al. [31] explored carbon composites of VS_2 as anode material for Na^+ batteries. Wu and coauthors synthesized the VSe_2 N-doped carbon nanofibers (NCNFs) by mixing vanadium acetylacetonate, graphene oxide, and PAN in DMF and blended at 40°C. After that, electrospinning process is employed to electro spun $C_{15}H_{21}O_6$-V@ PAN on aluminum foil. Then, muffle furnace is used to preoxidize the carbon fibers at 235°C in air for 3 hours and then these preoxidized carbon fibers are carbonized for 3 hours at 700°C in argon environment to obtain V_2O_3/CNFs. Then at the end, in H_2/Ar environment, synthesized V_2O_3-CNFs are placed downstream of VSe_2 powder in tube furnace for 10 hours at 300°C to get VSe_2/NCNFs. X-ray powder diffraction (XRD) is used to confirm the structure of as-prepared material. Also, field emission scanning electron microscope (FESEM) and TEM are used to study the morphology of the synthesized product. Results of FESEM image (Figure 17.5d) show that uniform nanofibers constitute the structure of VSe_2/NCNFs, and the diameter of the alone standing fiber is 150 nm. Energy dispersive spectroscopy (EDS) is used to map the individual elements present in the synthesized material. TEM images display that nanoflakes are embedded on the fiber surface and ultra-small VSe_2 nanoparticles distribution on nanofibers having diameter around 4 nm (Figure 17.5e).

Electrochemical performance of synthesized VSe_2/NCNFs is measured by assembling half coin cells in which counter electrode (cathode) is sodium. Full cells are also assembled with VSe_2/NCNFs as anode and $Na_3V_2(PO_4)_3$/C as cathode. Both half and full cells were assembled by using glove box filled with argon. Na-ion gel polymer is used as quasi-solid electrolyte and separator. Cycling performance of both half and full coin cells is displayed in Figure 17.5f and g.

Rate performance is measured by using various current densities such as 0.05, 0.1, 0.2, 0.5, 1, 2, and 5 A g^{-1}. The values of specific capacities at these densities are 420.8, 390.5, 374.4, 332.2, 312.5, and 278.1 mAh g^{-1}. When current density is returned to 0.05 from 5 A g^{-1}, VSe_2/NCNFs maintained the specific capacity. Long-term cycling performance is also evaluated for VSe_2/NCNFs used in half cell at current density of 5 A g^{-1}. VSe_2/NCNFs exhibit very good cycling stability and reversible capacity with only 0.002% reduction in capacity during each cycle. For 10,000 cycles, it shows overall capacity of 206.8 mAh g^{-1} at current density of 5 A g^{-1} (Figure 17.5f). Na gel polymer might be the reason to maintain the original shape of the electrode during long cycling performance. For the possible practical use, VSe_2/NCNFs electrode is tested for full battery where synthesized $Na_3V_2(PO_4)_3$/C is used as cathode. Rate capacities at different current densities are studied, where current densities are changing in the range 0.05–1 A g^{-1}. The full battery exhibits rate capacity of 74.8 mAh g^{-1} at 1 A g^{-1} and it gets higher when current density is returned to 0.05 A g^{-1}. The excellent electrochemical performance of VSe_2/NCNFs in half and full battery makes it a promising candidate to be used in practical quasi-solid NIBs.

Wu et al. [32] synthesized VSe_2 and investigated it for zinc-ion batteries (ZIBs). Synthesis of VSe_2 is carried out by using one-pot chemical liquid phases synthesis method. To explore the structure and morphology of the as-synthesized VSe_2, different characterizations like XRD, SEM, TEM, and EDS are carried out. Two-dimensional (2D) morphology of the as-prepared VSe_2 is revealed by TEM images. The resultant product contains many ultrathin nanosheets like objects. TEM and high-resolution transmission electron microscopy (HRTEM) images are displayed in Figure 17.6a and b. Pattern of selected area electron diffraction (SAED) is also given in the inset of Figure 17.6a, which depicts hexagonal arrangement of sharp diffraction spots revealing the single crystal nature of the nanosheet. HRTEM image revealed the thickness of the nanosheet, which is ~2.1 nm. EDS images revealed that V and Se are homogeneously distributed in individual nanosheet.

To study and understand the electrochemical performance of Zn^+ storage in VSe_2 nanosheets, half coin cells are assembled in which VSe_2 is used as cathode, Zn metal is used as counter anode, and 2M $ZnSO_4$ aqueous solution is used as electrolyte. The aqueous zinc battery exhibits excellent rate capability at different current densities 100, 200, 500, 1,000, and 2,000 mA g^{-1} with specific capacities of 131.8, 114.6, 105.2, 93.9, and 79.5 mAh g^{-1}. When the current density is restored to 100 mA g^{-1}, specific capacity recovers to 118.4 mAh g^{-1}. VSe_2 is also tested for long cycling performance (500 cycles) in ZIBs at 100 and 500 mA g^{-1} with 80.8% and 75.3% retention in the specific capacities, respectively (Figure 17.6c).

The mechanism of Zn^+ intercalation/de-intercalation is briefly given as follows:
At cathode

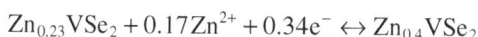

$$VSe_2 + 0.23Zn^+ + 0.46e^- \leftrightarrow Zn_{0.23}VSe_2$$

$$Zn_{0.23}VSe_2 + 0.17Zn^{2+} + 0.34e^- \leftrightarrow Zn_{0.4}VSe_2$$

At anode

$$0.4\,Zn \leftrightarrow 0.4\,Zn^{2+} + 0.8e^-$$

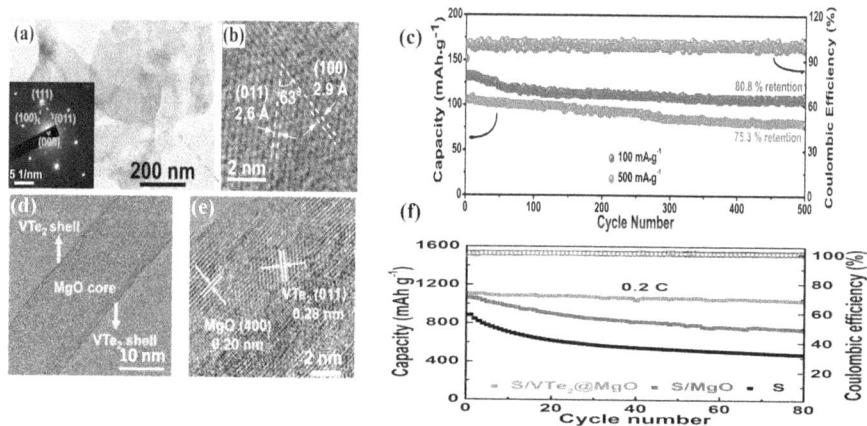

FIGURE 17.6 Displaying (a) TEM image with the inset of SAED; (b) HRTEM image; (c) long cycling performance of synthesized VSe_2 nanosheets; (d) TEM image and (e) HRTEM image of synthesized VTe_2@MgO heterostructure; and (f) cycling performance at 0.2 C of S/VTe_2@MgO and S/MgO. (Adapted with the permission from Refs. [32] and [33]. Copyright (2020), John Wiley & Sons and Copyright (2019), American Chemical Society.)

Results of Wu et al. revealed that due to excellent cyclic stability and high energy density, VSe$_2$ nanosheets are promising candidates for cathode material in ZIBs.

Wang et al. [33] investigated the electrochemical performance of VTe$_2$@MgO heterostructure for lithium–sulfur (Li–S) batteries. CVD system is employed for the fabrication of heterostructure VTe$_2$@MgO. Metallic VTe$_2$ over MgO is rarely fabricated by typical wet-chemical synthesis methods so far. After the synthesis of heterostructure VTe$_2$@MgO, different characterizations like SEM, TEM, SAED, STEM, and electron energy loss spectroscopy (EELS) are applied to examine the structure and morphology of the as-prepared heterostructure. SEM image reveals that the morphology of the as-prepared heterostructure is of tubular form. Further, TEM images show that VTe$_2$ shells are uniformly wrapped over the nanorods of MgO upon CVD (Figure 17.6d). Formation of in situ heterostructure VTe$_2$@MgO is confirmed by the HRTEM (Figure 17.6e) along with SAED which displays the coexistence of VTe$_2$ and MgO facets. Uniform distribution of Mg, V, Te, and O is revealed by STEM and EELS mapping which also confirms the VTe$_2$ incorporation on nanorods of MgO.

After the comprehensive study of the characterization, catalytic activity of VTe$_2$@MgO heterostructure is explored systematically to understand the redox reaction for sulfur. In this regard, CV curve, EIS plots, rate performance, galvanostatic charge/discharge profile, and cyclic performance are evaluated by assembling the half coin cells. In coin type cells, sulfur–carbon mixture, VTe$_2$@MgO along with aqueous binder is used to make slurry for fabricating cathode electrode on aluminum foil as current collector. Lithium foil is used as counter electrode (anode). Land CT2001A battery system is used to measure the cyclic performance in 1.7–2.8 V voltage range.

CV curve profile at scan rate of 0.05 mV s^{-1} exhibits that for S/VTe$_2$@MgO heterostructure, oxidation peak moves toward lower potential and reduction peaks move toward higher potential indicating the accelerated kinetics Li–S chemistry. Smaller charge transfer resistance R_{ct} is displayed by the S/VTe$_2$@MgO cathode electrode as compared to the other electrodes indicated by seemingly depressed semi-circle in the Nyquist plot. At small current density of 0.2 C, the cyclic stability of different cathodes is measured as shown in Figure 17.6f. S/TVe$_2$@MgO cathode at current density of 0.2 C displays an initial capacity of 1107 mAh g^{-1} and then it sustains to 1,034 mAh g^{-1} capacity after 80 cycles. Rate performance of different electrodes is compared at different current rates ranging from 0.2 to 3 C. At 0.2, 0.5, 1, 2, and 3 C, S/VTe$_2$@MgO exhibits 1,160, 1,020, 907, 783, and 705 mAh g^{-1} specific capacities, respectively. When the current density is again set to 0.2 C, capacity recovers to 1,150 mAh g^{-1} indicating superior rate capabilities in comparison to pure S and S/MgO electrodes. S/VTe$_2$@MgO cathode long cycling performance at 1 C is displayed with trivial loss in capacity (0.055%) for 1,000 cycles with 927 mAh g^{-1} initial capacity. Wang et al. smoothened the path for controllable and clean fabrication of heterostructure facilitator utilizing high electrocatalytic activities as a multifunctional promoter for Li–S batteries.

17.3 CONCLUSIONS

This chapter presented an overview about the computational approaches and experimental fabrication of various 2D vanadium dichalcogenides and their composites. Computational study can act as a bridge to perform experiments based on computational results and to verify them. Experimentally, different methods are used to synthesize VX_2 and their composites to be used as electrode in batteries. Results show that different morphologies help in the better transportation of ions in energy storage applications which deeply improve their performance. Electrochemical performance of VX_2 and their composites are investigated in potential innovative energy storage systems, i.e., metal-ion batteries such as LIBs, NIBs, KIBs, and ZIBs in this chapter. In nutshell, this inclusive review will trigger the interest in 2D VX_2-based nanomaterials with high power and energy density for the next-generation energy storage devices. This will ultimately lead to the expansion in the efficiency of energy systems, improve the stability of the grid, and enhance the penetration of green energy resources.

REFERENCES

1. M. Aneke, M. Wang, Energy storage technologies and real life applications: A state of the art review, *Applied Energy*, 179 (2016) 350–377.
2. M.-R. Gao, Y.-F. Xu, J. Jiang, S.-H. Yu, Nanostructured metal chalcogenides: Synthesis, modification, and applications in energy conversion and storage devices, *Chemical Society Reviews*, 42 (2013) 2986–3017.
3. Y.-J. Wang, D.P. Wilkinson, J. Zhang, Noncarbon support materials for polymer electrolyte membrane fuel cell electrocatalysts, *Chemical Reviews*, 111 (2011) 7625–7651.
4. G. Chen, J. Seo, C. Yang, P.N. Prasad, Nanochemistry and nanomaterials for photovoltaics, *Chemical Society Reviews*, 42 (2013) 8304–8338.
5. K. Kalyanasundaram, M. Grätzel, Themed issue: Nanomaterials for energy conversion and storage, *Journal of Materials Chemistry*, 22 (2012) 24190–24194.
6. A.J. Nozik, J. Miller, *Introduction to Solar Photon Conversion*. Washington, DC: ACS Publications, 2010.
7. J. Mao, J. Iocozzia, J. Huang, K. Meng, Y. Lai, Z. Lin, Graphene aerogels for efficient energy storage and conversion, *Energy & Environmental Science*, 11 (2018) 772–799.
8. W. Kang, N. Deng, J. Ju, Q. Li, D. Wu, X. Ma, L. Li, M. Naebe, B. Cheng, A review of recent developments in rechargeable lithium–sulfur batteries, *Nanoscale*, 8 (2016) 16541–16588.
9. Q. Yang, Z. Zhang, X.-G. Sun, Y.-S. Hu, H. Xing, S. Dai, Ionic liquids and derived materials for lithium and sodium batteries, *Chemical Society Reviews*, 47 (2018) 2020–2064.
10. R. Fang, S. Zhao, S. Sun, D.W. Wang, H.M. Cheng, F. Li, More reliable lithium-sulfur batteries: Status, solutions and prospects, *Advanced Materials*, 29 (2017) 1606823.
11. X. Chen, H. Lin, P. Chen, G. Guan, J. Deng, H. Peng, Smart, stretchable supercapacitors, *Advanced Materials*, 26 (2014) 4444–4449.
12. C. Jiang, J. Ma, G. Corre, S.L. Jain, J.T. Irvine, Challenges in developing direct carbon fuel cells, *Chemical Society Reviews*, 46 (2017) 2889–2912.
13. T. Cao, K. Huang, Y. Shi, N. Cai, Recent advances in high-temperature carbon–air fuel cells, *Energy & Environmental Science*, 10 (2017) 460–490.
14. S.K. Kurinec, *Emerging Photovoltaic Materials: Silicon & Beyond*. Hoboken, NJ: John Wiley & Sons (2018).
15. J. Chen, Z. Tang, Z. Pan, W. Shi, Y. Wang, Z.Q. Tian, P.K. Shen, Template-free growth of spherical vanadium disulfide nanoflowers as efficient anodes for sodium/potassium ion batteries, *Materials & Design*, 192 (2020) 108780.
16. X. Zhang, S.Y. Teng, A.C.M. Loy, B.S. How, W.D. Leong, X. Tao, Transition metal dichalcogenides for the application of pollution reduction: A review, *Nanomaterials*, 10 (2020) 1012.

17. B. Radisavljevic, A. Radenovic, J. Brivio, V. Giacometti, A. Kis, Single-layer MoS_2 transistors, *Nature Nanotechnology*, 6 (2011) 147–150.
18. Z. Yin, H. Li, H. Li, L. Jiang, Y. Shi, Y. Sun, G. Lu, Q. Zhang, X. Chen, H. Zhang, Single-layer MoS_2 phototransistors, *ACS Nano*, 6 (2012) 74–80.
19. K.F. Mak, K.L. McGill, J. Park, P.L. McEuen, The valley Hall effect in MoS_2 transistors, *Science*, 344 (2014) 1489–1492.
20. H. Li, X. Jia, Q. Zhang, X. Wang, Metallic transition-metal dichalcogenide nanocatalysts for energy conversion, *Chem*, 4 (2018) 1510–1537.
21. Y.-P. Gao, X. Wu, K.-J. Huang, L.-L. Xing, Y.-Y. Zhang, L. Liu, Two-dimensional transition metal diseleniums for energy storage application: A review of recent developments, *CrystEngComm*, 19 (2017) 404–418.
22. J. Lu, O. Zheliuk, I. Leermakers, N.F. Yuan, U. Zeitler, K.T. Law, J. Ye, Evidence for two-dimensional Ising superconductivity in gated MoS_2, *Science*, 350 (2015) 1353–1357.
23. Y. Wang, B. Qian, H. Li, L. Liu, L. Chen, H. Jiang, VSe_2/graphene nanocomposites as anode materials for lithium-ion batteries, *Materials Letters*, 141 (2015) 35–38.
24. H. Wang, H. Feng, J. Li, Graphene and graphene-like layered transition metal dichalcogenides in energy conversion and storage, *Small*, 10 (2014) 2165–2181.
25. R. Kappera, D. Voiry, S.E. Yalcin, B. Branch, G. Gupta, A.D. Mohite, M. Chhowalla, Phase-engineered low-resistance contacts for ultrathin MoS2 transistors, *Nature Materials*, 13 (2014) 1128–1134.
26. Y. Jing, Z. Zhou, C.R. Cabrera, Z. Chen, Metallic VS_2 monolayer: A promising 2D anode material for lithium ion batteries, *The Journal of Physical Chemistry C*, 117 (2013) 25409–25413.
27. N.S. Mikhaleva, M.A. Visotin, A.A. Kuzubov, Z.I. Popov, VS_2/graphene heterostructures as promising anode material for Li-ion batteries, *The Journal of Physical Chemistry C*, 121 (2017) 24179–24184.
28. D.B. Putungan, S.-H. Lin, J.-L. Kuo, Metallic VS_2 monolayer polytypes as potential sodium-ion battery anode via ab initio random structure searching, *ACS Applied Materials & Interfaces*, 8 (2016) 18754–18762.
29. J. Yang, J. Wang, X. Dong, L. Zhu, D. Hou, W. Zeng, J. Wang, The potential application of VS_2 as an electrode material for Mg ion battery: A DFT study, *Applied Surface Science*, 544 (2021) 148775.
30. L. Li, Z. Li, A. Yoshimura, C. Sun, T. Wang, Y. Chen, Z. Chen, A. Littlejohn, Y. Xiang, P. Hundekar, Vanadium disulfide flakes with nanolayered titanium disulfide coating as cathode materials in lithium-ion batteries, *Nature Communications*, 10 (2019) 1–10.
31. Y. Wu, W. Zhong, W. Tang, L. Zhang, H. Chen, Q. Li, M. Xu, S.-J. Bao, Flexible electrode constructed by encapsulating ultrafine VSe_2 in carbon fiber for quasi-solid-state sodium ion batteries, *Journal of Power Sources*, 470 (2020) 228438.
32. Z. Wu, C. Lu, Y. Wang, L. Zhang, L. Jiang, W. Tian, C. Cai, Q. Gu, Z. Sun, L. Hu, Ultrathin VSe_2 nanosheets with fast ion diffusion and robust structural stability for rechargeable zinc-ion battery cathode, *Small*, 16 (2020) 2000698.
33. M. Wang, Y. Song, Z. Sun, Y. Shao, C. Wei, Z. Xia, Z. Tian, Z. Liu, J. Sun, Conductive and catalytic VTe_2@ MgO heterostructure as effective polysulfide promotor for lithium–sulfur batteries, *ACS Nano*, 13 (2019) 13235–13243.

18 Emerging Applications of 2D Nanomaterials for Flexible Batteries (FBs)

Solen Kinayyigit
Gebze Technical University
NANOTerial Technology Corporation

Emre Bicer
Sivas University of Science and Technology

Abdullah Uysal
Gebze Technical University

CONTENTS

18.1 INTRODUCTION

Unprecedented demand for malleable and wearable electronics made the battery market alive and very dynamic in the last decade. Fast-changing industry urged the development of new generation batteries to power innovative portable and bendable devices for various applications ranging from electronic textiles to bio-electronics and more [1,2]. Following the current trend, high interest in research of these smart devices opened up the space for lightweight, ultra-thin, rollable, and stretchable

DOI: 10.1201/9781003178422-18

batteries since the conventional secondary batteries are not able to meet the expectation due to their heavy, rigid, and fragile nature [2,3]. In addition to the critical aspects in a high performance battery with balanced parameters, flexible batteries (FBs) should provide appreciable mechanical strength and operational safety while maintaining or, at least, not significantly sacrificing its electrochemical performance. Therefore, recent studies have leaned toward new designs, methods, and modified chemistries to provide optimum FBs [1,4].

In this regard, researchers focused on the emerging atomically thin 2D nanomaterials in recent years and employed them as additive, conductive support, and/or active materials. Decorating highly flexible, fiber-like structures with nanosheets or directly mixing with binders made a great progress on the improvement of mechanical and electrochemical properties of FBs. This chapter, therefore, discusses in detail the prominent examples of the featured 2D nanomaterials used in FBs as well as both advantages and disadvantages of each application presented herein.

18.2 CLASSIFICATION AND PRACTICAL APPLICATIONS IN FBs

In terms of composition and properties, 2D nanomaterials used in FBs can be classified into three main categories as monoelemental, 2D transition metal-based (transition metal dichalcogenides, TMDs), and polymer-based 2D nanomaterials (Figure 18.1). Emerging 2D mono-elemental materials (Xenes) such as graphene, phosphorene, germanene, silylene, borophene, and stanene have the potential to be employed in FBs. Yet, only graphene and phosphorene have been explored among Xenes [5–7]. 2D TMDs have been mostly utilized in flexible electrodes due to their unique properties [8,9]. In general, 2D polymer-based materials are used in electrolytes because of their low electrical conductivity. Still, there are few trials of polymer-based ones as active material in free-standing, flexible electrodes [10,11].

18.2.1 MONOELEMENTAL 2D NANOMATERIALS FOR FBs

As in rigid batteries, graphene and its derivatives are the 2D materials widely investigated in FBs due to their inherent mechanical flexibilities [12,13]. Although borophene, silylene, germanene, stanene, and bismuthene have already been proven to be promising candidates for FBs through density functional theory simulations, none of them have been tested in real applications. Metal oxides or metal sulfides with capacity >800 mAh g^{-1} are widely applied as conversion reaction-based electrodes in high energy-density flexible lithium-ion batteries (LIBs). Yet, most of the studies showed that their structure is unstable during deformation and long-term electrochemical cycling due to the sluggish reaction kinetics, severe lithiation-induced volume expansion, and applied stress. This superposition of stresses usually results in the pulverization of active material, poor electrochemical cycle, and irreversible deformation of electrode. Graphene-based scaffold is one possible solution for this kind of electrode but the need for well-constructed interspaces complicates the fabrication process and limits the active material content (usually <60 wt.%) [14]. Another strategy is to form elastic textures like wrinkles or origami on graphene to enhance intrinsic deformation ability. These structures bring a reversible strain ability in nanosheets through a

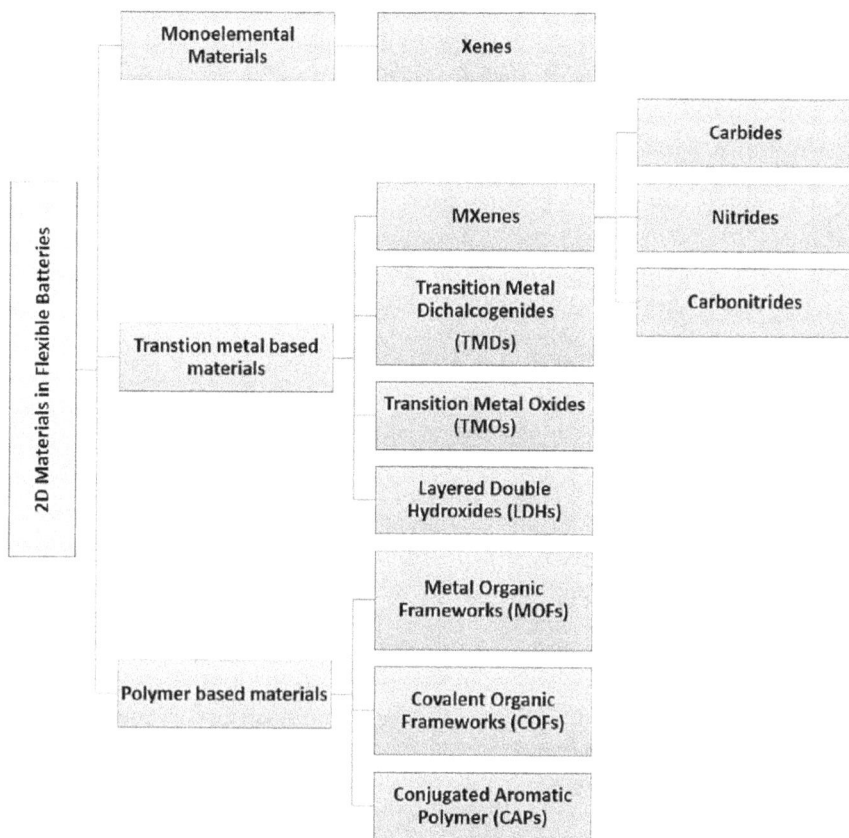

FIGURE 18.1 Schematic representation of the classification of emerging 2D nanomaterials used in FBs on material basis.

stress-release mechanism [4]. Chen et al. demonstrated that the hybrid structure of Fe_3O_4/graphene with the nested wrinkles delivered a specific capacity of 1,140 mAh g^{-1} and 99% capacity retention after 500 cycles at 10 C. Along with this outstanding performance, it showed a reversible stretchability over 100%. These extremely flexible Fe_3O_4 electrodes successfully absorbed lithiation-induced stress while relieving the tensile or compressive stresses of deformation by elastic folding/unfolding of wrinkles. More than tenfold increase in fracture toughness and twofold capacity increase proved the ability to accommodate the large deformations with cycling stability [14].

The use of porous and fibrous structures is a very efficient way to improve the specific surface area, the electrolyte penetration, and the ion accessibility in the electrodes of FBs [15,16]. Wrapping electrically conductive fiber with 2D nanomaterials is widely utilized to achieve higher energy density and power density in free-standing electrodes [16,17]. In this regard, N-doped reduced graphene oxide (rGO) wrapped carbon microfibers (CFs) were successfully synthesized by one-step high-temperature pyrolysis method [18]. The binder-free and current collector (metal foil)-free FBs delivered a reversible capacity of 400 mAh g^{-1} with capacity retention approaching 100% after 100 cycles.

Graphene is also involved in the studies to eliminate shuttling effect due to the dissolution of lithium polysulfides (LiPSs) in lithium–sulfur batteries, which is hard to achieve in flexible cathodes. Hybrid structures of sulfur substrates (oxides, sulfides, and nitrides) over the graphene nanosheets are very effective solutions for LiPS confinement during the charge/discharge process. In this system, polysulfides are bound by a polar transition metal through chemical interaction, catalysis redox, or polar absorption [19]. To give an example, CoS_2 is known to have the ability to trap and catalyze polysulfides via interface covalent bonding [12,20]. Li et al. took this advantage of the catalyzing ability of CoS_2 and combined with the confinement ability of graphene nanosheets to generate hybrid free-standing graphene/CoS_2/nano-sulfur hybrid papers for negative electrodes [12]. In this setup, CoS_2 nanoparticles were chemically anchored on graphene sheets uniformly through one-step hydrothermal synthesis. The strong interface covalent bonding between CoS_2 and graphene provided almost fivefold increase in tensile strength (15.3 MPa) and threefold increase in tensile strain (0.92%). In addition to the enhanced flexibility, one of the highest volumetric capacities (1,240 Ah L^{-1} at 1 C) in the literature was achieved with sulfur loading of 1.75 g cm³, which in return retained for 800 cycles with a low decay rate of 0.044% per cycle [12].

Similarly, Tian et al. used N-doped graphene as a conductive network and a flexible catalyst support to immobilize heterogenous CoS/CoO nanocrystals spatially in air cathode of the aqueous, flexible quasi-solid-state zinc air battery [13]. To do this, first 2D $Co(OH)_2$ nanosheets were sulfurized through hydrothermal treatment in the presence of N-doped graphene and turned into small CoS nanocrystals as in situ anchored on the surface of graphene nanosheets. Subsequent annealing treatment enabled the partial transformation of CoS to CoO at the external surface due to the small formation energy of CoO. The spatial distribution of resulting bifunctional electrocatalyst boosts the abundance of accessible surface areas by hindering the stacking of graphene nanosheets. As-prepared cathode exhibited a stable oxygen reduction reaction (ORR) and oxygen evolution reaction (OER) performance. Superior cycling performance was compared with the equivalent commercial noble-metal Pt/C + IrO_2 mixture catalysts. There was a slight voltage change observed after 100th cycle at 10 mA cm^{-2}; however, Pt/C + IrO^2 based battery faced a severe voltage gap just after 20 cycles. Moreover, its maximum power density was 137.8 mW cm^{-2}, whereas it was just 94.6 mW cm^{-2} for Pt/C cathode [13].

Other than graphene, novel flexible 2D carbon allotropes with in-plane nanopores such as graphyne or graphdiyne (GDY) having sp and sp^2 hybridized atoms gave new insights into FBs. Different from graphene, ion diffusion is not blocked by the aromatic carbon rings. Lower atomic density due to the higher porosity ratio grants extra flexibility to GDY-based electrode. Recently, He et al. produced hydrogen substituted graphdiyne (HsGDY) film on copper foil through an in situ cross-coupling reaction of triethynylbenzene [21]. In this extended π-conjugated carbon skeleton, the elimination of the oligomeric monomers enabled a range of pores with diameters ranging from 2 to 15 nm. The surface of copper substrate also functioned as a catalyst for the nucleation and the growth of the GDY film. The pore size and the thickness of the film were controlled by the initial amount of the monomer. Transmission electron microscopy (TEM) images revealed that as the thickness increases, pore size and porosity ratio also increase. At the same time, the tap density reduces, limiting the higher loading density of the film. A specific capacity of 650 mAh g^{-1} at a current density of 100 mAh g^{-1} is one of the best reversible capacities for sodium in the literature, which proves the facilitated transfer of ions with a larger diameter in the flexible electrode [21].

The dissimilar flexibility of batteries containing Xenes largely depends on their Young's modulus, which is usually related with π–π interactions and in-plane bond strength. Moreover, it is also related with the dihedral angles if the layered material has anisotropic crystal structure [22,23]. For instance, Young's modulus of black phosphorus (BP) varies in different directions and is much smaller compared to MoS$_2$, h-BN, and graphene [24,25] because the wavy structure of crystal lattice with dihedral angles endows BP with more tensile absorption ability and significant difference in stretchability [22,23].

18.2.2 TRANSITION METAL-BASED 2D NANOMATERIALS FOR FBS

18.2.2.1 MXenes in FBs

2D MXenes can be considered as one of the ideal substrates for anodes in alkali metal ion-based FBs. Transition metal-based carbides, nitrides, and carbonitrides have great potential as a scaffold to wrap active materials [26,27]. Flexible 2D MXene films can easily be fabricated by techniques such as vacuum filtration, sacrificial template, or clay rolling [4]. Although MXenes are not as cost-effective and environmentally friendly as graphene, they allow the intercalation of multivalent ions or ions with higher atomic radius than Li$^+$, providing more loading density in anodes. Besides, MXenes are very effective in achieving a robust solid electrolyte interface (SEI) due to their flat voltage profiles, low overpotentials, lithiophobic nature, and lower volume fluctuations during charge/discharge, which are the crucial factors for the cycle life and the safety of the FBs [28].

These features are demonstrated by constructing a pliable Ti$_3$C$_2$–Li film hybrid anode in a lamellar structure by a roll-to-roll mechanical approach. In this concept, lithiophobic Ti$_3$C$_2$ nanosheets confine the lithium plating at the nanogaps between the layers and act as an artificial SEI layer [26]. This approach prevents dendrite formation, enabling uniform nucleation during repeated plating/stripping of lithium metal. The overvoltage of the anode was extremely low (32 mV at 1.0 mA cm^{-2}) for metallic anode systems; even at the higher current density of 5.0 mA cm^{-2}, it was 103 mV. Under similar fabrication and testing conditions, the overvoltage of 2D boron nitride-Li metal anode was 106 mV (at 1.0 mA cm^{-2}) and 190 mV (at 1.0 mA cm^{-2}) for graphene-Li metal, respectively. Additionally, homogenous nucleation and parallel growth of Li metal on the Ti$_3$C$_2$ nanosheets were not observed in graphene. TEM and X-ray photoelectron spectroscopy (XPS) results demonstrated that its nucleation sites are located at the edges of the nanosheets only. The full cell performance of Ti$_3$C$_2$–Li–S was found to be remarkable with a reversible capacity of 948 mAh g^{-1} [26].

Aggregation, expansion, and restacking of nanosheets of the 2D nanomaterials are the general problems observed in the electrodes. In order to minimize these problems, a 3D porous structure from the assembly of 2D nanosheets may be used instead as in-plane van der Waals forces hold together the porous structure and eliminate the need for binder. Accordingly, Zhao et al. fabricated a flexible and highly conductive 3D MXene foam (Ti_3C_2/S) as anode material by using sulfur particles as a template for a porous structure which is tunable by the particle size (Figure 18.2) [27]. Sulfur was sublimated and removed below 300°C in order to achieve S-doped groups that facilitated the wetting of the electrolyte. This simple 3D foam structure eliminated the need for a binder, additives, and even a current collector. Nondiffusion capacitive contribution was 60.9% at 0.1 mV s^{-1} and 85.8% at 2 mV s^{-1}. The presence of substantial active sites and continuous electrolyte channels promoted infiltration and fast Li$^+$ ion transfer, yielding an initial reversible capacity of 455.5 mAh g^{-1} at 50 mA g^{-1} [27].

18.2.2.2 Transition Metal Dichalcogenides in FBs

Contrary to common rigidity of active materials, 2D TMDs are intrinsically flexible and possess elevated Young's modulus that enables the enduring of the strain up to 10% [29]. Therefore, TMD-based anodes need innovative strategies to eliminate the common volumetric expansion, peeling effect at the interfaces, and low electrical conductivity issues during intercalation/deintercalation process. The general approach for a design of a FB with such chalcogen-based active material is to support it with a conductive substrate having a high elastic modulus, such as carbon

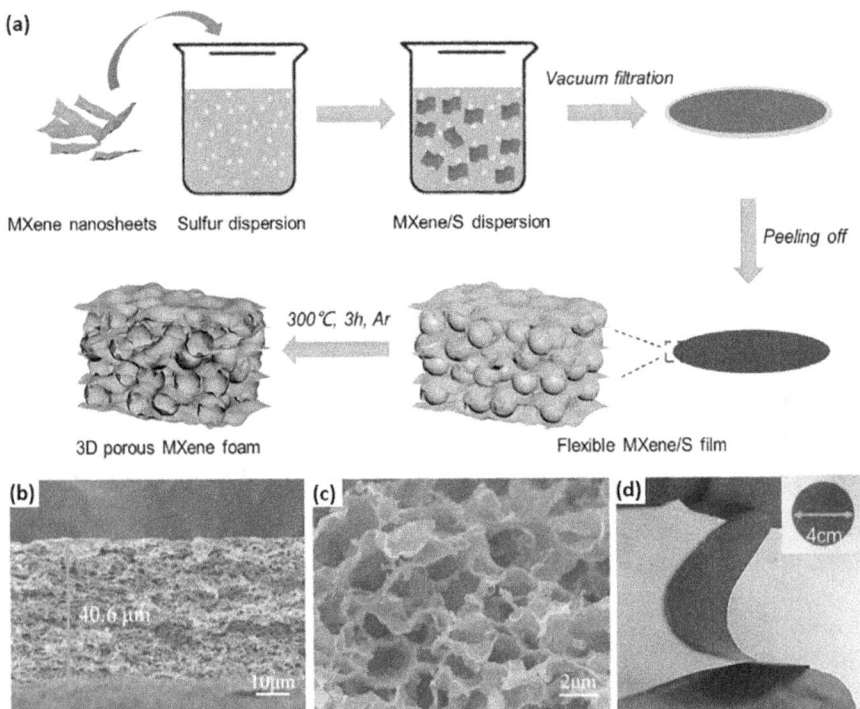

FIGURE 18.2 (a) Demonstration of the production process of the 3D flexible Ti_3C_2 MXene foam. (b and c) SEM images, and (d) digital photo of Ti_3C_2 MXene foam. (Adapted from Ref. [27] with permission from John Wiley & Sons, Copyright 2019, Wiley-VCH Verlag GmbH & Co. KGaA.)

or cellulose nanofibers (CNFs). To that end, Ni et al. compared 3D networks constructed from nitrogen-doped plain cellulose nanofibers (N-CNFs), nitrogen-doped multi-nanochannel cellulose nanofibers (NMCFs), and nitrogen-doped hollow-multi-nanochannel cellulose nanofibers (NHMCFs) decorated with vertical $MoSe_2$ nanosheets by hydrothermal synthesis method [8]. In this configuration, free electrons of nitrogen promoted the conductivity and the diffusion distance was shortened by vertically grown $MoSe_2$ nanosheets on both the inner and outer surface. Due to the extended surface area, 21.25% and 13.85% more mass loadings were obtained in NHMCFs and NMCFs, respectively. These channels provide enough room for the volume change during intercalation. Since $MoSe_2$@NHMCNFs have the highest surface area and more vertically anchored nanosheets with a mass loading of 71.55 wt.%, its capacity (586.7 mAh g^{-1} at 1 mA g^{-1}) was measured as 3.54-fold higher than that of $MoSe_2$@CNFs at the same conditions [8].

The flexible porous scaffold for TMDs can also be generated by the assembly of 2D nanoscale building blocks in the form of self-supporting hybrid structures [9,30,31]. The high electrical conductivity and intrinsic flexibility of MXenes make them potential candidates alternative to carbon-based networks. Recently, Bai et al. investigated flake-shaped folded $Ti_3C_2T_x$ nanosheets to encapsulate MoS_2/C nanosphere composites by vacuum filtration method [9]. The polyvinyl pyrrolidone (PVP)-derived carbon acts as both cushion to avoid stacking of nanosheets buffering the volume fluctuations and path for Li-ion/electron transfer hindering polysulfide dissolution during charge/discharge process. Furthermore, PVP-derived carbon not only covers MoS_2 nanosheets but also intercalates into those nanosheets and doubles the interlayer spacing, which provides extra storage capacity for lithium ions. The flexible, self-standing anode system (MoS_2/C@$Ti_3C_2T_x$) exhibited a capacity of 538.5 mAh g^{-1} at 0.05 A g^{-1} and also achieved a 96.7% of capacity retention after 150 cycles at 2 A g^{-1} [9].

18.2.2.3 Transition Metal Oxides (TMOs) in FBs

Mixed valence states, shortened ion diffusion distance, abundant active sites, and high redox activities in intercalation reactions make transition metal oxides (TMOs) ideal candidates for high-energy flexible electrodes. TMOs also have the ability to accommodate large ions such as sodium or potassium due to their high interlayer spacing distance and substantial lattice expansion [32]. It enables the advance of alternative rechargeable FBs beyond lithium chemistry. Mn-, Zn-, Fe-, Ni-, and Co-based oxides are the most investigated 2D active materials in FBs. To eliminate TMOs' irreversible restacking and aggregation tendencies, Zeng et al. developed binder-free anodes for Na/K storage in the form of sandwiched structured self-standing, flexible films by transforming 2D MXene ($Ti_3C_2T_x$) phase into sodium titanate (NTO)/potassium titanate integrated with graphene layers (Figure 18.3a) [33]. The ultrathin sandwiched structure of the few layered nanosheets was clearly observed in the HRTEM images (Figure 18.3c). Binder-free NTO/rGO anode with low working voltage exhibited a promising rate and cycling performance with 72 mAh g^{-1} at 5 A g^{-1} after 10,000 cycles. Long-term stability showed that graphene layers successfully buffered the expansion of $Ti_3C_2T_x$ lattice in the sandwiched structure. Although initial discharge capacity was 1,280 mAh g^{-1}, it lost 74% of its capacity

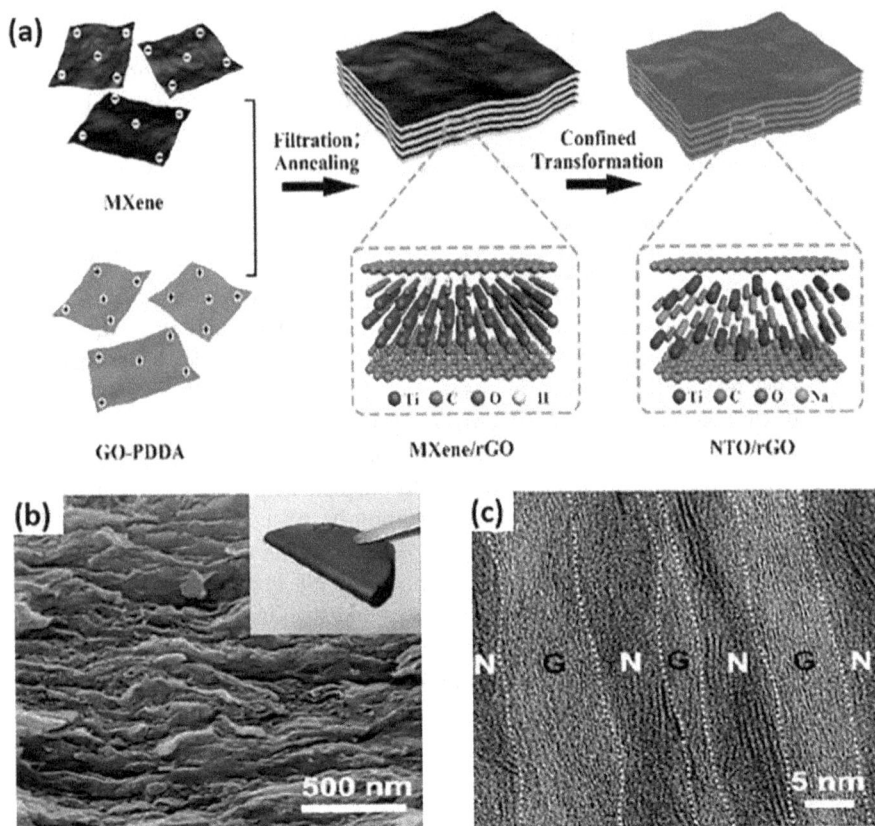

FIGURE 18.3 (a) Scheme illustrating preparation of flexible NTO/rGO films. (b) Sideview SEM image of NTO/rGO-10% hybrid film, digital photograph (inset). (c) Cross-sectional TEM image ("N" refers to NTO and "G" refers to rGO). (Adapted from Ref. [33] with permission from John Wiley & Sons, Copyright 2018, Wiley-VCH Verlag GmbH & Co. KgaA.)

at the first discharge cycle because of irreversible reactions and high amount of SEI formation [33]. New generation solid-state electrolytes may boost the cycling performance of this unique design.

Intrinsic conductivity, redox active sites, and wettability of electrodes can also be enhanced by coating or in situ growth of 2D metal oxides on fiber-like carbon or metal-based flexible structures [34]. Recently, Jin et al. managed to decorate Mn-doped Fe_2O_3 and NiO nanosheets on copper fibers to realize woven fabrics as flexible electrodes. Designing of 2D metal oxides into a 3D hierarchical structure demonstrated a stable performance up to 30,000 cycles even under various twisting and stress conditions [35]. Besides synthetic, nongreen, and costly inorganic scaffolds, there are also some robust, renewable, cheap, and biomass-derived alternative solutions to accommodate 2D nanomaterials in flexible electrodes. One of the most remarkable examples is the direct engineering of the reed straw waste as a 3D carbon host in both anode and cathode (Figure 18.4) [36]. Reed straw is naturally abundant on the surface of the ocean and possesses honeycomb-like cellular structure that aids in transfer of electrons. Thanks to the multi-level open and straight channels with enhanced wettability, quasi-solid-state Zn–MnO$_2$ battery with gel electrolyte delivered an initial specific capacity 306.7 mAh g^{-1} at 0.3 A g^{-1} and 95.2% of its capacity was retained at 3,000 cycles. Interestingly, Nyquist plots showed that the effect of impressive mass loading (weight percentage of 79.1% corresponding to 51.0 mg cm^3) on the Li$^+$ transfer rate was insignificant, which is favorable for maximizing the volumetric capacity in FBs. Moreover, extreme thermal and mechanical deformation did not cause any capacity fade and its performance of durability in harsh conditions was found to be outstanding in terms of safety aspects [36].

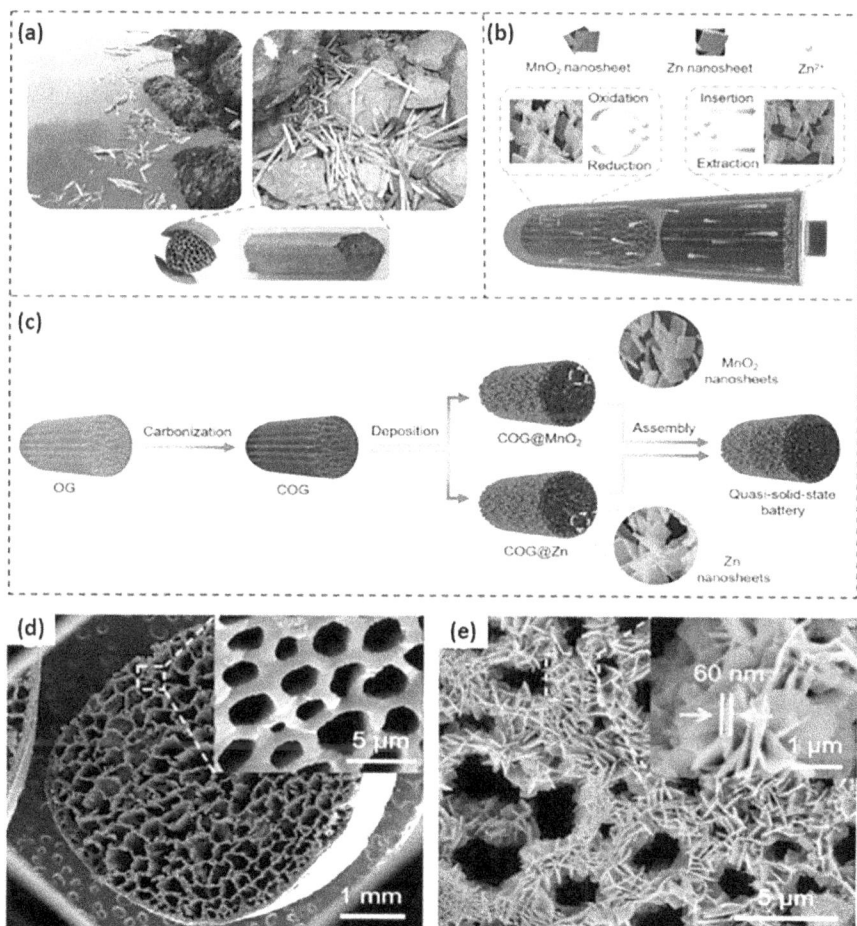

FIGURE 18.4 (a) Photographs of the ocean garbage (OG). (b) Schematic illustration of the structure and the working mechanism of the battery. (c) Schematic illustration of the preparation strategy for the battery from OG. The cross-sectional SEM images of (d) the raw carbonized-OG (COG) and (e) the COG@MnO$_2$ electrode. (Adapted from Ref. [49] with permission from Elsevier, Copyright 2020.)

18.2.2.4 Layered Double Hydroxides (LDHs) in FBs

Despite rapid progress on 2D material utilization and their investigation in FBs, layered double hydroxides (LDHs) are more investigated in conventional rechargeable battery components rather than FBs. Noble metals already offer satisfactory electrocatalytic activities, but their instability, rareness, and lack of multifunctionality restrict their practical application in metal-air batteries [37]. Regarding these issues, a bifunctional electrocatalyst (Co–CoO$_x$N–C) was directly synthesized from the precursor (CoAl-LDH@polydopamine) through hydrothermal and pyrolysis steps on a carbon cloth (CC) that functioned both as a current collector and a substrate [38]. The combination of OER active Co–CoO$_x$ and ORR active N-doped carbon in the form of vertically aligned nanosheet arrays in a single electrode gave a highly efficient bifunctional electrocatalyst with a small potential gap (0.678 V). As-prepared cathode was tested in an all-solid-state zinc-air battery and gave a power density of 20.7 mW cm^{-2} [38]. As known from previous reports, polymer-based solid-state electrolytes have relatively small lithium-ion transference numbers, instable at high potentials, and usually fail to prevent dendrite propagation at long cycles. At this point, 2D LDHs may offer substantial solutions for the above said issues by promoting flexibility as well. One of the latest studies evaluated the properties of 2D LDH/polyethylene oxide (PEO) composite solid-state electrolyte in flexible LiBs for the first time (Figure 18.5) [39]. Introduction of 5 wt.% 2D Mg-Al LDH nanosheets into PEO matrix gave Li$^+$ transference number and Li$^+$ conductivity of 0.42 and 1.1×10^{-5} S cm^{-1} at 30°C, respectively. The enhanced interface compatibility between the electrode and the electrolyte enabled a current density of 800 μA cm^{-2} at 60°C. This novel composite electrolyte performed a capacity of 138 mAh g^{-1} at 0.2 C and retained 88% of its capacity for 100 cycles with no dendrite occurrence. LDH nanosheets inhibited the motion of PEO chains by forming strong hydrogen bonds with those chains having high incidence of hydroxide radicals, promoting the migration of Li$^+$ ions by reducing the anion migration [39].

18.3 POLYMER-BASED 2D MATERIALS IN FBs

18.3.1 Metal Organic Frameworks (MOFs) in FBs

Generally, metal organic frameworks (MOFs) consist of covalently linked 2D organic ligands with heteroatoms like phosphorus, nitrogen, boron, or sulfur. Embedding these heteroatoms into the carbon backbone enhances their redox capability to such

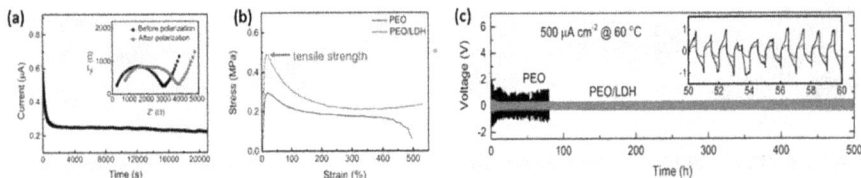

FIGURE 18.5 (a) AC impedance spectra of PEO/LDH composite polymer electrolyte (CPE). (b) Stress–strain curves of solid electrolytes. (c) Voltage profiles of the Li/CPE/Li symmetric cell. (Adapted from Ref. [39] with permission from Springer Nature, Copyright 2021.)

2D frameworks as potential active materials for the negative electrodes in FBs. Furthermore, polar functional groups at linkages contribute to the ion storage performance and/or block detrimental ion mobilities at the interface of 3D porous structure [6]. By taking advantage of these features in flexible lithium–sulfur batteries, Zheng et al. developed a layer involving ZIF-7 based MOF on carbon fiber to prohibit the shuttling effect without interruption of ZIF-7 electrolyte infiltration in flexible Li–S batteries (Figure 18.6) [10]. The interlayer ZIF-7@CNF was fabricated by chemically processing in situ grown ZnO nanorods into vertically aligned ZIF-7 nanosheets on a carbon fiber cloth. An excellent discharge specific capacity of 1226.9 mAh g^{-1} at 0.2 C may be attributed to the polar and sulfiphilic nature of the ZIF-7@CNF interlayer to successfully inhibit LiPS dissolution [10].

In another example, a zinc-based MOF derivation on a CC was used as an active material in the flexible cathode for the aluminum-air battery [11]. Zn-MOF was functionalized with sublimated ferrocene to decorate the inner/outer surface by iron atoms. With the incorporation of ferrocene molecules with Zn-MOF as a host as well as the rich nitrogen source from pyridinic N sites enabled open-circuit voltage of 2.1 V at bending state and a stable discharge voltage of ≈1.5 V over 8 h at 1 mA cm^{-2} [11].

Zhou et al. utilized LDH sheets to physically block LiPS by modification of polypropylene separator [40]. These cationic sheets (Mg$_2$Al-LDH) acted as a barrier layer to confine the crossover of polysulfides and enabled the use of pure sulfur in flexible cathodes. In comparison to the performance of bare separator, Mg$_2$Al-LDH (density: ~0.018 mg cm^{-2}, thickness: ~20–30 nm) improved reversible capacity decay from 0.29% to 0.18% per cycle.

18.3.2   COVALENT ORGANIC FRAMEWORKS (COFs) IN FBs

Different from MOFs, 2D layered covalent organic frameworks (COFs) do not involve metal ions/clusters but instead covalently bonded pure organic building blocks. High chemical stability, tunable pore size, long-range order, and rigid backbones make 2D COFs promising networks to support polymer-based solid-state electrolytes in FBs [6]. Recently, 2D COFs containing hydrazone linkages were modified with the condensed PEO through a solvothermal process and evolved into lithiophilic imine

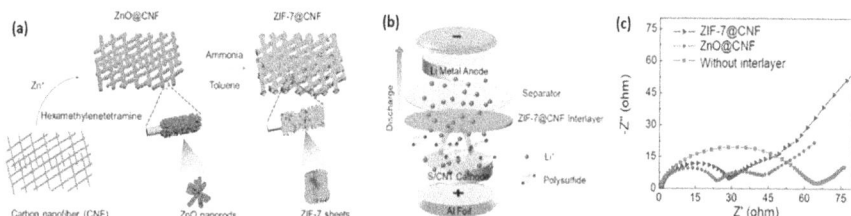

FIGURE 18.6   (a) Schematic of ZIF-7@CNF interlayer fabrication process. (b) Schematic of the ZIF-7@CNF interlayer in Li–S batteries. (c) Electrochemical impedance spectroscopy (EIS) of working electrodes. (Adapted from Ref. [10] with permission from The Royal Society of Chemistry.)

bonds which were beneficial for the uniform deposition and Li^+ conductivity [41]. Both crystalline 2D COF structure and the accumulated flexible PEO groups in glassy phase provided plenty of ion hopping sites and conduction pathways to enhance Li^+ conductivity (1.33×10^{-3} S cm^{-1} at 200°C) [41]. Availability of lithiophilic imine bonds through multifunctional materials containing COFs can also be an advantage for achieving stable electrodes in alkali-based FBs. In this regard, the limited wettability of $Ti_3C_2T_x$ substrate was improved by constructing a freestanding MXene/COF nanocomposite film for dendrite-free lithium anodes [42]. $Ti_3C_2T_x$ nanosheets were decorated with Li-reactive 2D Schiff-base COF-LZU1 microspheres to increase the ratio of lithiophilic functional groups. Here, COF microspheres worked as nucleation seeds to regulate the morphology of the deposited Li metal. However, COFs incorporation over 5 wt.% was found to have a reducing effect on both conductivity and flexibility. Owing to these merits, hybrid anode performed a Coulombic efficiency of 97.6% at 0.5 mA cm^{-2} in the flexible Li–S full cell [42].

18.3.3 CONJUGATED AROMATIC POLYMERS (CAPS) IN FBS

As a subclass of emerging porous polymers, 2D conjugated aromatic polymers (CAPs) have been extensively explored within the last few years, especially for electrodes in alkali metal ion FBs [43–46]. The reason is their superior charge–discharge voltage varying in the range of 3–4 V, which is crucial for higher energy density [47,48]. Despite having polarization losses, withstanding ability to structural alterations gave redox-active conjugated polymers an edge on FBs [49]. Recently, Duan et al. synthesized a crystalline few-layer 2D polyimide (2DPI) as anode for flexible sodium-ion batteries [43]. Few-layer π-conjugated 2DPI nanosheets with a thickness of ~1.5 to 2.6 nm and an average pore size of 1.3 nm were generated by organizing internal order and in-plane periodicity by means of hydrogen bonding and imidization. Planar hydrogen-bonded networks with a number of redox-active sites (imide functional groups and triazine rings) delivered a specific capacity of 312 mAh g^{-1} at 0.1 A g^{-1}. Furthermore, it performed an outstanding stability by retaining 95% of initial capacity for 1,100 cycles [43]. In another study, a novel 2D bipolar CAP with porous-honeycomb structure was developed as cathode active material for sodium storage in FBs [44]. A new material platform was synthesized through ionothermal method to achieve short-range-ordered 2D structures, 300 nm wide and less than ten stacking layers. The presence of benzene and triazine rings is featured in p-/n-dopable regions together for a continuous discharge process, enabling a wide working potential of 1.3–4.1 V vs. Na/Na^+ and thus a higher energy density. Negative bipolar porous organic electrode provided a specific power of 10 $kWkg^{-1}$, which was one of the best among Na-ion batteries. Although its self-discharge characteristic was not comparable to Na-ion or other intercalation-based electrodes, the retention of 80% of initial capacity for 7,000 cycles at 1.0 A g^{-1} was a remarkable and promising performance for the future [44].

In another example, Liu et al. coated the copper foil by polyporphyrin (TThPP)-based 2D COF through chemical oxidative polymerization technique [50]. The TThPP nanosheets linked via the thiophenephenyl groups to the foil and generated a conductive film. As-prepared anode exhibited a reversible capacity of 666 mAh g^{-1} at a current density of 200 mA g^{-1}. The same group further investigated sulfur storage capability of TThPP-based 2D COFs in flexible Li–S batteries [51]. The galvanostatic charge/discharge test at a current rate of 0.5 C resulted with a capacity of 633 mAh g^{-1} after 200 cycles. The capacity decay rate of 0.16% per cycle and 96% of Coulombic efficiency proved that the proposed cathode (COF/S) system was highly effective to alleviate the shuttle effect of Li–S intermediates.

18.4   CONCLUSION AND FUTURE PERSPECTIVES

Emerging technologies such as internet of things or wearable electronics necessitate high performance, innovative, and cost-effective strategies on FBs. However, vast majority of the research efforts agreed that traditional energy storage materials are incapable to meet all the urgent demands for FBs. Although emerging 2D nanomaterials with unique features have great contribution to elastic properties, their performances solely are not enough for practical applications due to the complex chemistry in rechargeable batteries. 2D materials used in rigid batteries can be rendered suitable for FBs by a flexible matrix involving binders and additives. As illustrated with recent studies in this chapter, the use of intrinsically flexible emerging 2D materials can provide the same properties without compromising from the specific power and the energy densities. Rising the tap density is of paramount importance to elevate the volumetric capacity.

Hybrid 2D nanomaterials or multiphase mixtures offer relatively simple but effective solutions to achieve multifunctional properties [48,52]. For instance, in sulfur-based FBs, oppositely charged metal hydroxide layers can immobilize LiPS by electrostatic interactions and act as advanced sulfur host, while graphene nanosheets can boost electron transfer rate and hinder the irreversible restacking of individual LDH layers [53]. The incorporation of nanotubes are also an effective way to bridge sheets and prevent agglomeration. In many cases, 2D active material should be supported with a conductive substrate in order to achieve flexibility and retain structural integrity for longer lifetimes. Nanofibers are widely utilized as conductive scaffolds for 2D nanomaterials in FBs [54,55]. In situ growth of 2D nanosheets on freestanding papers is one of the prospective engineering applications to fabricate pliable electrodes. Other functionalization techniques like chemical doping, surface modification, or structural modulation are also widely investigated to improve both electrochemical and mechanical properties of 2D nanomaterials used in FBs [56–58].

In addition to robust and rational design, extra attention must be paid on the low-cost and eco-friendly synthesis of 2D nanomaterials to realize large-scale application of FBs. Establishing universal evaluation standards for characterization in addition to further advances of in situ characterization techniques will be helpful to have a better understanding on the reaction kinetics, ion transport mechanisms, and/or catalytic activities of emerging 2D nanomaterials in FBs. It is sure that FBs are in the early stages of development and there is still room for more revolutionary 2D nanomaterials possible to use in FBs.

REFERENCES

1. Bocchetta, P.; Frattini, D.; Ghosh, S.; Mohan, A. M. V.; Kumar, Y.; Kwon, Y. Soft materials for wearable/flexible electrochemical energy conversion, storage, and biosensor devices. *Materials (Basel)*, **2020**, *13* (12), 2733. doi: 10.3390/ma13122733.

2. Yi, F.; Ren, H.; Shan, J.; Sun, X.; Wei, D.; Liu, Z. Wearable energy sources based on 2D materials. *Chem. Soc. Rev.*, **2018**, *47* (9), 3152–3188. doi: 10.1039/c7cs00849j.

3. Fu, K. K.; Cheng, J.; Li, T.; Hu, L. Flexible batteries: From mechanics to devices. *ACS Energy Lett.*, **2016**, *1* (5), 1065–1079. doi: 10.1021/acsenergylett.6b00401.

4. Liu, Y.; Sun, Z.; Tan, K.; Denis, Di. K.; Sun, J.; Liang, L.; Hou, L.; Yuan, C. Recent progress in flexible non-lithium based rechargeable batteries. *J. Mater. Chem. A*, **2019**, *7* (9), 4353–4382. doi: 10.1039/c8ta10258a.

5. Wang, R.; Dai, X.; Qian, Z.; Zhong, S.; Chen, S.; Fan, S.; Zhang, H.; Wu, F. Boosting lithium storage in free-standing black phosphorus anode via multifunction of nanocellulose. *ACS Appl. Mater. Interfaces*, **2020**, *12* (28), 31628–31636. doi: 10.1021/acsami.0c08346.

6. Zhang, C.; Lu, C.; Zhang, F.; Qiu, F.; Zhuang, X.; Feng, X. Two-dimensional organic cathode materials for alkali-metal-ion batteries. *J. Energy Chem.*, **2018**, *27* (1), 86–98. doi: 10.1016/j.jechem.2017.11.008.

7. Zhang, R.; Palumbo, A.; Kim, J. C.; Ding, J.; Yang, E. H. Flexible graphene-, graphene-oxide-, and carbon-nanotube-based supercapacitors and batteries. *Ann. Phys.*, **2019**, *531* (10). doi: 10.1002/andp.201800507.

8. Ni, X.; Cui, Z.; Luo, H.; Chen, H.; Liu, C.; Wu, Q.; Ju, A. Hollow multi-nanochannel carbon nanofibers@MoSe$_2$ nanosheets composite as flexible anodes for high performance lithium-ion batteries. *Chem. Eng. J.*, **2021**, *404*. doi: 10.1016/j.cej.2020.126249.

9. Bai, Z.; Yang, Y.; Zhang, D.; Wang, Y.; Guo, Y.; Yan, H.; Chu, P. K.; Luo, Y. Carbon-encapsulated nanosphere-assembled MoS$_2$ nanosheets with large interlayer distance for flexible lithium-ion batteries. *J. Solid State Electrochem.*, **2021**, *25* (5), 1657–1665. doi: 10.1007/s10008-021-04936-8.

10. Zheng, S.; Sun, D.; Wu, L.; Liu, S.; Liu, G. Carbon fiber supported two-dimensional ZIF-7 interlayer for durable lithium-sulfur battery. *J. Alloys Compd.*, **2021**, *870*. doi: 10.1016/j.jallcom.2021.159412.

11. Huang, L.; Zang, W.; Ma, Y.; Zhu, C.; Cai, D.; Chen, H.; Zhang, J.; Yu, H.; Zou, Q.; Wu, L.; et al. In-situ formation of isolated iron sites coordinated on nitrogen-doped carbon coated carbon cloth as self-supporting electrode for flexible aluminum-air battery. *Chem. Eng. J.*, **2021**, *421*. doi: 10.1016/j.cej.2021.129973.

12. Li, H.; Wen, X.; Shao, F.; Zhou, C.; Zhang, Y.; Hu, N.; Wei, H. Interface covalent bonding endowing high-sulfur-loading paper cathode with robustness for energy-dense, compact and foldable lithium-sulfur batteries. *Chem. Eng. J.*, **2021**, *412*. doi: 10.1016/j.cej.2021.128562.

13. Tian, Y.; Xu, L.; Li, M.; Yuan, D.; Liu, X.; Qian, J.; Dou, Y.; Qiu, J.; Zhang, S. Interface engineering of CoS/CoO@N-doped graphene nanocomposite for high-performance rechargeable Zn-air batteries. *Nano-Micro Lett.*, **2021**, *13* (1). doi: 10.1007/s40820-020-00526-x.

14. Chen, J.; Wen, L.; Fang, R.; Wang, D. W.; Cheng, H. M.; Li, F. Stress release in high-capacity flexible lithium-ion batteries through nested wrinkle texturing of graphene. *J. Energy Chem.*, **2021**, *61*, 243–249. doi: 10.1016/j.jechem.2021.03.021.

15. Lei, W.; Xiao, W.; Li, J.; Li, G.; Wu, Z.; Xuan, C.; Luo, D.; Deng, Y. P.; Wang, D.; Chen, Z. Highly nitrogen-doped three-dimensional carbon fibers network with superior sodium storage capacity. *ACS Appl. Mater. Interfaces*, **2017**, *9* (34), 28604–28611. doi: 10.1021/acsami.7b08704.

16. Liu, M.; Zhang, P.; Qu, Z.; Yan, Y.; Lai, C.; Liu, T.; Zhang, S. Conductive carbon nano-fiber interpenetrated graphene architecture for ultra-stable sodium ion battery. *Nat. Commun.*, **2019**, *10* (1). doi: 10.1038/s41467-019-11925-z.

17. Fu, L.; Tang, K.; Song, K.; Van Aken, P. A.; Yu, Y.; Maier, J. Nitrogen doped porous carbon fibres as anode materials for sodium ion batteries with excellent rate perfor-mance. *Nanoscale*, **2014**, *6* (3), 1384–1389. doi: 10.1039/c3nr05374a.

18. Islam, M. S.; Faisal, S. N.; Tong, L.; Roy, A. K.; Zhang, J.; Haque, E.; Minett, A. I.; Wang, C. H. N-doped reduced graphene oxide (RGO) wrapped carbon microfibers as binder-free electrodes for flexible fibre supercapacitors and sodium-ion batteries. *J. Energy Storage*, **2021**, *37*. doi: 10.1016/j.est.2021.102453.

19. Tao, H.; Fan, Q.; Ma, T.; Liu, S.; Gysling, H.; Texter, J.; Guo, F.; Sun, Z. Two-dimensional materials for energy conversion and storage. *Prog. Mater. Sci.*, **2020**, *111*. doi: 10.1016/j.pmatsci.2020.100637.

20. Ma, Z.; Li, Z.; Hu, K.; Liu, D.; Huo, J.; Wang, S. The enhancement of polysulfide absorbsion in Li-S batteries by hierarchically porous CoS_2/carbon paper interlayer. *J. Power Sources*, **2016**, *325*, 71–78. doi: 10.1016/j.jpowsour.2016.04.139.

21. He, J.; Wang, N.; Cui, Z.; Du, H.; Fu, L.; Huang, C.; Yang, Z.; Shen, X.; Yi, Y.; Tu, Z.; et al. Hydrogen substituted graphdiyne as carbon-rich flexible electrode for lithium and sodium ion batteries. *Nat. Commun.*, **2017**, *8* (1). doi: 10.1038/s41467-017-01202-2.

22. Zhu, J. P.; Xiao, G. S.; Zuo, X. X. Two-dimensional black phosphorus: An emerging anode material for lithium-ion batteries. *Nano-Micro Lett.* 2020. doi: 10.1007/s40820-020-00453-x.

23. Ju, J.; Ma, J.; Wang, Y.; Cui, Y.; Han, P.; Cui, G. Solid-state energy storage devices based on two-dimensional nano-materials. *Energy Storage Mater.*, **2019**, *20*, 269–290. doi: 10.1016/j.ensm.2018.11.025.

24. Peng, Q.; Hu, K.; Sa, B.; Zhou, J.; Wu, B.; Hou, X.; Sun, Z. Unexpected elastic isotropy in a black phosphorene/TiC_2 van der Waals heterostructure with flexible Li-ion battery anode applications. *Nano Res.*, **2017**, *10* (9), 3136–3150. doi: 10.1007/s12274-017-1531-5.

25. Peng, Q.; Wang, Z.; Sa, B.; Wu, B.; Sun, Z. Blue phosphorene/MS_2 (M = Nb, Ta) heterostructures as promising flexible anodes for lithium-ion batteries. *ACS Appl. Mater. Interfaces*, **2016**, *8* (21), 13449–13457. doi: 10.1021/acsami.6b03368.

26. Li, B.; Zhang, D.; Liu, Y.; Yu, Y.; Li, S.; Yang, S. Flexible Ti_3C_2 MXene-lithium film with lamellar structure for ultrastable metallic lithium anodes. *Nano Energy*, **2017**, *39*, 654–661. doi: 10.1016/j.nanoen.2017.07.023.

27. Zhao, Q.; Zhu, Q.; Miao, J.; Zhang, P.; Wan, P.; He, L.; Xu, B. Flexible 3D porous MXene foam for high-performance lithium-ion batteries. *Small*, **2019**, *15* (51). doi: 10.1002/smll.201904293.

28. Sun, Y.; Chen, D.; Liang, Z. Two-dimensional MXenes for energy storage and conversion applications. *Mater. Today Energy*, **2017**, *5*, 22–36. doi: 10.1016/j.mtener.2017.04.008.

29. Mukherjee, S.; Singh, G. Two-dimensional anode materials for non-lithium metal-ion batteries. *ACS Appl. Energy Mater.*, **2019**, *2* (2), 932–955. doi: 10.1021/acsaem.8b00843.

30. Ren, J.; Wang, Z.; Yang, F.; Ren, R. P.; Lv, Y. K. Freestanding 3D single-wall carbon nanotubes/WS_2 nanosheets foams as ultra-long-life anodes for rechargeable lithium ion batteries. *Electrochim. Acta*, **2018**, *267*, 133–140. doi: 10.1016/j.electacta.2018.01.167.

31. Wen, H.; Kang, W.; Liu, X.; Li, W.; Zhang, L.; Zhang, C. Two-phase interface hydrother-mal synthesis of binder-free SnS_2/graphene flexible paper electrodes for high-perfor-mance Li-ion batteries. *RSC Adv.*, **2019**, *9* (41), 23607–23613. doi: 10.1039/c9ra03397a.

32. Li, Y.; Wang, R.; Guo, Z.; Xiao, Z.; Wang, H.; Luo, X.; Zhang, H. Emerging two-dimensional noncarbon nanomaterials for flexible lithium-ion batteries: Opportunities and challenges. *J. Mater. Chem. A*, **2019**, *7* (44), 25227–25246. doi: 10.1039/c9ta09377j.

33. Zeng, C.; Xie, F.; Yang, X.; Jaroniec, M.; Zhang, L.; Qiao, S. Z. Ultrathin titanate nanosheets/graphene films derived from confined transformation for excellent Na/K ion storage. *Angew. Chemie Int. Ed.*, **2018**, *57* (28), 8540–8544. doi: 10.1002/anie.201803511.

34. Zhu, J.; Yin, Z.; Yang, D.; Sun, T.; Yu, H.; Hoster, H. E.; Hng, H. H.; Zhang, H.; Yan, Q. Hierarchical hollow spheres composed of ultrathin Fe_2O_3 nanosheets for lithium storage and photocatalytic water oxidation. *Energy Environ. Sci.*, **2013**, *6* (3), 987–993. doi: 10.1039/c2ee24148j.

35. Jin, Z.; Li, P.; Jin, Y.; Xiao, D. Superficial-defect engineered nickel/iron oxide nanocrystals enable high-efficient flexible fiber battery. *Energy Storage Mater.*, **2018**, *13*, 160–167. doi: 10.1016/j.ensm.2018.01.010.

36. Zhao, J.; Wu, W.; Jia, X.; Xia, T.; Li, Q.; Zhang, J.; Wang, Q.; Zhang, W.; Lu, C. High-value utilization of biomass waste: From garbage floating on the ocean to high-performance rechargeable $Zn-MnO_2$ batteries with superior safety. *J. Mater. Chem. A*, **2020**, *8* (35), 18198–18206. doi: 10.1039/d0ta05926a.

37. Oncu, A.; Cetinkaya, T.; Akbulut, H. Enhancement of the electrochemical performance of free-standing graphene electrodes with manganese dioxide and ruthenium nanocatalysts for lithium-oxygen batteries. *Int. J. Hydrogen Energy*, **2021**, *46* (33), 17173–17186. doi: 10.1016/j.ijhydene.2021.02.154.

38. Li, S.; Xie, W.; Song, Y.; Shao, M. Layered double hydroxide@polydopamine core-shell nanosheet arrays-derived bifunctional electrocatalyst for efficient, flexible, all-solid-state zinc-air battery. *ACS Sustain. Chem. Eng.*, **2020**, *8* (1), 452–459. doi: 10.1021/acssuschemeng.9b05754.

39. Wang, Q.; Wu, J. F.; Yu, Z. Y.; Guo, X. Composite polymer electrolytes reinforced by two-dimensional layer-double-hydroxide nanosheets for dendrite-free lithium batteries. *Solid State Ionics*, **2020**, *347*. doi: 10.1016/j.ssi.2020.115275.

40. Zhou, Y.; Hu, G.; Zhang, W.; Li, Q.; Zhao, Z.; Zhao, Y.; Li, F.; Geng, F. Cationic two-dimensional sheets for an ultralight electrostatic polysulfide trap toward high-performance lithium-sulfur batteries. *Energy Storage Mater.*, **2017**, *9*, 39–46. doi: 10.1016/j.ensm.2017.06.005.

41. Zhang, G.; Hong, Y. L.; Nishiyama, Y.; Bai, S.; Kitagawa, S.; Horike, S. Accumulation of glassy poly(ethylene oxide) anchored in a covalent organic framework as a solid-state Li+ electrolyte. *J. Am. Chem. Soc.*, **2019**, *141* (3), 1227–1234. doi: 10.1021/jacs.8b07670.

42. Wei, C.; Wang, Y.; Zhang, Y.; Tan, L.; Qian, Y.; Tao, Y.; Xiong, S.; Feng, J. Flexible and stable 3D lithium metal anodes based on self-standing MXene/COF frameworks for high-performance lithium-sulfur batteries. *Nano Res.*, **2021**. doi: 10.1007/s12274-021-3433-9.

43. Duan, H.; Lyu, P.; Liu, J.; Zhao, Y.; Xu, Y. Semiconducting crystalline two-dimensional polyimide nanosheets with superior sodium storage properties. *ACS Nano*, **2019**. doi: 10.1021/acsnano.8b09416.

44. Sakaushi, K.; Hosono, E.; Nickerl, G.; Gemming, T.; Zhou, H.; Kaskel, S.; Eckert, J. Aromatic porous-honeycomb electrodes for a sodium-organic energy storage device. *Nat. Commun.*, **2013**, *4*. doi: 10.1038/ncomms2481.

45. Xia, S. B.; Li, F. S.; Shen, X.; Li, X.; Cheng, F. X.; Sun, C. K.; Guo, H.; Liu, J. J. A photochromic zinc-based coordination polymer for a Li-ion battery anode with high capacity and stable cycling stability. *Dalt. Trans.*, **2018**, *47* (37), 13222–13228. doi: 10.1039/c8dt02930j.

46. Yoshizawa-Fujita, M.; Horiuchi, S.; Uemiya, T.; Ishii, J.; Takeoka, Y.; Rikukawa, M. Polyether-based supramolecular electrolytes with two-dimensional boroxine skeleton. *Front. Energy Res.*, **2021**, *9*. doi: 10.3389/fenrg.2021.663270.

47. Kuila, B. K. Nanoheterostructured materials based on conjugated polymer and two-dimensional materials: Synthesis and applications. In S. Jit and S. Das (Eds) *2D Nanoscale Heterostructured Materials*. Elsevier, 2020, pp. 91–124. doi: 10.1016/b978-0-12-817678-8.00004-x.

48. Ren, Y.; Yu, C.; Chen, Z.; Xu, Y. Two-dimensional polymer nanosheets for efficient energy storage and conversion. *Nano Res.*, **2021**, *14* (6), 2023–2036. doi: 10.1007/s12274-020-2976-5.

49. Xie, J.; Gu, P.; Zhang, Q. Nanostructured conjugated polymers: Toward high-performance organic electrodes for rechargeable batteries. *ACS Energy Lett.*, **2017**, *2* (9), 1985–1996. doi: 10.1021/acsenergylett.7b00494.

50. Yang, H.; Zhang, S.; Han, L.; Zhang, Z.; Xue, Z.; Gao, J.; Li, Y.; Huang, C.; Yi, Y.; Liu, H.; et al. High conductive two-dimensional covalent organic framework for lithium storage with large capacity. *ACS Appl. Mater. Interfaces*, **2016**, *8* (8), 5366–5375. doi: 10.1021/acsami.5b12370.

51. Liao, H.; Wang, H.; Ding, H.; Meng, X.; Xu, H.; Wang, B.; Ai, X.; Wang, C. A 2D porous porphyrin-based covalent organic framework for sulfur storage in lithium-sulfur batteries. *J. Mater. Chem. A*, **2016**, *4* (19), 7416–7421. doi: 10.1039/c6ta00483k.

52. Chua, S.; Fang, R.; Sun, Z.; Wu, M.; Gu, Z.; Wang, Y.; Hart, J. N.; Sharma, N.; Li, F.; Wang, D. W. Hybrid solid polymer electrolytes with two-dimensional inorganic nanofillers. *Chem. Eur. J.*, **2018**, *24* (69), 18180–18203. doi: 10.1002/chem.201804781.

53. Cao, Y.; Lin, Y.; Wu, J.; Huang, X.; Pei, Z.; Zhou, J.; Wang, G. Two-dimensional MoS_2 for Li–S batteries: Structural design and electronic modulation. *ChemSusChem*, **2020**, *13* (6), 1392–1408. doi: 10.1002/cssc.201902688.

54. Lin, L.; Ning, H.; Song, S.; Xu, C.; Hu, N. Flexible electrochemical energy storage: The role of composite materials. *Compos. Sci. Technol.*, **2020**, *192*. doi: 10.1016/j.compscitech.2020.108102.

55. Jamesh, M. I. Recent advances on flexible electrodes for Na-ion batteries and Li–S batteries. *J. Energy Chem.*, **2019**, 15–44. doi: 10.1016/j.jechem.2018.06.011.

56. Shen, X.; Zheng, Q.; Kim, J. K. Rational design of two-dimensional nanofillers for polymer nanocomposites toward multifunctional applications. *Prog. Mater. Sci.*, **2021**, *115*. doi: 10.1016/j.pmatsci.2020.100708.

57. Jana, M.; Xu, R.; Cheng, X. B.; Yeon, J. S.; Park, J. M.; Huang, J. Q.; Zhang, Q.; Park, H. S. Rational design of two-dimensional nanomaterials for lithium-sulfur batteries. *Energy Environ. Sci.*, **2020**, *13* (4), 1049–1075. doi: 10.1039/c9ee02049g.

58. Yang, Y.; Liu, X.; Zhu, Z.; Zhong, Y.; Bando, Y.; Golberg, D.; Yao, J.; Wang, X. The role of geometric sites in 2D materials for energy storage. *Joule*, **2018**, *2* (6), 1075–1094. doi: 10.1016/j.joule.2018.04.027.

19 Nanocomposites of 2D Materials for Flexible Li-Ion Batteries

Demet Ozer and Zeliha Ertekin
Hacettepe University

CONTENTS

19.1 INTRODUCTION

Lithium-ion batteries (LIBs) have gained significant interest as a major alternative for energy storage tools due to their great energy density, extended cycle life, environmentally friendly, low-cost energy supply, etc. [1]. Unfortunately, conventional LIBs cannot satisfy the requirements of flexible electronics because traditional electrode materials are not sufficiently bendable, and it is desirable to prevail them with innovative alternatives that meet the following criteria: (i) to transfer electrons quickly, they should have high electrical conductivity; (ii) the active components should be firmly fixed on a high-loading-capacity supporting substrate; (iii) they should be both flexible and robust, and stay stable even when deformed; and (iv) they should have excellent electrochemical properties [2]. Nowadays, research focuses more on the enhancement of flexible Li-ion batteries (FLIBs). FLIBs are made up of the same components as traditional LIBs. Each component must be flexible, lightweight, stretchable, and implantable to provide the demands of portable and wearable electronics. The flexibility of FLIBs is typically analyzed according to their stretchability and bendability. An ideal flexible battery should possess great flexibility, high energy and power density, outstanding stability, and dynamically stable power output.

DOI: 10.1201/9781003178422-19

The FLIBs can be implemented in various ranges of energy storage devices. Significantly, they can also be used for portable medical devices during the COVID-19, which can also help to increase the flexible battery market. FLIBs can provide consistent, dependable power for a long time while maintaining high reliability and performance. As a result, implanted medical devices such as new pacemakers, medical patches, medical diagnostic sensors, drug delivery systems, disposable medical devices, and biosensors are appropriate. The portable and flexible devices have also emerged as fast-rising attention from many companies. The key market actors are Samsung SDI (Korea), STMicroelectronics (Switzerland), Blue Spark Technologies (United States), NEC Corp. (Japan), LG Chem (South Korea), Enfucell Oy Ltd. (Finland), Apple Inc. (United States), etc. For example, the NEC Corp. company produced a flexible organic radical battery with 0.3 mm thickness. Panasonic Corporation created FLIBs with a 0.55 mm thickness appropriate for use in card-type and wearable devices [3]. The global market for FLIBs is expected to reach over 296 million USD by 2025 [4].

To date, many researchers have considerable efforts to build up new FLIBs with bendable and stretchable properties. However, several challenges need to improve, such as electrode degradation, low battery performance, and flexible cell architecture design. The most promising strategy to overcome the fabrication challenge of FLIBs is to prepare electrodes using nanostructure materials. The formation of nanostructures has changed remarkable properties toward bulk materials, and the specific capacity, rate performance, life cycle, stability, etc., have enhanced the efficiency in electrochemical applications. In this chapter, innovations in the research and manufacture of FLIBs relevant to nanocomposites of 2D materials are discussed in detail. In addition, the design of the complete battery system and the materials (electrodes, substrates/collectors, and separators/electrolytes) are expressed as promising candidates for the next-generation FLIBs. Besides, future research directions for FLIBs are presented. It is discussed how important it is to design and manufacture FLIBs to achieve high flexibility, high power, and energy density all at the same time.

19.2   DESIGN OF FLEXIBLE LI-ION BATTERIES

FLIBs, the same as conventional LIBs, have three main components: substrates, electrolytes, and electrodes (anode and cathode) together with the current collector and battery shell (Figure 19.1a). All major components of FLIBs are bendable and stretchable. The recent research is mainly focused on improving the electrode materials properties such as high specific capacity, better cycle life, mechanical stability, and fast Li-ion diffusion. The chemical processes take place between the anode and cathode to provide the electrons. The intercalation and deintercalation of Li-ion take place between the anode and the cathode, as shown in Figure 19.1b. During charging, lithium ions deintercalate from cathodes and intercalate into anodes, while lithium ions move from the anodes to the cathodes during the discharge process. The cell architecture is an important factor as well as electrode materials to develop fully flexible LIBs. To improve the flexibility of the battery, each component should be portable, bendable, and wearable. All correlations between mechanical flexibility and the electrochemical performance of LIBs should be evaluated as well.

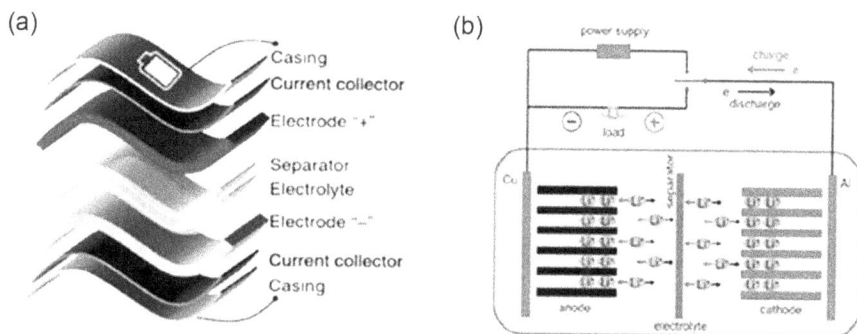

FIGURE 19.1 (a) A typical flexible Li-ion battery (FLIB) design. (Adapted with permission from Ref. [5]. Copyright (2016). Copyright the Authors, some rights reserved; exclusive licensee (Springer Nature). Distributed under a Creative Commons Attribution License 4.0 (CC BY).) (b) Schematic representation of FLIBs. (Adapted with permission from Ref. [6]. Copyright (2019) World Scientific Publishing Company.)

The most important focus in portable electronic devices is to design practical and esthetic devices as well as high efficiency. Although smaller, thinner, and lighter batteries are being produced, the design of FLIBs is a limiting issue. Current battery technology mainly offers design in fixed shapes such as cylindrical, prism, or pouch. The electrodes used in commercial LIBs have a cofacial configuration, but this design is not suitable for flexible batteries due to the deformation of the cell. More importantly, this configuration can cause frictional force between the electrodes when the cells are bent. To overcome this problem, three different cell types have been proposed in the literature. These are classified as coplanar cells, noble type cells, and cable/wire type cells [7]. The cathode and anode electrodes are positioned in the coplanar arrangement, which is in the same plane. The kinetics of ions transport is slower than in traditional designs since the electrodes are in the same plane. However, the flexible coplanar cell shows high flexibility upon buckling or stretching conditions and has a thickness of <0.5 mm. In the noble type of cell, the slurry containing active material is spread on the electrode, and then the prepared electrodes are wrapped in metal straps. Because the flexing component of the battery contains metal straps, this sort of cell provides cycle stability during flexing testing. Cable/wire type batteries differ from traditional flat and rigid batteries, and they are generally designed for wearable systems. This structure preserves its mechanical and electrochemical properties even in highly deformable shapes. Figure 19.2 shows the SEM image of the hollow helical anode and the application of flexibility to the cable cell in various folding and bending forms, as well as a schematic representation of the flexible cable cell's construction. An electroplated hollow helical anode has 12 strips of Ni–Sn coated Cu wire. It has been shown that the cathode composite thickness and the quantification of Ni–Sn coated wires were easily adjusted to improve the battery capacity. By putting a needle into a small gap within the hollow helix anode, the electrolyte was easily injected. The hollow, spring-like shape enables mechanical flexibility as well as efficient electrolyte transfer to the electrodes [8]. Soon, the cable-type battery might play an essential role in

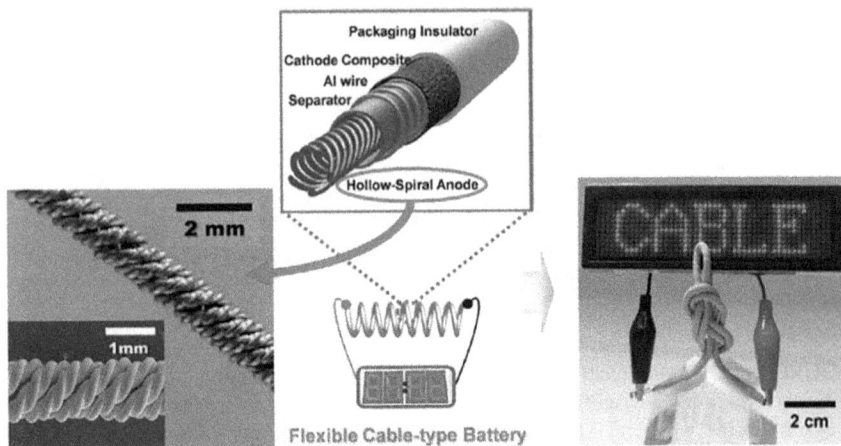

FIGURE 19.2 Schematic design of flexible cable battery with SEM image of the anode and flexibility application of the cable battery in a variety of folding and twisting situations. (Adapted with permission from Ref. [8]. Copyright (2012) Wiley.)

the portable, wearable, and flexible electronics sectors, overcoming the constraints of present flexible energy storage technologies.

19.2.1 SUBSTRATES/COLLECTORS FOR FLIBS

The current collectors and binders are generally the first components in the construction of flexible electrodes. Cu and Al foils are common substrates for LIBs. Despite its flexibility, mechanical fatigue makes it tough for them to withstand bending to a tiny diameter, like smaller than 2 mm, or repetitive folding hundreds of times [3]. Stress mitigators include porous constructions with pores that allow for the release of various types of stresses. Porous current collectors are commonly made of carbon compounds like graphene, metals, and conductive polymers. For flexible batteries, establishing a dependable network of current collectors is critical. Two distinct approaches may be utilized to combine active materials with flexible substrates. The first of these ways entails utilizing the conductive and flexible matrix as a current collector or embedding it in surfaces with active components that are securely attached by coating, sputtering deposition, or other new procedures. The second option is to utilize a non-conductive substrate with good mechanical qualities, such as polydimethylsiloxane, plastic, or cotton. The mechanically robust substrate must be adopted in the whole system.

The fabrication of a flexible and thinner packaging is also required for applications in flexible devices. Also, the battery packaging must be made of stable and corrosion-resistant material. The soft aluminum pouches or tough polymers can be used as ideal package materials for FLIBs. Recently, different kinds of polymers have been carried out as current collectors. They can be classified as polymer substrates (scaffolds), conductive polymers, and polymer composites together with conductive materials [7].

19.2.2   SEPARATORS/ELECTROLYTES FOR FLIBs

The electrolytes are an important part of FLIBs and are used to obtain diffusion medium and prevent contacts of electrodes as an insulator. Excellent conductivity, high electronic insulation, self-flexibility, excellent contact with active materials, and low interface resistance are all desirable characteristics for FLIBs [9]. Liquid electrolytes are widely used in flexible systems because of their unique electrical conductivity. Conventional organic liquid electrolytes, as is well known, have excellent conductivity and good compatibility with electrodes, but they can pose a serious safety threat when used in flexible electronics. However, recently to improve the limitations of liquid electrolytes such as safety, toxicity, and inability to design, solid electrolytes can be utilized as an alternative electrolyte. The current research has also been focusing on the fabrication of eco-friendly FLIBs. As a result, polymer solid electrolytes used as both an electrolyte and a separator must have strong ionic conductivity, mechanical flexibility, electrochemical stability, and low toxicity. According to their physical composition, polymer electrolytes can be categorized as solid-state polymer electrolyte, gel polymer electrolyte, or composite polymer electrolyte. They can also contain either solvent or nanoparticles. The usual polymers utilized in FLIBs are as follows: poly(ethylene oxide), polyimide, poly(ethyl cyanoacrylate), poly(vinylidene fluoride), poly(arylene ether), and poly(vinylidene fluoride-cohexafluoropropylene). Porthault groups prepared an efficient gel polymer electrolyte made up of PVDF-HFP, LiFSI, Pyr13FSI, and Li-MMT clay for FLIBs. This material has high ionic conductivity (0.48 mS cm^{-1}) as well as good thermal stability (up to 140°C). The new battery design decreasing cell thickness (760 μm for doubled-sided architecture) is performed more than 800 cycles with 80% of the initial capacity at room temperature [10]. The research about FLIBs started to use ionic liquids doped gel polymer as an electrolyte to enhance safety and thermal stability due to their non-flammability, low vapor pressure, and high chemical and thermal stability. Li et al. reported electrolyte made up of ionic liquid (1-Ethyl-3 methylimidazolium dicyanamide), and the polymer is composed of poly(vinylidene fluoride-co-hexafluoropropylene), which has good thermal stability (non-volatile up to 300°C) and slightly higher ionic conductivity (6×10^{-4} S cm^{-1}, the ratio of 1:2 polymer and ionic liquid). The flexible thin-film LIB shows excellent electrochemical performance under flat and bent conditions [11].

Although the separators have not attracted intensive interest as much as electrodes in FLIBs, it takes a role to determine parameters such as internal cell resistance, operating temperature range, and long cycle stability during the lithiation process [12]. Hence, it is vital to optimizing the separators to improve both the safety requirements and electrochemical performance of LIBs. In addition, it should provide a homogenous transport of ions between cathode and anode together with minimizing the degradation of the cell. Traditional polyethylene-based separators have a porous structure that enhances mechanical compatibility and electrochemical stability between separators/electrodes in addition to ionic conductivity. The battery community faces a challenge in incorporating such hard and delicate materials into flexible batteries. The designs of polymer-supported porous ceramic electrolytes allow the solution of this problem with improved mechanical properties.

Therefore, most of the research on FLIBs has been conducted using polymer-based electrolytes because of their mechanical suitability and technological readiness.

19.2.3 2D Electrodes for FLIBs

A wide variety of electrodes have been fabricated and designed for FLIBs to improve their electrochemical performance and stability so far. These electrodes, which are categorized as anode or cathode, are often made of soft and twistable inorganic or organic materials contained in a flexible layer or composites of these materials. Anodes are one of the critical components of LIBs and should not be brittle materials. Conductive additives such as carbon-based materials, non-conductive flexible substrates, or 2D nanostructured active materials are preferred to prepare flexible anodes. Generally, carbon nanofibers, graphene, carbon cloth (CC), and carbon nanotubes (CNTs) are used as carbon-based materials, while polymers, cotton, cloth, cellulose paper, etc., are preferred as non-conductive materials. Another vital component of FLIBs is the cathode materials. They are extremely important in determining the capacity and energy density of FLIBs. These materials with various matrixes have been applied to increase the conductivity, lightweight, stability, and charge/discharge performance of flexible batteries. For example, carbon-based materials like CNT, graphene, carbon nanofibers, CC, ultrathin graphite foam or metal materials, or organic polymer materials are primarily used for cathode materials [13].

The components or compounds utilized in electrode materials must be carefully chosen to achieve good electrochemical performance. Various approaches have been successfully applied to improve the energy storage performance of electroactive materials, including (i) downscale active materials, especially nanosize; (ii) changing composition of the active materials; (iii) embedding materials or adding functional groups; (iv) adjusting particles size and surface; (v) preventing the corrosion using coatings; and (vi) changing the electrolyte [14].

The state-of-the-art technology of flexible batteries is the utilization of nanostructured materials as electrodes. Two-dimensional (2D) nanomaterials are extremely useful in many components of FLIBs, including anodes, cathodes, separators, electrolytes, catalysts, and current collectors, among other dimensions. It is well known that 2D materials can provide higher energy storage capabilities by improving Li diffusion between electrodes. Therefore, these materials are commonly used as electrodes to improve cyclability, volumetric and gravimetric capacities [15]. The highlighted advantage of 2D materials is illustrated in Figure 19.3. 2D materials offer an opportunity to increase their electrical conductivity, which improves their electrochemical performance. In addition to these properties, 2D materials can enable uniform Li^+ ion electrodeposition and prevent dendrite formation.

The atomic structures of 2D nanomaterials differ from those of bulk materials, with variations in atomic arrangement, chemical valence, coordination number, and bond length. They have sheet-like structures with lateral sizes more than 100 nm but thicknesses of only a single or a few atoms (typically <5 nm). The exposed surface area of 2D nanosheets is greater, allowing for more Li-insertion channels and shorter pathways for quicker lithium-ion diffusion. Besides this, they have a small weight, more active sites, and sensible distribution for the construction of flexible batteries.

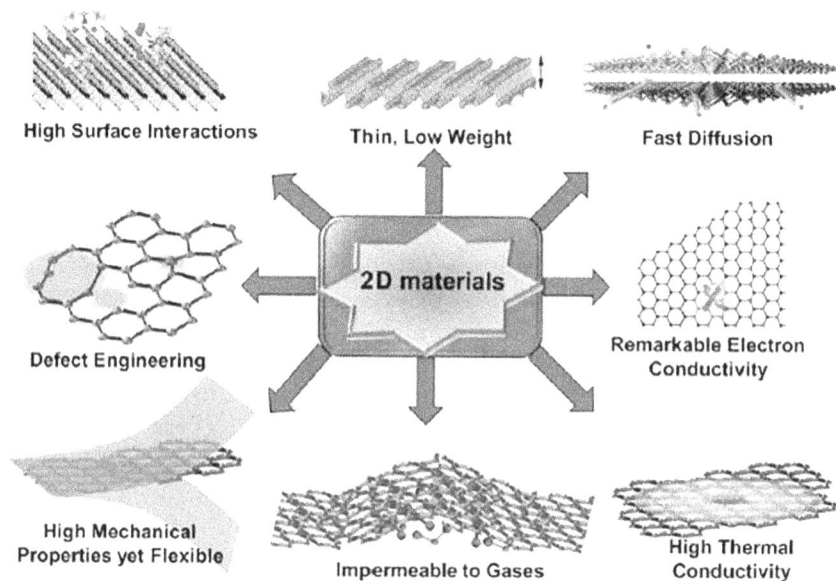

FIGURE 19.3 Some of the highlights of 2D materials that offer versatile properties in energy storage devices. (Adapted with permission from Ref. [15]. Copyright (2020) ACS.)

The porous structures have a lower bending stiffness and are therefore more flexible than solid structures.

Up to now, the layered structure 2D nanomaterials have been widely employed as anode and cathode due to their high specific surface area and a vast number of active sites, which provide great convenience to the intercalation of lithium ions [6]. Alleviating the interfacial tension between the active particles and the substrate and increasing the adhesion strength and areas of the active materials to be mixed with the current collector of the substrates are important ways to increase the efficiency of flexible batteries. The adaptation of various materials in the interlocked composite is generally used for FLIBs to provide good electrochemical performance, great mechanical flexibility as well as electrical stability in the case of stretching and bending situations. The most used 2D nanocomposites-based electrodes can be classified into three categories as adapted materials: inorganic, organic, and organic–inorganic materials.

19.2.3.1 2D Inorganic Material-Based Electrodes

Carbon-based materials are widely applied as porous and high surface area electrodes to design and fabricate flexible batteries. The formation of composites of other active materials with carbon-based materials produces synergistic effects to form a flexible network and improve the efficiency of batteries [16]. When carbon-based materials increase the functionality, flexibility, stability, and conductivity of materials, active materials enhance the specific capacity, density, and decrease costs. Since 2004, the two-dimensional carbon sheet graphene and its derivatives have become popular due to its unrivaled properties like mechanical flexibility and strength, high surface area, excellent thermal and chemical stability, great electrical conductivity $(4.8 \times 10^2 \text{ S cm}^{-1})$, and charge-carrier mobility that act as a current collector and a conducting agent. The CNTs are used as both a current collector, because of their strong adhesion, excellent mechanical durability due to Young's modulus, low contact resistance, and a conductive additive to increase the electrochemical performance and mechanical stability. For example, the 2D paper-like graphite/CNT composite was prepared, and the graphite layer remained firmly attached to the CNT film after folding. It had stronger structural integrity and lower contact resistance than a traditional Cu current collector, allowing for more efficient electron transfer at the composite interface. The graphite-CNT electrode exhibited good cycling stability and rate capability, with a specific capacity of 335 mAh g^{-1} at 0.1 C and capacity retention of 99.1% after 50 cycles (328 mAh g^{-1} at 2 C, 20.3% higher than the graphite-Cu electrodes). In comparison, the graphite-Cu showed a slightly lower cycling performance with a high specific capacity (318 mAh g^{-1} at 0.1 C) and excellent capacity retention (96.2%) after 50 cycles. The relatively low weight of the graphite-CNT electrodes resulted in a higher than 180% increase in gravimetric energy density when compared to the graphite-Cu electrodes [17]. The silicon as a nonmetal has a high theoretical specific capacity (4,200 mAh g^{-1}, ten times more than commercial graphite) [18]. The vacuum filtering process was used to combine 2D graphene with a bendable hybrid anode for FLIBs. The obtained flexible graphene-Si composite electrodes showed much higher discharge capacity than that of the graphene electrodes. Wang and colleagues developed a binder-free anode made up of silicon nanowires@graphene sheaths@reduced graphene oxide sheet (SiNW@G@RGO), which has a high reversible specific capacity of 1,600 mAh g^{-1} at a current density of 2.1 A g^{-1}, 80% capacity retention after 100 cycles, and superior rate capability of 500 mAh g^{-1} at 8.4 A g^{-1} [19]. The addition of silicon increased the lithium storage capacity. When the graphene support changed with graphitic structure CNT, the CNT/Si sheet composite was fabricated and demonstrated both high specific energy capacity (1,494 mAh g^{-1}) and good cyclic stability with a capacity retention of over 94% after 45 cycles [20].

The formation of 2D nanocomposite can be done in two different ways: The various oxides or sulfides can be combined with 2D carbon-based materials, or 2D oxides or sulfides can be combined with carbon-based materials. The various types of transition metal oxides (TiO_2, SnO_2, Co_3O_4, etc.) embed 2D carbon-based materials using different synthesis approaches (hydrothermal, solvothermal, impregnation, filtration, etc.). Changing the surface characteristics of the produced materials enhances the electrochemical performance of the electrodes when different production techniques are used. In addition, as compared to solo metal oxide electrodes, the creation of hybrid electrodes increased capacity and cycle stability. In a theoretical study, the formation of composite between 2D MoO_2 and graphene enhances the electrical conductivity, increases the theoretical specific capacity of anode material to 1,411 mAh g^{-1}, and decreases the diffusion barrier to 77 meV for LIBs. The 2D van der Waals heterostructure and the flexibility of graphene can provide electrochemically active areas and prevent deformations during charge–discharge processes [21]. Xiong and coworkers experimentally used binary metal oxide $ZnMn_2O_4$-graphene (ZMO-G) hybrid nanosheet synthesized through a two-step synthesis and $LiFePO_4$ nanosheet prepared by the solvothermal method as anode and cathode, respectively, for preparation of flexible pouch cell (Figure 19.4a). The formation of nanosheets was

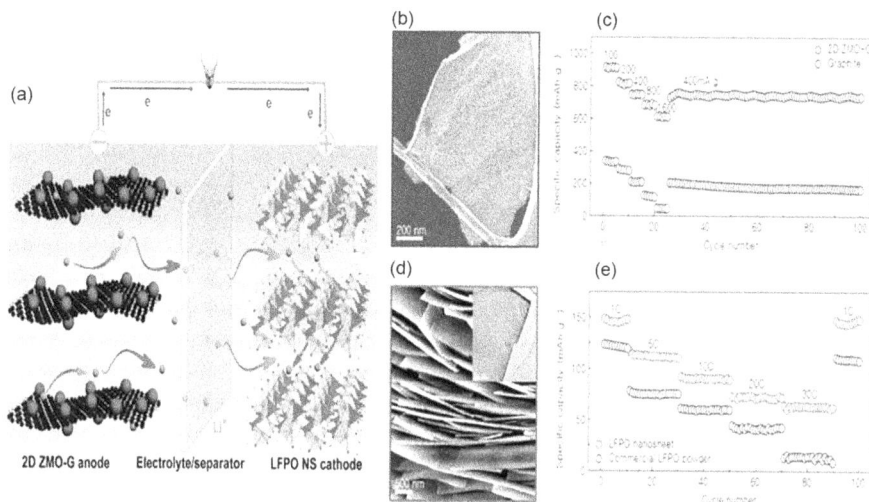

FIGURE 19.4 (a) Schematic view of LIBs with 2D ZMO-G hybrid anode and LFPO cathode, (b) SEM image of ZMO-G anode, (c) rate capabilities of 2D ZMO-G hybrid nanosheets and conventional graphite anodes, (d) SEM image of LFPO cathode, and (e) rate capabilities of LFPO nanosheets and commercial LFPO powder cathode. (Adapted with permission from Ref. [22] Copyright (2015) Elsevier.)

shown in Figure 19.4b–d. These two-dimensional nanosheets offer fast charge and discharge rates due to their short Li$^+$ ion diffusion length and active charge transport channel [22]. They show excellent charge and discharge rate and cycling stability (Figure 19.4c), compared to the conventional battery that includes graphite anode and commercial LiFePO$_4$ cathode. The creation of ZMO-G hybrid nanosheets increases the electrical conductivity of the graphene nanosheets while providing additional electrochemically active surfaces for electrolyte ion absorption and desorption. They can also significantly accelerate charge transfer during electrochemical processes. The ZMO-G-based half-cell has a high discharge capacity of 1,130 mAh g^{-1} at 400 mA g^{-1}. In the first cycle, it maintained a reversible charge capacity of 714 mAh g^{-1} with a Coulombic efficiency of 65%. Figure 19.4e compares the rate capacities of LFPO nanosheet with commercial LFPO powder cathodes, with nanosheet having a higher rate capacity than powder.

The 2D metal chalcogenides (sulfur and selenide) have gained significant efforts due to their cheap cost, abundance, and high theoretical specific capacity for FLIBs. They have been combined with carbon nanosheets and show superior rate capability. For example, the nickel sulfide@carbon nanotube paper-like composite (NiS@CNT) was prepared by a facile electrodeposition method using nickel sulfide nanoparticles and CNT thin films as an anode using a binder-free flexible high-capacity material for LIBs. Despite the high theoretical capacity and price efficiency, the low electronic conductivity of nickel sulfide reduces the electrochemical performance when used alone. The combination with carbon-based materials improves the capacity and besides tolerates the mechanical stress. It shows 845 mAh g^{-1} specific capacity at 60 mA g^{-1}. Furthermore, its specific capacity was maintained at 644 mAh g^{-1}

after 100 cycles (current density of 300 mA g^{-1}), demonstrating the composite's high cycling stability [23]. The MoSe$_2$ and single-walled carbon nanotube (SWCNT) hybridized film composite was synthesized by solvothermal method, and strong chemical bonding formed between the two-dimensional octahedral metallic MoSe$_2$ nanosheet. Conductive SWCNT film provides good adhesion and structural stability via a C–O–Mo bond, and these prevent a structural breakdown during the charging/discharging procedure. The obtained 2D layer MoSe$_2$/SWCNTs composite shows high-rate performance and possessed a capacity of 630 mAh g^{-1} (current of 3,000 mA g^{-1}) as a flexible and light binder-free LIB anode material. After 100 cycles, the strong bonding of composite accommodates volume change throughout the repeated charge and discharge process, resulting in 89% capacity retention and cyclic durability with a reversible capacity of 971 mAh g^{-1} (current density of 300 mA g^{-1}) [24]. The 2D MXenes (metal carbide) nanosheets serve as a conductive component that facilitates electron and ion movement as well as a substrate that prevents the additive nanostructures from aggregating. Metallic conductivity, hydrophilic surfaces, and good mechanical characteristics are all features of MXenes. They also have a higher capacity for reversible intercalation of a wide range of organic molecules and metal cations (Li$^+$, Na$^+$, Mg^{2+}, and Al^{3+}). In addition, MXenes, another remarkable family of 2D nanomaterials, are preferred as electrode materials in LIBs studies due to their low diffusion barrier and high lithium storage capacity [25]. Two-dimensional free-standing and flexible titanium carbide-CNT composite (Ti$_3$C$_2$T$_x$/CNT) paper as an efficient cathode material was prepared by a simple vacuum-assisted filtration method. It has about 105 mAh g^{-1} at 0.1 C, 50 mAh g^{-1} at 10 C capacity, and the discharge capacity (at 1 C) was maintained for up to 500 cycles at 80 mAh g^{-1} with Coulombic efficiencies close to 100% for LIBs. The CNT acts as a spacer and prevents electrolyte ions from re-stacking, preventing rapid diffusion between the MXene flakes [26]. In another example, the delaminated Nb$_2$CT$_x$ and CNT as an interlayer spacer at 0.5 C formed nanocomposite paper as electrode material in coin cell for LIB. From cyclic voltammetry analysis, the anode material had a first-cycle capacity of 780 mAh g^{-1} and a reversible capacity of 420 mAh g^{-1}, with a Coulombic efficiency ~100% at 0.5 C. The reversible capacity at 2.5 C enhanced from 320 to 370 mAh g^{-1} after 100 cycles and to 430 mAh g^{-1} after 300 cycles [27].

19.2.3.2   2D Organic Material-Based Electrodes

The cost and weight of the inorganic materials used in commercial batteries, the ease of reusability, features such as high mechanical flexibility, and molecular-level tunable electrochemical properties increase the significance of organic materials for battery applications. The conductive polymers, carbonyl compounds, radical polymers, and organosulfides have been used as organic electrodes for FLIBs. Organic polymers have emerged as attractive, flexible electrode components for LIBs. They have a high mechanical tolerance limit, high theoretical capacity, excellent thermal and chemical stability, low density, ease of manufacturing, low cost, reversibility, and adaptability for LIBs. Moreover, polymeric structures are nearly identical even after hundreds to thousands of cycles.

Although there is significant work to develop flexible inorganic electrode materials, inorganic materials are brittle and have an inadequate elastic deformation.

For this reason, composites of polymers with conductive materials have been studied to obtain flexible electrodes. For instance, polymer active materials such as polymerized polyimide with high theoretical capacity, high conductivity, and inherent flexibility have been offered [13]. Organic materials, especially polymers, are used as electrode materials and have a naturally flexible system. Generally, these materials are prepared together with carbon-based materials to form 2D materials with high electrical conductivity and specific surface area. For instance, Chang and coworkers used polyimide as a conjugated carbonyl polymer and prepared two-dimensional graphene@polyimide@reduced graphene composite film through solvothermal reaction, vacuum filtration, and carbonization methods as an excellent flexible cathode material for LIBs [28]. The binder-free composite formation enhances the efficiency of electrode material. It has an ultrahigh reversible capacity of 198 mAh g^{-1} at a current density of 30 mA g^{-1} and outstanding rate performance of 100 mAh g^{-1} at high current densities of 6,000 mA g^{-1}, as well as approximately 70% capacitance retention after 2,500 cycles (current density of 1,000 mA g^{-1}). Wu et al. reported cathode electrodes with a binder-free flexible material using a highly conductive SWNT film by in situ polymerized polyimide (PI) (Figure 19.5a) [29]. Also, the prepared PI/SWNT film was shown in Figure 19.5b and c. The synthesized flexible polyimide@SWCNT film shows better electronic conductivity (Figure 19.5d) and rate performance than the pure PI electrode (Figure 19.5e). PI/SWNT film shows a capacity of 226 mAh g^{-1} at 0.1 C, while pure PI has a capacity of 170 mAh g^{-1} at 0.1 C. When a sulfur-linked carbonyl-based poly(2,5-dihydroxyl-1,4-benzoquinonyl sulfide) compound was combined with a SWCNT through vacuum filtration method without any metal current collector, the flexible binder-free composite cathode was formed. It delivered a capacity of 75 mAh g^{-1} at a current density of 5,000 mA g^{-1} (0.47 mAh cm^{-2}), which affirms its excellent rate performance at high current density. It has a discharge capacity of 182 mAh g^{-1} (0.9 mAh cm^{-2}) at a current rate of 50 mA g^{-1} and a potential window of 1.5–3.5 V and retains 89% of its initial capacity after 500 charge–discharge cycles (current density of 250 mA g^{-1}) [30]. The SWCNTs can improve performance and cyclic stability in addition to functioning as current collectors and conductive additives. SWCNTs create a conductive network in composite electrodes, enhancing the overall conductivity and the cathode electrode performance.

As another 2D support, CCs have been widely used for the formation of electrode materials. Wu and coworkers prepared composite electrodes for FLIBs using perylene diimide (3,4,9,10-perylene tetracarboxylic diimide or PID) as an organic electroactive material and CC. CC is used as a current collector. The PID/CC composite was produced with a high specific capacity of 136 mAh g^{-1} at a current density of 50 mA g^{-1} as an efficient cathode material with both lithium foil and complete cells with prelithiated CC as an anode. The PDI/CC cathodes demonstrate high cycling stability in both types of batteries, keeping 84% of their initial capacities after 300 charge–discharge cycles at 500 mA g^{-1} [31].

One of the difficulties in electrode applications of organic materials is their low conductivity. Organic electrode materials with poor electronic conductivity have a lower kinetic reaction rate. The conductivity of organic-based polymers can be improved by adding a conductive substance. Furthermore, the design variety of

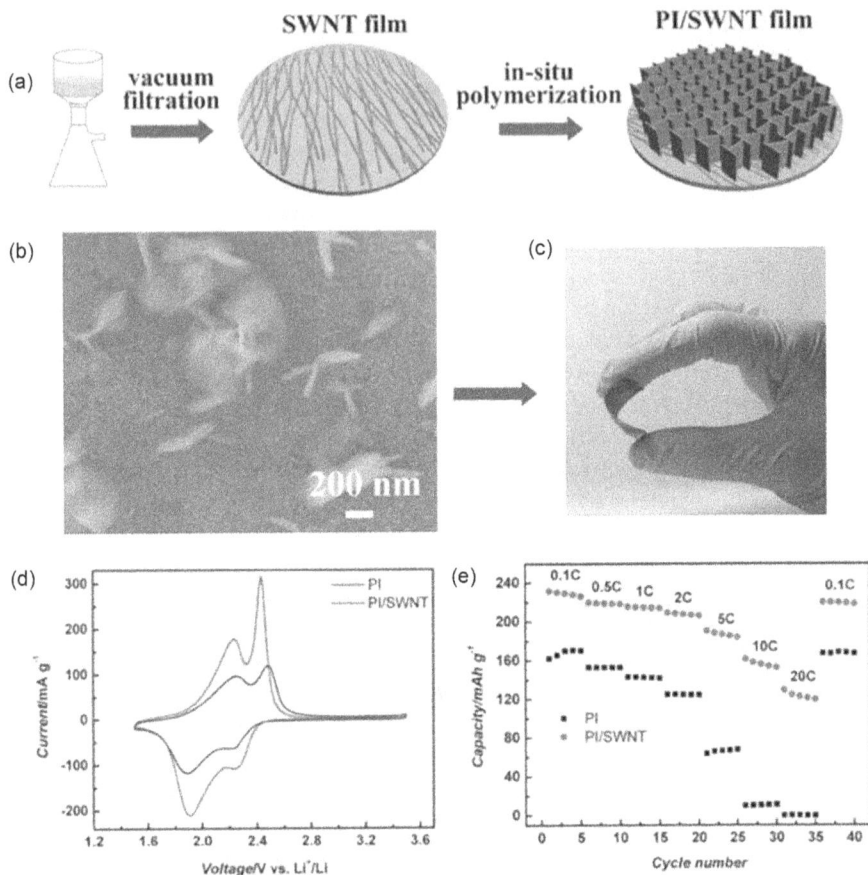

FIGURE 19.5 (a) In situ preparation of polyimide/single-wall carbon nanotube (PI/SWNT) film, (b) SEM image of PI/SWNT film, (c) photographs of prepared PI/SWNT film electrode, (d) cyclic voltammetry curves of PI and PI/SWNT electrodes at 0.1 mV s^{-1}, and (e) rate performance of PI and PI/SWNT electrodes at different current rates. (Adapted with permission from Ref. [29]. Copyright (2014) John Wiley & Sons.)

organic materials, which allows for the creation of 2D or 3D materials with large-conjugated structures, improves electron transport and conductivity [32]. As a consequence of the formation of organic material composites with conductive materials, large capacity, high-rate performance, outstanding cycle performance, and strong flexibility make them a viable option with low cost and high sustainability for future FLIB applications.

19.2.3.3 2D Organic–Inorganic Material-Based Electrodes

Metal-organic frameworks (MOFs), also known as porous coordination polymers, are a well-organized mixture of inorganic units (metal ions, metal nodes, or metal clusters) and organic ligands linked by coordination bonds. Due to their unique properties, such as ultrahigh porosity, structural tailorability, large surface area, functional diversity, crystallinity, and adaptability, MOFs have gained significant attention for energy applications. In alkali ion batteries, the ease of tailoring the flexibility and the pore size procures space for their investigation as efficient anodes. Many publications on the use of MOFs in rechargeable batteries have been published; however, only a few studies have been found about flexible batteries.

MOFs have better thermal stability compared to pure organic materials. Metal cations in the MOF structure form active sites for redox reactions in LIBs. The reason why MOF-related materials are considered the most promising candidate as electrode materials for LIBs is briefly explained as follows: (i) the ultra-high porosity of MOFs can tolerate volume expansion during lithium ions storage; (ii) electrode materials with a larger specific surface area can be obtained by developing MOFs with various structures and coordination metal centers; and (iii) due to their adjustable pore structures, they provide an advantage for interface charge transport [33].

Composites are known as a synergistic combination of two or more materials and show different properties than their types. Mostly, MOFs do not have high conductivity. Therefore, researchers try to increase the conductivity by introducing alternative conductive materials like carbon nanomaterials, graphene, and reduced graphene oxide (rGO). For example, Jin's group synthesized Fe-MOF/rGO composite as an anode material with high electrically conductive, good rate capability, and excellent cycling stability. The synergistic impact of MOFs with high theoretical capacity and rGO compounds with high electrical conductivity was responsible for the increase in electrochemical performance [34]. An et al. successfully prepared flexible layered nickel-based MOF with a 2-dimensional layered structure using 3,30,5,50-tetramethyl-4,40-bipyrazole (H_2Me_4bpz) ligand, as given in Figure 19.6a. The electrochemical efficiency of the Ni–Me_4bpz anode electrode was determined using cyclic voltammetry. As shown in Figure 19.6b, the three cathodic peaks obtained after the first cycle disappeared. This change is explained by the reduction of the initial discharge specific capacity from 320 to 140 mAh g^{-1}, as shown in the charge/discharge cycle tests (Figure 19.6c) [35]. It has been indicated that a 2D layered structure with flexible rotational properties can improve lithium intercalation/deintercalation and electrochemical performance for LIBs. However, the new design MOFs with better functionality are still in their infancy, and more investigation is urgently needed for the complete FLIBs.

MOF-derived composites, unlike the electrode materials described above, have a broad variety of derivatives and controlled architectures, making them attractive as electrode materials for LIBs. For example, using terephthalic acid (H_2BDC) and tetra-n-butyl titanate, MIL 125 framework was synthesized by solvothermal method and then calcined at 380°C for 5 hours in the air. The most important result of the derivation of metal oxide from the MOF template is to improve and change the surface properties of obtained materials because of MOF's high surface area and porosity. The MIL 125 turned to mesoporous anatase phase of TiO_2, which is widely used as an important material due to unique properties such as natural abundance, band-gap efficiency, and photo- and electrocatalytic activity [36]. The MIL 125 derived TiO_2 has a high reversible capacity, good rate capacity, and long-term cycle stability; however, mechanical flexibility is a great challenge. With the construction of 2D flexible graphene nanosheets (GS) using a two-step vacuum filtration process,

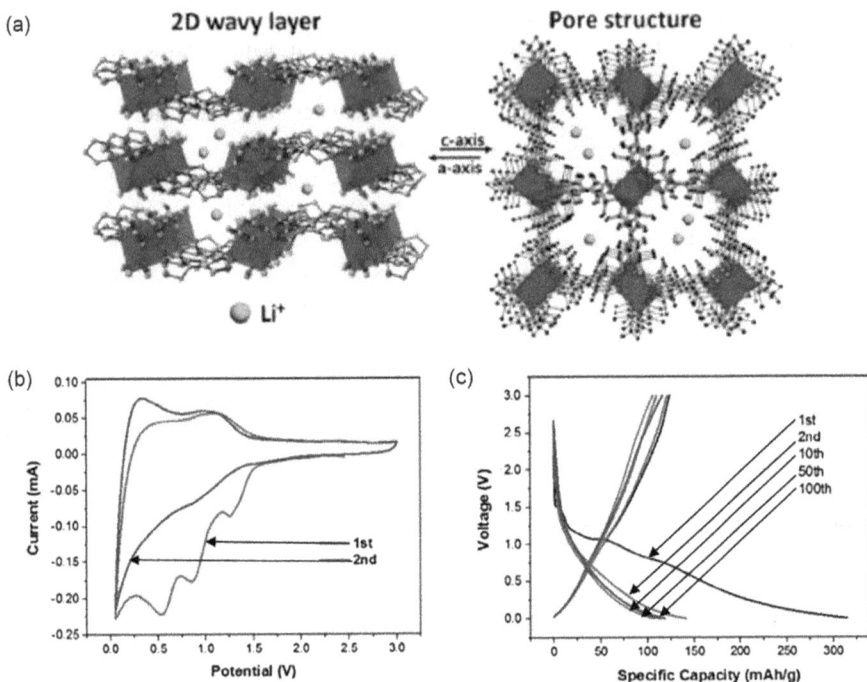

FIGURE 19.6 (a) The structure of Ni-Me$_4$bpz framework, (b) cyclic voltammetry for Ni-Me$_4$bpz, and (c) charge–discharge curves of Ni-Me$_4$bpz electrode. (Adapted with permission from Ref. [35]. Copyright (2015) Elsevier.)

the highly conductive TiO$_2$@GS electrode was developed as a free-standing anode material for flexible batteries with a high specific capacity of 205 mAh g^{-1} at 0.5C, great rate capability (76 mAh g^{-1} at 20C), and satisfying cycling stability (70.5% capacity retention over 3,500 cycles at 5C). Consequently, the formation of composite enhanced the flexibility, reversible capacity, durability, and structural and cyclic stability of electrodes for LIBs.

19.3   CHALLENGES AND PERSPECTIVES FOR FUTURE RESEARCH

Developing technology, the rapid increase in the use of electric vehicles and mobile electronic devices, consisting of intermittent systems of energy sources like wind and sun, which are planned to replace fossil fuels that are starting to run out, has made the development of renewable energy storage systems important. The batteries, especially LIBs, are extremely promising energy storage devices. The increment of small-size, portable and wearable smart devices changed the direction of research to flexible batteries. The recent advances of FLIBs with a focus on 2D nanomaterials were highlighted in this chapter. To improve the electrochemical performances, the design of 2D nanocomposites for each component of batteries is a great choice. For the key component for FLIBs, carbon materials, including graphene and their composites, were widely used as a flexible electrode material. Most of them are intrinsically rigid materials, and the design and development of novel electrode materials is still a challenge to improve electrochemical performance, safety, and mechanical properties. Another significant problem for flexible batteries is the dynamic stability of power output during deformation, in addition to great flexibility and high energy density at the same time.

The supported architecture and novel electrode materials help to increase the specific capacity of the FLIBs, and besides this, the bendability and mechanical flexibility of the electrodes have been enhanced. The increased demand for miniaturized products used in electronic applications has led to an increase in the adoption of thin-film FLIBs. From the overall discussion of this chapter, the parameters highlighted in Figure 19.7 should be improved for the commercial use of promising FLIBs. Especially, research approaches for FLIBs should focus mainly on the development of flexible active materials considering the parameters. The electrodes, the electrolyte, and the electrochemical reactions in the battery system are affected by the external environment. Therefore, alternative packaging materials with the shape deformable should also be investigated.

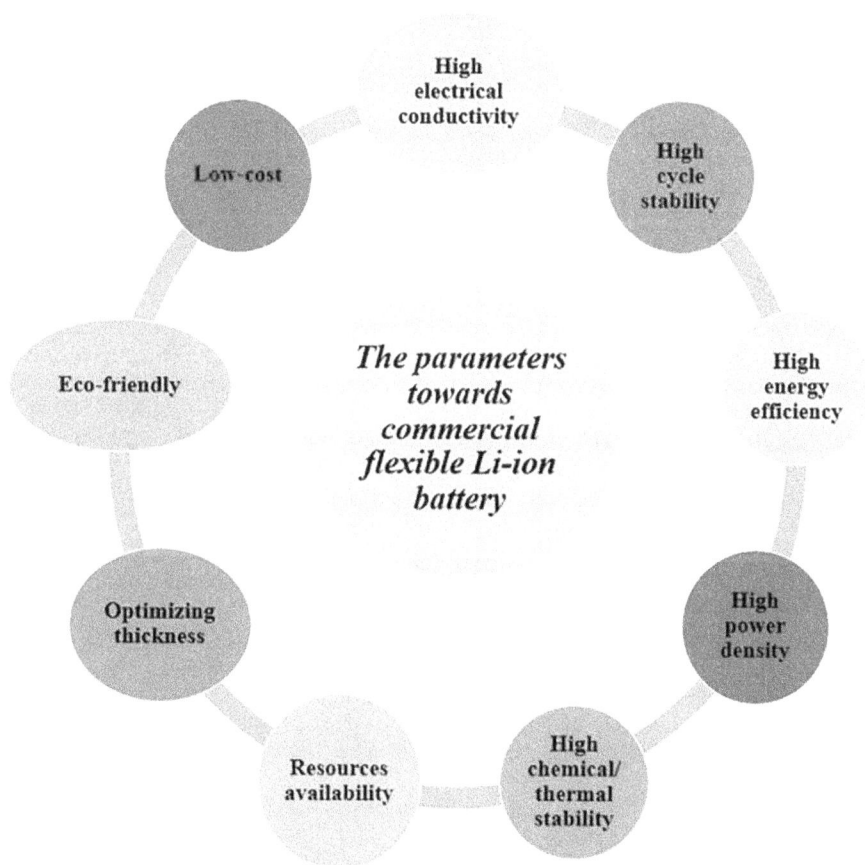

FIGURE 19.7   The factors toward commercialization of flexible Li-ion batteries.

19.4 CONCLUSIONS

Detailed research about electrodes based on 2D nanocomposite materials for FLIBs is presented. 2D nanocomposite materials have been extensively studied in many research and applications due to their unique physicochemical and structural characteristics. Using these materials to produce electrodes for thinner, lighter, and more stable FLIBs is one of the most promising techniques. Recently, many existing 2D nanocomposites, including MOFs, MXene, metal oxides, chalcogenides, graphene, and CNTs, have great potential as anode/cathode materials for FLIBs due to their huge surface area, high electrical conductivity, and adjustable active sites for the lithiation/delithiation process. 2D nanocomposite materials with outstanding lithium storage performance and flexibility offer great potency to prepare electrodes for FLIBs. However, novel 2D nanocomposite materials are required to enhance the electrochemical capacity of FLIBs under bending and stretching conditions. Until far, most research has concentrated on the design of a single flexible electrode for FLIBs. On the other hand, more significant effort is needed to fabricate complete FLIBs. Hopefully, this chapter will contribute to research for the next-generation FLIBs based on nanocomposites of 2D materials.

REFERENCES

1. D. Ozer, Z. Ertekin, K. Pekmez, N.A. Oztas, Fuel effects on $Li_2CuP_2O_7$ synthesized by solution combustion method for lithium-ion batteries, *Ceramics International*, 2019, 45, 4626–4630.
2. Y. Zhang, Y. Jiao, M. Liao, B. Wang, H. Peng, Carbon nanomaterials for flexible lithium ion batteries, *Carbon*, 2017, 124, 79–88.
3. G. Qian, X. Liao, Y. Zhu, F. Pan, X. Chen, Y. Yang, Designing flexible lithium-ion batteries by structural engineering, *ACS Energy Letters*, 2019, 4, 690–701.
4. https://www.marketsandmarkets.com/Market-Reports/flexible-battery-market-190884508.html (accessed June 28, 2021).
5. M.R. Lukatskaya, B. Dunn, Y. Gogotsi, Multidimensional materials and device architectures for future hybrid energy storage, *Nature Communications*, 2016, 7, 1–13.
6. X. Zang, T. Wang, Z. Han, L. Li, X. Wu, Recent advances of 2D nanomaterials in the electrode materials of lithium-ion batteries, *Nano*, 2019, 14, 1930001.
7. H. Cha, J. Kim, Y. Lee, J. Cho, M. Park, Issues and challenges facing flexible lithium-ion batteries for practical application, *Small*, 2018, 14, 1702989.
8. Y.H. Kwon, S.W. Woo, H.R. Jung, H.K. Yu, K. Kim, B.H. Oh, S. Ahn, S.Y. Lee, S.W. Song, J. Cho, Cable-type flexible lithium ion battery based on hollow multi-helix electrodes, *Advanced Materials*, 2012, 24, 5192–5197.
9. Z. Fang, J. Wang, H. Wu, Q. Li, S. Fan, J. Wang, Progress and challenges of flexible lithium ion batteries, *Journal of Power Sources*, 2020, 454, 227932.
10. H. Porthault, C. Calberg, J. Amiran, S. Martin, C. Páez, N. Job, B. Heinrichs, D. Liquet, R. Salot, Development of a thin flexible Li battery design with a new gel polymer electrolyte operating at room temperature, *Journal of Power Sources*, 2021, 482, 229055.
11. Q. Li, H. Ardebili, Flexible thin-film battery based on solid-like ionic liquid-polymer electrolyte, *Journal of Power Sources*, 2016, 303, 17–21.
12. P. Chaturvedi, A.B. Kanagaraj, A. Alhammadi, H. Al Shibli, D.S. Choi, Fabrication of PVDF-HFP-based microporous membranes by the tape casting method as a separator for flexible Li-ion batteries, *Bulletin of Materials Science*, 2021, 44, 1–7.

13. T. Tao, S. Lu, Y. Chen, A review of advanced flexible lithium-ion batteries, *Advanced Materials Technologies*, 2018, 3, 1700375.
14. A. Noori, M.F. El-Kady, M.S. Rahmanifar, R.B. Kaner, M.F. Mousavi, Towards establishing standard performance metrics for batteries, supercapacitors and beyond, *Chemical Society Reviews*, 2019, 48, 1272–1341.
15. R. Rojaee, R. Shahbazian-Yassar, Two-dimensional materials to address the lithium battery challenges, *ACS Nano*, 2020, 14, 2628–2658.
16. J. Islam, F.I. Chowdhury, J. Uddin, R. Amin, J. Uddin, Review on carbonaceous materials and metal composites in deformable electrodes for flexible lithium-ion batteries, *RSC Advances*, 2021, 11, 5958–5992.
17. K. Wang, S. Luo, Y. Wu, X. He, F. Zhao, J. Wang, K. Jiang, S. Fan, Super-aligned carbon nanotube films as current collectors for lightweight and flexible lithium ion batteries, *Advanced Functional Materials*, 2013, 23, 846–853.
18. M. Phadatare, R. Patil, N. Blomquist, S. Forsberg, J. Örtegren, M. Hummelgård, J. Meshram, G. Hernández, D. Brandell, K. Leifer, Silicon-nanographite aerogel-based anodes for high performance lithium ion batteries, *Scientific Reports*, 2019, 9, 1–9.
19. B. Wang, X. Li, X. Zhang, B. Luo, M. Jin, M. Liang, S.A. Dayeh, S. Picraux, L. Zhi, Adaptable silicon–carbon nanocables sandwiched between reduced graphene oxide sheets as lithium ion battery anodes, *ACS Nano*, 2013, 7, 1437–1445.
20. K. Fu, O. Yildiz, H. Bhanushali, Y. Wang, K. Stano, L. Xue, X. Zhang, P.D. Bradford, Aligned carbon nanotube-silicon sheets: A novel nano-architecture for flexible lithium ion battery electrodes, *Advanced Materials*, 2013, 25, 5109–5114.
21. J. Ma, J. Fu, M. Niu, R. Quhe, MoO_2 and graphene heterostructure as promising flexible anodes for lithium-ion batteries, *Carbon*, 2019, 147, 357–363.
22. P. Xiong, L. Peng, D. Chen, Y. Zhao, X. Wang, G. Yu, Two-dimensional nanosheets based Li-ion full batteries with high rate capability and flexibility, *Nano Energy*, 2015, 12, 816–823.
23. P. Fan, H. Liu, L. Liao, J. Fu, Z. Wang, G. Lv, L. Mei, H. Hao, J. Xing, J. Dong, Flexible and high capacity lithium-ion battery anode based on a carbon nanotube/electrodeposited nickel sulfide paper-like composite, *RSC Advances*, 2017, 7, 49739–49744.
24. T. Xiang, S. Tao, W. Xu, Q. Fang, C. Wu, D. Liu, Y. Zhou, A. Khalil, Z. Muhammad, W. Chu, Stable 1T-$MoSe_2$ and carbon nanotube hybridized flexible film: Binder-free and high-performance Li-ion anode, *ACS Nano*, 2017, 11, 6483–6491.
25. Y. Li, R. Wang, Z. Guo, Z. Xiao, H. Wang, X. Luo, H. Zhang, Emerging two-dimensional noncarbon nanomaterials for flexible lithium-ion batteries: Opportunities and challenges, *Journal of Materials Chemistry A*, 2019, 7, 25227–25246.
26. A. Byeon, M.-Q. Zhao, C.E. Ren, J. Halim, S. Kota, P. Urbankowski, B. Anasori, M.W. Barsoum, Y. Gogotsi, Two-dimensional titanium carbide MXene as a cathode material for hybrid magnesium/lithium-ion batteries, *ACS Applied Materials and Interfaces*, 2017, 9, 4296–4300.
27. O. Mashtalir, M.R. Lukatskaya, M.Q. Zhao, M.W. Barsoum, Y. Gogotsi, Amine-assisted delamination of Nb_2C MXene for Li-ion energy storage devices, *Advanced Materials*, 2015, 27, 3501–3506.
28. B. Chang, J. Ma, T. Jiang, L. Gao, Y. Li, M. Zhou, Y. Huang, S. Han, Reduced graphene oxide promoted assembly of graphene@ polyimide film as a flexible cathode for high-performance lithium-ion battery, *RSC Advances*, 2020, 10, 8729–8734.
29. H. Wu, S.A. Shevlin, Q. Meng, W. Guo, Y. Meng, K. Lu, Z. Wei, Z. Guo, Flexible and binder-free organic cathode for high-performance lithium-ion batteries, *Advanced Materials*, 2014, 26, 3338–3343.
30. K. Amin, Q. Meng, A. Ahmad, M. Cheng, M. Zhang, L. Mao, K. Lu, Z. Wei, A carbonyl compound-based flexible cathode with superior rate performance and cyclic stability for flexible lithium-ion batteries, *Advanced Materials*, 2018, 30, 1703868.

31. D.-Q. Wu, D. Lu, P. Yang, L. Ma, B. Jiang, X. Xi, F.-C. Meng, W.-B. Zhang, F. Zhang, Q.-Q. Zhong, An organic solvent-free approach towards PDI/carbon cloth composites as flexible lithium ion battery cathodes, *Chinese Journal of Polymers Science*, 2020, 38, 540–549.

32. H. Wang, C.-J. Yao, H.-J. Nie, K.-Z. Wang, Y.-W. Zhong, P. Chen, S. Mei, Q. Zhang, Recent progress in carbonyl-based organic polymers as promising electrode materials for lithium-ion batteries (LIBs), *Journal of Materials Chemistry A*, 2020, 8, 11906–11922.

33. Y. Jiang, H. Zhao, L. Yue, J. Liang, T. Li, Q. Liu, Y. Luo, X. Kong, S. Lu, X. Shi, Recent advances in lithium-based batteries using metal organic frameworks as electrode materials, *Electrochemistry Communications*, 2020, 122, 106881.

34. Y. Jin, C. Zhao, Z. Sun, Y. Lin, L. Chen, D. Wang, C. Shen, Facile synthesis of Fe-MOF/RGO and its application as a high performance anode in lithium-ion batteries, *RSC Advances*, 2016, 6, 30763–30768.

35. T. An, Y. Wang, J. Tang, Y. Wang, L. Zhang, G. Zheng, A flexible ligand-based wavy layered metal-organic framework for lithium-ion storage, *Journal of Colloid and Interface Science*, 2015, 445, 320–325.

36. H. Luo, C. Xu, B. Wang, F. Jin, L. Wang, T. Liu, Y. Zhou, D. Wang, Highly conductive graphene-modified TiO_2 hierarchical film electrode for flexible Li-ion battery anode, *Electrochimica Acta*, 2019, 313, 10–19.

20 Graphene-Based 2D Nanomaterials for Supercapacitors

Kaushalya Bhakar and Naresh A. Rajpurohit
Central University of Gujarat

Meena Nemiwal
Malaviya National Institute of Technology

Dinesh Kumar
Central University of Gujarat

CONTENTS

20.1 INTRODUCTION

Energy storage devices are important in today's world to meet the growing need for reliable and portable power sources. Supercapacitors are also known as ultra-capacitors, consider as the most promising electrochemical energy storage device. Supercapacitors are important in supporting the voltage of the device throughout the

DOI: 10.1201/9781003178422-20

current fluctuation in everything from compact equipment to rechargeable vehicles. In recent years, supercapacitors attract significant interest because of their distinctive properties, including high power density, long-life cycle performance, remarkable safety standards, and rapid charge/discharge rates [1]. The supercapacitor can be considered a future of energy storage devices. It may replace the battery for energy storage applications. Supercapacitors store a higher energy density than traditional dielectric capacitors. Supercapacitors have a similar cell construction as traditional dielectric capacitors, but highly porous material electrodes are used in the place of metal electrodes [2]. Supercapacitors can be divided into two types depending upon their energy storage process. The first is known as electrochemical double-layer capacitance (EDLC), and the other is pseudocapacitor. Both types of mechanisms must be assimilated to enhance the efficiency of the supercapacitor on a single electrode material. The hybrid energy storage system can bring high power and energy density compared to EDLC and pseudocapacitance material alone. In EDLC, carbon-based nanomaterial including activated carbon, carbon nanotube (CNT), and graphene are used as active electrode material. For pseudocapacitor, transition metal oxides like ruthenium oxide, manganese oxide, nickel oxide, cobalt oxide, ferric oxide, and conducting polymers such as polypyrrole (PPY) and polyaniline (PANI) are used as conductor material. These carbonaceous materials have been used as a good supportive material for a different metal nanoparticle type, which improves catalytic activity. Scientists are continuously developing innovative electrode materials at low cost with eco-friendly nature and ingenious device design. Since graphene was discovered, it led to great changes in material science. Graphene is the well-known 2D carbon single layer comprising all-sp^2-hybridized carbons considered a constituent material for carbon material of other dimensionalities. Graphene-based material has appeared as a suitable candidate for energy generation and storage application because of its superior properties, i.e., lightweight, high thermal (5,000 W m^{-1} K^{-1}), and electrical conductivity (108 S m^{-1}), highly tuneable surface area (up to 2,675 m^{2}g^{-1}). This is higher than 1D (CNT) and 3D (graphite) carbon materials, high transparency, strong mechanical strength (~1 TPa), and chemical stability [3]. Graphene-based materials combine superior physical, chemical, and mechanical properties, enhancing their availability and applications in various fields. For example, electronics, highly efficient structural nanocomposite, and sustainable energy production and storage—supercapacitor lithium-ion batteries, solar cell, fuel cell, photovoltaic cell, etc. [4]. The theoretical specific capacitance of monolayer graphene is 21 μF/cm^2, and for complete surface area utilization, it reached 550 F g^{-1}. However, practically the rate of specific capacitance is low compared to the theoretical rate because of the major aggregation of graphene sheets during the synthesis and application processes [3]. Therefore, to improve the electrochemical activity, graphene-based materials are incorporated with various transition metal oxides, hydroxide, and conducting polymer. Several techniques synthesize graphene using graphite as a starting material. Graphene and its composite nanostructures have been recommended as desirable and substitute electrode materials for energy storage because of their tremendous stability during chemical reactions and improved thermal and electrical conductivities.

20.2 SYNTHESIS OF GRAPHENE-BASED MATERIALS

The flexible graphite can thermally expand and consequently exfoliated to acquire graphene. Graphene is a monolayer arrangement of carbons with thickness at the atomic level. The unique electrical and optical properties of graphene make it more popular at the industrial level. Using liquid-phase exfoliation, pristine graphite can directly convert to graphene, which can oxidize to obtain graphene oxide (GO) [5]. The GO can further reduce graphene oxide (rGO), and mainly, this pathway is used to synthesize GO-based nanocomposite in energy generation and storage applications. GO is a layered structure. Depending upon the synthesis procedure, graphene surfaces contain various oxygen-based functional groups such as carboxyl (C=O), a hydroxyl group (OH), and epoxy group (CO) for stabilization (Figure 20.1). GO consists of phenol epoxy and epoxide group on the planar surface and carboxylic group at the edges of the graphene sheet. Compared to graphene, GO and rGO can easily disperse in various solvents [6]. The GO dispersion in an aqueous medium forms a monolayer sheet and stabilizes through a weak van der Waals and dipole interaction in an aqueous medium. The reduction of GO to rGO employing urea as a reducing agent is another environment-friendly approach for the synthesis of rGO. The resultant material was successfully utilized as an effective electrode material for energy storage devices. The urea reduced the oxygen-containing functional group present on the surface and the edges of GO. The subsequent rGO displayed a tremendous change in electrochemical properties, including specific capacitances near 255 and 100 F g^{-1} at a different electrical current flow rate of 0.5 and 30 A g^{-1}, correspondingly, along with a keeping in the capacity of 93% [7].

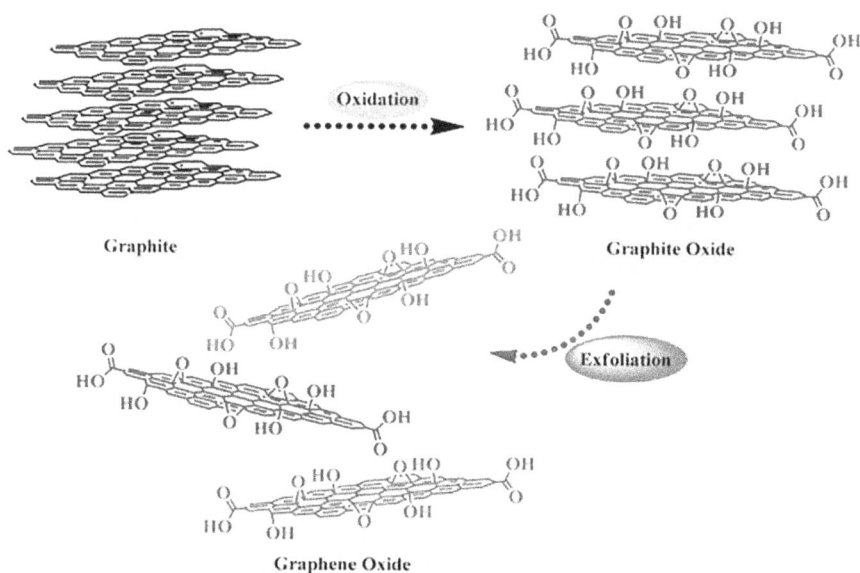

FIGURE 20.1   Schematic diagram of graphene oxide synthesis via oxidation. (Adapted with permission from Ref. [8]. Copyright (2021) MDPI.)

Several physical, chemical, and mechanical techniques are available to develop and modify graphene-based materials to change their efficiency and properties. Graphene has good flexibility, strength, and high porosity to affect composite when binding together positively. The synthesis of graphene can be achieved by following a mechanical and chemical exfoliation, epitaxial growth, chemical vapor deposition (CVD), chemical synthesis, pyrolysis, etc. [9]. The microwave synthesis and unzipping of CNT are also used for graphene development. In mechanical exfoliation, the graphene flakes extract on the surface of the substrate through a single layer formulation. A single graphene layer extraction can be done using adhesive tapes and atomic force microscopy [6]. However, mechanical exfoliation is related to low yield and not suitable for mass production. The deposition of graphene on various substrates like silicon carbide (SiC) and metal surface takes place through epitaxial growth and CVD method [10]. CVD offers better properties, including single-layer structure, large crystal area, well-defined basal and edge plane enhanced, and less defect in graphene layer, improving the charge carrier ability in the electronic application. The graphene sheets, which CVD of hydrocarbon develops, possess unique characteristics with fewer defects. The expensive instrumentation restricts large-scale production and makes it a poor candidate for EDLC electrode material development. The chemical exfoliation method considers as an ideal method for the synthesis of graphene-based material for energy storage application. In the chemical exfoliation method, the exfoliation of graphite is done by adding alkali solution and inserting the alkali ions between the graphite sheets [11]. The formation of graphene via chemical synthesis takes place through exfoliation of graphite to GO and further reduction of GO to graphene using a suitable reducing agent such as hydrazine hydrate [12]. The high conductivity with no defect in graphene sheet and large-scale production at low cost make this method more demandable. While the partial agglomeration of graphene sheets during the reduction process increases their diameter. The synthesized graphene can potentially be fused with other organic or inorganic materials to acquire new graphene nanohybrid structures, including graphene with polymers, metals, transition metal oxides, and CNTs for supercapacitors. The handling of graphene-based material like graphene papers and thin films is hindered due to restacking and agglomeration of graphene sheets, which reduce the specific surface area and diffusion of electrolytes. Novel approaches have been used throughout the years to resolve the restacking and agglomeration-related issues, including intercalation of spacer between graphene sheets, template-assisted growth, and crumpling of graphene sheets [13].

20.3 GRAPHENE-BASED ELECTROCHEMICAL DOUBLE LAYER SUPERCAPACITOR

An electric double-layer supercapacitor is also recognized as a supercapacitor and ultra-capacitor. This is an energy storage structure based on the charge–discharge mechanism in an electric double layer on a porous electrode where cation and anion of electrolyte solution form a Helmholtz layer on the surface of the electrode (Figure 20.2). Two highly porous electrodes that are electrochemically stable and do not go under faradic processes developed EDLC. These electrodes are immersed in an electrolyte solution to develop a highly active surface area and minimal distance between

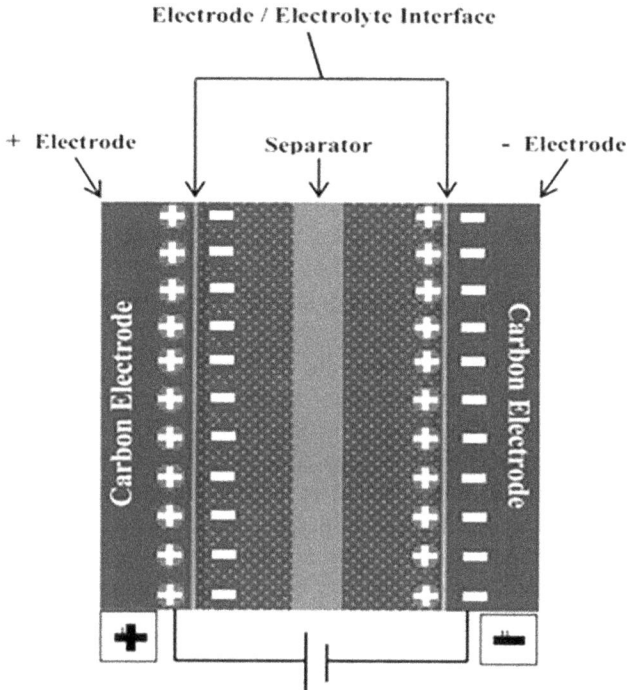

FIGURE 20.2 Schematic EDLC showing charge storage mechanism. (Adapted with permission from Ref. [19]. Copyright (2021) Elsevier.)

the two electrodes. Graphene-based composite is utilized as an active material for EDLC because of its good mechanical and electrical properties. Earlier research suggests that chemically changed graphene has good electrical conductivity. The theoretical value of specific capacitance for graphene is very low. The chemically modified graphene electrode possesses a suitable specific capacitance of about 130 F g^{-1} in an aqueous solution of KOH electrolyte, and for organic electrolytes, it is 99 F g^{-1} [14]. The specific capacitance of particularly hydrazine hydrate reduce GO is also very poor. The major reason for the lower specific capacitance of graphene material is the strong π-π interaction, resulting in the clustering and restacking of graphene sheets. The clustering in single-layer graphene sheets is avoided by intercalation of hard species known as spacer-like nanoparticle, polymer, and CNT between the layers [15]. The intercalation of carbon black nanoparticles in graphene enhances the charge storage capacity and provides an open structure for ion diffusion on the electrode, increasing electrochemical efficiency [16]. Li et al. synthesized CNT intercalated graphene composite to prevent restacking of the graphene layer. This multilayer graphene/CNT-based hybrid electrode possesses a high surface capacitance of 140 F g^{-1} at a current density of 0.1 A g^{-1} in 1 M H_2SO_4 solution [17].

Similarly, various heteroatoms like boron, nitrogen, and sulfur are doped in graphene, altering the electronic composition and density of state drastically, thus adjusting the quantum capacitance, leading to better interfacial capacitance rates. Xu et al. [18] developed a self-assembled nanostructure through the hydrothermal method with a 3D network structure. The 3D network structure of graphene hydrogel possesses high electric conductivity and no structural changes and shows specific capacitance values 175 and 152 F g^{-1} at a scan rate of 10 and 20 mV s^{-1}.

20.4 GRAPHENE-BASED PSEUDOCAPACITOR

Pseudocapacitor works on energy storage, mainly dependent on fast and reversible charge transfer reactions between an electrolyte solution and electrode. In pseudocapacitors, the reaction occurs both on the surface and bulk in the electrode materials, which enhance the electric capacity and higher specific capacitance compared to carbon material-based EDLC (Figure 20.3) [20]. For pseudocapacitor, transition metal oxide, including MnO$_2$, NiOx, conducting polymer, and oxygen-containing functional group compound is used as an active material. The storage device is known as pseudocapacitors, but they are the combination of EDLC and pseudocapacitance supported by the surface of the active material.

The volume expansion and shrinkage of material during the electrosorption process are major drawbacks of pseudocapacitor, reducing cyclic stability [21]. The graphene-based electrode could be considered a good candidate for improving the pseudocapacitors' performance. The GO has a high pseudocapacitance compared to rGO due to many oxygen functionalized groups with high surface area [22]. But the functional groups present on the surface of GO can negatively affect the performance and cyclic stability of pseudocapacitor. Graphene can be used as a supportive template for many nanostructures to enhance specific surface area and electrical performance. Metal oxide is extensively used for energy storage applications. The pseudocapacitor uses the surface of metal oxide for redox reactions and stores energy. MnO$_2$ has a high theoretical value of specific capacitance and ease in synthesis at a low cost, but practically the specific capacitance is very low due to poor conductivity. To enhance MnO$_2$ specific capacitance rate combined with carbenoids material.

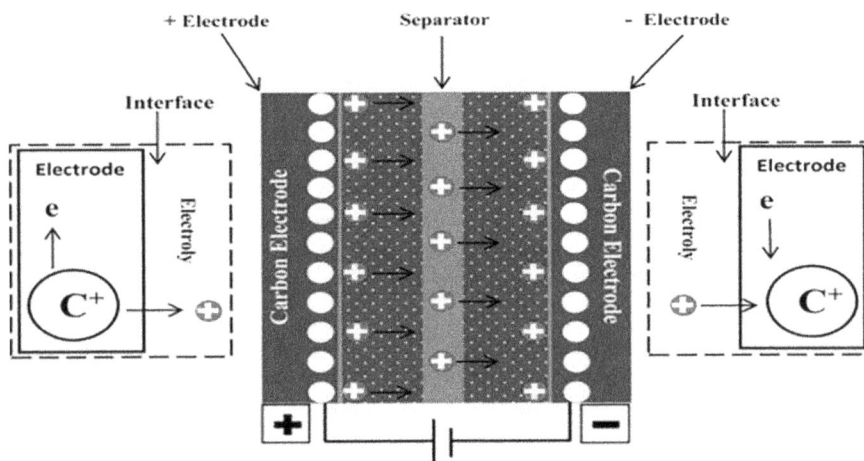

FIGURE 20.3 Representation of the charge storage mechanism of pseudocapacitor. (Adapted with permission from Ref. [19]. Copyright (2021) Elsevier.)

The fabrication of rGO/MnO$_2$ textile electrode using electrodeposition method where MnO$_2$ particles deposit on the surface of the textile cover with rGO. The specific capacitance of rGO/MnO$_2$-based nanostructure is 315 F g^{-1} at a scan rate 2 mV s^{-1}, which is higher than the only textile-covered rGO [23]. The conductive polymer is another material for good specific capacitance, high charge storage capacity, and environmental stability. The low mechanical stability of conductive polymer restricts its application in energy storage.

The combination of a conductive polymer with mechanically stable material like graphene can boost their charge storage capacity. Graphene-PANI nanocomposite has great potential as a supercapacitor electrode and exhibits improved electrochemical performance. Wu et al. prepared GO/PANI composite via in situ polymerization of aniline with chemically modified GO in an acidic medium. The GO/PANI composite was also reduced with hydrazine with reduced PANI and formed graphene/PANI composite. The nanocomposite shows a good specific capacitance of 480 F g^{-1} at a current density of 0.1 A g^{-1} [24]. The results show that good cyclic stability and specific capacitance could be achieved by depositing PANI with chemically modified graphene. The assimilation of the polymer chain in graphene/conducting polymer composite can also enhance the flow of electrons and increase the charge–discharge rate.

20.5 GRAPHENE-BASED 2D MATERIALS FOR SUPERCAPACITORS

Pristine GO is not sufficiently applicable for an efficient super-capacitance electrode since it has random orientation and high aggregation of graphene sheets. This issue can be overcome by doping various components, which are described in the below section.

20.5.1 HETEROATOM-DOPED GRAPHENE AND HYBRIDS FOR SUPERCAPACITORS

Nanocomposites of graphene possess high specific capacitance. Still, their efficiency can be improved by doping heteroatoms such as boron (B), nitrogen (N), phosphorus (P), and sulfur (S). Relatively highly oxygen functionalized groups on GO restrict the diffusion of ions, occurring in bad electrochemical reactions in organic electrolytes. Heteroatoms are doped with GO to overcome this problem since they can easily tune the electronic properties, mechanical and chemical stability of carbon frameworks [25,26].

20.5.1.1 N-Doped Heteroatom

The N-doped structures have a large surface area and low density of oxygen-containing groups resulting that it is a favorable electrode material for EDLCs in an organic electrolyte solution. The composite of NiSe$_2$@N-rGO has increased the specific surface area and electric conductivity of materials resulting in NiSe$_2$@N-rGO composite producing a large specific capacitance of 2451.4 F g^{-1} at 1 A g^{-1} current density. It is stable under the 0–1.6 V range. Moreover, energy density and power density are recorded in 40.5 and 841.5 W kg^{-1}, respectively [27]. Unfortunately, the nitrogen-doping will deteriorate the conductivity of the nanocomposites due to intensive p-p interaction. This p-p interaction is resulting in low surface area and low capacitance. To avoid these following problems an effective strategy is designed by the addition of CNT and conducting polymer such as PANI (polyaniline), polyphenol in between the graphene sheets [28]. Wang et al. reported N-doped GO/Ni-ferrite@ PANI hybrid materials in which their specific capacitance was found out 667.0 at A g^{-1} and obtained good power and energy density 110.8 and 927 Wh kg^{-1}. Their retention capacity was 90% over 10,000 cycles [29].

20.5.1.2 B-Doped Heteroatom

Doping of chemicals in the GO is one of the effective methods to tune the electronic properties. B-doped GO is synthesized by the reduction of GO under borane tetrahydrofuran (B-rGO). The B-doped rGO had a large specific surface area near about 466 $m^2 g^{-1}$, which is including excellent specific capacitance of 200 F g^{-1} in an aqueous electrolyte solution. It also shows good cyclic stability after 4,500 cycles [30]. Balaji S. et al. prepared B-doped rGO through supercritical fluid processing as well as hydrothermal route. In which specific capacitance is 286 F g^{-1} with 96% retention capacitance at 20 A g^{-1} over 10,000 cycles [31]. Similarly, hierarchically porous B-doped carbon material is synthesized that exhibits high specific capacitance 379.9 F g^{-1} at 1 A g^{-1}, in which cycling performance retention capacitance is near about 96.3% over 10,000 cycles [32]. Prakash D. et al. developed N-B doped GO, including excellent specific capacitance 885 F g^{-1} at 10 A g^{-1} with retention capacitance of 77.8% over 10,000 cycles [33].

20.5.1.3 S-Doped Heteroatom

S-doped graphene electrodes also exhibited improved capacitive like nitrogen. S-doped nanocomposites are generally prepared with dominant sulfur sources such as thiophene. S-doped carbon nanocomposites showed efficient energy storage that was prepared simple flame pyrolysis technique using cow margarine and carbon disulfide. They obtained maximum specific capacitance 337 F g^{-1} at 1 A g^{-1}, in which retaining capacitance was found out 89% over 20,000 cycles with higher current density ~15 A g^{-1} [34].

20.5.2 Metal-Organic Frameworks-Doped Graphene Electrode

Metal-organic frameworks (MOFs) are new types of crystalline materials, in which metal clusters are coordinated with organic linkers through a covalent bond. They possessed some remarkable properties such as tuneable porosity, have a large specific surface area, and a well-arranged crystalline structure. Therefore, they are extensively used in electrochemical sensing, catalytic activity, drug delivery, and energy storage system [35–37]. The major issue of MOFs regarding materials is their electronic conductivity and stability is poor in aqueous media, which affects the energy storage capacity; to overcome this problem, MOFs are incorporated with rGO. Recently MOFs-rGO combination was proposed, which boosts the electrical conductivity and mechanical stability of the composite. Simple ultrasonication method, which delivers high energy storage capacity, synthesizes Cu-MOFs (benzene 1,3,5-tricarboxylic acid)/rGO hybrid material at 685.33 F g^{-1} at 1.6 A g^{-1}. It shows retention capacity 91.91% after 1,000 cycles at 8 A g^{-1} [38].

Rajpurohit et al. [39] investigated for super capacitance performance of bimetallic MOFs using Cu and Fe as metal centers with 1,3,5-tricarboxylic acid as ligand with

S-doped GO. The capacitance behavior of CuFe@BTS/S-GNS has been examined under 1 M Na_2SO_4 electrolyte solution in three and two-electrode system assembly in which specific capacitance was found 1164.3 F g^{-1} at 0.5 A g^{-1} and deliver remarkable power and energy density 195.1 W kg^{-1} and 96.57 Wh kg^{-1}, respectively, over 1.8 V. It shows an excellent retention capacity 92.5% over 10,000 cycles at 1.0 A g^{-1} current density. In recent times, sulfide-based MOFs have been prepared, and rGO-$NiMoO_4$@Ni-Co-S hybrid nanocomposite possessed a high specific capacity of 318 mAh g^{-1} at 1 A g^{-1} remarkable cyclic performance of 88.87% after 10,000 cycles. The hybrid materials also possessed satisfactory energy density and power density of 57.2 WhK g^{-1} and 801.8 WK g^{-1}, respectively [40].

20.5.3 GRAPHENE-BASED NANOCOMPOSITES WITH CONDUCTING POLYMERS (PSEUDOCAPACITORS)

Current scenario conductive polymers are extensively used for energy store application due to their high pseudocapacitance. Many conducting polymers (CPs) are investigated for the super capacitance application. Among them, PANI, polypyrrole (PPY), and polythiophene are widely used (PTH). PANI is considering the most promising material due to its low cost, ease of preparation, and high specific capacitance. Hao and co-workers [41] synthesized a new type of conductive material based on fibrillar polyaniline incapacitated with GO layers, which was prepared via in situ polymerizations of the aniline in the presence of GO. The specific capacitance of synthesized PANI/GO composite up to 531 F g^{-1} is higher than PANI alone, showing the synergetic effect between PANI and GO. In another work, Hao et al. [42] reported the variation in electrochemical properties of PANI/GO by changing the size and feeding ratios of graphene material. At 12,500 mesh size, graphene sheets have a high specific capacitance of 746 F g^{-1} with retention capacitance of 73% over 500 cycles. Similar at 500 mesh size-specific capacitance was obtained 627 F g^{-1} with 64% retention capacitance over 500 cycles. The improved specific capacitance and cyclic stability show a synergetic effect among the two components. Similar polypyrrole has also been used as the conducting polymer. Lee and co-workers synthesized PPY/GO-based super capacitance with satisfactory specific capacitance 267 F g^{-1} at scan rate 100 mV s^{-1} [43]. Zhao and co-workers also work based on PPY/GO materials in which has remarkable specific capacitance was obtained near about 510 F g^{-1} with high capacitance. Their retention ratio found out near 70% near 10,000 cycles [44].

20.5.4 METAL OXIDE-BASED GRAPHENE NANOCOMPOSITES FOR SUPERCAPACITORS

The sheet-like construction of graphene-based material enhances their application as a matrix for metal oxide. Incorporating GO enhances the flexible capacity of metal oxide. The composites of rGO-metal oxide can be synthesized via various methods, such as electrodeposition, hydrothermal, co-precipitated, and spray pyrolyzing. RuO_2 metal oxide was anchored onto rGO. The specific capacitance of RuO_2/rGO is higher than pure RuO_2 or rGO because of the synergistic effect. In this effect, RuO_2 separates the rGO sheets, resulting in an increment of EDLC of rGO; similarly, dispersed rGO prevented the agglomeration of the RuO_2 [45,46]. In super-capacitance materials, the metal with a broad range of bandgap is widely used. Zn and Cu oxide are remarkable applications in super-capacitance materials due to suitable bandgap. The GO-CuO for specific capacitance was obtained at 790 F g^{-1} at a scan rate of 5 mV s^{-1} in KCl. Similarly, ZnO Specific capacitance was found out near about (332–445) F g^{-1} in KCl electrolyte solution [47]. Various kinds of metal oxide and GO-based supercapacitor electrodes are studied by Mishra and Ramaprabhu [48]; from their study, it is found that hybrid nanomaterials have higher specific capacitance compared to pristine GO and metal oxide alone. It results from a significant interaction between rGO and metal oxide, indicating that rGO-Metal hybrid is suitable for energy storage purposes. Wang et al. [49] reported that the hybrid materials of iron oxide and GO have excellent specific capacitance 908 F g^{-1} at 2 A g^{-1} in 1 M KOH electrolyte solution. Kuila et al. also worked on iron oxide and rGO hybrid material in which specific capacitance was found out 782 F g^{-1} at 3 A g^{-1}, and retention ratio is also high 99.99% after 1,000 cycles. Energy density was also found at 39.1 WhK g^{-1} in 6 M KOH electrolyte solution. Zhao et al. reported the growth of CNTs on Ni-ion/rGO plates, which show excellent electrochemical resulting in which specific capacitance is found out 1,235 F g^{-1} at 1 A g^{-1} in 6 M KOH electrolyte solution [50].

20.6   CONCLUSION

This chapter summarizes the present development in graphene-based 2D material for energy storage, such as supercapacitors. Combining graphene and its derivative with the traditional electrode materials has carried about many extraordinary developments in energy storage systems, including supercapacitors. The high specific surface area, porosity, good electrical conductivity, no structural change with chemical reaction, and external stress of graphene enhance the electron's movement and increase the specific capacitance. The chemical exfoliation method considers as an excellent pathway for graphene synthesis at a low cost. Recently, graphene is combined with various other compositions, including heteroatoms, polymer, MOFs, to develop EDLC and pseudocapacitor for energy storage. One of the major drawbacks of the graphene-based electrode material is the gap between laboratory research and practical application, which needs to be lost.

ACKNOWLEDGMENT

Kaushalya Bhakar, Naresh A. Rajpurohit, and Dinesh Kumar are thankful to the Central University of Gujarat, Gandhinagar, for supporting this work.

REFERENCES

1. Miller, John R., and Patrice Simon. "Electrochemical capacitors for energy management." *Science Magazine* 321, no. 5889 (2008): 651–652.
2. Vivekchand, S. R. C., Chandra Sekhar Rout, K. S. Subrahmanyam, Achutharao Govindaraj, and C. N. R. Rao. "Graphene-based electrochemical supercapacitors." *Journal of Chemical Sciences* 120, no. 1 (2008): 9–13.
3. Xia, Jilin, Fang Chen, Jinghong Li, and Nongjian Tao. "Measurement of the quantum capacitance of graphene." *Nature Nanotechnology* 4, no. 8 (2009): 505–509.

4. Huang, Yi, Jiajie Liang, and Yongsheng Chen. "An overview of the applications of graphene-based materials in supercapacitors." *Small* 8, no. 12 (2012): 1805–1834.

5. Hirata, Masukazu, Takuya Gotou, Shigeo Horiuchi, Masahiro Fujiwara, and Michio Ohba. "Thin-film particles of graphite oxide 1: High-yield synthesis and flexibility of the particles." *Carbon* 42, no. 14 (2004): 2929–2937.

6. Ke, Qingqing, and John Wang. "Graphene-based materials for supercapacitor electrodes: A review." *Journal of Materiomics* 2, no. 1 (2016): 37–54.

7. Lei, Zhibin, Li Lu, and X. S. Zhao. "The electrocapacitive properties of graphene oxide reduced by urea." *Energy & Environmental Science* 5, no. 4 (2012): 6391–6399.

8. Garg, Bhaskar, Tanuja Bisht, and Yong-Chien Ling. "Graphene-based nanomaterials as heterogeneous acid catalysts: a comprehensive perspective." *Molecules* 19, no. 9 (2014): 14582–14614.

9. Olabi, A. G., Mohammad Ali Abdelkareem, Tabbi Wilberforce, and Enas Taha Sayed. "Application of graphene in energy storage device: A review." *Renewable and Sustainable Energy Reviews* 135(2021): 110026.

10. Yu, Qingkai, Jie Lian, Sujitra Siriponglert, Hao Li, Yong P. Chen, and Shin-Shem Pei. "Graphene segregated on Ni surfaces and transferred to insulators." *Applied Physics Letters* 93, no. 11 (2008): 113103.

11. Yu, X. Z., C. G. Hwang, Chris M. Jozwiak, A. Kohl, Andreas K. Schmid, and Alessandra Lanzara. "New synthesis method for the growth of epitaxial graphene." *Journal of Electron Spectroscopy and Related Phenomena* 184, no. 3–6 (2011): 100–106.

12. Hirata, Masukazu, Takuya Gotou, Shigeo Horiuchi, Masahiro Fujiwara, and Michio Ohba. "Thin-film particles of graphite oxide 1: High-yield synthesis and flexibility of the particles." *Carbon* 42, no. 14 (2004): 2929–2937.

13. Yan, Jun, Tong Wei, Bo Shao, Fuqiu Ma, Zhuangjun Fan, Milin Zhang, Chao Zheng, Yongchen Shang, Weizhong Qian, and Fei Wei. "Electrochemical properties of graphene nanosheet/carbon black composites as electrodes for supercapacitors." *Carbon* 48, no. 6 (2010): 1731–1737.

14. Stoller, Meryl D., Sungjin Park, Yanwu Zhu, Jinho An, and Rodney S. Ruoff. "Graphene-based ultracapacitors." *Nano letters* 8, no. 10 (2008): 3498–3502.

15. Zhu, Jixin, Dan Yang, Zongyou Yin, Qingyu Yan, and Hua Zhang. "Graphene and graphene-based materials for energy storage applications." *Small* 10, no. 17 (2014): 3480–3498.

16. Ji, Liwen, Praveen Meduri, Victor Agubra, Xingcheng Xiao, and Mataz Alcoutlabi. "Graphene-based nanocomposites for energy storage." *Advanced Energy Materials* 6, no. 16 (2016): 1502159.

17. Qiu, Ling, Xiaowei Yang, Xinglong Gou, Wenrong Yang, Zi-Feng Ma, Gordon G. Wallace, and Dan Li. "Dispersing carbon nanotubes with graphene oxide in water and synergistic effects between graphene derivatives." *Chemistry: A European Journal* 16, no. 35 (2010): 10653–10658.

18. Bai, Hua, Chun Li, Xiaolin Wang, and Gaoquan Shi. "A pH-sensitive graphene oxide composite hydrogel." *Chemical Communications* 46, no. 14 (2010): 2376–2378.

19. Abbas, Qaisar, Rizwan Raza, Imran Shabbir, and A. G. Olabi. "Heteroatom doped high porosity carbon nanomaterials as electrodes for energy storage in electrochemical capacitors: A review." *Journal of Science: Advanced Materials and Devices* 4, no. 3 (2019): 341–352.

20. Li, Qi, Michael Horn, Yinong Wang, Jennifer MacLeod, Nunzio Motta, and Jinzhang Liu. "A review of supercapacitors based on graphene and redox-active organic materials." *Materials* 12, no. 5 (2019): 703.

21. Lu, Qi, Jingguang G. Chen, and John Q. Xiao. "Nanostructured electrodes for high-performance pseudocapacitors." *Angewandte Chemie International Edition* 52, no. 7 (2013): 1882–1889.

22. Xu, Bin, Shufang Yue, Zhuyin Sui, Xuetong Zhang, Shanshan Hou, Gaoping Cao, and Yusheng Yang. "What is the choice for supercapacitors: graphene or graphene oxide?" *Energy & Environmental Science* 4, no. 8 (2011): 2826–2830.

23. Yu, Guihua, Liangbing Hu, Michael Vosgueritchian, Huiliang Wang, Xing Xie, James R. McDonough, Xu Cui, Yi Cui, and Zhenan Bao. "Solution-processed graphene/MnO$_2$ nanostructured textiles for high-performance electrochemical capacitors." *Nano Letters* 11, no. 7 (2011): 2905–2911.

24. Zhang, Kai, Li Li Zhang, X. S. Zhao, and Jishan Wu. "Graphene/polyaniline nanofiber composites as supercapacitor electrodes." *Chemistry of Materials* 22, no. 4 (2010): 1392–1401.

25. Li, Wenrong, Dehong Chen, Zheng Li, Yifeng Shi, Ying Wan, Guan Wang, Zhiyu Jiang, and Dongyuan Zhao. "Nitrogen-containing carbon spheres with very large uniform mesopores: The superior electrode materials for EDLC in organic electrolyte." *Carbon* 45, no. 9 (2007): 1757–1763.

26. Li, Shin-Ming, Shin-Yi Yang, Yu-Sheng Wang, Hsiu-Ping Tsai, Hsi-Wen Tien, Sheng-Tsung Hsiao, Wei-Hao Liao, Chien-Liang Chang, Chen-Chi M. Ma, and Chi-Chang Hu. "N-doped structures and surface functional groups of reduced graphene oxide and their effect on the electrochemical performance of supercapacitor with organic electrolyte." *Journal of Power Sources* 278(2015): 218–229.

27. Maurya, Oshnik, Somnath Khaladkar, Michael R. Horn, Bhavesh Sinha, Rajendra Deshmukh, Hongxia Wang, TaeYoung Kim, Deepak P. Dubal, and Archana Kalekar. "Emergence of Ni-based chalcogenides (S and Se) for clean energy conversion and storage." *Small* 17(2021): 2100361.

28. You, Bo, Lili Wang, Li Yao, and Jun Yang. "Three-dimensional N-doped graphene–CNT networks for supercapacitor." *Chemical Communications* 49, no. 44 (2013): 5016–5018.

29. Mei, Jun, and Long Zhang. "Facile and economic synthesis of nitrogen doped graphene/manganese dioxide composites in aqueous solution for energy storage devices." *Materials Letters* 143(2015): 163–166.

30. Han, Jongwoo, Li Li Zhang, Seungjun Lee, Junghoon Oh, Kyoung-Seok Lee, Jeffrey R. Potts, Junyi Ji, Xin Zhao, Rodney S. Ruoff, and Sungjin Park. "Generation of B-doped graphene nanoplatelets using a solution process and their supercapacitor applications." *ACS Nano* 7, no. 1 (2013): 19–26.

31. Balaji, S. Suresh, M. Karnan, P. Anandhaganesh, Shaikh Mohammad Tauquir, and M. Sathish. "Performance evaluation of B-doped graphene prepared via two different methods in symmetric supercapacitor using various electrolytes." *Applied Surface Science* 491 (2019): 560–569.

32. Du, Pengcheng, Luohua Liu, Yiwei Dong, Wangzu Li, Jinmei Li, Zhenlin Liu, and Xue Wang. "Synthesis of hierarchically porous boron-doped carbon material with enhanced surface hydrophobicity and porosity for improved supercapacitor performance." *Electrochimica Acta* 370(2021): 137801.

33. Prakash, Duraisamy, and Sellaperumal Manivannan. "N, B co-doped and crumpled graphene oxide pseudocapacitive electrode for high energy supercapacitor." *Surfaces and Interfaces* 23(2021): 101025.

34. Bondarde, Mahesh P., Pravin H. Wadekar, and Surajit Some. "Synthesis of sulfur doped carbon nanoparticle for the improvement of supercapacitive performance." *Journal of Energy Storage* 32 (2020): 101783.

35. Manivel, Perumal, Vembu Suryanarayanan, Noel Nesakumar, David Velayutham, Kanagaraj Madasamy, Murugavel Kathiresan, Arockia Jayalatha Kulandaisamy, and John Bosco Balaguru Rayappan. "A novel electrochemical sensor based on a nickel-metal organic framework for efficient electrocatalytic oxidation and rapid detection of lactate." *New Journal of Chemistry* 42, no. 14 (2018): 11839–11846.

36. Yang, Dong, Manuel A. Ortuno, Varinia Bernales, Christopher J. Cramer, Laura Gagliardi, and Bruce C. Gates. "Structure and dynamics of Zr_6O_8 metal-organic framework node surfaces probed with ethanol dehydration as a catalytic test reaction." *Journal of the American Chemical Society* 140, no. 10 (2018): 3751–3759.

37. Horcajada, Patricia, Tamim Chalati, Christian Serre, Brigitte Gillet, Catherine Sebrie, Tarek Baati, Jarrod F. Eubank et al. "Porous metal–organic-framework nanoscale carriers as a potential platform for drug delivery and imaging." *Nature Materials* 9, no. 2 (2010): 172–178.

38. Saraf, Mohit, Richa Rajak, and Shaikh M. Mobin. "A fascinating multitasking Cu-MOF/rGO hybrid for high performance supercapacitors and highly sensitive and selective electrochemical nitrite sensors." *Journal of Materials Chemistry A* 4, no. 42 (2016): 16432–16445.

39. Rajpurohit, Anuja S., Ninad S. Punde, and Ashwini K. Srivastava. "A dual metal organic framework based on copper-iron clusters integrated sulphur doped graphene as a porous material for supercapacitor with remarkable performance characteristics." *Journal of Colloid and Interface Science* 553(2019): 328–340.

40. Acharya, Jiwan, Gunendra Prasad Ojha, Byoung-Suhk Kim, Bishweshwar Pant, and Mira Park. "Modish designation of hollow-tubular $rGO–NiMoO_4$@ Ni-Co-S hybrid core–shell electrodes with multichannel superconductive pathways for high-performance asymmetric supercapacitors." *ACS Applied Materials & Interfaces* 13, no. 15 (2021): 17487–17500.

41. Wang, Hualan, Qingli Hao, Xujie Yang, Lude Lu, and Xin Wang. "A nanostructured graphene/polyaniline hybrid material for supercapacitors." *Nanoscale* 2, no. 10 (2010): 2164–2170.

42. Wang, Hualan, Qingli Hao, Xujie Yang, Lude Lu, and Xin Wang. "Effect of graphene oxide on the properties of its composite with polyaniline." *ACS Applied Materials & Interfaces* 2, no. 3 (2010): 821–828.

43. Bose, Saswata, Tapas Kuila, Thi Xuan Hien Nguyen, Nam Hoon Kim, Kin-tak Lau, and Joong Hee Lee. "Polymer membranes for high temperature proton exchange membrane fuel cell: Recent advances and challenges." *Progress in Polymer Science* 36, no. 6 (2011): 813–843.

44. Zhang, Li Li, Shanyu Zhao, Xiao Ning Tian, and X. S. Zhao. "Layered graphene oxide nanostructures with sandwiched conducting polymers as supercapacitor electrodes." *Langmuir* 26, no. 22 (2010): 17624–17628.

45. Zhi, Mingjia, Chengcheng Xiang, Jiangtian Li, Ming Li, and Nianqiang Wu. "Nanostructured carbon–metal oxide composite electrodes for supercapacitors: A review." *Nanoscale* 5, no. 1 (2013): 72–88.

46. Wu, Zhong-Shuai, Guangmin Zhou, Li-Chang Yin, Wencai Ren, Feng Li, and Hui-Ming Cheng. "Graphene/metal oxide composite electrode materials for energy storage." *Nano Energy* 1, no. 1 (2012): 107–131.

47. Chakraborty, Sohini, and N. L. Mary. "A carbon nanotube reinforced functionalized styrene–maleic anhydride copolymer as an advanced electrode material for efficient energy storage applications." *New Journal of Chemistry* 44, no. 11 (2020): 4406–4416.

48. Mishra, Ashish Kumar, and Sundara Ramaprabhu. "Functionalized graphene-based nanocomposites for supercapacitor application." *The Journal of Physical Chemistry C* 115, no. 29 (2011): 14006–14013.

49. Wang, Huanwen, Zijie Xu, Huan Yi, Huige Wei, Zhanhu Guo, and Xuefeng Wang. "One-step preparation of single-crystalline Fe_2O_3 particles/graphene composite hydrogels as high-performance anode materials for supercapacitors." *Nano Energy* 7(2014): 86–96.

50. Zhang, Li Li, Zhigang Xiong, and X. S. Zhao. "A composite electrode consisting of nickel hydroxide, carbon nanotubes, and reduced graphene oxide with an ultrahigh electrocapacitance." *Journal of Power Sources* 222 (2013): 326–332.

21 MXene-Based 2D Nanomaterials for Supercapacitors

Felipe M. de Souza
Pittsburg State University

Anuj Kumar
GLA University

Ram K. Gupta
Pittsburg State University

CONTENTS

21.1 INTRODUCTION

The potential applications of 2D materials came into the spotlight mostly after the 2010 Nobel Prize for the development of single-layer graphene by Andre Geim and Kostya Novoselov. Along that other materials such as double-layered hydroxides, phosphorenes, transition metals dichalcogenides, silicene, germancene, and hexagonal boron nitride have also been recently received tremendous attention. In 2011, titanium carbide (Ti_3C_2) was successfully synthesized by Naguib et al. as a 2D nanostructure that presented a conductivity of 6,500 S cm^{-1} and decent optical properties as it transmitted over 97% of light per nanometer thickness [1]. The Ti_3C_2 nanosheet structure had individual aligned planes and a high density of free electrons, which prompted the remarkable conductivity properties and showed a quantum confinement effect. This property arises when a material is sufficiently small (a few nanometers) to cause the energy bandgap to increase as a function of the particle size. Such

DOI: 10.1201/9781003178422-21

intriguing properties sparked the research on transition metal carbides and nitrides, collectively named as MXenes. The notable and yet partially uncharted features of MXenes such as electrochemical, morphological, conductivity, semiconductivity, and optical properties led to applications in saturable absorbers, biomedicine, nonlinear photonics, field-effect transistors, conductive electrodes, batteries, and supercapacitors. Based on the plethora of utilizations, the understanding of MXene's electrochemical properties and concepts is vital. The general formula of MXenes can be expressed as $M_{n+1}X_nT_x$, where $n + 1$ describes the layers of transition metals (M), in between the layer of C or N (X). Lastly, T_x describes the functional groups like hydroxyl, chlorine, or oxygen fluorine located in the MXene surface and/or edge [2]. Various types of MXenes are possible because of the use of many transition metals, surface functionalities, and up to three different atomic structures, which sum to more than 30 MXenes. Besides, the surface can be tuned through chemical modification to introduce specific functional moiety or doping with heteroatoms to increase the number of active sites that can store charge for energy applications [3–5].

MXenes can be also synthesized using methods applied for the synthesis of graphenes such as the top-bottom approach where exfoliation of bulk starting material can provide nanosheets with a thickness of a single or few atoms layers. This structure provides high surface area and conductivity, which are core requirements for batteries and supercapacitor applications as they lead to stable charge/discharge cycles and rapid power delivery. The charge/discharge mechanism occurs mainly through the adsorption/desorption of ions in the electric double layer (Helmholtz double layer) formed in between the electrode and electrolyte. This phenomenon defines the traditional electric double-layer capacitors. Such electrochemical properties are influenced by the type of materials used and surface area. However, these devices often suffer from low energy density (around 10 Wh kg^{-1}), because of the limited surface area that can be achieved. A known way to improve the performance of capacitors lies in employing pseudocapacitive materials because of their redox process, which allows more charge to be stored. For that purpose, metal oxides have been used; however, their low conductivity imposes a challenge. MXenes became a viable option as these materials are often conductive and can vary their oxidation number due to the transition metal that facilitates charge transfer. Also, the 2D structure with a few atomic layers provides a high surface area. Lukatskaya et al. synthesized $Ti_3C_2T_x$ that achieved 210 F g^{-1} and showed a volumetric capacitance of 1,500 F cm^{-3}, which surpassed the capacitance of state-of-the-art RuO_2 [6]. It was observed that lowering film thickness provided a better alignment of 2D structures, which improved the specific capacitance. In addition, pseudocapacitance played a major role in improving the electrochemical properties. Despite the appealing properties of 2D MXenes, there are some challenges such as the tendency to restack the nanolayers, low flexibility, prone to oxidation, and difficulties in scaling up production. As an alternative, some approaches such as integration of MXenes with other materials like graphene, carbon nanotubes, metals, metal sulfides, nanomaterials, and polymers have been performed [7–9]. The latter strategy of combining polymers with MXenes has been proven to be quite successful. Approaches such as preparing a polymer separately and blending with an MXene (ex situ) are worthwhile options since several polymers can be used by implementing this approach. Also, organized structures and accurate control composition can be obtained. Such effects arise due to the attractive inter-molecular forces between the polymer and MXene. Such approaches also provide enhanced tensile strength and improve cation intercalation in between the polymer/MXene, which increases volumetric capacitance, attenuates the tendency of MXenes films to restack, and improves stability against oxygen. Such factors were observed by Ling et al. after combining $Ti_3C_2T_x$ with poly diallyl dimethylammonium chloride (PDDA) and separately with poly(vinyl alcohol) (PVA) [8]. Polymers and MXenes composites can also be prepared in situ, as the polymeric film is formed over the surface of the MXene. This strategy is seldomly employed compared with the ex situ approach due to the limitation of monomers that can be used. Yet, the synergy effect between polymers and MXenes is beneficial for the improvement of electrochemical, mechanical, and chemical stability of electrodes, which improves their applicability. Some techniques for the fabrication of these composite electrodes include hot press, vacuum-assisted filtration (VAF), and drop-casting [10–12]. The following sessions discuss the approaches for the synthesis of MXenes and their use in supercapacitors regarding approaches, morphology effects, and properties.

21.2 APPROACHES FOR THE SYNTHESIS OF MXene

MXenes have shown promising results and possibilities of further enhancing their energy storage properties by tuning their synthesis approaches. The common approach used for their synthesis lies on the chemical or physical etching of their bulk counterpart in the form of $M_{n+1}AX_n$ where M is the transition metal, A represents the metals or semimetals from the groups IIIA and IVA, and X can be C or N. Among those, Ti_3AlC_2 has been widely studied while many other MAX phase compounds have been proposed theoretically [13,14]. Through that, mainly two approaches have been adopted: (i) top-down approaches such as exfoliation of bulk MXenes or (ii) bottom-up approaches such as chemical vapor deposition (CVD) form a 2D film through gasses reactants. The following session discusses in detail the synthetic approaches for MXenes.

The chemical etching is a viable approach for the synthesis of MXenes starting from a MAX compound. Instead of physical exfoliation that is used to obtain nano-flakes of graphene or black phosphorus, MXenes can be obtained through a top-bottom method, which usually requires a selective etching that cannot be performed only by sonication or physical exfoliation. Instead, specific etching is required usually by hydrofluoric acid (HF), other fluorine-based etching agents, or heating. Such a process is required as the MAX layers have strong metallic bonds. However, the M–A bonds are more susceptible to break in comparison to M–X bonds. Thus, selective etching can be performed on the A species from the MAX to form the 2D layer MXene. For example, Naguib et al. obtained 2D nanolayers of Ti_3C_2 from the selective etching of Ti_3AlC_2 with a 50 wt.% HF solution for 2 hours at room temperature [1]. The proposed reaction for the synthesis of Ti_3C_2 is shown in reactions (21.1) to (21.3). The reaction showed that after the chemical etching (reaction 21.1), Ti_3C_2 further reacts with water to add hydroxyl groups on the MXene surface and edges to form $Ti_3C_2(OH)_2$ (reaction 21.2). Also, the excess of HF can covalently bond to the Ti_3C_2 surface and form a fluorine termination (reaction 21.3). This facile approach can also be applied for the synthesis of other MXenes such as Nb_4C_3, Nb_2C, V_2C, Ti_2C, and Ta_4C_3. Along with that mixed addition of two transition metals can also be performed, which changes the distribution of metals into the 2D arrangement, occasioning a change in properties. Also, Mo layers can be introduced in between the MXene 2D structure as in the case of Mo_2TiC_2 or $Mo_2Ti_2C_3$. It has been observed that in some scenarios a lower concentration of HF can lead to better capacitance as it provides the optimal allocation of ions in between the layers. On the other hand, the case in which the ions diffuse too deep in between the layers leads to higher activation energy [15].

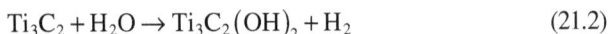

$$Ti_3AlC_2 + 3HF \rightarrow AlF_3 + 3/2H_2 + Ti_3C_2 \qquad (21.1)$$

$$Ti_3C_2 + H_2O \rightarrow Ti_3C_2(OH)_2 + H_2 \qquad (21.2)$$

$$Ti_3C_2 + 2HF \rightarrow Ti_3C_2F_2 + H_2 \qquad (21.3)$$

Munir et al. studied the influence of crucial parameters such as HF concentration, different solvents, and sonication time to optimize the overall yield of Ti_3C_2 from the starting Ti_3AlC_2 [16]. The most satisfactory results were obtained under the condition of a 30 wt.% HF solution, dimethyl sulfoxide (DMSO) as a solvent, and 135 minutes of sonication time. The schematics for the addition of solvents, sonication time, effect of increasing concentration of HF etching, and scanning electron microscopic (SEM) images for the samples are shown in Figure 21.1. In the SEM images, it is notable that the pristine Ti_3AlC_2 presented an uneven morphology surface. On the other hand, with the increase of HF concentration, there was the appearance of an accordion-like structure. As the concentration of HF went up to 30 wt.%, the most predominant accordion-shaped morphology was observed (Figure 21.1g), which suggested a higher surface area in comparison to the other samples. On top of that, DMSO was the solvent that provided the highest interlayer d-spacing, which is an important parameter to avoid restacking of flakes. This effect was observed due to the size and chemical stability of this solvent in comparison to water, ammonia, and urea. Also, when water was used as intercalant other peaks likely related to Al_2O_3, AlF_3, and $TiOF_2$ were observed, which may be undesirable for the system.

Despite the efficiency of HF as an etching agent, this compound raises concerns regarding its toxicity. Because of this, other approaches have been developed as a safer option for the chemical etching, which consists of the in situ formation of HF. Based on that, a single-step and high-efficiency method for the synthesis of Ti_3C_2 through Ti_3AlC_2 has been reported by Ghidiu et al., where HCl and LiF solutions were added into the reaction system to form HF to perform the etching [17]. They obtained a clay that could be molded into several shapes before drying and achieved a volumetric capacitance of around 900 F cm^{-3}. The chemical reaction for the in situ fabrication of HF by using LiF and HCl is shown in reaction (21.4).

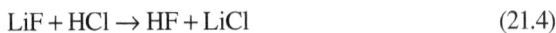

$$LiF + HCl \rightarrow HF + LiCl \qquad (21.4)$$

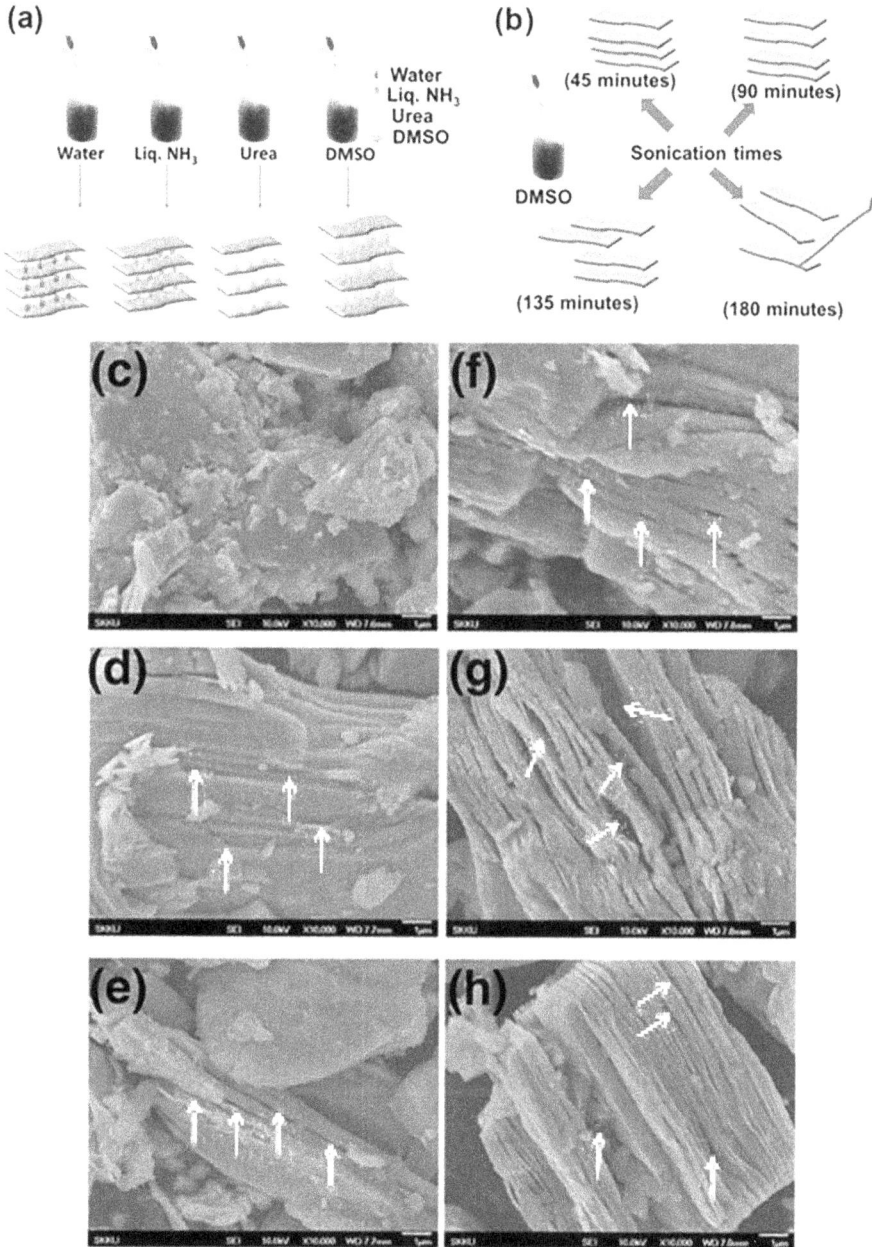

FIGURE 21.1 (a) Effect of layer distancing through the addition of different solvents. (b) Effect of sonication time into 2D layer exfoliation. SEM images for (c) Pristine Ti$_3$AlC$_2$ and Ti$_3$C$_2$ obtained using increasing concentrations of HF: (d) 5, (e) 10, (f) 20, (g) 30, and (h) 40 wt.%. (Adapted with permission from Ref. [16], Copyright 2021, American Chemical Society.)

Another in situ fabrication of HF was performed by Wang et al. who utilized the traditional hydrothermal method with NH_4F solution and Ti_3AlC_2 powder to obtain $Ti_3C_2T_x$, where T is F, OH, and O [18]. The sample's capacitance was measured to be 141 F cm^{-3} in 3 M KOH electrolyte. The chemical reaction for the in situ fabrication of HF is shown in reaction (21.5).

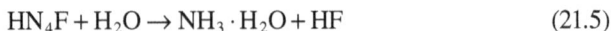

$$HN_4F + H_2O \rightarrow NH_3 \cdot H_2O + HF \tag{21.5}$$

In situ preparation of HF offers a safer procedure, Li^+ and NH_4^+ cations for intercalation in 2D nanoflakes, decreases the attraction forces between the layers and thus prevents restacking. Also, several fluoride salts such as LiF, NaF, KF, CsF, or CaF_2 along with HCl or H_2SO_4 can be used to perform the exfoliation process, giving some versatility to the procedures [17,19]. The MXenes obtained from the in situ HF method often present higher hydrophilicity due to the presence of OH and O in the terminations, along with the range of zeta-potential in between -30 and -80 mV showing a negative charge that interacts with water. These features allow several processing techniques such as spray and spin coating, writing, printing, and electrophoretic deposition.

Fluorine-free approaches have also been developed [20]. The process consisted of a hydrothermal method assisted with NaOH to obtain a $Ti_3C_2T_x$, where T is either hydroxyl groups or oxygen. The relatively aggressive parameters such as high concentration of a NaOH solution (27.5 mol L^{-1}) and high temperature (270°C) yielded around 92% of $Ti_3C_2T_x$. Satisfactory electrochemical properties were obtained under these conditions as the scan rate of 2 mV s^{-1} and 1 mol L^{-1} H_2SO_4 electrolyte provided the gravimetric capacitance of 314 F g^{-1}, which was over 214% higher than $Ti_3C_2T_x$ obtained through the traditional HF etching process. However, to achieve these values the high alkali concentration and temperature parameters must be met otherwise under low temperatures the Al can form oxides that cause a jamming effect and therefore are not removed from the interlayer, jeopardizing its electrochemical properties by hindering the formation of well-spaced 2D layers. When the system is exposed to a high temperature but lower NaOH concentration, there was an enhancement of Al dissolution. It occurred because $Al(OH)_3$ reacts endothermically with alkalis, hence high temperature favors this process, which leads to better etching. However, part of the Ti_3C_2 was oxidized forming Na-Ti-O compounds (NTOs) such as $Na_2Ti_3O_7$ and $Na_2Ti_5O_{11}$, which suffer from poor cycling stability. This side reaction leads to poor sodium accommodation between the layers due to its size (0.97 Å) and electrode volume change due to sodium alloying or conversion causing cell deterioration as reported for Ti, Sn, Sb, and MoS_2-based electrodes [21–23].

Other top-bottom approaches to synthesize MXenes include scotch tape mechanical exfoliation, high-power ultrasonication, and chemical intercalation. Scotch tape mechanical exfoliation leads to high-quality layers with almost no imperfections in the structure. However, its low yield makes it challenging to scale up, despite its use for the fabrication of transistors. On the other hand, high-power ultrasonication leads to quite the opposite features, as the yield to obtain MXenes is higher with a larger number of defects, which can disrupt the 2D layer pattern. Chemical intercalation relies on the use of several solvents to promote the peeling of 2D layers. Solvents such as *N, N*-dimethylformamide, *n*-butylamine, choline, tetrabutylammonium hydroxide (TBAOH), isopropylamine, and DMSO are used for this. The latter is specific to $Ti_3C_2T_x$, while the former two are general to MXenes [24,25]. It is important to mention that this approach can demand a day to a week to take full effect. Also, the solvent plays some influence on the MXene's electrochemical properties. Similarly, ionic cations such as NH_4^+, Na^+, K^+, Mg^{2+}, or even Al^{3+} can also be intercalated through an electrochemical approach or spontaneously [26]. Electrochemical etching is another facile approach that can be performed by using HCl under a mild temperature of 50°C. It leads to terminations of O, OH, and Cl, possibly leading to some defects in the structure due to unselective etching of Ti.

Bottom-up techniques such as CVD and atomic layer deposition can also be employed for the synthesis of MXenes. It usually yields materials with low defects, the low thickness of the order of few nanometers, and large lateral active areas of around 100 μm. However, these methods have been seldomly explored by the scientific community, requiring further research for these techniques [27].

21.3 MXene FOR SUPERCAPACITORS

21.3.1 EFFECT OF INTERLAYER DISTANCE ON CHARGE-STORAGE CAPACITY

Finding materials that can effectively store energy with reasonable cost, high efficiency, and facile synthetic process are some of the core aspects to develop the sustainable future need of energy. To reach a satisfactory capacitance, these materials are required to be conductors or semiconductors, perform stable redox reactions to introduce pseudocapacitance, and endure several charge/discharge cycles without losing their overall capacitance. 2D materials offer practical solutions for that as their ultrathin morphology improves both electron flow and ionic transportation. Yet, further research is required to enhance their cycling stability. The interlayer spacing plays a major role as larger interlayer distancing allows higher and better ionic charge carrier transportation, and hence the higher capacitance. A small interlayer spacing leads to the tendency of irreversible restacking, which decreases the overall capacitance and permeation paths for intercalants and charge carriers. Many research aims to increase the interlayer spacing to decrease the energy barriers that arise from ionic diffusion. Also, the consistent sorption and desorption of cations can lead to volumetric changes that decrease the electrode's stability, which is more noticeable in layered 2D structures [28]. Another important factor regarding the increase of interlayer spacing is that it allows the use of other cations besides Li^+, such as Na^+, K^+, Mg^{2+}, Al^{3+}, and NH_4^+, enabling a wider set of abundant species that can be used to develop supercapacitors. Some strategies used to improve interlayer spacing are further discussed to provide an insight into how to optimize experiment designs.

The first commonly used strategy lies in using organic solvents as intercalants to increase the interlayer spacing. Peng et al. synthesized an ultrathin vanadyl phosphate ($VOPO_4$) 2D layered that was further intercalated with triethylene glycol (TEG) and tetrahydrofuran (THF) to provide improved sodium intercalation [29]. The possibility to adjust the interlayer spacing by selecting optimal intercalants able to coordinate with the 2D $VOPO_4$ nanolayers is important to improve ion transport as well as structure integrity when the electrode is submitted through many electrochemical cycles [30]. Isopropanol (IPA) and intercalants were used to promote the separation of the layers in $VOPO_4$ by removing water and replacing it with TEG and THF as demonstrated in Figure 21.2. The bond formed between

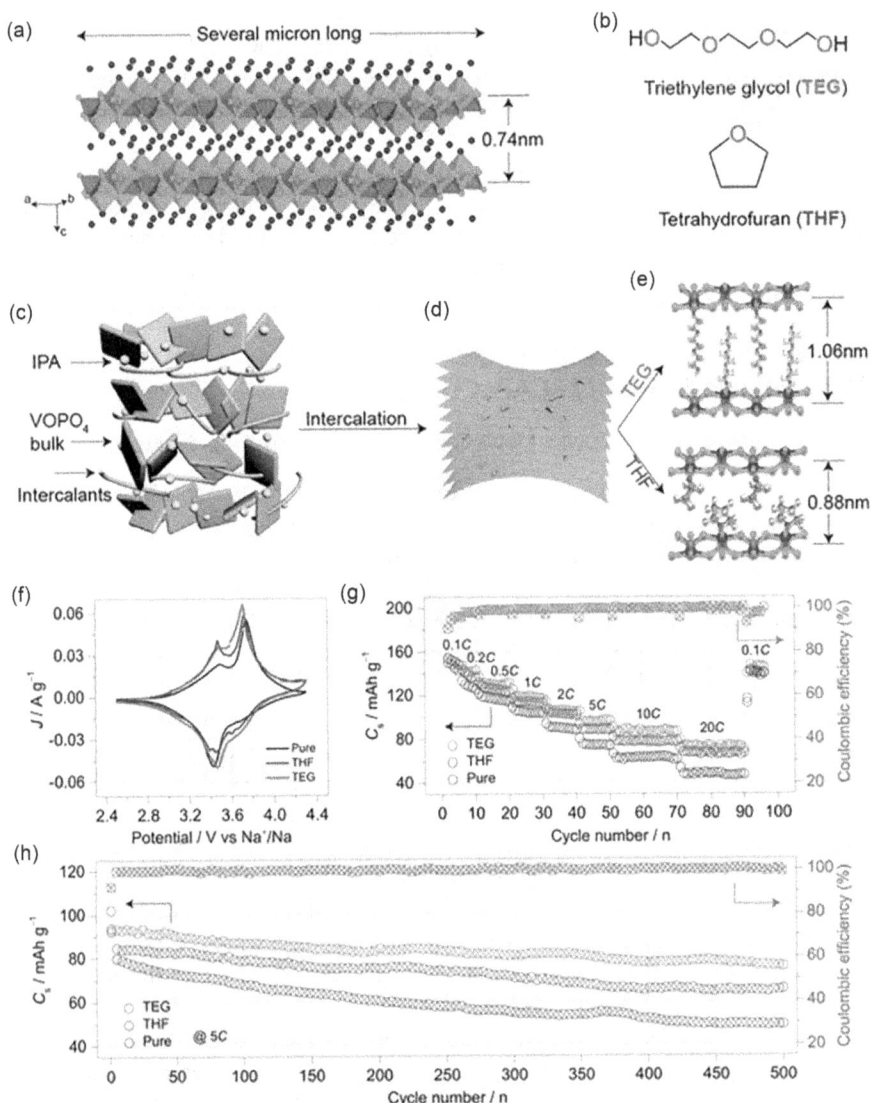

FIGURE 21.2   Scheme for the exfoliation of 2D $VOPO_4$ nanolayer. (a) Bulky $VOPO_4 \cdot 2H_2O$ structure. (b) Structures of TEG and THF are used as intercalants. (c and d) Intercalation process. (e) Interlayer spacing by using TEG and THF. Electrochemical analysis for the pure $VOPO_4$ and intercalated with THF and TEG. (f) CV. (g) Rate performance from the C range from 0.1 to 20. (h) Cycling stability as well as Coulombic efficiency. (Adapted with permission from Ref. [29], Copyright (2017), American Chemical Society.)

TEG and $VOPO_4$ provided the most stable alignment, which was responsible for increasing the distance between the layers in comparison to the starting material and THF, as confirmed by X-ray diffraction. To properly analyze the influence of space interlayer on the capacity, cyclic voltammetry (CV), rate capability as well as stability tests were performed. It was noticeable that the overall CV area increased by the order of $VOPO_4$ Pure $<$ THF $<$ TEG. It occurred because the larger space between the nanolayers allowed more Na^+ to allocate in between it as it is represented by the two sharp redox peaks. The same trend was observed for the rate performance as $VOPO_4$-TEG had higher capacitance values for all the current values, followed by $VOPO_4$-THF intercalated nanosheets and the pure material with no intercalants. The stability test for the electrodes was studied and again the TEG intercalated with 2D $VOPO_4$ displayed the highest capacity retention after 500 cycles of 88% followed by 76% for the THF electrode and 51% for the pure $VOPO_4$. Such an approach is relatively facile and variable for other intercalants such as alcohols, amines as well as other 2D layered structured materials. Hence, these conditions allowed the electrode to effectively adsorb Na^+, which is a desired as this cation is more abundant and cheaper than Li^+.

Another interesting strategy to increase the interlayer distance is pillaring. It consists of introducing a surfactant that can diffuse within the layers and increase the space between them. Luo et al. fabricated pillared Ti_3C_2 where the MXene preparation consisted of pre-pillaring with cetyltrimethylammonium bromide (CTAB), which is a cationic surfactant. The positive charge at the end of the alkyl chain from CTAB provides attractive electrostatic interaction with the Ti_3C_2 surface as the latter is negatively charged. The structural analysis demonstrated a shift in the Ti_3C_2 peaks to a lower angle, implying an increase in the interlayer spacing. It was observed that CTAB yielded the largest distancing between the nanoflakes as it contains an alkyl chain of 16 carbons compared with dodecyltrimethylammonium bromide (DTAB, which contains only 12 carbons. It occurred due to the intercalant's larger alkyl chain length. However, a further increase in the carbon chain caused a decrease in interlayer space because the intercalation energy was not enough to overcome the steric hindrance from the alkyl chain. Hence, there was an optimum chain length for intercalants to be introduced between the nanolayers. Another relevant aspect is to effectively introduce the surfactant in between the layer in the temperature range of 40°C–60°C. However, this condition can partially oxidize the surface of Ti_3C_2, which may decrease its conductivity. By performing this pre-pillaring step, a satisfactory interlayer spacing of around 2.23 nm was created. The pillaring was successfully performed using tin cations (Sn^{4+}) because of the large volume expansion that occurs along with the alloying with lithium, ultimately leading to extra capacity. It is important to mention that the electrochemical mechanism between Li and Sn is slow, which may decrease the electrode's power density. However, when these species reach 10 nm size, this drawback is eliminated as the slow cation diffusion is overcome by the pseudocapacitance that takes place in the electrode's surface. Based on that, $SnCl_4$ emulsion nanoparticles smaller than 10 nm were used as they formed Sn^{4+} cations that intercalated within the CTAB@Ti_3C_2 that had 2.230 nm through ion exchange leading to the nanocomposite [CTAB-Sn(IV)@Ti_3C_2]. The summary of this process is demonstrated in Figure 21.3a.

(a)

MAX   →(HF)→   MXene   +   CTAB

◉ Sn(IV) nanocomplex   ⊙H ●C ◐N ●Br

CTAB-Sn(IV)@MXene   ←(Sn⁴⁺)←   CTAB pre-pillaring MXene

(b)

(c)

(d)

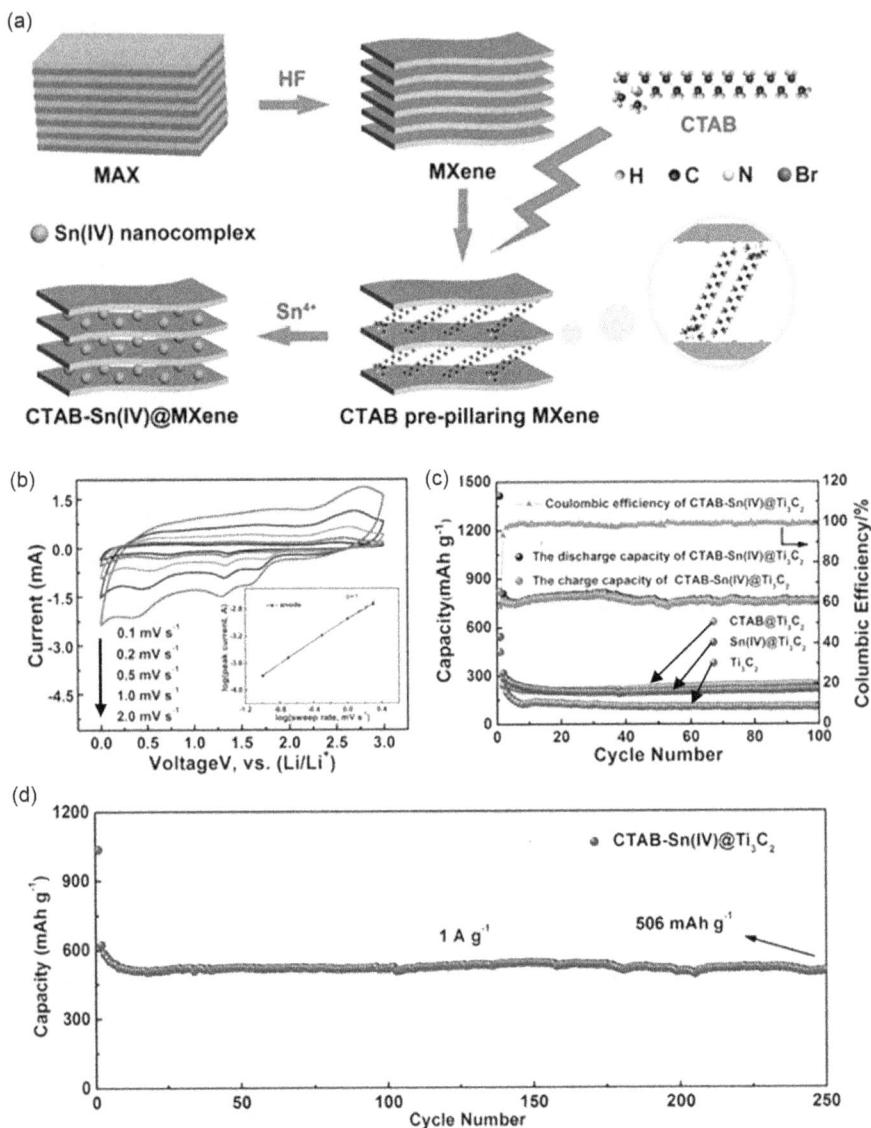

FIGURE 21.3 (a) Schematics for the fabrication of CTAB-Sn(IV)@Ti$_3$C$_2$. (b) CV for the CTAB-Sn(IV)@Ti$_3$C$_2$. (c) Charge/discharge capacity and Coulombic efficiency of CTAB-Sn(IV)@Ti$_3$C$_2$. (d) Cycling stability of CTAB-Sn(IV)@Ti$_3$C$_2$ at 1 A g^{-1}. (Adapted with permission from Ref. [31], Copyright (2016), American Chemical Society.)

The CV graph demonstrated a cathodic and anodic current signal related to the Li$^+$ complexed Sn(IV) along with the electrochemical activity of Ti$_3$C$_2$ (Figure 21.3b). The highest capacity reversibility was achieved for the nano-composed CTAB-Sn(IV)@ Ti$_3$C$_2$ yielding 765 mAh g^{-1} in contrast with Sn(IV)@Ti$_3$C$_2$ and CTAB@Ti$_3$C$_2$ that yielded 218 and 248 mAh g^{-1}, respectively (Figure 21.3c). Besides the higher values for reversible capacity for the nanocomposite CTAB-Sn(IV)@Ti$_3$C$_2$, it also yielded a stable Coulombic efficiency in the range of 95%–98%. Also, the cycling stability for the nanocomposites was high, suggesting that the intercalating agents were effectively introduced in between the layers providing a 506 mAh g^{-1} capacity after 250 cycles (Figure 21.3d). Notably, interlayer spacing plays an important role in improving the electrochemical properties, mostly related to improving the amounts of cations that can be intercalated in between the 2D layers, which leads to a considerable gain in capacity.

21.3.2 SUPERCAPACITORS BASED ON BULK MXENE

The MXenes by themselves have demonstrated superb properties due to their high surface area and conductivity, as for the case of Ti$_3$C$_2$, it has been reported to reach values in between 6,000 and 10,000 S cm^{-1} [32, 33]. On top of that, its hydrophilicity improves its dispersion in water that facilitates the diffusion of electrolytes. Such properties yield applications in electromagnetic interference shielding, sensors, conversion of light to heat, catalysis, water treatment, biomedicine, batteries, and supercapacitors. Values around 1,220 F cm^{-3} have been reported for MXenes due to the high-density structure in combination with an efficient pseudocapacitance mechanism [34]. However, a high capacitance can be only reached at low scan rates, otherwise, the MXene film may deteriorate under high scan rates or fast charge/discharge processes. Another important aspect is that the optimal properties are acquired when the MXene film has the dimension of few nanometers thick since if it reaches the micrometer scale, the layers can restack due to van der Waals interactions, covering the active pores, hindering the electrolyte permeation and, decreasing the redox activity and charge transfer. Naguib et al. proposed a novel approach to developing liquid crystal Ti$_3$C$_2$ fibers that presented enhanced electrochemical and mechanical properties in comparison to its layered counterpart [35]. The MXene flakes went through a probe sonication process, which decrease its lateral dimension from 3 µm (L-Ti$_3$C$_2$) to 0.3 µm (S-Ti$_3$C$_2$). Through that, the liquid crystal structures of L-Ti$_3$C$_2$ and S-Ti$_3$C$_2$ went from isotropic to nematic, respectively. The Ti$_3$C$_2$ fibers were obtained through a wet-spinning process, at which the MXene was extruded through a syringe and entered in a solution of an organic solvent such as chitosan or acetic acid. Due to the MXene's hydrophilicity, its structure coagulated into a fiber due to the repulsion with the organic solvent. It was observed that the fiber structure with a larger lateral dimension synthesized in chitosan (L-Ti$_3$C$_2$-Chitosan) presented higher tensile strength. It occurred because the level of coagulation for the larger fiber in chitosan was higher, leading to a denser structure and therefore expected to be mechanically stronger. However, the highest electrochemical performance was obtained for the S-Ti$_3$C$_2$ coagulated in chitosan, which was likely due to its higher surface area, which allowed more intercalation with ions, thus increasing its capacitance. On top of that, the capacitance retention and Coulombic efficiency for S-Ti$_3$C$_2$ coagulated in chitosan were also analyzed reaching values around 98% for both. This study demonstrated a viable strategy to further improve the pristine MXenes' electrochemical properties by varying the processing method.

As discussed, MXenes are mostly obtained by chemical etching from HF in the MAX phase starting material, in which the most notable example is the case of

Ti$_3$AlC$_2$ (MAX) that is converted to Ti$_3$C$_2$ (MXene). The removal of Al atoms from the structure leads to an accordion-like morphology that is held by van der Waals interactions. On top of that, negatively charged groups such as O, F, and OH are introduced into the edges and the surface of MXenes. The combination of MXenes with other materials that are currently studied for the development of high-performance supercapacitors composites is an interesting path for researchers. A combination between MXenes and layered double hydroxides (LDH) has been pro- posed. The LDH are inorganic two-dimensional materials composed of transition metal hydroxides. These materials have presented high energy storage properties due to their fast-redox rate and high theoretical specific capacitance. However, the major concern with LDH is their poor conduction and tendency to agglomerate. Thus, the combination with other materials can increase their conductivity and ion transfer to improve electrochemical properties. Previous studies were made to analyze the synergy between LDH such as NiFe with graphene [36,37]. It was observed that the composite obtained by the combination of LDH and graphene had more active sites and chemical stability against oxidative environments. Composites were also obtained by using activated carbon, porous carbon, reduced graphene oxide (RGO), and carbon nanotubes along with the LDH. Hence, in pursuit of developing high-performance supercapacitors, MXenes were also incorporated with LDH due to their high conductivity that improves charge transfer rate, and its layered structures expose more active sites for more adsorption of cations such as Li$^+$ to improve stability over extended charge/distance cycles. Zhang et al. synthesized a NiMn-LDH with Ti$_3$C$_2$ in a 2:1 mass ratio (NM/M 21), which provided a specific capacitance of 1,330 mAh g^{-1} at 0.1 A g^{-1} [38]. Also, it maintained 88.6% of its capacitance efficiency after 3,000 cycles. The procedure to obtain the composite was based on the selective etch- ing of Ti$_3$AlC$_2$ into Ti$_3$C$_2$ with HF and sonication. Then Ni^{2+} and Mn^{2+} were added into the system with Ti$_3$C$_2$. The electrostatic attraction between the negative charges from the MXene's surface (O, OH, and F) and the metallic cation's positive charge formed an organized arrangement. Then, Ni^{2+} and Mn^{2+} were reduced to form the 2D LDH structure (NiMn-LDH).

The in situ crystallization promoted the binding (covalent bonding) between the oxygen from the MXene with the transition metals from the LDH in NM/M 21, which showed to be a crucial factor for the improvement in electrochemical prop- erties. On the other hand, when the mechanical mixture of NiMn-LDH and Ti$_3$C$_2$ (NM+M 21) were analyzed, the specific capacitance was 875 mAh g^{-1} at 0.1 A g^{-1} and capacitance efficiency dropped to 41.8% after 3,000 cycles. Such difference was observed because of the chemical bond in the NM/M 21 composite, which stabi- lized the interface against the electrolyte and prevented structural degradation. Also, the chemical attachment between these two 2D layered materials led to satisfactory improvements in electrochemical behavior. The CV for the composites showed the largest area compared to individual materials due to the high dispersity of composite to expose more active sites and consequently more intercalation of Li$^+$. Also, the NM/M 21 demonstrated high stability as it maintained the highest Coulombic effi- ciency of 90.4%. In the case of the physical mixture, NM+M 21, there was no chemi- cal bond between them, suggesting that the main reason for the improvement in electrochemical properties is the electrostatic interaction between MXene and LDH.

21.3.3 SUPERCAPACITORS BASED ON THIN FILMS OF MXENE

High surface area and enhancement of active sites to further improve the electro-chemical properties are, in essence, the main reasons for the synthesis of ultra-thin MXene films. When the thickness of these materials reaches the dimension of a few nanometers, some of their properties change due to quantum confinement leading to a general improvement of conductibility and pseudocapacitance efficiency. In consequence of that, the morphology of thin films can be further modified by introducing hetero atoms that can be non-metals or metals. Also, porosity and crystallinity can be adjusted. Gogotsi et al. demonstrated the synthetic procedure to obtain Mo_2C thin films [39]. It consisted of dissolving metallic Mo in molten Cu at a temperature around 1,090°C forming a Mo-Cu alloy. Then methane was blown into the metallic solution to react with Mo in the interface and converted into Mo_2C 2D layer. An important parameter for this approach is the low methane concentration; otherwise, it could lead to the formation of graphene instead. Also, rapid cooling of the system is important to obtain clean Mo_2C crystals. The Cu substrate can be etched by $(NH_4)_2S_2O_8$. Other parameters such as the increase in temperature lead to an increase in the nucleation density, whereas lateral size increases by increasing the reaction time [40].

Xu et al. synthesized high-performance Mo_2C crystals that presented around 10 μm of lateral size and 3 nm of thickness [41]. A different approach to obtain thin films has been performed by Kim et al. through the self-assembly method that yielded 10 nm thick Ti_3C_2 films under a few minutes [42]. First, Ti_3C_2 was diluted in DI water. Ethyl acetate was added (10% v/v) to form a solvent on the water surface. Due to the higher ethyl acetate's vapor pressure, it evaporates creating a natural convection chain that developed regular-sized convection cells, a process known as Rayleigh–Bénard convection. The evaporation leads to an increase of water's surface tension below where the ethyl acetate layer was. This effect leads to the MXene flakes to self-assemble on the surface in consequence of Marangoni's effect, which is the mass transfer through an interface between two liquids (fluids) because of the surface tension gradient. After the complete evaporation of ethyl acetate, the MXene films in different spots of the solution's surface were formed. Then, hydrochloric acid was added to disrupt the electrostatic repulsion and assemble the MXene film that was floating. The formed film can then be transferred to the desired substrate; in this case, the authors used polyethylene terephthalate.

Besides different approaches to fabricate MXenes in a scalable and efficient way, scientists must also aim for other ways to further optimize the properties of the material. One viable strategy is the combination of different electrode materials in a way that one is used in supercapacitors and the other in batteries, which form a hybrid energy storage device. These hybrid supercapacitors (hybrid SC) aim to further improve power and energy density, charge/discharge capability, and cycling stability [43]. In correlation to that, the combination of MXenes with carbon-based materials, conductive polymers, and transition metal oxides (TMO) became a possible way to obtain high-performance supercapacitors. Even though the latter carries an inherent drawback of low electrical conductivity, MnO_2 as one of the TMO has been explored widely for its applications in supercapacitors. The advantages of using

TMO are related to low cost and toxicity along with facile synthesis. It also presents stable redox activity for the fast charge/discharge process due to the variation from Mn^{3+} and Mn^{4+}. Thus, the combination of MnO_2 with MXenes should suppress its inherent low conductivity. Associating these parameters Mahmood et al. fabricated a composite based on thin MXene (>50 nm) with MnO_2 nanowires (NWR) [44]. The synthesis consisted of traditional etching of Ti_3AlC_2 using HF to obtain Ti_3C_2 sonicated with DMSO. The MnO_2 NWR were synthesized through a modified annealing method that consisted of the precursor $KMnO_4$ dissolved in water followed by the addition of $(NH_4)_2S_2O_8$ and HNO_3 to obtain MnO_2 NWR. Further, a DMSO solution containing Ti_3C_2 and MnO_2 NWR was prepared. The MnO_2 NWRs served the purpose of improving the connection between MXene and MnO_2 as well as avoiding the MXene's restacking. Besides, the combination of these two materials into one electrode surpasses its electrochemical activity when they are used separately, suggesting an advantageous synergy effect between MXenes and MnO_2. Hence, the high conductibility of MXenes and fast redox capability of MnO_2 provide a boost into the composite electrochemical properties. Such synergy was reflected in the results of a specific capacitance of 611 F g^{-1} in contrast to Ti_3C_2 and MnO_2 NWR by themselves that yielded 527 and 337 F g^{-1}, respectively.

There was a considerable improvement in electrochemical properties of Ti_3C_2/MnO_2 composite compared to individual materials (Figure 21.4a). It suggests that intercalation/deintercalation of K^+ and H^+ cations in the composite's surface is describing its faradaic behavior. Also, the large loop from the CV suggests an increased surface area as the presence of Ti_3C_2 nanoflakes provides a faster pathway that leads to higher mobility of charge carrier species, hence decreasing the Ti_3C_2/MnO_2 electrode's internal resistance. Figure 21.4b shows the CV cycles of Ti_3C_2/MnO_2 composite at various scan rates. The electrode's cycling stability was also analyzed through CV (Figure 21.4c–d). The high capacity retention of 96% based on 1,000 cycles was observed with a minor decrease in capacitance from 528 to 507 F g^{-1}.

21.3.4 FLEXIBLE SUPERCAPACITORS BASED ON MXENE

The scaling use of electronic devices led to the introduction of stretchable materials that were used to the buildup of several high technology devices such as wearable sensors, implantable devices, artificial skin, among others, which often require the use of supercapacitors or batteries into its fabrication [45,46]. The long cycle stability, high power, and energy density offered by supercapacitors made room for these materials to be introduced in stretchable devices. Currently, materials such as metal oxides, conductive polymers, and carbon materials are the most present in commercial supercapacitors. However, some drawbacks require attention such as the inherent low specific capacitance of carbon materials and relatively low conductibility and stability of some conductive polymers and metal oxides. Hence, a promising approach to counter these obstacles relies on the use of MXenes. These 2D layered materials are known for their high conductibility, high surface accompanied to high redox activity (pseudocapacitance), and cycling stability. Thus, MXenes combined with carbon nanotubes or graphene can prompt satisfactory properties due to synergy

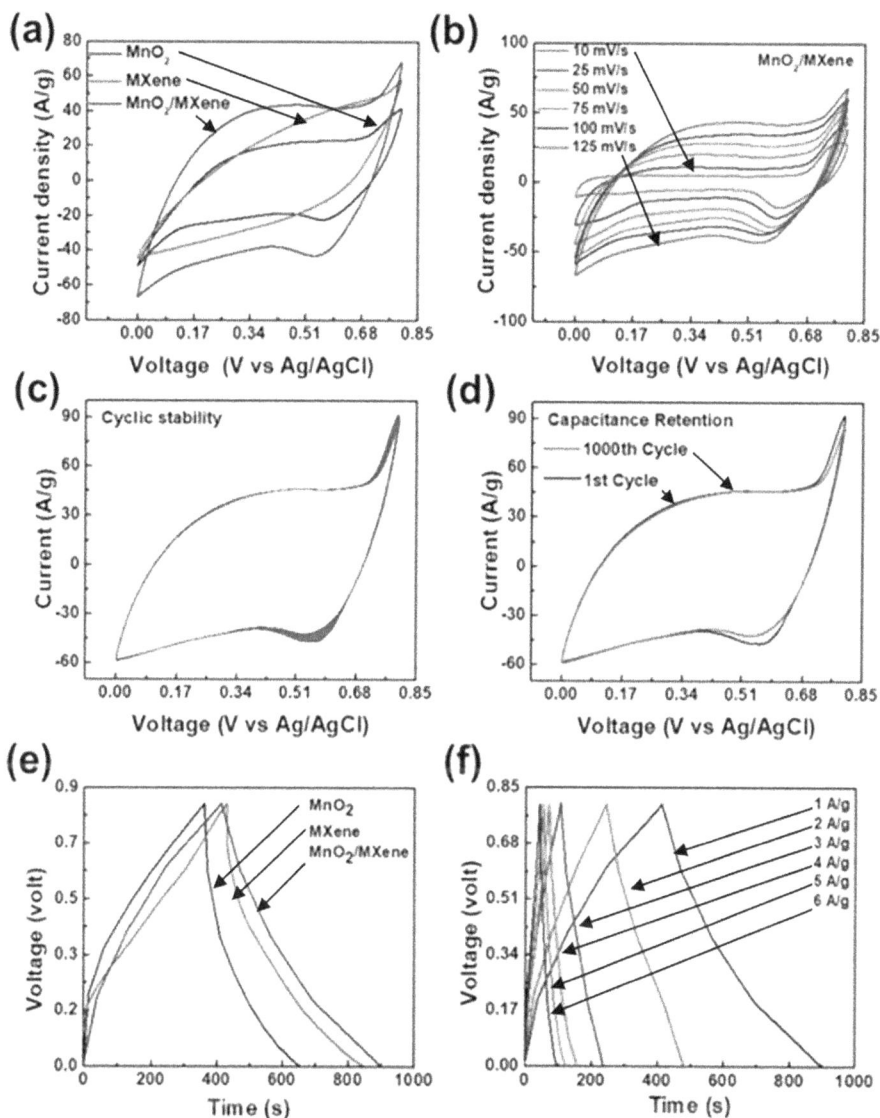

FIGURE 21.4 (a) CV curves for MnO_2, MXene (Ti_3C_2), and the composite. (b) CV curves at different scan rates for MnO_2/MXene. (c, d) First and 1,000th CV cycles to analyze stability and restacking tendency of the composite. (e) Charge–discharge curves for MnO_2, MXene (Ti_3C_2), and the composite at 1 A g^{-1}. (f) Charge–discharge curves at different current densities. (Adapted with permission from Ref. [44], Copyright (2021), American Chemical Society.)

between these two materials as graphene, for example, can introduce stretchability besides its electrochemical performance and surface area. Also, nanosized materials like those have a positive influence on the exposure of active sites, in comparison to bulkier materials while introducing bendability and a wider range of mechanical properties that MXenes by themselves usually lack.

MXenes were adequately applied to flexible and rigid micro-supercapacitors, metal-ion batteries, and others [47,48]. However, the high elasticity modulus in the range of 3–75 GPa in a dry state can make these materials brittle [49]. Despite that, a study showed the synthesis of an MXene/poly (diallyl dimethylammonium chloride) composite that was able to stretch up to 40% of its initial length. However, when the strain was applied the resistance increased [50]. Hence, to find a middle ground between facile procedure and a good balance of properties, Zhou et al. developed a Ti_3C_2/RGO composite that remained lasting conductivity after several strain cycles [34]. To fabricate the composite, a solution containing MXene and RGO in water was submitted to VAF. Then a substrate of acrylic elastomer was stretched up to 300% in one axis and the composite film was transferred to its surface. A hand press was used to adhere MXene/RGO composite film to the substrate surface and left it to dry. After that, the substrate was slowly returned to its original length.

The effect of applying uniaxial strains of 100%, 200%, and 300% at the Ti_3C_2/RGO composites with increasing concentrations of RGO is demonstrated in the SEM images (Figure 21.5a–c). It was notable that even after largely applied strain, the composite films maintained their morphology in a wrinkled form due to lower Young's modulus, higher elongation at break, and relatively stronger interlayer interactions. On the other hand, SEM images for the pure Ti_3C_2 films demonstrated fragmentation and cracks after applying uniaxial strains (Figure 21.5d–f). The small flake size along with the high mechanical modulus of pure MXene films was responsible for the cracked structure. Hence, the large and flexible RGO flakes combined with the small MXene flakes introduced stretchability and structural alignment from the former and higher surface area that improves the electrochemical properties from the latter. The film's ultimate strength increased with an increase in RGO concentration (Figure 21.5g). The Ti_3C_2/RGO composite, despite presenting higher inherent resistance in comparison to pure Ti_3C_2, the cyclic stretching and relaxation of the composite film did not influence the resistance when higher concentrations of RGO were added (Figure 21.5h–i). On the other hand, a considerable variation of resistance was observed for the pure Ti_3C_2 films.

As noted, the increase in RGO concentration into the composite film leads to an increment of electrode robustness as well as electrical resistance. Thus, an optimal concentration of RGO should be proposed to achieve the best balance between mechanical and electrochemical properties. It was observed that the composite film with 50% RGO yielded the most satisfactory results. The 50 wt.% Ti_3C_2/RGO composite film had its electrochemical properties analyzed (Figure 21.6a–c). Meanwhile, the pure Ti_3C_2 film was also analyzed for comparison (Figure 21.6d–f). The CV for the composite displayed a wider squared area compared with the MXene film, suggesting the charging/discharging process occurs through an electrical double-layer, leading to a higher capacitance. Figure 21.6b shows that the composite film was barely influenced by the strain, indicating that the structure maintained its electrochemical properties regardless of mechanical stretching. On the other hand, MXene film deteriorated its electrochemical properties after being submitted into mechanical strain (Figure 21.6e). It occurred because the structure cracking led to a bigger edge-to-edge spacing between the MXene, which increases the electrode's resistance. Also, the large strain caused the MXene to peel off, decreasing the amount of active

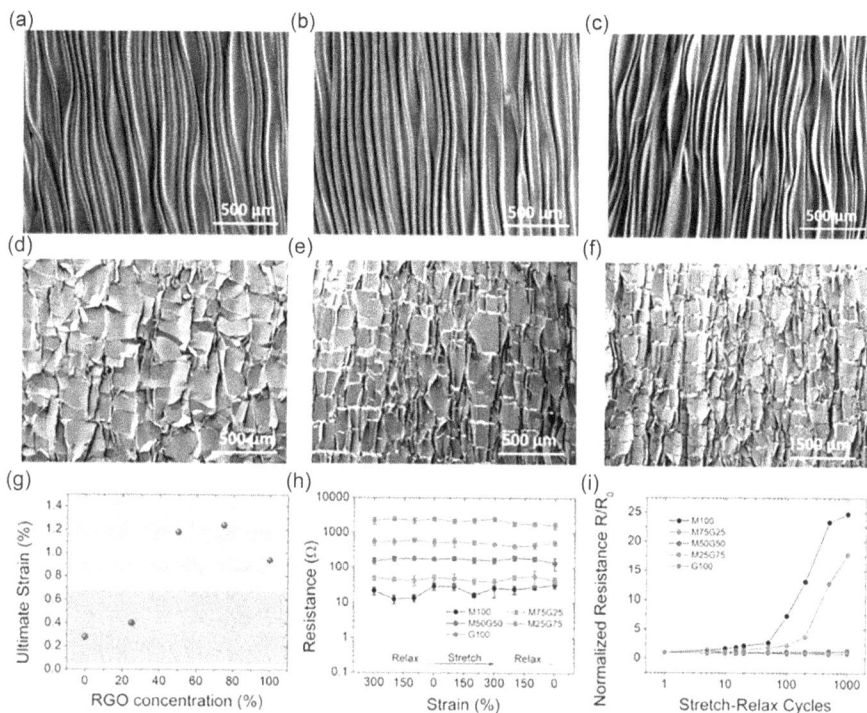

FIGURE 21.5 SEM for the Ti_3C_2/RGO composite films after applying a strain of (a) 100%, (b) 200%, and (c) 300%. SEM for the pure Ti_3C_2 films after applying a strain of (d) 100, (e) 200, and (f) 300%. (g) Ultimate strain as a function of RGO concentration. (h) Influence of mechanical cycles on the resistance of the film. (i) Normalized resistance of the prepared films as a function of stretch-relax cycles. (Adapted with permission from Ref. [34], Copyright (2020), American Chemical Society.)

material due to poor adherence and inherent brittleness. Consequently, the capacitance for the composite was higher than the MXene film. This study shows a viable example for the development of a flexible supercapacitor through the combination of an MXene (Ti_3C_2) with a conductive layered component (RGO), which yielded a composite film with better mechanical properties at the expense of an increase in electrical resistance. Despite that, the electrochemical properties remained nearly constant regardless of the strain applied in the composite electrode.

Sun et al. fabricated a compressive and flexible MXene/N-doped carbon foam (NCF)-based supercapacitor [51]. The hollow and interconnected structure of carbon foam was obtained through single-step pyrolysis. The NCF worked synergistically with the MXene nanosheets by further improving its cycling stability, enhanced ion diffusion due to the networked and hollowed structure combined with the inherent MXene's high conductibility and hydrophilicity. The results obtained from the free-standing composite were promising as the gravimetric capacitance was around 330 $F\ g^{-1}$ and volumetric capacitance was 3,160 $mF\ cm^{-3}$, accompanied to a capacity retention of 99.2% after 10,000 cycles. These properties were presented after 60% strain in any direction.

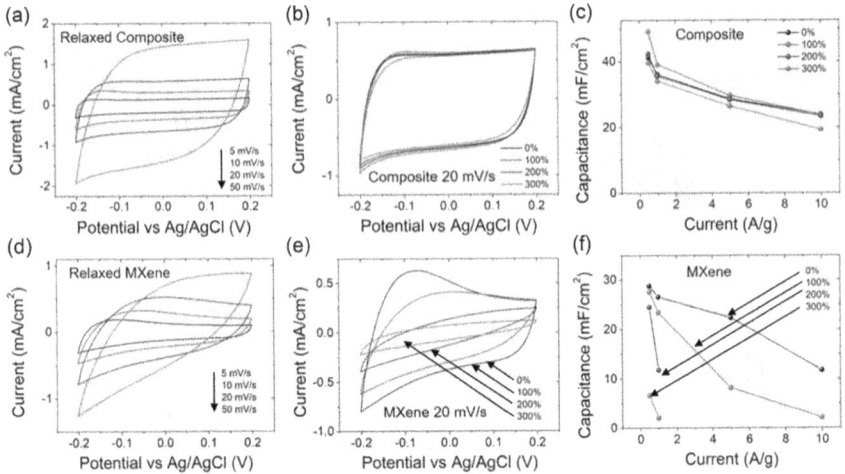

FIGURE 21.6 Electrochemical analysis for the Ti_3C_2/RGO composite film through (a) CV, (b) influence of strain in CV, and (c) Specific capacitance obtained by galvanostatic charge/ discharge at different current densities and strains varying from 0% to 300%. Electrochemical analysis for the pure Ti_3C_2 film through (d) CV, (e) influence of strain in CV, and (f) specific capacitance obtained by galvanostatic charge/discharge at different current densities and strains varying from 05 to 300%. (Adapted with permission from Ref. [34], Copyright (2020), American Chemical Society.)

21.4 CONCLUSION

Based on the discussions and analysis provided in this chapter, it was notable that MXenes hold great potential due to facile techniques that yield 2D materials with the thickness of a few atoms. Such morphology is achieved usually through HF etching (ex situ or in situ) that selectively removes the "A" atom from a starting MAX material leading to the general formula of MXenes. This process gives advantages such as inherently high conductivity and surface area, which allows fast charge transfer and eases the intercalation of ions, which are crucial aspects for the general enhancement of electrochemical behavior. Yet, several strategies for further improvement are available. The increase of interlayer spacing is a strategy to increment the intercalation/deintercalation of several cations such as Li^+, K^+, Na^+, NH_4^+, Mg^{2+}, or Al^{3+}; while simultaneously countering one of the drawbacks of MXenes, which is their tendency to restack. It can be performed by the addition of organic solvents or promoting covalent bonds that form pillared structures in between the 2D MXene layers that prevent them from restacking. These materials can also be fabricated into fibers, which improve both electrochemical stability and added considerable tensile strength for the MXenes fibers. Composites based on MXenes and LDH have yielded applicable results as MXenes provide higher conduction and surface area increases, whereas LDH provides faster redox pseudocapacitance and chemical stability. Facile methods to obtain ultra-thin MXene films have also been reported showing potential large-scale processes. Finally, flexible supercapacitors have also been fabricated by using MXenes and carbon-based materials adhered to flexible substrates, which yielded electronic devices able to withstand several cycles of mechanical folding without losing their properties. Hence, despite the low cycling stability of MXenes the scientific community is succeeding in overcoming these drawbacks and purifying the advantages of these 2D layered materials through different fabrication techniques and development of composites showing a promising path for future research.

REFERENCES

1. Naguib M, Kurtoglu M, Presser V, Lu J, Niu J, Heon M, Hultman L, Gogotsi Y, Barsoum MW (2011) Two-dimensional nanocrystals produced by exfoliation of Ti_3AlC_2. *Adv Mater* 23:4248–4253.
2. Gao L, Li C, Huang W, Mei S, Lin H, Ou Q, Zhang Y, Guo J, Zhang F, Xu S, Zhang H (2020) MXene/polymer membranes: Synthesis, properties, and emerging applications. *Chem Mater* 32:1703–1747.
3. Nan J, Guo X, Xiao J, Li X, Chen W, Wu W, Liu H, Wang Y, Wu M, Wang G (2019) Nanoengineering of 2D MXene-based materials for energy storage applications. *Small* n/a:1902085.
4. Naguib M, Mochalin VN, Barsoum MW, Gogotsi Y (2014) 25th anniversary article: MXenes: A new family of two-dimensional materials. *Adv Mater* 26:992–1005.
5. Gogotsi Y, Anasori B (2019) The rise of MXenes. *ACS Nano* 13:8491–8494.
6. Lukatskaya MR, Kota S, Lin Z, Zhao M-Q, Shpigel N, Levi MD, Halim J, Taberna P-L, Barsoum MW, Simon P, Gogotsi Y (2017) Ultra-high-rate pseudocapacitive energy storage in two-dimensional transition metal carbides. *Nat Energy* 2:17105.
7. Xu C, Song S, Liu Z, Chen L, Wang L, Fan D, Kang N, Ma X, Cheng H-M, Ren W (2017) Strongly coupled high-quality graphene/2D superconducting Mo_2C vertical heterostructures with aligned orientation. *ACS Nano* 11:5906–5914.
8. Ling Z, Ren CE, Zhao M-Q, Yang J, Giammarco JM, Qiu J, Barsoum MW, Gogotsi Y (2014) Flexible and conductive MXene films and nanocomposites with high capacitance. *Proc Natl Acad Sci* 111:16676 LP–16681.
9. Li X, Dai Y, Ma Y, Liu Q, Huang B (2015) Intriguing electronic properties of two-dimensional MoS_2/TM_2CO_2(TM = Ti, Zr, or Hf) hetero-bilayers: type-II semiconductors with tunable band gaps. *Nanotechnology* 26:135703.
10. Wu X, Hao L, Zhang J, Zhang X, Wang J, Liu J (2016) Polymer-$Ti_3C_2T_x$ composite membranes to overcome the trade-off in solvent resistant nanofiltration for alcohol-based system. *J Memb Sci* 515:175–188.
11. Shahzad F, Alhabeb M, Hatter CB, Anasori B, Man Hong S, Koo CM, Gogotsi Y (2016) Electromagnetic interference shielding with 2D transition metal carbides (MXenes). *Science* 353:1137 LP–1140.
12. Carey M, Hinton Z, Sokol M, Alvarez NJ, Barsoum MW (2019) Nylon-6/$Ti_3C_2T_z$ MXene nanocomposites synthesized by in situ ring opening polymerization of ε-caprolactam and their water transport properties. *ACS Appl Mater Interfaces* 11:20425–20436.
13. Naguib M, Mashtalir O, Carle J, Presser V, Lu J, Hultman L, Gogotsi Y, Barsoum MW (2012) Two-dimensional transition metal carbides. *ACS Nano* 6:1322–1331.
14. Hu M, Hu T, Li Z, Yang Y, Cheng R, Yang J, Cui C, Wang X (2018) Surface functional groups and interlayer water determine the electrochemical capacitance of $Ti_3C_2T_x$ MXene. *ACS Nano* 12:3578–3586.
15. Levi MD, Lukatskaya MR, Sigalov S, Beidaghi M, Shpigel N, Daikhin L, Aurbach D, Barsoum MW, Gogotsi Y (2015) Solving the capacitive paradox of 2D MXene using electrochemical quartz-crystal admittance and in situ electronic conductance measurements. *Adv Energy Mater* 5:1400815.

16. Munir S, Rasheed A, Rasheed T, Ayman I, Ajmal S, Rehman A, Shakir I, Agboola PO, Warsi MF (2020) Exploring the influence of critical parameters for the effective synthesis of high-quality 2D MXene. *ACS Omega* 5:26845–26854.

17. Ghidiu M, Lukatskaya MR, Zhao M-Q, Gogotsi Y, Barsoum MW (2014) Conductive two-dimensional titanium carbide 'clay' with high volumetric capacitance. *Nature* 516:78–81.

18. Wang L, Zhang H, Wang B, Shen C, Zhang C, Hu Q, Zhou A, Liu B (2016) Synthesis and electrochemical performance of $Ti_3C_2T_x$ with hydrothermal process. *Electron Mater Lett* 12:702–710.

19. Yang J, Naguib M, Ghidiu M, Pan L-M, Gu J, Nanda J, Halim J, Gogotsi Y, Barsoum MW (2016) Two-dimensional Nb-based M_4C_3 solid solutions (MXenes). *J Am Ceram Soc* 99:660–666.

20. Li T, Yao L, Liu Q, Gu J, Luo R, Li J, Yan X, Wang W, Liu P, Chen B, Zhang W, Abbas W, Naz R, Zhang D (2018) Fluorine-free synthesis of high-purity $Ti_3C_2T_x$ (T=OH, O) via alkali treatment. *Angew Chemie Int Ed* 57:6115–6119.

21. Zhu H, Jia Z, Chen Y, Weadock N, Wan J, Vaaland O, Han X, Li T, Hu L (2013) Tin anode for sodium-ion batteries using natural wood fiber as a mechanical buffer and electrolyte reservoir. *Nano Lett* 13:3093–3100.

22. He M, Kravchyk K, Walter M, Kovalenko MV (2014) Monodisperse antimony nanocrystals for high-rate Li-ion and Na-ion battery anodes: Nano versus bulk. *Nano Lett* 14:1255–1262.

23. Hu Z, Wang L, Zhang K, Wang J, Cheng F, Tao Z, Chen J (2014) MoS_2 nanoflowers with expanded interlayers as high-performance anodes for sodium-ion batteries. *Angew Chemie Int Ed* 53:12794–12798.

24. Mashtalir O, Lukatskaya MR, Zhao M-Q, Barsoum MW, Gogotsi Y (2015) Amine-assisted delamination of Nb_2C MXene for Li-ion energy storage devices. *Adv Mater* 27:3501–3506.

25. Naguib M, Unocic RR, Armstrong BL, Nanda J (2015) Large-scale delamination of multilayers transition metal carbides and carbonitrides "MXenes." *Dalt Trans* 44:9353–9358.

26. Lukatskaya MR, Mashtalir O, Ren CE, Dall'Agnese Y, Rozier P, Taberna PL, Naguib M, Simon P, Barsoum MW, Gogotsi Y (2013) Cation intercalation and high volumetric capacitance of two-dimensional titanium carbide. *Science* 341:1502 LP–1505.

27. Hu M, Zhang H, Hu T, Fan B, Wang X, Li Z (2020) Emerging 2D MXenes for supercapacitors: Status, challenges and prospects. *Chem Soc Rev* 49:6666–6693.

28. Zhu Y, Qian Y, Ju Z, Ji Y, Yan Y, Liu Y, Yu G (2020) Understanding charge storage in hydrated layered solids $MOPO_4$ (M = V, Nb) with tunable interlayer chemistry. *ACS Nano* 14:13824–13833.

29. Peng L, Zhu Y, Peng X, Fang Z, Chu W, Wang Y, Xie Y, Li Y, Cha JJ, Yu G (2017) Effective interlayer engineering of two-dimensional $VOPO_4$ nanosheets via controlled organic intercalation for improving alkali ion storage. *Nano Lett* 17:6273–6279.

30. Lagaly G (1986) Interaction of alkylamines with different types of layered compounds. *Solid State Ionics* 22:43–51.

31. Luo J, Zhang W, Yuan H, Jin C, Zhang L, Huang H, Liang C, Xia Y, Zhang J, Gan Y, Tao X (2017) Pillared structure design of MXene with ultralarge interlayer spacing for high-performance lithium-ion capacitors. *ACS Nano* 11:2459–2469.

32. Dillon AD, Ghidiu MJ, Krick AL, Griggs J, May SJ, Gogotsi Y, Barsoum MW, Fafarman AT (2016) Highly conductive optical quality solution-processed films of 2D titanium carbide. *Adv Funct Mater* 26:4162–4168.

33. Sang X, Xie Y, Lin M-W, Alhabeb M, Van Aken KL, Gogotsi Y, Kent PRC, Xiao K, Unocic RR (2016) Atomic defects in monolayer titanium carbide ($Ti_3C_2T_x$) MXene. *ACS Nano* 10:9193–9200.

34. Zhou Y, Maleski K, Anasori B, Thostenson JO, Pang Y, Feng Y, Zeng K, Parker CB, Zauscher S, Gogotsi Y, Glass JT, Cao C (2020) $Ti_3C_2T_x$ MXene-reduced graphene oxide composite electrodes for stretchable supercapacitors. *ACS Nano* 14:3576–3586.

35. Naguib M (2020) Multifunctional pure MXene fiber from liquid crystals of only water and MXene. *ACS Cent Sci* 6:344–346.

36. Zhou G, Wang D-W, Yin L-C, Li N, Li F, Cheng H-M (2012) Oxygen bridges between NiO nanosheets and graphene for improvement of lithium storage. *ACS Nano* 6:3214–3223.

37. Jiang Y, Song Y, Li Y, Tian W, Pan Z, Yang P, Li Y, Gu Q, Hu L (2017) Charge transfer in ultrafine LDH nanosheets/graphene interface with superior capacitive energy storage performance. *ACS Appl Mater Interfaces* 9:37645–37654.

38. Zhang D, Cao J, Zhang X, Insin N, Wang S, Zhao Y, Qin J (2020) NiMn-layered double hydroxides chemically anchored on Ti_3C_2 MXene for superior lithium ion storage. *ACS Appl Energy Mater* 3:11119–11130.

39. Gogotsi Y (2015) Transition metal carbides go 2D. *Nat Mater* 14:1079–1080.

40. Li X, Cai W, An J, Kim S, Nah J, Yang D, Piner R, Velamakanni A, Jung I, Tutuc E, Banerjee SK, Colombo L, Ruoff RS (2009) Large-area synthesis of high-quality and uniform graphene films on copper foils. *Science* 324:1312 LP–1314.

41. Xu C, Wang L, Liu Z, Chen L, Guo J, Kang N, Ma X-L, Cheng H-M, Ren W (2015) Large-area high-quality 2D ultrathin Mo_2C superconducting crystals. *Nat Mater* 14:1135–1141.

42. Kim SJ, Choi J, Maleski K, Hantanasirisakul K, Jung H-T, Gogotsi Y, Ahn CW (2019) Interfacial assembly of ultrathin, functional MXene films. *ACS Appl Mater Interfaces* 11:32320–32327.

43. Hong Ng VM, Huang H, Zhou K, Lee PS, Que W, Xu JZ, Kong LB (2017) Recent progress in layered transition metal carbides and/or nitrides (MXenes) and their composites: Synthesis and applications. *J Mater Chem A* 5:3039–3068.

44. Mahmood M, Rasheed A, Ayman I, Rasheed T, Munir S, Ajmal S, Agboola PO, Warsi MF, Shahid M (2021) Synthesis of ultrathin MnO_2 nanowire-intercalated 2D-MXenes for high-performance hybrid supercapacitors. *Energy & Fuels* 35:3469–3478.

45. Li S, Peele BN, Larson CM, Zhao H, Shepherd RF (2016) A stretchable multicolor display and touch interface using photopatterning and transfer printing. *Adv Mater* 28:9770–9775.

46. Lipomi DJ, Vosgueritchian M, Tee BC-K, Hellstrom SL, Lee JA, Fox CH, Bao Z (2011) Skin-like pressure and strain sensors based on transparent elastic films of carbon nanotubes. *Nat Nanotechnol* 6:788–792.

47. Naguib M, Come J, Dyatkin B, Presser V, Taberna P-L, Simon P, Barsoum MW, Gogotsi Y (2012) MXene: A promising transition metal carbide anode for lithium-ion batteries. *Electrochem Commun* 16:61–64.

48. Kurra N, Ahmed B, Gogotsi Y, Alshareef HN (2016) MXene-on-paper coplanar microsupercapacitors. *Adv Energy Mater* 6:1601372.

49. Come J, Xie Y, Naguib M, Jesse S, Kalinin S V, Gogotsi Y, Kent PRC, Balke N (2016) Nanoscale elastic changes in 2D $Ti_3C_2T_x$ (MXene) pseudocapacitive electrodes. *Adv Energy Mater* 6:1502290.

50. An H, Habib T, Shah S, Gao H, Radovic M, Green MJ, Lutkenhaus JL (2018) Surface-agnostic highly stretchable and bendable conductive MXene multilayers. *Sci Adv* 4:eaaq0118.

51. Sun L, Song G, Sun Y, Fu Q, Pan C (2020) MXene/N-doped carbon foam with three-dimensional hollow neuron-like architecture for freestanding, highly compressible all solid-state supercapacitors. *ACS Appl Mater Interfaces* 12:44777–44788.

22 2D Nanomaterials for Flexible Supercapacitors

Yamin Zhang, Jinyang Zhang,
Linrui Hou, and Changzhou Yuan
University of Jinan

CONTENTS

22.1 INTRODUCTION

With the rapid development of wearable electronic devices (folding mobile phones, smart bracelets, flexible displays, etc.), the manufacturing of flexible energy storage devices has become an emerging field [1]. Due to the advantages of high power density, medium energy density, rapid charge and discharge, long cycle life, and mechanical flexibility, FSCs have become the most promising energy supply equipment in flexible electronic equipment [2]. 2D nanomaterials are ultra-thin layered crystals with only one or a few atomic layers being thick. The ultra-thin properties provide a larger surface area for charge storage, and excellent mechanical properties, including bending/folding strength and excellent flexibility, make them promising applications in FSCs. The earliest 2D nanomaterials were proposed in 2004 by Andre Geim and Kostya Novoselov, two physicists from the University of Manchester, who prepared single-layer graphene. After more than 10 years of development, various 2D nanomaterials have been gradually discovered and studied (Figure 22.1). We will describe this in detail in Section 22.2.

Typical FSCs are composed of flexible electrodes, flexible electrolytes, separators, and flexible packaging materials [3]. Flexible electrodes can be roughly divided into two categories: One is a flexible substrate electrode that grows the active material in situ on the substrate. Different substrates are grown by different methods, such as foam nickel, carbon paper, conductive paper, steel mesh, and other conductive

DOI: 10.1201/9781003178422-22

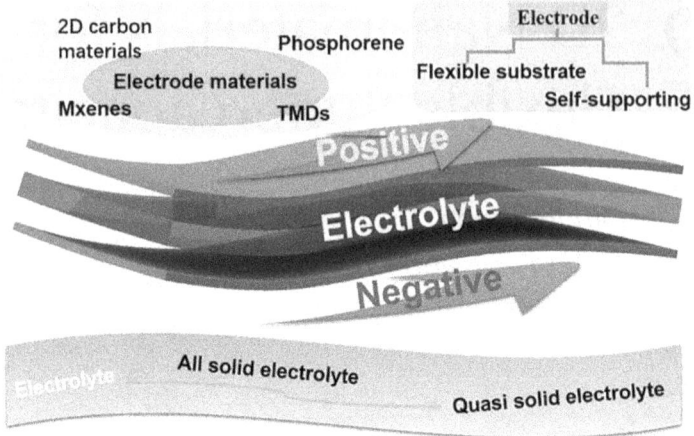

FIGURE 22.1   Schematic diagram of the composition of 2D nanomaterials FSCs. -Ł

substrates. In general, active materials are loaded on the surface by hydrothermal deposition and electrochemical deposition. For insulating substrates such as pet, sponge, and non-woven fabric, the electrical properties of insulating substrates were changed firstly by sputtering deposition, atomic layer deposition, and photolithography, and then, the active materials were grown. The other is self-supporting electrodes (such as graphene and carbon nanofibers) with flexibility and excellent electrochemical activity, which integrate current collector and active materials [4].

In addition, the choice of electrolyte is very important for flexible devices to exhibit excellent electrochemical performance under bending, torsion, folding, and other external forces [5]. Traditional electrolytes are in a liquid state (organic or water-soluble electrolytes), which are prone to leakage, volatilization, chemical corrosion, and other problems when subjected to external force, especially organic electrolytes. It not only causes irreversible damage to devices but also pollutes the environment. A solid electrolyte can avoid the above problems. At the same time, when the device is assembled, the solid electrolyte takes into account the dual functions of ion transport and the separator between positive and negative electrodes, which simplifies the device structure and reduces the cost.

According to the current research, there are two kinds of solid electrolytes: all solid electrolytes and quasi-solid electrolytes. All solid electrolytes are divided into inorganic solid electrolytes, solid polymer electrolytes, and composite polymer electrolytes. They are solid at room temperature, and the movement of ions occurs in the solid state. Their main advantage is that they can completely remove any liquid components, thus improving the safety of the whole equipment. But, the main limitation is that the ionic conductivity is often much lower than that of liquid electrolytes. Quasi solid-state electrolyte is also known as gel polymer electrolyte. The ion conduction mechanism is mainly in the solvent or plasticizer. They are composed of polymer networks that swell in solvents containing active ions, so they have both the mechanical properties of solids and the high transport properties of liquids. Therefore, a quasi-solid electrolyte is the best choice for FSCs.

This chapter systematically introduces the intelligent design and construction of 2D nanomaterials for FSCs, including 2D carbomaterials (graphite, graphene, and graphyne), phosphorus, MXenes, and transition metal disulfide compounds. In addition, the application prospect and the challenge are prospected. At the same time, the assembly and practical application of FSCs are discussed.

22.2 DESIGN AND CONSTRUCTION OF 2D NANOMATERIALS FOR FLEXIBLE SUPERCAPACITORS

22.2.1 2D CARBON MATERIALS (GRAPHENE, GRAPHYNE)

2D carbon nanomaterials have been widely used in various energy storage devices due to their excellent conductivity, thermal conductivity, and stable chemical properties, including FSCs [6]. Graphene is the most classic 2D carbon nanomaterial. Graphene is a two-dimensional material stripped from graphite material. Its structure is composed of a single layer of carbon atoms arranged. Each carbon atom and three neighboring carbon atoms form three σ bonds through SP^2 hybrid orbitals, forming a solid hexagonal honeycomb structure, which makes it have excellent mechanical properties. In addition, the P orbitals of each carbon atom perpendicular to the plane of the layer form large π bonds of multiple atoms throughout the whole layer, and the electrons can move freely, thus graphene has excellent electrical conductivity (electron mobility up to $200,000 \, cm^2 V^{-1} s^{-1}$) [7]. These excellent properties make graphene an attractive electrode material for FSCs. When graphene is applied to FSCs, its macro structure includes one-dimensional graphene (graphene fiber), two-dimensional graphene (graphene film), and three-dimensional graphene (graphene hydrogel, aerogel).

Graphene fiber has high electrical conductivity, mechanical properties, rigidity, strong stability, and a high aspect ratio (aspect ratio). Therefore, it has high flexibility and good weaving ability and is suitable for microelectronic devices, wearable electronic products, and intelligent textiles [8]. However, pure graphene fibers have low specific surface area and poor capacitance storage, because of the π–π interaction between graphene sheets during the preparation process, resulting in an irreversible agglomeration phenomenon [9]. Therefore, the structural design of graphene fibers and the preparation of graphene fibers with porous structure and large specific surface area have become an important topic for FSCs [10]. Cheng et al. [11] adopted the method of confined gas self-driven assembly and used graphene and poly(3,4-ethylenedioxythiophene):poly(styrenesulfonate) (PEDOT:PSS) conducting polymer as raw materials to successfully develop a composite electrode of fibrous supercapacitor with hollow structure (Figure 22.2a-e). Compared with the solid composite fiber electrode, the hollow composite fiber electrode not only has a tensile strength of up to 631 MPa and 4,700 S m^{-1} conductivity (Figure 22.2f and g). With both inner and outer double-layer electrode/electrolyte contact interface, it also shows a significant increase in specific capacitance and energy density. The FSCs composed of hollow composite fiber electrode show excellent electrochemical performance and flexibility. The area-specific capacitance and specific energy density of the FSCs are up to 304.5 mF cm^{-2} and 27.1 μWh cm^{-2} (Figure 22.2h), respectively, at the current density of 0.08 mA cm^{-2} which also exhibits excellent long-term cycle stability (96% after

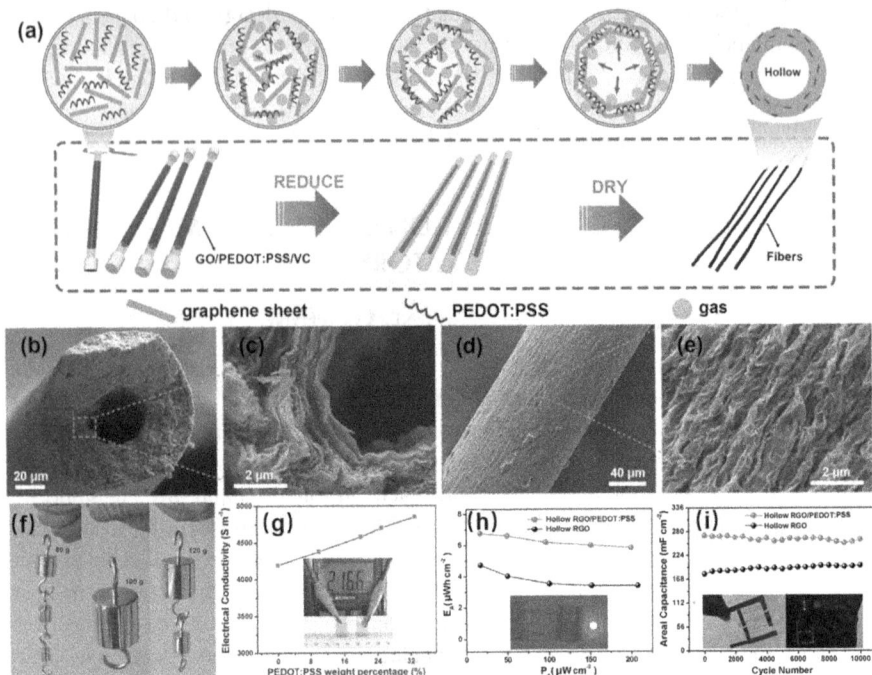

FIGURE 22.2 (a) Diagram of hollow structural fiber formation. (b) and (c) SEM images of fiber sections at low and high magnifications. (d) and (e) SEM images of the fiber side at low and high magnifications. (f) Loading test for fibers. (g) Different percentages of PEDOT:PSS conductivity. (h) Ragone diagram of a fiber-optic supercapacitor. The inset shows a serial fiber optic supercapacitor lights up a red LED. (i) Long-term cycling stability of fiber optic supercapacitors at 0.48 mA cm^{-2}. The inset shows three fiber-optic supercapacitors connected in series, attached to clothing, powering three LED lights. (Adapted with permission from Ref. [11]. Copyright (2016) Wiley-VCH.)

10,000 cycles) (Figure 22.2i) and has no significant change in charge–discharge performance at different bending angles and for 500 times of continuous bending.

Graphene film, also known as graphene paper, is a self-supporting graphene paper formed by multilayer graphene paper. It has distinguished electrical conductivity and flexibility and can be directly used as an electrode without binder, conductive agent, and current collector. When used as the electrode material of FSCs, it has excellent electrochemical performance [12]. Wang and his colleagues [13] fabricated graphene paper by vacuum filtration assembly method and first studied them as flexible electrode materials. However, due to the interaction between graphene layers, the irreversible agglomeration of graphene sheets occurs. This agglomeration reduces the effective contact area of graphene paper and limits ion diffusion, which shows unsatisfactory electrochemical performance. In order to solve this problem, the agglomeration of graphene can be avoided by adding a spacer. For example, the introduction of carbon black nanoparticles as a spacer greatly hindered the self-healing of graphene in the process of vacuum filtration [14]. In addition, graphene paper is formed by using graphene with a porous structure, which can provide more channels for ions and accelerate ion migration [15]. In addition to the above methods, template-assisted growth and wrinkling of graphene plates can also be used to prevent the aggregation of graphene plates, increase the surface area, and promote the transport of electrolyte ions in order to improve the electrochemical performance of FSCs [16].

Three-dimensional graphene is a 3D porous network macrostructure based on graphene, which has many advantages, such as porous structure, large specific surface area, strong flexibility, and fast electron transfer rate, including aerogels, hydrogels, foam, and sponge. Graphene gel is the best electrode material for FSCs. At the same time, there are some obstacles, such as the defects in the structure of graphene gel and the stacking of disorder. The uneven force between graphene sheets leads to low overall strength [17]. The composite of carbon materials (carbon nanotubes, porous carbon) with good conductivity can not only hinder the aggregation of graphene nanosheets but also improve the conductivity. Niu et al. [18] prepared porous carbon and graphene composite aerogels using the self-loading method. This material has the advantages of multi-pore distribution, excellent conductivity, large effective contact area, rapid ion diffusion/transmission channel, high rate performance, energy density, and cycling stability. Recently, Li et al. [19] prepared novel metal-free FSCs based on redox-active lignosulfonate functionalized graphene hydrogel. It has excellent mechanical strength and can be directly used as a self-supporting electrode without the need for any other adhesives or additives, so the lignin graphene hydrogel electrode shows good rate capability, long period stability, and high specific capacitance.

Scientists have been working on developing new ways to synthesize new carbon allotropes and explore their new properties. In 1987, Baughman et al. first predicted that graphyne is a series of stable crystalline carbon allotropes with high SP hybridization [20]. In 2010, a team led by Li Yu Liang, Academician of the Key Laboratory of Institute of Organic Solids, successfully synthesized graphyne film, a new allotrope of large-area carbon, on the surface of copper sheet by a chemical method using the coupling reaction of hexaacetylene benzene under the catalysis of copper sheet [21]. Graphyne is a 1,3-diacetylene bond conjugated benzene ring to form a two-dimensional plane network structure of all carbon molecules, with a wealth of carbon chemical bonds, large conjugated system wide surface space excellent chemical stability, and excellent electrical properties. However, due to the difficulty in obtaining high-quality graphyne acetylene samples suitable for the separation process, the relevant separation studies are mainly based on theory and calculation methods. The theoretical calculation proves that graphyne acetylene has a theoretical capacity of $2,233\,mA\,h\,g^{-1}$ when applied to lithium-ion batteries, which is six times that of graphite ($372\,mAh\,g^{-1}$) [22]. Chen and his colleagues [23] calculated the electrochemical properties of the elemental mixed graphitic alkyne applied to supercapacitors. The element-doped graphyne has a higher theoretical capacity as the presence of acetylene bonds increases the pore size of graphene compared to graphene, reducing the average coordination number of C atoms and plane bulk density. Graphyne aperture than graphene is more advantageous to ion contact with the electrolyte. Therefore, graphyne has a higher theoretical capacity than graphene, easier diffusion ability, larger pore diameter, etc., and has a greater prospect in the application of FSCs. Graphyne has attracted great interest and attention from researchers since its successful synthesis.

So far, the basic and application research of graphyne is on the rise, showing a huge and vast space. The controllable preparation methods, system characterization methods, and controllable structure of graphyne still need to be explored continuously. Information technology and other aspects of the research in the future will show unlimited space for innovation [24].

22.2.1.1 Phosphorene

Phosphorene, also known as black phosphosene or two-dimensional black phosphorus, is a two-dimensional nanomaterial composed of folded monolayer phosphorus atoms separated from black phosphorus, and SP^3 hybridized between phosphorus atoms. In 2014, Chen Xianhui's group from the University of Science and Technology of China and Zhang Yuanbo's group from Fudan University cooperated to prepare monolayer pholene for the first time [25]. They used conventional mechanical stripping to strip thin sheets of black phosphorus onto a degenerately doped silicon wafer with a layer of heat-growing silicon dioxide. Transistors made of black phosphoene had electron mobility of $1,000\,cm^2\,V^{-1}\,s^{-1}$. Two structures with different edge configurations can be obtained by cutting pholene along two different crystal phases: sawtooth type and armchair type. Due to the complex electronic structure, phosphoenes show remarkable anisotropy and have excellent properties in mechanics, optics, electron mobility efficiency, etc. [26].

The two-dimensional phosphosene in the environment is very unstable due to easy oxidation and degradation in the air and multilayer phosphosene can be stable for several days, but single phosphosene olefinic corrosion will be completely stable a few hours later. The main reason is that phosphosene reacts with oxygen for the oxidation of phosphorus and phosphorus oxide can cause a strong reaction with water, which leads to phosphorus degradation phenomenon. It can be modified by chemical functionalization and physical modification to prevent the degradation of phosphoene. Chemical functionalization usually introduces specific functional groups with different properties on the surface of phosphoene to prevent the reaction of phosphoene with oxygen and realize its passivation. Physical Modification: Usually, pholene is physically coated (with relatively stable chemicals) by modification to passivate the BP to form a protective film on the pholene structure [27].

Pholene with excellent electrical conductivity, larger specific surface area, and excellent mechanical properties, such as making phosphorus olefine to be applied to flexible supercapacitor, has high prospects. In addition, the substrate supporting electrode prepared by the composite carbon nanotubes can improve the electrical conductivity of the composite material, enhance the infiltration of electrolyte, and limit the stack of phosphoene nanosheets to assemble the flexible device to obtain $41\ F\ cm^{-3}$ volume specific capacity [28]. Self-supported electrodes are another mode of FSCs, which can be achieved by composite polymers such as polypyrrole/phosphoene self-supported electrodes synthesized by electrochemical polymerization. At the current density of $0.5\ A\ g^{-1}$, a high specific capacity of $497\ F\ g^{-1}$ was obtained, and there was almost no attenuation after 10,000 cycles. Or, carbon nanotubes/black phosphoene/carbon nanotubes (CNTs/BP-CNTs) non-woven self-supporting electrode was prepared by mixing with carbon nanotubes through microfluidic spinning technology (Figure 22.3a–c). Its porous structure forms a conductive network with abundant ion transport paths, which can rapidly diffuse ions [29].

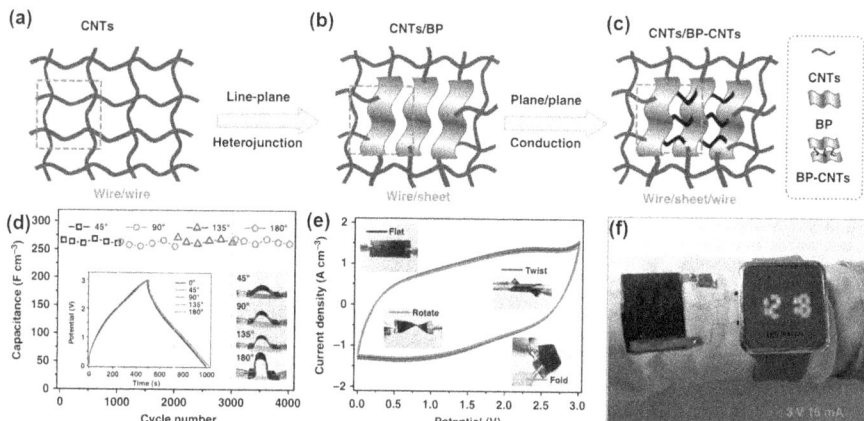

FIGURE 22.3 (a)–(c) Schematic illustration of the designed electrode structure. CNTs/BP refers to the physical mixture of carbon nanotubes and black phosphorous, CNTs/BP-CNTs refer to the physical mixture of carbon nanotubes and carbon nanotubes chemical-bridged black phosphorous. (d) Bending cycle stability of physical hybrid supercapacitors based on CNTs/BP/CNTs at 0.4 A cm^{-3} current density. The inset shows constant current charge–discharge curves and photos at different bending angles. (e) CNTs/BP -CNTs electrochemical performance test at distortion. (f) CNTs/BP -CNTs practical application. (Adapted with permission from Ref. [29]. Copyright (2018) Copyright The Authors, some rights reserved; exclusive licensee [Springer Nature].)

They bend at any angle to maintain excellent area specific capacity (Figure 22.3d), there is no change in their curriculum vitae curves (Figure 22.3e), and their flexible supercapacitor was assembled on the clothes to power the watch (Figure 22.3f). In conclusion, although phosphoene has become a new and promising two-dimensional nanomaterial due to its unique structure and excellent electrochemical performance, there are still some problems to be solved. For example, phosphoene structure is unstable and easy to degrade, so it still needs to find appropriate materials to inhibit it. The preparation of high-quality phospholene with controllable thickness and large size still needs more effort to meet the market demand.

22.2.1.2 MXenes

MXenes material is a kind of two-dimensional layered metal carbon/nitrogen compounds, which has been widely concerned for its excellent properties such as diversity of elemental composition, high electron mobility, unique in-plane anisotropy, high electrical conductivity, and excellent mechanical properties. The precursor phase of MXenes is MAX phase, where M is the transition metal element, A is the III or the IV main group element, and X is N or C. MXenes are usually obtained by the selective etching of a layer by MAX using hydrofluoric acid (HF), fluoric acid/hydrochloric acid mixture, and other methods [30]. In 2011, Prof. Gogotsi and Barsoum from Drexel University, USA, obtained Ti_3C_2 by hydrofluoric acid etching Al in Ti_3AlC_2 for the first time and named it MXenes. Up to now, 155 species of precursor MAX phase have been reported, and according to the theoretical prediction, more than 80 species of MXenes can be found, and a variety of solid solutions can be formed by changing the composition, so it can be predicted that MXenes will be a large family [31].

Through first-principle calculation, it is proved that the elastic modulus and bending strength of MXenes during stretching are higher than those of graphene, and MXenes have excellent mechanical properties [32]. It has been reported that through wet spinning, the synergistic effect between graphene oxide liquid crystal and MXenes sheet was used to continuously prepare MXenes-based fibers and assemble them into fiber flexible supercapacitors (FSCs). The results show that the obtained hybrid fibers have excellent electrical conductivity (29,000 S m^{-1}) and excellent volume-specific capacitance (586.4 F cm^{-3}) [33]. It has also been reported that an ultra-thin film is formed by inserting polypyrrole into layered Ti_3C_2 by an electrophoresis deposition method, which can be used as a self-supporting electrode for flexible all-solid supercapacitors [34]. Li and colleagues [35] used the hydrothermal method to grow Fe_2O_3 nanorod arrays on carbon cloth as conductive substrates and then immersed the prepared Fe_2O_3/carbon cloth in MXenes solution to build the MXenes layer, which not only improved the area capacitance of the original Fe_2O_3/carbon cloth (725 mF cm^{-2} at a current density of 1 mA cm^{-2}), but also provided a stable conductive carrier for pseudocapacitive Fe_2O_3 by providing an efficient electron transport pathway. The results show that the MXenes films have excellent electrochemical performance. However, although the MXenes films have excellent electrical conductivity, their internal ion kinetics retardation becomes the fundamental factor restricting their electrochemical performance. To solve this problem, Wang et al. [36] combined MXenes and bacterial cellulose (BC) with a loading capacity of up to 5 mg cm^{-2} to form a porous three-dimensional network film. The porous three-place network not only provides abundant electron transport channels, a larger active surface but also enhances the mechanical properties of the film. They combined polyaniline/BC cathode with MXenes/BC anode to produce a flexible asymmetric pseudocapacitor that provides a high area capacitance of up to 925 mF cm^{-2}.

As the energy storage devices in addition to the above research, to improve the MXenes electrode electrochemical performance, researchers have taken a lot of effort, such as formation of flexible 3D long-short MXenes bubble, to promote the electrode surface of ions or electron transfer rate. Using capillary forces assembly strategy makes MXenes with the three-dimensional structure of reunion to improve the specific surface area. Similarly, inhibiting MXenes aggregation can also allow particles (including carbon nanotubes, graphene, and polymers) to be inserted between the layers to act as baffles to prevent MXenes from stacking [37].

22.2.2 TRANSITION METAL DICHALCOGENIDES (TMDS)

Transition metal dichalcogenides (TMDs) are two-dimensional layered material, and the basic chemical formula can be written as MX_2. It consists of two layers of sulfur group elements (X = S, Se, Te, etc.) in which the transition metal elements (M = Ti, V, Ta, Mo, W Re, etc.) form a layered sandwich structure. TMDs have band gap structures, including direct band gap and indirect band gap. This solves the problem that graphene has no band gap. Therefore, it has been widely used in catalysts, biosensors, FSCs, and other aspects [38].

Since 2007, Soon and Loh [39] have for the first time synthesized single-layer edge-guided TMDs high-density nano-walled MS_2 films by chemical vapor deposition. These MS_2 films have excellent electrochemical storage properties (excellent specific capacitance of 100 F g^{-1}), which are superior to many carbon materials, due to the energy storage mechanism of the Ferrari process. The unique layered structure of TMDs has excellent mechanical properties, as well as large specific surface area and highly exposed active edges, making it more suitable for flexible devices. Liu et al. [40] used the impregnation casting method to soak MoS_2/CNTs mixed ink slowly on the A4 paper and dry it to obtain the continuous film with porous structure. MoS_2 is connected with CNTs to improve the conductivity and promote the electron migration rate. The interconnected structure has an open porous structure, which is conducive to the full infiltration of electrolyte.

To further improve the stability of TMDs, it has been reported that one-dimensional single crystal WO_3 nanowires were formed on W foil, then vulcanized on the nanowires, and the surface of single-crystal WO_3 was transformed into two-dimensional WS_2 by chemical conversion (Figure 22.4a). The whole electrode material is seamless (Figure 22.4b), with a large specific surface area, a synergistic effect of different structural properties of one-dimensional and two-dimensional components, and strong mechanical properties (Figure 22.4c). The FSCs not only achieve excellent capacitance performance but also have very long cycle stability in the process of charge–discharge cycle (no obvious attenuation in 30,000 cycles) [41].

The above studies have fully proved that 2D TMDs have an efficient application in FSCs. However, the research on 2D TMDs energy storage is still in its early stage, and there are still many shortcomings. The two-dimensional TMDs of most metallic phases have high conductivity but poor thermal stability. The stable 1H/2H phase TMDs conduct very poorly, and the flexible electrodes prepared are not thin enough to be manufactured on a large scale.

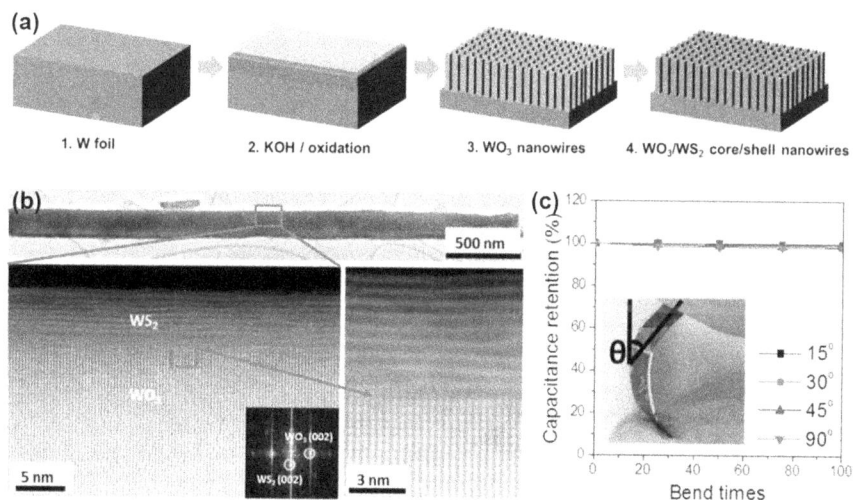

FIGURE 22.4 (a) Schematic for the fabrication process of WO_3/WS_2 core/shell nanowires on a W foil. (b) Low and high magnification TEM images of WO_3/WS_2 core/shell nanowire. (c) Capacitance retention under repeated mechanical deformations at various bending angles. (Adapted with permission from Ref. [41]. Copyright (2016) Copyright The Authors, some rights reserved; exclusive licensee [American Chemical Society].)

22.2.3 ASSEMBLY AND PRACTICAL APPLICATIONS OF FLEXIBLE SUPERCAPACITORS

In recent years, along with the people of portable and wearable flexible device with wide attention, various flexible inoculation, and supercapacitor, the FSCs assembly type and functional characteristics vary, according to the different assembly structures that can be mainly divided into three types [42]: one-dimensional fibrous FSCs, two-dimensional planar FSCs, and three-dimensional porous FSCs.

The structure of one-dimensional fibrous FSCs is similar to a sandwich, including electrode, electrolyte, and electrode. According to the assembly method, it can be divided into three types [43]: coaxial device composed of inner and outer electrodes (single coaxial fiber), a device composed of parallel fiber electrodes (two parallel fibers), and a device composed of two-stage fiber winding (two twisted fibers) (Figure 22.5a–c). For single coaxial fiber electrodes, the inner layer of the device is the collector (usually conductive metal wire) which covers the electrode materials. Then, the gel or solid electrolyte is evenly wrapped between the inner electrode and the outer electrode, and the outermost layer is a current collector. This kind of assembly structure makes the contact area between the two electrodes larger, provides more active sites for the ion desorption process, and can maintain the integrity even in the bending state, resulting in excellent electrochemical performance. However, it is difficult to assemble the device, and the inner and outer electrodes are hard to be tightly wrapped, resulting in a large air attack. As a result, the internal resistance becomes larger and the device loses its energy storage properties. Two parallel fiber electrodes are assembled in such a way that the electrode

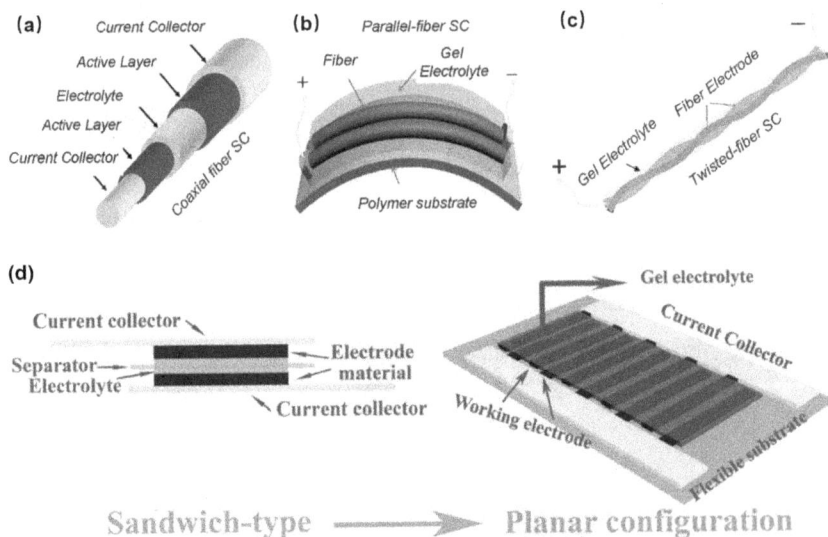

FIGURE 22.5 Three typical device configurations of fiber SCs: (a) single coaxial fiber, (b) two parallel fibers, and (c) two twisted fibers. (Adapted with permission from Ref. [43]. Copyright (2014) Copyright The Authors, some rights reserved; exclusive licensee [The Royal Society of Chemistry].) (d) Schematic comparison of the sandwich-type supercapacitors (left) and planar supercapacitors (right). (Adapted with permission from Ref. [1]. Copyright (2014) Copyright The Authors, some rights reserved; exclusive licensee [The Royal Society of Chemistry].)

fibers are placed parallel on a planar substrate. This structure can assemble multiple parallel electrodes to obtain higher energy density and power density. But, the existence of planar substrate limits its development in wearable devices. Finally, a device consisting of two levels of fibers intertwined, the structure of which is to intertwine two fiber electrodes, evenly covers the gel or solid electrolyte, not only serves as electrolyte but also functions as a diaphragm. This assembly method can be easily woven into clothes and has a good application prospect in wearable energy storage devices. Whereas, the contact area of the two fiber electrodes with this structure is small when they are wound, which is not conducive to the effective transmission of electrons and reduces the electro-chemical performance of energy storage devices.

The above three kinds of fibrous FSCs have their advantages and disadvantages, but they have good application prospects in flexible devices, especially single coaxial fiber, and two twisted fiber electrodes, which are widely used in flexible devices such as microsensors, micro-cameras, and wearable smart clothing. Qin et al. [44] obtained flexible films of carbon nanofibers by wet spinning and activated them with potassium hydroxide to make them rich in nitrogen oxides and produce uniformly distributed pore structures. To make the flexible devices have high flexibility, stretch-compression, and high ionic conductivity, the selection of flexible electrolytes is also the focus of attention.

In addition, fiber FSCs can also be combined with various electronic devices, such as photoelectric converters and sensors, so that their applications are more extensive. However, there are still some problems that hinder the development of fibrous FSCs. For example, the high-tensile fiber can be realized by winding, but the conductive polymer covering the surface is easy to dehydrate in the air, resulting in poor flex-ibility. Therefore, appropriate packaging methods are still needed to be explored. Compared with two-dimensional planar FSCs, the electrochemical performance is poor, so it is still necessary to choose appropriate materials to improve their energy density and power density.

A two-dimensional planar FSCs are composed of a fluid collector, a membrane, and a positive and negative electrode. To achieve flexible flexibility, the appropriate electrode materials, substrates, electrolytes, and packaging materials need to be selected. It can be divided into the conventional sandwich-type supercapacitors and planar super-capacitors (interdigitated structure) (Figure 22.5d) according to the different ways of assembly [1]. First of all, the sandwich type usually consists of three parts filled with sufficient electrolytes between the positive and negative poles (which also act as a membrane). The structure is simple and easy to prepare and suitable for mass production. Nonetheless, its structure has a small electrode area, increasing its thickness can improve its charge storage, but at the same time, it will increase the ion transport distance and reduce its power density. Due to its special structure, the thickness could not be ultra-thin, which limits the development of this sandwich structure supercapacitors. To solve this problem, another two-dimensional planar FSCs interdigitated structure is proposed. It is assembled by loading two pairs of fingerlike microelectrodes on a flexible substrate, which are arranged in equidistant pairs. The structure of the superca-pacitors can be ultra-thin, giving the device excellent flexibility. At the same time, this microstructure can be connected in series with multiple interdigitated structure super-capacitors to improve their energy density. However, the preparation of interdigitated structure is difficult and the cost is high, which limits its commercial development.

The flexible electrode materials can be divided into two kinds. One is that the electrode material itself forms a flexible film, that is, the self-supporting electrode material; the other is that when the electrode material is in powder form, it needs to be coated on the flexible substrate (such as metal foil fiber cloth paper), that is, the flexible substrate electrode material.

For the application of a flexible substrate electrode material, Yu et al. [45] prepared an ultra-thin transparent and flexible graphene film with a capacitance of 135 F g^{-1} when used as a supercapacitor electrode on a PET substrate and a glass sheet. In the same study, MnO$_2$ nanosheets were deposited on transparent indium tin oxide-polyethylene terephthalate substrates by screen printing, which achieved high specific capacitance (774 F g^{-1}) and stable long cycle performance. In order to improve the electrical conductivity of the electrode material, the electrode material can be attached to the conductive metal foil. It is necessary to select the appropriate metal foil to deposit the electrode material so that it will not easily peel off when bending and deforming. Shah and his colleagues [46] used air-assisted chemical vapor deposition to grow vertical flexible substrate electrodes of carbon nanotubes directly on conductive substrates to produce ultracapacitors that showed very small equivalent series resistance (about a few hundred milliohms), a direct result of the combination of CNTs with metal foil.

The self-supporting electrode material is both an electrode material and a fluid collector, which greatly simplifies the structure of the energy storage device, reduces the cost, and improves the electrochemical performance. Niu et al. [47] assembled FSCs using independently flexible single-walled carbon nanotubes (SWCNTs) films as both positive and negative electrodes. The prepared SWCNTs thin-film compact supercapacitor with small equivalent series resistance has higher energy density and power density (43.7 Wh kg^{-1} and 197.3 kW kg^{-1}, respectively). These results clearly show the potential applications of stand-alone SWCNTs films in compact design supercapacitors, which have the potential to improve performance and significantly increase energy and power density.

The above one-dimensional fibrous and two-dimensional FSCs have excellent prospects in various electronic devices, wearable smart clothing, foldable mobile phones, and other aspects. However, it is limited to small or miniature electronic devices and could not meet the requirements for large equipment. To greatly improve the device's energy output, researchers have recently developed a thick three-dimensional porous electrode material, including aerogel, foam, and sponge. This thicker structure (mostly more than 100 μm) can accommodate more active materials, resulting in higher energy output. The porous structure increases ion transport channels, increasing the transport rate, and resulting in a high power density.

Flexible aerogel electrode material has the advantages of large electric storage capacity, small internal resistance, lightweight, developed pores, and good flexibility. Song et al. [48] designed nitrogen-doped biomass-derived carbon/rGO aerogel networks (N-C-RGO-Networks) (Figure 22.6a–d) by the in situ generation of three-dimensional interlinked graphene aerogel interfaces. This structure (Figure 22.6e and f) exhibited significantly enhanced energy storage capacity in binderless systems, reaching 320 and 200 at 0.1 and 10 A g^{-1}, respectively (Figure 22.6g). In the same study, carbon fiber reinforced nanocellulose and multi-walled carbon nanotubes were used as a hybrid aerogel that acts as a highly flexible electrode. Its structure has

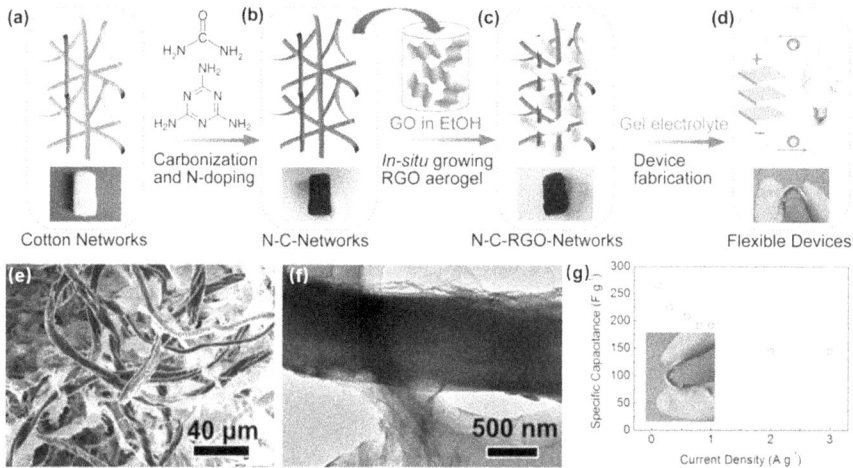

FIGURE 22.6 (a) Initial cotton. (b) N-C-Networks after carbonization and nitrogen doping. (c) N-C-RGO-Networks via establishing graphene aerogel interfaces. (d) The flexible devices and supercapacitor power. (e) SEM images and (f) TEM image of N-C-RGO-Networks. (g) Specific capacitance of the flexible all-solid-state device with the inset of the optical photo. (Adapted with permission from Ref. [48]. Copyright (2016) Elsevier.)

the advantages of three-dimensional porous structure, excellent electrical conductivity, binder-free properties, etc. The assembled supercapacitor also has excellent mechanical strength and the maximum stress can reach 27.9 MPa. Different from aerogels, flexible foam electrode materials usually use CVD or electrode positioning methods to directly grow electrode materials on foam substrate (including nickel foam and copper foam). Zhang et al. [49] grew ultrathin $NiCo_2O_4$ nanosheets in situ on Ni foam by optimizing the hydrothermal reaction based on crystal growth kinetics. The ultrathin $NiCo_2O_4$ structured nanosheet array can expose more active sites, provide a rich diffusion channel, and buffer the stress caused by phase transition during the charge and discharge process of the supercapacitor.

In summary, three kinds of flexible ultracapacitors with different assembly structures and their application range are summarized, including one-dimensional fibrous flexible ultracapacitors, two-dimensional planar flexible ultracapacitors, and three-dimensional porous flexible ultracapacitors. These three device configurations have their own advantages and disadvantages. Therefore, further research is needed to find suitable flexible electrode materials, electrolyte, diaphragm, and packaging materials to obtain flexible devices with high energy output and meet commercial requirements.

22.2.4 CONCLUSION AND PERSPECTIVE

In this chapter, we summarize various 2D nanomaterials, including 2D carbon nanomaterials, phosphoenes, MXenes, and transition metal sulfides, as well as the advantages and disadvantages of FSCs, assembly methods, and applications. How to select electrode materials with high conductivity, high specific capacity, and high mechanical flexibility is the problem that FSCs urgently need to solve at present. In addition, multi-function system integration will be the focus of the future development of flexible devices, such as biosensors, photoelectric converters, pressure sensors, and other integrated multifunction devices. For example, micro-supercapacitors, pressure sensors, photodetectors, and gas sensors are integrated to mimic the functions of human skin and sensory organs.

There are still many unsolved problems in the development of FSCs, which are not only the focus and hotspot of future research but also the opportunities in the research process: (i) At present, FSCs could not meet the requirements of high energy output equipment, and electrode materials still need to be deeply explored and new innovations to meet the requirements of commercialization. (ii) It is very important to study the mechanical strength and flexibility of FSCs, which directly affects their practical application. Therefore, it is very necessary to study the tensile properties of the device and the electrochemical properties of bending in different states during the research. (iii) Commonly used electrolytes are generally toxic and corrosive, which can cause harm to the human body. Therefore, to avoid electrolyte leakage, high safety factor packaging technology is needed to ensure the normal operation of wearable FSCs.

At present, FSCs are still under development, but industrialization requires further research and selection of electrode materials to achieve superior charge storage. At the same time, the preparation process is simple and the cost is low, which provides strong support for the large-scale production of FSCs.

REFERENCES

1. Peng X, Peng L, Wu C, Xie Y (2014) Two dimensional nanomaterials for flexible supercapacitors. *Chem. Soc. Rev.* 43:3303–3323.
2. Kumar KS, Choudhary N, Jung Y, and Thomas J (2018) Recent advances in two-dimensional nanomaterials for supercapacitor electrode applications. *ACS Energy Lett.* 3:482–495.
3. Han Y, Ge Y, Chao Y, Wang C, Wallace GG (2018) Recent progress in 2D materials for flexible supercapacitors. *J. Energy Chem.* 27:57–72.
4. Zhang L, Yu X, Zhu P, Zhou F, Li G, Sun R, and Wong C-P (2018) Laboratory filter paper as a substrate material for flexible supercapacitors. *Sutain. Energ. Fuels.* 2:147–154.
5. Dai H, Zhang G, Rawach D, Fu C, Wang C, Liu X, Dubois M, Lai C, Sun S (2021) Polymer gel electrolytes for flexible supercapacitors: Recent progress, challenges, and perspectives. *Energy Storage Mater.* 34:320–355.
6. Huang L, Santiago D, Loyselle P, and Dai L (2018) Graphene-based nanomaterials for flexible and wearable supercapacitors. *Small.* 14:1800879.
7. Chee WK, Lim HN, Zainal Z, Huang NM, Harrison I, Andou Y (2016) Flexible graphene-based supercapacitors: A review. *J. Phys. Chem. C.* 120:4153–4172.
8. Karim N, Afroj S, Tan S, He P, Fernando A, Carr C, Novoselov KS (2017) Scalable production of graphene-based wearable e-textiles. *ACS Nano.* 11:12266–12275.
9. Wang S, Liu N, Su J, Li L, Long F, Zou Z, Jiang X, Gao Y (2017) Highly stretchable and self-healable supercapacitor with reduced graphene oxide based fiber springs. *ACS Nano.* 11:2066–2074.
10. Lim T, Ho BT, Suk JW (2021) High-performance and thermostable wire supercapacitors using mesoporous activated graphene deposited on continuous multilayer graphene. *J. Mate. Chem. A.* 9:4800–4809.

11. Qu G, Cheng J, Li X, Yuan D, Chen P, Chen X, Wang B, Peng H (2016) A fiber super-capacitor with high energy density based on hollow graphene/conducting polymer fiber electrode. *Adv. Mater.* 28:3646–3652.

12. Xie P, Yuan W, Liu X, Peng Y, Yin Y, Li Y, Wu Z (2021) Advanced carbon nanomaterials for state-of-the-art flexible supercapacitors. *Energy Storage Mater.* 36:56–76.

13. Angelopoulou P, Vrettos K, Georgakilas V, Avgouropoulos G (2020) Graphene aerogel modified carbon paper as anode for lithium-ion batteries. *Chem. Select.* 5:2719–2724.

14. Wang G, Sun X, Lu F, Sun H, Yu M, Jiang W, Liu C, Lian J (2012) Flexible pillared gra-phene-paper electrodes for high-performance electrochemical supercapacitors. *Small* 8:452–459.

15. Li Q, Guo X, Zhang Y, Zhang W, Ge C, Zhao L, Wang X, Zhang H, Chen J, Wang Z, Sun L (2017) Porous graphene paper for supercapacitor applications. *J. Mate. Sc. Technol.* 33:793–799.

16. El-Kady MF, Shao Y, Kaner RB (2016) Graphene for batteries, supercapacitors and beyond. *Nat. Rev. Mater.* 1:16033.

17. Du Y, Xiao P, Yuan J, Chen J (2020) Research progress of graphene-based materials on flexible supercapacitors. *Coatings* 10:892.

18. Xu M, Wang A, Xiang Y, Niu J (2021) Biomass-based porous carbon/graphene self-assembled composite aerogels for high-rate performance supercapacitor. *J. Clean. Prod.* 315:128110.

19. Li F, Wang X, Sun R (2017) A metal-free and flexible supercapacitor based on redox-active lignosulfonate functionalized graphene hydrogels. *J. Mater. Chem. A.* 5:20643–20650.

20. Malko D, Neiss C, Viñes F, Görling A (2012) Competition for graphene: Graphynes with direction-dependent dirac cones. *Phys. Rev. Lett.* 108:086804.

21. Li G, Li Y, Liu H, Guo Y, Li Y, Zhu D (2010) Architecture of graphdiyne nanoscale films. *Chem. Commun.* 46:3256–3258.

22. Mortazavi B, Shahrokhi M, Madjet ME, Hussain T, Zhuang X, Rabczuk T (2019) N-, B-, P-, Al-, As-, and Ga-graphdiyne/graphyne lattices: First-principles investigation of mechanical, optical and electronic properties. *J. Mater. Chem.* 7:3025–3036.

23. Chen X, Xu W, Song B, He P (2020) First-principles study of stability, electronic struc-ture and quantum capacitance of B-, N- and O-doped graphynes as supercapacitor elec-trodes. *J. Phys. Condens.* 32:215501.

24. Kang J, Wei Z, Li J (2019) Graphyne and its family: Recent theoretical advances. *ACS. Appl. Mater. Inter.* 11:2692–2706.

25. Li L, Yu Y, Ye GJ, Ge Q, Ou X, Wu H, Feng D, Chen XH, Zhang Y (2014) Black phos-phorus field-effect transistors. *Nat. Nanotechnol.* 9:372–377.

26. Qiao J, Kong X, Hu Z-X, Yang F, Ji W (2014) High-mobility transport anisotropy and linear dichroism in few-layer black phosphorus. *Nat. Commun.* 5:4475.

27. Wu Y, Yuan W, Xu M, Bai S, Chen Y, Tang Z, Wang C, Yang Y, Zhang X, Yuan Y, Chen M, Zhang X, Liu B, Jiang L (2021) Two-dimensional black phosphorus: Properties, fabrication and application for flexible supercapacitors. *Chem. Eng. J.* 412:128744.

28. Yang B, Hao C, Wen F, Wang B, Mu C, Xiang J, Li L, Xu B, Zhao Z, Liu Z, Tian Y (2017) Flexible black-phosphorus nanoflake/carbon nanotube composite paper for high-performance all-solid-state supercapacitors. *ACS. App. Mater. Inter.* 9:44478–44484.

29. Wu X, Xu Y, Hu Y, Wu G, Cheng H, Yu Q, Zhang K, Chen W, Chen S (2018) Microfluidic-spinning construction of black-phosphorus-hybrid microfibres for non-woven fabrics toward a high energy density flexible supercapacitor. *Nat. Commun.* 9:4573.

30. Hu M, Zhang H, Hu T, Fan B, Wang X, Li Z (2020) Emerging 2D MXenes for superca-pacitors: Status, challenges and prospects. *Chem. Soc. Rev.* 49:6666–6693.

31. Wang K, Zheng B, Mackinder M, Baule N, Garratt E, Jin H, Schuelke T, Fan QH (2021) Efficient electrophoretic deposition of MXene/reduced graphene oxide flexible electrodes for all-solid-state supercapacitors. *J. Energy Storage* 33:102070.

32. Borysiuk VN, Mochalin VN, Gogotsi Y (2015) Molecular dynamic study of the mechanical properties of two-dimensional titanium carbides $Ti_{n+1}C_n$ (MXenes). *Nanotechnology* 26:265705.

33. Yang Q, Xu Z, Fang B, Huang T, Cai S, Chen H, Liu Y, Gopalsamy K, Gao W, Gao C (2017) MXene/graphene hybrid fibers for high performance flexible supercapacitors. *J. Mater. Chem. A* 5:22113–22119.

34. Zhu M, Huang Y, Deng Q, Zhou J, Pei Z, Xue Q, Huang Y, Wang Z, Li H, Huang Q, Zhi C (2016) Highly flexible, freestanding supercapacitor electrode with enhanced performance obtained by hybridizing polypyrrole chains with MXene. *Adv. Energy Mater.* 6:1600969.

35. Li F, Liu Y-L, Wang G-G, Zhang H-Y, Zhang B, Li G-Z, Wu Z-P, Dang L-Y, Han J-C (2019) Few-layered $Ti_3C_2T_x$ MXenes coupled with Fe_2O_3 nanorod arrays grown on carbon cloth as anodes for flexible asymmetric supercapacitors. *J. Mater. Chem. A* 7:22631–22641.

36. Wang Y, Wang X, Li X, Bai Y, Xiao H, Liu Y, Liu R, Yuan G (2019) Engineering 3D ion transport channels for flexible MXene films with superior capacitive performance. *Adv. Funct. Mate.* 29:1900326.

37. Zhao Z, Wang S, Wan F, Tie Z, Niu Z (2021) Scalable 3D self-assembly of MXene films for flexible sandwich and microsized supercapacitors. *Adv. Funct. Mater.* 31:2101302.

38. Tanwar S, Arya A, Gaur A, Sharma AL (2021) Transition metal dichalcogenide (TMDs) electrodes for supercapacitors: A comprehensive review. *J. Phys. Condens. Mat.* 33:303002.

39. Soon JM, Loh KP (2007) Electrochemical double-layer capacitance of MoS_2 nanowall films. *Electrochem. Solid. St.* 10:A250.

40. Liu A, Lv H, Liu H, Li Q, Zhao H (2017) Two dimensional MoS_2/CNT hybrid ink for paper-based capacitive energy storage. *J. Mater. Sci. Mater. Electron.* 28:8452–8459.

41. Choudhary N, Li C, Chung H-S, Moore J, Thomas J, Jung Y (2016) High-performance one-body core/shell nanowire supercapacitor enabled by conformal growth of capacitive 2D WS_2 layers. *ACS Nano* 10:10726–10735.

42. Dong L, Xu C, Li Y, Huang Z-H, Kang F, Yang Q-H, Zhao X (2016) Flexible electrodes and supercapacitors for wearable energy storage: A review by category. *J. Mater. Chem. A* 4:4659–4685.

43. Yu D, Qian Q, Wei L, Jiang W, Goh K, Wei J, Zhang J, Chen Y (2015) Emergence of fiber supercapacitors. *Chem. Soc. Rev.* 44:647–662.

44. Qin T, Peng S, Hao J, Wen Y, Wang Z, Wang X, He D, Zhang J, Hou J, Cao G (2017) Flexible and wearable all-solid-state supercapacitors with ultrahigh energy density based on a carbon fiber fabric electrode. *Adv. Energy Mater.* 7:1700409.

45. Yu A, Roes I, Davies A, Chen Z (2010) Ultrathin, transparent, and flexible graphene films for supercapacitor application. *Appl. Phys. Lett.* 96:253105.

46. Shah R, Zhang X, Talapatra S (2009) Electrochemical double layer capacitor electrodes using aligned carbon nanotubes grown directly on metals. *Nanotechnology* 20:395202.

47. Niu Z, Zhou W, Chen J, Feng G, Li H, Ma W, Li J, Dong H, Ren Y, Zhao D, Xie S (2011) Compact-designed supercapacitors using free-standing single-walled carbon nanotube films. *Energy Environ. Sci.* 4:1440–1446.

48. Song W-L, Li X, Fan L-Z (2016) Biomass derivative/graphene aerogels for binder-free supercapacitors. *Energy Storage Mater.* 3:113–122.

49. Zhang X, Yang F, Chen H, Wang K, Chen J, Wang Y, Song S (2020) In situ growth of 2d ultrathin $NiCo_2O_4$ nanosheet arrays on Ni foam for high performance and flexible solid-state supercapacitors. *Small* 16:2004188.

Index

For Product Safety Concerns and Information please contact our EU
representative GPSR@taylorandfrancis.com
Taylor & Francis Verlag GmbH, Kaufingerstraße 24, 80331 München, Germany

www.ingramcontent.com/pod-product-compliance
Lightning Source LLC
Chambersburg PA
CBHW060754220326
41598CB00022B/2434